AMERICAN MATHEMATICAL SOCIETY
COLLOQUIUM PUBLICATIONS

VOLUME XXV

LATTICE THEORY

BY

GARRETT BIRKHOFF

PUBLISHED BY THE
AMERICAN MATHEMATICAL SOCIETY
PROVIDENCE, RHODE ISLAND
1940

International Standard Serial Number 0065-9258
International Standard Book Number 0-8218-1025-1
Library of Congress Catalog Card Number 66-23707

Revised edition, 1948
Third edition, 1967
Third edition, third printing, 1979
Reprinted with corrections, 1984

Printed in the United States of America

PREFACE TO THE THIRD EDITION

The purpose of this edition is threefold: to make the deeper ideas of lattice theory accessible to mathematicians generally, to portray its structure, and to indicate some of its most interesting applications. As in previous editions, an attempt is made to include current developments, including various unpublished ideas of my own; however, unlike previous editions, this edition contains only a very incomplete bibliography.

I am summarizing elsewhere† my ideas about the role played by lattice theory in mathematics generally. I shall therefore discuss below mainly its logical structure, which I have attempted to reflect in my table of contents.

The beauty of lattice theory derives in part from the extreme simplicity of its basic concepts: (partial) ordering, least upper and greatest lower bounds. In this respect, it closely resembles group theory. These ideas are developed in Chapters I–V below, where it is shown that their apparent simplicity conceals many subtle variations including for example, the properties of modularity, semimodularity, pseudo-complements and orthocomplements.

At this level, lattice-theoretic concepts pervade the whole of modern *algebra*, though many textbooks on algebra fail to make this apparent. Thus lattices and groups provide two of the most basic tools of "universal algebra", and in particular the structure of algebraic systems is usually most clearly revealed through the analysis of appropriate lattices. Chapters VI and VII try to develop these remarks, and to include enough technical applications to the theory of groups and loops with operators to make them convincing.

A different aspect of lattice theory concerns the foundations of set theory (including general topology) and real analysis. Here the use of various (partial) orderings to justify transfinite inductions and other limiting processes involves some of the most sophisticated constructions of all mathematics, some of which are even questionable! Chapters VIII–XII describe these processes from a lattice-theoretic standpoint.

Finally, many of the deepest and most interesting applications of lattice theory concern (partially) ordered mathematical structures having also a binary addition or multiplication: lattice-ordered groups, monoids, vector spaces, rings, and fields (like the real field). Chapters XIII–XVII describe the properties of such systems,

† G. Birkhoff, *What can lattices do for you?*, an article in *Trends in Lattice Theory*, James C. Abbot, ed., Van Nostrand, Princeton, N.J., 1967.

and also those of positive linear operators on partially ordered vector spaces. The theory of such systems, indeed, constitutes the most rapidly developing part of lattice theory at the present time.

The labor of writing this book has been enormous, even though I have made no attempt at completeness. I wish to express my deep appreciation to those many colleagues and students who have criticized parts of my manuscript in various stages of preparation. In particular, I owe a very real debt to the following: Kirby Baker, Orrin Frink, George Grätzer, C. Grandjot, Alfred Hales, Paul Halmos, Samuel H. Holland, M. F. Janowitz, Roger Lyndon, Donald MacLaren, Richard S. Pierce, George Raney, Arlan Ramsay, Gian-Carlo Rota, Walter Taylor, and Alan G. Waterman.

My thanks are also due to the National Science Foundation for partial support of research in this area and of the preparation of a preliminary edition of notes, and to the Argonne National Laboratory and the Rand Corporation for support of research into aspects of lattice theory of interest to members of their staffs.

Finally, I wish to thank Laura Schlesinger and Lorraine Doherty for their skillful typing of the entire manuscript.

TABLE OF CONTENTS

CHAPTER I

TYPES OF LATTICES

1. Posets; Chains

Pure lattice theory is concerned with the properties of a single undefined binary relation $\leqq$, to be read "is contained in", "is a part of", or "is less than or equal to". This relation is assumed to have certain properties, the most basic of which lead to the following concept of a "partially ordered set", alias "partly ordered set" or "poset".

DEFINITION. A *poset* is a set in which a binary relation $x \leqq y$ is defined, which satisfies for all x, y, z the following conditions:

P1. For all x, $x \leqq x$. (Reflexive)
P2. If $x \leqq y$ and $y \leqq x$, then $x = y$. (Antisymmetry)
P3. If $x \leqq y$ and $y \leqq z$, then $x \leqq z$. (Transitivity)

If $x \leqq y$ and $x \neq y$, one writes $x < y$, and says that x "is less than" or "properly contained in" y. The relation $x \leqq y$ is also written $y \geqq x$, and read y contains x (or includes x). Similarly, $x < y$ is also written $y > x$. The above notation and terminology are standard.

There are countless familiar examples of partly ordered sets—i.e., of mathematical relations satisfying P1–P3. Three of the simplest are the following.

Example 1. Let $\Sigma(\mathrm{I})$ consist of all subsets of any class I, including I itself and the void class $\varnothing$; and let $x \leqq y$ mean x is a subset of y.

Example 2. Let $\mathbf{Z}^+$ be the set of positive integers; and let $x \leqq y$ mean that x divides y.

Example 3. Let $\dot{F}$ consist of all real single-valued functions $f(x)$ defined on $-1 \leqq x \leqq 1$; and let $f \geqq g$ mean that $f(x) \geqq g(x)$ for *every* x with $-1 \leqq x \leqq 1$.

We now state without proof two familiar laws governing inclusion relations, which follow from P1–P3.

LEMMA 1. *In any poset, $x < x$ for no x, while $x < y$ and $y < z$ imply $x < z$. Conversely, if a binary relation $<$ satisfies the two preceding conditions, define $x \leqq y$ to mean that $x < y$ or $x = y$; then the relation $\leqq$ satisfies* P1–P3.

In other words, strict inclusion is characterized by the anti-reflexive and transitive laws.

It is easily shown that a poset P can contain at most one element a which satisfies $a \leqq x$ for all $x \in P$. For if a and b are two such elements, then $a \leqq b$ and also $b \leqq a$, whence $a = b$ by P2. Such an element, if it exists, is denoted by

O, and is called the *least* element of X. The dual *greatest* element of P, if it exists, is denoted by I. The elements O and I, when they exist, are called *universal bounds* of P, since then $O \leqq x \leqq I$ for all $x \in P$.

LEMMA 2. *If* $x_1 \leqq x_2 \leqq \cdots \leqq x_n \leqq x_1$, *then* $x_1 = x_2 = \cdots = x_n$. (*Anti-circularity*)

Example 4. Let $\mathbf{R}$ be the set of real numbers, and let $x \leqq y$ have its usual meaning for real numbers.

The relation of inclusion in this and other important posets satisfies:

P4. Given x and y, either $x \leqq y$ or $y \leqq x$.

DEFINITION. A poset which satisfies P4 is said to be "simply" or "totally" ordered and is called a *chain*.

In other words, of any two distinct elements in a chain, one is less and the other greater. Clearly, the posets of Examples 1–3 are not chains: they contain pairs of elements x, y which are *incomparable*, in the sense that neither $x \leqq y$ nor $y \leqq x$ holds.

From Examples 1–4 above, many other posets can be constructed as subsets. More precisely, let P be any poset, and let S be any subset of P; define the relation $x \leqq y$ in S to mean that $x \leqq y$ in P. Conditions P1–P3 being satisfied by $\leqq$ in P, they are satisfied *a fortiori* in S. A similar observation holds for P4. There follows, trivially,

THEOREM 1. *Any subset S of a poset P is itself a poset under the same inclusion relation* (*restricted to S*). *Any subset of a chain is a chain.*

Thus, the set $\mathbf{Z}^+$ of positive integers is a chain under the relation $\leqq$ of relative magnitude of Example 4, though a poset which is not a chain under the partial ordering of Example 2.

Example 5. (a) The set $\{1, 2, \cdots, n\}$ forms a chain $\mathbf{n}$ (the *ordinal* number n) in its natural order. (b) When unordered, so that no two elements are comparable, it forms another poset (the *cardinal* number n).

The family of all subsets of any class distinguished by any given special property forms a poset under set-inclusion. Thus this is true of the subgroups of any group, the vector subspaces of any vector space, the Borel subsets of any T_0-space etc. For instance, an ideal is a subset H of a ring R distinguished by the properties (i) $a, b \in R$ imply $a - b \in R$, and (ii) $a \in H$ and $b \in R$ imply $ab \in H$ and $ba \in H$. Specialized to this case, the principle stated above yields an important example, which will be studied more deeply in Chapters VII and XIV below.

Example 6. Let P consist of the ideals $H, J, K, \cdots$ of any ring R; let $H \leqq K$ mean that H is a subset of K (i.e., that $H \subset K$).

2. Isomorphism; Duality

A function θ: $P \to Q$ from a poset P to a poset Q is called *order-preserving* or *isotone* if it satisfies

$$(1) \qquad x \leqq y \quad \text{implies} \quad \theta(x) \leqq \theta(y).$$

An isotone function which has an isotone two-sided inverse is called an *isomorphism*. In other words, an isomorphism between two posets P and Q is a bijection which satisfies (1) and also

$$\theta(x) \leqq \theta(y) \quad \text{implies} \quad x \leqq y. \tag{1'}$$

An isomorphism from a poset P to itself is called an *automorphism*.

Two posets P and Q are called *isomorphic* (in symbols, $P \cong Q$), if and only if there exists an isomorphism between them.

The *converse* of a relation ρ is, by definition, the relation $\breve{\rho}$ such that $x\breve{\rho}y$ (read, "x is in the relation $\breve{\rho}$ to y") if and only if $y\rho x$. Thus the converse of the relation "includes" is the relation "is included in"; the converse of "greater than" is "less than". It is obvious from inspection of conditions P1–P3 that

THEOREM 2 (DUALITY PRINCIPLE). *The converse of any partial ordering is itself a partial ordering.*

DEFINITION. The *dual* of a poset X is that poset $\breve{X}$ defined by the converse partial ordering relation on the same elements.

Since $X \cong \breve{\breve{X}}$, this terminology is legitimate: the relation of duality is symmetric.

DEFINITION. A function $\theta: P \to Q$ is *antitone* if and only if

$$x \leqq y \quad \text{implies} \quad \theta(x) \geqq \theta(y), \tag{2}$$

$$\theta(x) \leqq \theta(y) \quad \text{implies} \quad x \geqq y. \tag{2'}$$

A bijection (one–one correspondence) θ which satisfies (2)-(2′) is called a *dual isomorphism*.

We shall refer to systems isomorphic with $\breve{X}$ as "dual" to X. Obviously posets are dual in pairs, whenever they are not self-dual. Similarly, definitions and theorems about posets are dual in pairs whenever they are not self-dual; and if any theorem is true for all posets, then so is its dual.

As we shall see later, this Duality Principle applies to algebra, to projective geometry, and to logic.

Many important posets are self-dual (i.e., anti-isomorphic with themselves). Thus Example 1 of §2 is self-dual; the correspondence which carries each subset into its complement is one-to-one and inverts inclusion. Similarly the set of all linear subspaces of n-dimensional Euclidean space which contains the origin is self-dual: the correspondence carrying each subspace into its orthogonal complement is one-to-one and inverts inclusion.

In these cases the self-duality is of period two: the image $(x')'$ of the image x' of any x is again x. Such self-dualities (dual automorphisms) are called *involutions*.

Exercises for §§1–2:

1. Prove Lemma 1.
2. Prove Lemma 2.
3. Show that there are exactly three different ways of partly ordering a set of two elements.
4. (a) Show that there are just two nonisomorphic posets of two elements, both of which are self-dual.

(b) Show that there are five nonisomorphic posets of three elements, three of which are self-dual.

*5. (a) Let $G(n)$ denote the number of nonisomorphic posets of n elements. Show that $G(4) = 16$, $G(5) = 63$, $G(6) = 318$. (I. Rose – R. T. Sasaki)

(b) $G^*(n)$ denote the number of different partial orderings of n elements. Show $G^*(2) = 3$, $G^*(3) = 19$, $G^*(4) = 219$, $G^*(5) = 4231$, $G^*(6) = 130{,}023$, $G^*(7) = 6{,}129{,}859$.

(c) How many of the preceding give self-dual posets?

(d) Is $G^*(n)$ odd for all n? Justify.

3. Diagrams; Graded Posets

The notion of "immediate superior" in a hierarchy can be defined in any poset, as follows.

DEFINITION. By "a covers b" in a poset P, it is meant that $a > b$, but that $a > x > b$ for no $x \in P$.

By the *order* $n(P)$ of a poset P is meant the (cardinal) number of its elements. When this number is finite, P is called a "finite" poset. Using the covering relation, one can obtain a graphical representation of any finite poset P as follows.

Draw a small circle to represent each element of P, placing a higher than b whenever $a > b$. Draw a straight segment from a to b whenever a covers b. The resulting figure is called a *diagram* of P: examples are shown in Figures 1a–1e below.

Since $a > b$ if and only if one can move from a to b downward along some broken line, it is clear that any finite poset is defined up to isomorphism by its diagram. It is also clear that the diagram of the dual P of a poset P is obtained from that of P by turning the latter upside down.

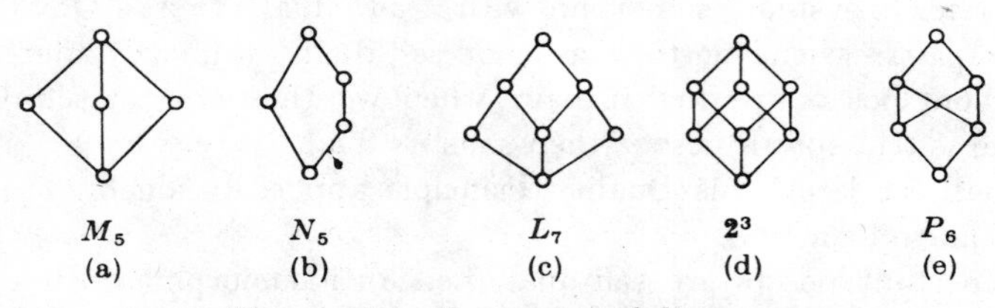

FIGURE 1. Examples of diagrams

DEFINITION. By a *least* element of any subset X of P, we mean an element $a \in X$ such that $a \leqq x$ for all $x \in X$. By a *greatest* element of X, we mean an element $b \in X$ such that $b \geqq x$ for all $x \in X$.

The preceding concepts are not to be confused with the concepts of *minimal* and *maximal* elements. A minimal element of a subset X of a partly ordered set P is an element a such that $a < x$ for no $x \in X$; maximal elements are defined dually. Clearly, a least element must be minimal and a greatest element maximal, but the converse is not true.

THEOREM 3. *Any finite nonempty subset X of a poset has minimal and maximal members.*

PROOF. Let X consist of $x_1, \cdots, x_n$. Define $m_1 = x_1$, and m_k as x_k if $x_k < m_{k-1}$ and m_{k-1} otherwise. Then m_n will be minimal. Similarly, X has a maximal element.

THEOREM 4. *In chains, the notions of minimal and least (maximal and greatest) element of a subset are effectively equivalent. Hence any finite chain has a least (first) and greatest (last) element.*

PROOF. If $x < a$ for no $x \in X$, then by P4 $x \geqq a$ for every $x \in X$.

THEOREM 5. *Every finite chain of n elements is isomorphic with the ordinal number* $\mathbf{n}$ *(the chain of integers* $1, \cdots, n$*).*

That is, there is a bijection ϕ from the chain X onto $\{1, \cdots, n\}$ such that $x_1 \leqq x_2$ if and only if $\phi(x_1) \leqq \phi(x_2)$; finite chains are finite ordinal numbers.

PROOF. Let ϕ map the least $x \in X$ into 1, and the least of the remaining $x \in X$ into 2, etc.

The length of a finite chain $\mathbf{n}$ is defined to be $n - 1$ (see its diagram). More generally, the *length* $l(P)$ of a poset P is defined as the least upper bound of the lengths of the chains in P. When $l(P)$ is finite, P is said to be of *finite length.* Any poset of finite length is defined up to isomorphism by its covering relation: $a > b$ if and only if a finite sequence $x_0, x_1, \cdots, x_n$ exists, such that $a = x_0$, $b = x_n$, and x_{i-1} covers x_i for $i = 1, \cdots, n$.

The isomorphism or nonisomorphism of two finite posets can often be tested most simply by drawing their diagrams. Any isomorphism must be one-to-one onto between lowest elements, between next lowest elements, and so on. Corresponding elements must be covered by equal numbers of elements, and the covering elements must also correspond. These principles make it easy to enumerate the different (i.e., nonisomorphic) posets having $n = 4$ elements (say); there are exactly 16 of them.

In a poset P of finite length with O, the *height* or dimension $h[x]$ of an element $x \in P$ is, by definition, the l.u.b. of the lengths of the chains $O = x_0 < x_1 < \cdots < x_l = x$ between O and x. If P has a universal upper bound I, then clearly $h[I] = l[P]$. Clearly also, $h[x] = 1$ if and only if x covers O; such elements are called "atoms" or "points" of P.

The height function is especially important in *graded posets.* These are defined as posets P with a function $g: P \to \mathbf{Z}$ from P to the chain of all integers (in their natural order), such that:

G1. $x > y$ implies $g[x] > g[y]$ (strict isotonicity).

G2. If x covers y, then $g[x] = g[y] + 1$.

Any graded poset satisfies the following

JORDAN–DEDEKIND CHAIN CONDITION. *All maximal chains between the same endpoints have the same finite length.*

LEMMA 1. *Let P be any poset with O in which all chains are finite. Then P satisfies the Jordan–Dedekind chain condition if and only if it is graded by* $h[x]$.

PROOF. If P is graded by $h[x]$, then the Jordan–Dedekind chain condition follows obviously: the length of any maximal chain joining the endpoints a and $b > a$ is $h[b] - h[a]$. Conversely, if the Jordan–Dedekind chain condition holds, then $h[x]$ is the length of every maximal chain from O to x, from which G1 and G2 follow immediately.

Exercises for §3:

1. (a) Show that the diagram of a poset is an oriented graph,† if we draw an arrow from x to y if and only if x covers y.

(b) Show that a finite oriented graph is associated with a poset if and only if $\overrightarrow{a_0a_1}, \overrightarrow{a_1a_2}, \cdots, \overrightarrow{a_{n-1}a_n}$ is incompatible with $\overrightarrow{a_0a_n}$.

(*c) Show that any oriented graph defines a quasi-ordered set, if $a \geqq b$ is defined to mean $a = b$, $\overrightarrow{ab}$, or $\overrightarrow{aa_1}, \overrightarrow{a_1a_2}, \cdots, \overrightarrow{a_nb}$ for suitable $a_1, \cdots, a_n$.

2. Show that the "covering relations" of any poset form a new poset if ("x covers y") > ("u covers v") means that $y \geqq u$.

3. Show that any isotone transformation $P \to P_1$ of one poset P onto another P_1 carries connected components of the graph of P into connected components of the graph of P_1.

4. Which of the diagrams of Figure 1 represent self-dual posets? Define two new self-dual posets by means of diagrams.

5. Show that a chain can be defined as a set of elements in which a transitive relation $x > y$ is defined, such that for any elements u, v, one and only one of the relations $u > v$, $u = v$, $v > u$ holds.

6. Show that chains are those posets, all of whose subsets are lattices.

7. Show that no finite poset of more than two elements is defined to within isomorphism by its graph.

8. Let P be a poset of finite length. Show that any two elements of P have an upper bound if and only if P has a universal upper bound I.

9. Show that, in a poset P of finite length, a chain from a to b is maximal if and only if it is connected in the graph of P.

4. Lattices

An *upper bound* of a subset X of a poset P is an element $a \in P$ containing every $x \in X$. The *least upper bound* is an upper bound contained in every other upper bound; it is denoted l.u.b. X or sup X. By P2, sup X is unique if it exists. The notions of *lower bound* of X and greatest lower bound (g.l.b. X or inf X) of X are defined dually. Again by P2, inf X is unique if it exists.

DEFINITION. A *lattice*‡ is a poset P any two of whose elements have a g.l.b. or "meet" denoted by $x \wedge y$, and a l.u.b. or "join" denoted by $x \vee y$. A lattice L is *complete* when each of its subsets X has a l.u.b. and a g.l.b. in L.

Setting $X = L$, we see that any nonvoid complete lattice contains a least element O and a greatest element I. Evidently, the dual of any lattice is a lattice, and the dual of any complete lattice is a complete lattice, with meets and joins interchanged. Any finite lattice or lattice of finite length is complete. More sophisti-

† For the concepts of graph and oriented graph ("digraph"), cf. O. Ore, *Theory of graphs*, Amer. Math. Soc., 1962.

‡ The concept of a lattice ("Dualgruppe") was first studied in depth by Dedekind [**1**, pp. 113–14]. That of a complete lattice was introduced by the author in [**1**, p. 442].

cated "chain conditions" implying completeness for lattices will be discussed in Chapter VIII.

Any chain is a lattice, in which $x \wedge y$ is simply the smaller and $x \vee y$ is the larger of x and y. Not every lattice is complete: thus the rational numbers are not complete, and the real numbers (in their natural order) are not complete unless $-\infty$ and $+\infty$ are adjoined as "universal bounds".

The lattice of all subsets of a given set X (Example 1, §1) is complete; the empty set $\varnothing$ is O, X itself is I. For any family A of subsets $S_\alpha \subset X$, inf A is just the intersection $\bigcap_A S_\alpha$ of the S_α, and sup A is the set-union $\bigcup_A S_\alpha$.

Definition. A *sublattice* of a lattice L is a subset X of L such that $a \in X$, $b \in X$ imply $a \wedge b \in X$ and $a \vee b \in X$.

A sublattice is a lattice in its own right with the same join and meet operations. The empty subset is a sublattice; so is any one-element subset. More generally, given $a \leqq b$ in a lattice L, the *(closed) interval* $[a, b]$ of all elements $x \in L$ which satisfy $a \leqq x \leqq b$ is a sublattice. A *convex* subset of a poset P is a subset which contains $[a, b]$ whenever it contains a and b, $a \leqq b$. A subset S of a lattice L is a *convex sublattice* when $a, b \in S$ imply $[a \wedge b, a \vee b] \subset S$.

A subset of a lattice L can be itself under the same (relative) order without being a sublattice. The following example is typical of a wide class of such (complete) lattices.

Example 7. Let Σ consist of the subgroups of any group G, and let $\leqq$ mean set-inclusion. Then Σ is a complete lattice, with $H \wedge K = H \cap K$ (set-intersection), and $H \vee K$ the least subgroup in Σ containing H and K (which is not their set-theoretical union).

In the preceding example, the set-union of two incomparable subgroups is never a subgroup; hence the lattice of Example 7 is *not* a sublattice of the lattice of all subsets of G. Examples 6 and 7 are typical of a wide class of complete lattices, characterized in terms of the following concept.

Definition. A property of subsets of a set I is a *closure* property when (i) I has the property, and (ii) any intersection of subsets having the given property itself has this property.

Closure properties will be systematically investigated in Chapters V and VIII. For the present, we note only the following result.

Theorem 6. *Let L be any complete lattice, and let S be any subset of L such that* (i) $I \in S$, *and* (ii) $T \subset S$ *implies* inf $T \in S$. *Then S is a complete lattice.*

Proof. For any (nonvoid) subset T of S, evidently inf T (in L) is a member of S by (ii), and is the greatest lower bound of T in S. Dually, let $U \subset S$ be the set of all upper bounds of T in S; it is nonvoid since $I \in S$. Then inf $U \in S$ is also an upper bound of T, moreover it is a *least* upper bound since inf $U \leqq u$ for all $u \in U$. This proves that S is a complete lattice.

Corollary. *Those subsets of any set which have a given closure property form a complete lattice, in which the lattice meet of any family of subsets S_α is their intersection, and their lattice join is the intersection of all subsets T_β which contain every S_α.*

Direct products. Besides occurring naturally in all branches of mathematics, new lattices can also be constructed from given ones by various processes. One such process consists in forming direct products. This is analogous to the processes of forming direct products of groups and direct sums of rings.

DEFINITION. The *direct product*† PQ of two posets P and Q is the set of all couples (x, y) with $x \in P$, $y \in Q$, partially ordered by the rule that $(x_1, y_1) \leqq (x_2, y_2)$ if and only if $x_1 \leqq x_2$ in P and $y_1 \leqq y_2$ in Q.

THEOREM 7. *The direct product LM of any two lattices is a lattice.*

PROOF. For any two elements (x_i, y_i) in LM $(i = 1, 2)$, the element $(x_1 \vee x_2, y_1 \vee y_2)$ contains both of the (x_i, y_i)—hence is an upper bound for the pair. Moreover every other upper bound (u, v) of the two (x_i, y_i) satisfies $u \geqq x_i$ $(i = 1, 2)$ and hence (by definition of l.u.b.) $u \geqq x_1 \vee x_2$; likewise, $v \geqq y_1 \vee y_2$, and so $(u, v) \geqq (x_1 \vee x_2, y_1 \vee y_2)$. This shows that

$$(x_1 \vee x_2, y_1 \vee y_2) = (x_1, y_1) \vee (x_2, y_2), \tag{3}$$

whence the latter exists. Dually,

$$(x_1 \wedge x_2, y_1 \wedge y_2) = (x_1, y_1) \wedge (x_2, y_2), \tag{3'}$$

which proves that L is a lattice.

5. Lattice Algebra

The binary operations $\wedge$ and $\vee$ in lattices have important algebraic properties, some of them analogous to those of ordinary multiplication and addition ($\cdot$ and $+$). First, one easily verifies:

LEMMA 1. *In any poset P, the operations of meet and join satisfy the following laws, whenever the expressions referred to exist:*

L1. $x \wedge x = x, \quad x \vee x = x.$ *(Idempotent)*

L2. $x \wedge y = y \wedge x, \quad x \vee y = y \vee x.$ *(Commutative)*

L3. $x \wedge (y \wedge z) = (x \wedge y) \wedge z, \quad x \vee (y \vee z) = (x \vee y) \vee z.$ *(Associative)*

L4. $x \wedge (x \vee y) = x \vee (x \wedge y) = x.$ *(Absorption)*

Moreover $x \leqq y$ is equivalent to each of the conditions

$$x \wedge y = x \quad \textit{and} \quad x \vee y = y. \qquad \textit{(Consistency)}$$

PROOF. The idempotent and commutative laws are evident. The associative laws L3 are also evident since $x \wedge (y \wedge z)$ and $(x \wedge y) \wedge z$ are both equal to g.l.b. $\{x, y, z\}$ whenever all expressions referred to exist. The equivalence between $x \geqq y$, $x \wedge y = y$, and $x \vee y = x$ is easily verified, and implies L4.

† Direct products are also called "cardinal products", for reasons to be explained in §III.1.

LEMMA 2. *If a poset P has an O, then*

$$O \wedge x = O \quad \textit{and} \quad O \vee x = x \quad \textit{for all } x \in P.$$

Dually, if P has a universal upper bound I, then

$$x \wedge I = x \quad \textit{and} \quad x \vee I = I \quad \textit{for all } x \in P.$$

The proofs will be left to the reader.

LEMMA 3. *In any lattice, the operations of join and meet are isotone:*

$$\textit{If } y \leqq z, \quad \textit{then } x \wedge y \leqq x \wedge z \quad \textit{and} \quad x \vee y \leqq x \vee z. \tag{4}$$

PROOF. By L1–L4 and consistency, $y \leqq z$ implies

$$x \wedge y = (x \wedge x) \wedge (y \wedge z) = (x \wedge y) \wedge (x \wedge z),$$

whence $x \wedge y \leqq x \wedge z$ by consistency. The second inequality of (4) can be proved dually (Duality Principle).

LEMMA 4. *In any lattice, we have the distributive inequalities*

$$x \wedge (y \vee z) \geqq (x \wedge y) \vee (x \wedge z), \tag{5}$$

$$x \vee (y \wedge z) \leqq (x \vee y) \wedge (x \vee z). \tag{5'}$$

PROOF. Clearly $x \wedge y \leqq x$, and $x \wedge y \leqq y \leqq y \vee z$; hence $x \wedge y \leqq x \wedge (y \vee z)$. Also $x \wedge z \leqq x, x \wedge z \leqq z \leqq y \vee z$; hence $x \wedge z \leqq x \wedge (y \vee z)$. That is, $x \wedge (y \vee z)$ is an upper bound of $x \wedge y$ and $x \wedge z$, from which (5) follows. The distributive inequality (5′) follows from (5) by duality.

LEMMA 5. *The elements of any lattice satisfy the modular inequality:*

$$x \leqq z \quad \textit{implies} \quad x \vee (y \wedge z) \leqq (x \vee y) \wedge z. \tag{6}$$

PROOF. $x \leqq x \vee y$ and $x \leqq z$. Hence $x \leqq (x \vee y) \wedge z$. Also $y \wedge z \leqq y \leqq x \vee y$ and $y \vee z \leqq z$. Therefore $y \wedge z \leqq (x \vee y) \wedge z$, whence $x \vee (y \wedge z) \leqq (x \vee y) \wedge z$, Q.E.D.

It is evident from Lemmas 1–5 that lattice theory has a strong algebraic flavor. We now prove that it can be regarded as a branch of algebra: *identities* L1–L4 *completely characterize lattices*.† To prove this fact, and in various other connections, the following notion is helpful.

DEFINITION. A system with a single binary, idempotent, commutative and associative operation is called a *semilattice*.

Lemma 1 has the following immediate corollary; a dual corollary for joins also holds (cf. §II.2).

† Actually, Dedekind used L1–L4 to define lattices. The use of inf and sup in posets was first studied by C. S. Peirce, Amer. J. Math. **3** (1880), 15–57 (esp. p. 33). E. Schröder [**1**, p. 197] corrected the erroneous impression of Peirce that all lattices were distributive.

COROLLARY. *Let P be any poset in which any two elements have a meet. Then P is a semilattice with respect to the binary operation* $\wedge$.

Such semilattices are called *meet*-semilattices. Conversely:

LEMMA 6. *Under the relation defined by*

$$(*) \qquad x \leqq y \quad \text{if and only if} \quad x \circ y = x,$$

any semilattice with binary operation $\circ$ *is a poset in which* $x \circ y = \text{g.l.b.}\ \{x, y\}$.

PROOF. The idempotent law $x \circ x = x$ implies the reflexive law $x \leqq x$. The commutative law $x \circ y = y \circ x$ makes $x \leqq y$ (i.e., $x \circ y = x$) and $y \leqq x$ (i.e., $y \circ x = y$) imply $x = x \circ y = y \circ x = y$, the antisymmetric law P1. The associative law makes $x \leqq y$ and $y \leqq z$ imply

$$x = x \circ y = x \circ (y \circ z) = (x \circ y) \circ z = x \circ z, \quad \text{whence } x \leqq z,$$

proving the transitive law P3. We leave it to the reader to prove that $x \circ y \leqq x$ and $x \circ y \leqq y$. Finally, if $z \leqq x$ and $z \leqq y$, then $z \circ (x \circ y) = (z \circ x) \circ y = z \circ y = z$, whence $z \leqq x \circ y$, proving that $x \circ y = \text{g.l.b.}\ \{x, y\}$.

THEOREM 8. *Any system L with two binary operations which satisfy* L1–L4 *is a lattice, and conversely.*

PROOF. First, by Lemma 6, any L which satisfies L1–L4 is a poset in which $x \wedge y = \text{g.l.b.}\ \{x, y\}$, so that $x \leqq y$ means $x \wedge y = x$. Second, by L4, $x \wedge y = x$ implies $x \vee y = (x \wedge y) \vee y = y$ and (by duality) conversely. Hence $x \leqq y$ is also equivalent to $x \vee y = y$. By duality, it follows that $x \vee y = \text{l.u.b.}\ \{x, y\}$; hence L is a lattice. The converse was proved in Lemma 1, completing the proof.

Exercises for §§4–5:

1. Prove that $(a \vee b) \wedge (c \vee d) \geqq (a \wedge c) \vee (b \wedge d)$ for all a, b, c, d in any lattice.
2. Prove (4) directly from L1–L4.
3. Prove that any interval of a lattice is a sublattice, and that so is any intersection of intervals.
4. (a) Draw diagrams of five nonisomorphic lattices of five elements; three are self-dual.
   (b) Show that every lattice of five elements is isomorphic to one of these.
   (c) Show that there are just four nonisomorphic nonvoid lattices of less than five elements.
   (d) Show that there are just 15 lattices of six elements, of which seven are self-dual. (*Hint.* Add O, I to posets of four elements.)
   (e) How many nonisomorphic lattices have seven elements? (*Hint.* Add O, I to posets of five elements. How many are lattices?)
5. Show that the "join" of any two sets, in the lattice of all sets closed under any closure operation, is the closure of their set-union.
6. Show that the following form complete lattices: (a) The normal subgroups of any group, (b) the characteristic subgroups of any group, (c) the right-ideals of any ring, (d) the ideals of any lattice, (e) the invariant subalgebras of any linear algebra.
7. Let Φ be any class of single-valued transformations ϕ of a set I. Show that the subsets X of I with the property that $\phi(X) \subset X$ for all $\phi \in \Phi$ form a complete lattice.
8. Show that the convex subsets of Euclidean space form a complete lattice.

*9. A subset S of a vector space V with scalars in an ordered field F is called *convex* if $x, y \in S$, $\lambda, \mu \geqq 0$ and $\lambda + \mu = 1$ imply $\lambda x + \mu y \in S$. Prove that the convex subsets of V form a complete lattice.

10. Show that if $n(P) \leqq 5$ and P contains universal bounds O and I, then P is a lattice.

11. Show that, for any subset S of a lattice L, the set of all lower bounds of S is a sublattice of L.

12. Prove that the complete lattice of all ideals of **Z** is isomorphic with the lattice of all nonnegative integers under divisibility. Describe the universal bounds of the latter.

13. (a) Show that the eight-element lattice of all subsets of a set of three points contains no seven-element sublattice.

(*b) Show that any lattice of $n > 6$ elements contains a sublattice of exactly six elements.

6. Distributivity

In many lattices, the analogy between the lattice operations $\wedge$, $\vee$ and the arithmetic operations $\cdot$, $+$ includes the distributive law $x(y + z) = xy + xz$. In such lattices, the distributive inequalities (5)–(5′) can be sharpened to identities. These identities do not hold in all lattices; for instance, they fail in the lattices M_5 and N_5 of Figures 2a and 2b.

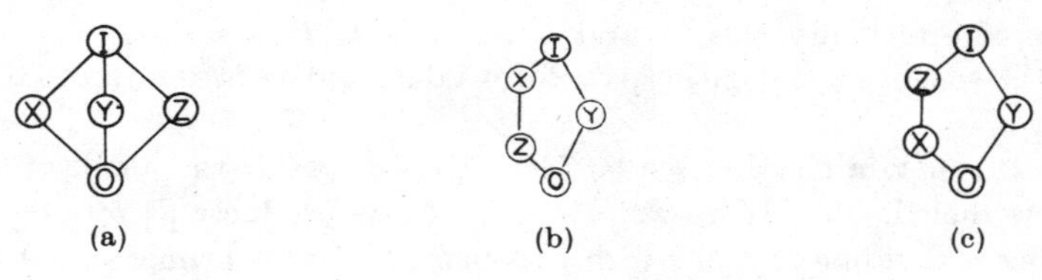

FIGURE 2

We now study distributivity, first proving a result which has no analog in ordinary algebra (since $a + (bc) \neq (a + b)(a + c)$ in general).

THEOREM 9. *In any lattice, the following identities are equivalent:*

L6′. $\quad x \wedge (y \vee z) = (x \wedge y) \vee (x \wedge z) \quad$ *all* x, y, z,

L6″. $\quad x \vee (y \wedge z) = (x \vee y) \wedge (x \vee z) \quad$ *all* x, y, z.

Caution. The truth of L6′ does *not* imply that of L6″ for *individual* elements x, y, z of a lattice, as Figures 2b-2c show.

PROOF. We shall prove that L6′ implies L6″. The converse implication L6″ $\Rightarrow$ L6′ will then follow by duality.

We have, for any x, y, and z,

$$\begin{aligned}
(x \vee y) \wedge (x \vee z) &= [(x \vee y) \wedge x] \vee [(x \vee y) \wedge z] && \text{by L6}' \\
&= x \vee [z \wedge (x \vee y)] && \text{by L4, L2} \\
&= x \vee [(z \wedge x) \vee (z \wedge y)] && \text{by L6}' \\
&= [x \vee (z \wedge x)] \vee (z \wedge y) && \text{by L3} \\
&= x \vee (z \wedge y) && \text{by L4.}
\end{aligned}$$

DEFINITION. A lattice is *distributive* if and only if the identity L6′ (and hence L6″) holds in it.

The lattices of Examples 1–5 of §1 are all distributive; those of Examples 6–7, however, are not distributive in general. Thus, the real numbers (Example 4) form a distributive lattice because of the following easily proved result.

LEMMA. *Any chain is a distributive lattice.*

In fact, $x \wedge y$ is the lesser of x and y; $x \vee y$ is the greater of x and y; $x \wedge (y \vee z)$ and $(x \wedge y) \vee (y \wedge z)$ are both equal to x if x is smaller than y or z; and both equal to $y \vee z$ in the alternative case that x is bigger than y and z.

The dual of any distributive lattice is distributive, and any sublattice of a distributive lattice is distributive. Any direct product of distributive lattices is distributive. The familiar fact that the lattice of Example 1 is distributive can be generalized as follows.

DEFINITION. A *ring* of sets is a family Φ of subsets of a set I which contains with any two sets S and T also their (set-theoretic) intersection $S \cap T$ and union $S \cup T$. A *field* of sets is a ring of sets which contains with any S also its set-complement S'.

Any ring of sets under the natural ordering $S \subset T$ is a distributive lattice. Thus the open sets in a topological space form a distributive lattice; so do the closed sets.

Example 2 is also a distributive lattice. Here $x \vee y$ is the g.c.d. of x and y, and $x \wedge y$ is their l.c.m. If we write x, y, and z as products $\prod p_i^{e_i}$ of powers of all the primes p_i dividing x, y, or z with exponent $e_i = 0$ for primes p_i not dividing the number, then $e_i(x \wedge y) = \min \{e_i(x), e_i(y)\}$, $e_i(x \vee y) = \max \{e_i(x), e_i(y)\}$, and so for all i, $e_i(x \wedge (y \vee z)) = e_i((x \wedge y) \vee (x \wedge z))$ is the *median* of the three exponents $e_i(x)$, $e_i(y)$, $e_i(z)$.

An important property of distributive lattices is the following.

THEOREM 10. *In a distributive lattice, if* $c \wedge x = c \wedge y$ *and* $c \vee x = c \vee y$, *then* $x = y$.

PROOF. Using repeatedly the hypotheses, L4, L2, and L6, we have

$$\begin{aligned} x &= x \wedge (c \vee x) = x \wedge (c \vee y) = (x \wedge c) \vee (x \wedge y) \\ &= (c \wedge y) \vee (x \wedge y) = (c \vee x) \wedge y = (c \vee y) \wedge y = y. \end{aligned}$$

This completes the proof. (The converse will be proved as a Corollary of Theorem II.13.)

7. Modularity

Setting $x \leqq z$ in the distributive law L6, so that $z = x \vee z$, we get the self-dual "modular" identity:

L5. If $x \leqq z$, then $x \vee (y \wedge z) = (x \vee y) \wedge z$.

Thus any distributive lattice satisfies L5.

Definition. A lattice is *modular* when it satisfies the modular identity† L5.

Not every lattice is modular: the five-element lattice N_5 of Figure 2b is nonmodular. Again, though every distributive lattice is modular, the five-element lattice M_5 of Figure 2a is modular but not distributive. The lattice M_5 is isomorphic with the lattice of all normal subgroups of the four-group. That M_5 is modular is a corollary of the following theorem.

Theorem 11. *The normal subgroups of any group G form a modular lattice.*

Proof. The normal subgroups of G certainly form a lattice, in which $M \wedge N = M \cap N$ is the intersection of M and N and $M \vee N = MN \neq M \cup N$ is the set of products xy with $x \in M$, $y \in N$. To prove that the lattice is modular, it suffices by the modular inequality (6) to show that $L \subset N$ implies $(L \vee M) \cap N \subset L \vee (M \cap N)$. To show this, suppose $a \in (L \vee M) \cap N$. Then, *if* $LM = ML$, so that $L \vee M = LM$, we have $a = bc$ where $b \in L$, $c \in M$, and $bc \in N$. Hence $c = b^{-1}a$, where $b^{-1} \in L \subset N$ and $a \in (L \vee M) \cap N \subset N$, proving $c \in N$. Since $c \in M$ as above, $c \in M \cap N$, and so $a = bc \in L \vee (M \cap N)$. This proves that $(L \vee M) \cap N \subset L \vee (M \cap N)$, as claimed.

Remark. The preceding shows also that if subgroups L, M, N of a group G satisfy $LM = ML$, and $L \subset N$, then $L \vee (M \cap N) = (L \vee M) \cap N$.

Any sublattice of a modular lattice is modular. Therefore the subspaces of any vector space and the ideals of any ring (Example 6 of §1) form modular lattices, being subgroups of the modular lattice of all (normal) subgroups of the relevant additive group. Likewise, any direct product of modular lattices is modular.

The lattice of Figure 2b is easily verified to be nonmodular. We now show that the lattice of Figure 2b is the only nonmodular lattice of five elements. Indeed we shall prove much more.

Theorem 12. *Any nonmodular lattice L contains the lattice N_5 of Figure 2b as a sublattice.*

Proof. By definition, L contains elements x, y, and z such that $x < z$ and $x \vee (y \wedge z) < (x \vee y) \wedge z$. Then the lattice formed by y, $x \vee y$, $y \wedge z$, $(x \vee y) \wedge z$ and $x \vee (y \wedge z)$ is isomorphic to N_5. For, trivially, $y \wedge z \leqq x \vee (y \wedge z) < (x \vee y) \wedge z \leqq x \vee y$, while $[x \vee (y \wedge z)] \vee y = x \vee y$ and dually. Moreover $y \wedge z = x \vee (y \wedge z)$ is impossible, since it would imply $x \leqq y \wedge z$, whence $(x \vee y) \wedge z = x \vee (y \wedge z)$, contrary to hypothesis.

A basic property of modular lattices is the following "transposition principle" due to Dedekind [**2**, p. 245].

Theorem 13. *In any modular lattice M, the mappings $\phi_a: x \to x \wedge a$ and $\psi_b: y \to y \vee b$ are inverse isomorphisms from $[b, a \vee b]$ to $[a \wedge b, a]$ and vice-versa.*

† The basic properties of modular lattices ("Dualgruppen von Modultypus") and distributive lattices were discovered by R. Dedekind [**1**]; see also E. Schröder [**1**].

PROOF. If $x \in [b, a \vee b]$, then $x\phi_a \in [a \wedge b, a]$ by the isotonicity of ϕ_a. Moreover $(x \wedge a) \vee b = x \wedge (a \vee b)$ by L5, since $x \in [b, a \vee b]$. That is, $x\phi_a\psi_b = x$; the proof that $y\psi_b\phi_a = y$ for all $y \in [a \wedge b, a]$ is dual.

COROLLARY. *In any modular lattice L:*

(ξ) *If $a \neq b$ and both a and b cover c, then $a \vee b$ covers a as well as b.*

(ξ') *Dually, if $a \neq b$ and c covers both a and b, then a and b both cover $a \wedge b$.*

(In Theorem II.16, it will be shown that conditions (ξ)–(ξ') are sufficient as well as necessary for modularity, in lattices of finite length.)

PROOF. If a and $b \neq a$ cover c, then $c = a \wedge b$. Hence, by the theorem, $[a, a \vee b] \cong [a \wedge b, b] \cong \mathbf{2}$, and so $a \vee b$ covers a. By the same argument, $a \vee b$ covers b. The proof of (ξ') is dual, Q.E.D.

Theorem 13 has many other consequences, whose formulation is much simplified by two further definitions.

DEFINITION. Two intervals of a lattice are called *transposes* when they can be written as $[a \wedge b, a]$ and $[b, a \vee b]$ for suitable a, b. Likewise, two intervals $[x, y]$ and $[x', y']$ are called *projective* (in symbols $[x, y] \sim [x', y']$) if and only if there exists a finite sequence $[x, y], [x_1, y_1], [x_2, y_2], \cdots, [x', y']$ in which any two successive quotients are transposes.

COROLLARY 1. *In Theorem* 13, *subintervals of $[a \wedge b, a]$ and $[a, a \vee b]$ are mapped onto (isomorphic) transposed intervals by ψ_b and ϕ_a, respectively.*

COROLLARY 2. *Projective intervals are isomorphic, in any modular lattice.*

Exercises for §§6–7:

1. Show that the two lattices of Figures 2a–2b are the only nondistributive lattices of five elements.
2. Show that L6 and L1–L3 imply $x \vee (x \wedge y) = x \wedge (x \vee y)$.
3. Show that if we add to a distributive lattice L new elements O, I satisfying $O < x < I$ for all $x \in L$, the result is a distributive lattice.
4. Show that the "Riemann" partitions of an interval into a finite number of nonoverlapping subintervals form a distributive lattice.
5. Show that L5 is equivalent to: if $x < z$, then $x \vee (y \wedge z) \nless (x \vee y) \wedge z$.
6. (a) Show that the lattice N_5 of Figure 1b (§3) is not modular.
   (b) Show that the lattice of Figure 1a is modular.
   (c) Show that N_5 is the only nonmodular five-element lattice.
7. (a) Show that the modular law is self-dual.
   (b) Show that any lattice of length two is modular.
8. In Figure 2c, $x \wedge (y \vee z) = x = (x \wedge y) \vee (x \wedge z)$, yet the dual is false since $x \vee (y \wedge z) = x < z = (x \vee y) \wedge (x \vee z)$. Why does this not contradict Theorem 9?
9. Show that the submodules of any R-module (over a ring R) form a modular lattice.
10. In a modular lattice, let $a \wedge b = O$, $a \vee b = I$, and let $O < c < a$, $O < d < b$. Show that $\{a, b, c, d\}$ generates a sublattice isomorphic with $\mathbf{3} \times \mathbf{3}$, where $\mathbf{3}$ is a chain with 3 elements.
11. Given nonvoid posets P and Q, show that the group of automorphisms of PQ contains a subgroup isomorphic with (Aut P) $\times$ (Aut Q).

8. Semimodularity

Lattices of finite length which satisfy (ξ) *or* (ξ') are called semimodular.† More precisely, a lattice of finite length which satisfies (ξ) is called (upper) *semimodular*; one which satisfies (ξ') is called *lower semimodular*, or "dually" semimodular.

It is easy to show that any interval sublattice or direct product of (upper) semimodular lattices is itself (upper) semimodular. However, since every nonmodular lattice contains a sublattice isomorphic with the lattice N_5 of Figure 1b, which does not satisfy (ξ) or (ξ'), a sublattice of an (upper) semimodular lattice need not be semimodular. In particular, this is true of the seven-element lattice of Figure 1c. Actually (Dilworth), *every* finite lattice is isomorphic with a sublattice of a semimodular lattice; this will be shown in Chapter IV.

The following two examples define typical upper semimodular lattices of finite length.

Example 8. Let F be any field, and let $A(F; n)$ be the set of all *subspaces* (or "flats") of the n-dimensional affine space over F (subsets which contain, with any two points, the entire straight line joining them). Then $A(F; n)$ is an upper semimodular lattice of length $n + 1$, in which $d[x]$ is one less than the geometric dimension. Figure 3a depicts the diagram of $A(\mathbf{Z}_2, 2)$.

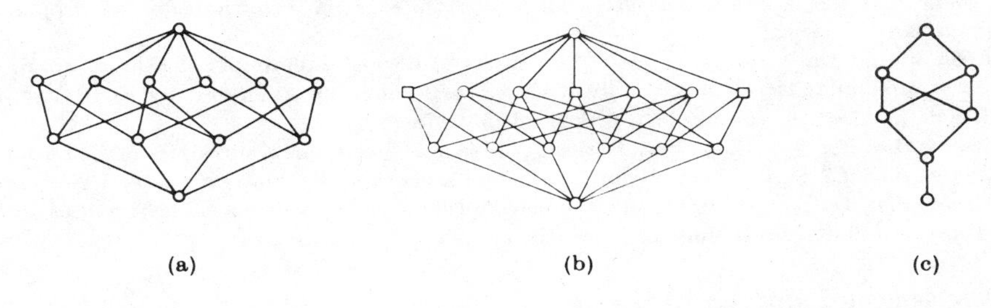

(a) (b) (c)

FIGURE 3

Example 9. Let S be a set of n elements. The symmetric *partition lattice* of length $n - 1$ is the poset Π_n of equivalence relations on (partitions of) S, partially ordered by making $\rho \leqq \tau$ mean that $x\rho y$ implies $x\tau y$—i.e., that the partition $\pi(\rho)$ is a refinement of $\pi(\tau)$.

In Π_n, the *meet* $\rho \wedge \tau$ has for equivalence classes the intersections $R_i \cap T_j$ of those for ρ (the R_i) and those for τ (the T_j), so that $x(\rho \wedge \tau)y$ if and only if $x\rho y$ and $x\tau y$. The *join* $\rho \vee \tau$ is the meet of all equivalence classes which contain both ρ and τ. In Π_n, O is the equality relation, while I is the degenerate partition having S for its only equivalence class.

Further, τ *covers* ρ in Π_n if and only if $\pi(\tau)$ is obtainable from $\pi(\rho)$ by uniting two equivalence classes. Finally, $h[\rho] = n - \nu(\rho)$, where $\nu(\rho)$ is the number of

† Semimodular lattices were apparently first considered by the author [**1**, p. 446]; see also Fr. Klein, Math. Z. **42** (1936), 58–81. Example 9 was first studied by the author in [**3**, pp. 446–52].

equivalence classes into which ρ partitions S. Hence Π_n is a *graded lattice*, for any finite n. Figure 3b depicts Π_4.

A closely related poset is the following.

Example 10. Let $\mu: N = m_1 + \cdots + m_r$ and $\nu: N = n_1 + \cdots + n_s$ be any two partitions of the positive integer N into positive integers m_i resp. n_j. Define $\mu \leqq \nu$ to mean that the partition ν can be obtained from μ (after possible rearrangement) by grouping suitable summands.

The resulting poset P_N cannot satisfy (ξ) for $N > 4$, because it is not a lattice (Figure 3c depicts P_5). However, it does satisfy the Jordan–Dedekind chain condition; see §II.8.

Exercises:

1. Show that any nonmodular lattice contains a non-semimodular sublattice.
2. Let M be any modular lattice, and $s < I$ any maximal proper element of M. Show that if we delete the $x \leqq s$ in M, we get a semimodular lattice L. Show that if $p > O$ in L, the $y \geqq p$ form a modular lattice.
3. (a) Show that if, in any semimodular lattice L, we equate to I all elements with $d[x] \geqq n$ (n any fixed integer), we get a semimodular lattice.

   (b) Show that this is a join-epimorphic image of L. (S. Mac Lane)
4. Show that the symmetric partition lattice of degree n is nonmodular if $n > 2$.
5. Show that the lattice of all partitions of any finite graph into connected subgraphs is semimodular.

*6. Show that the "normal cosets" (i.e., cosets of normal subgroups) of a finite group G form a semimodular lattice if and only if the normal subgroup generated by each element $a \in G$ (not the identity) is a minimal proper normal subgroup.

7. Show that if $k \leqq 5$, then every lattice of order $n > k$ contains a sublattice of k elements.
8. Show that if $2 \leqq n \leqq 7$, every lattice of order n contains a sublattice of $n - 1$ elements.
9. Show that if $k \leqq 6$, every modular lattice of order $n > k$ contains a sublattice of order k.

(Examples 7–9 describe results of Fr. Klein.)

9. Complemented Modular Lattices

By a *complement* of x in a lattice L with O and I is meant an element $y \in L$ such that $x \wedge y = O$ and $x \vee y = I$; L is called *complemented* if all its elements have complements. A lattice is called *relatively complemented* if all its (closed) intervals are complemented. Theorem 10 asserts that, in a given interval $[a, b]$ of a distributive lattice, an element c can have at most one relative complement.†

In §1, only Example 1 is necessarily complemented. The modular lattice of all subspaces of the n-dimensional vector space $F^n = V_n(F)$ over any field (or division ring) F is complemented. The case $V_2(\mathbf{Z}_2)$ gives the modular lattice M_5 of Figure 1a.

THEOREM 14. *Any complemented modular lattice M is relatively complemented.*

PROOF. First, if $O \leqq x \leqq b$ in M, then $x \wedge (x' \wedge b) = (x \wedge x') \wedge b = O$ trivially, while by L5

$$x \vee (x' \wedge b) = (x \vee x') \wedge b = I \wedge b = b.$$

† For more general definitions, see G. Szasz [Sz, §18], and Publ. Math. Math. Debrecen **3** (1953), 9–16.

Hence $[O, b]$ is a complemented modular sublattice B of M. Dually, $[a, b] \subset B$ is a complemented modular sublattice of B, Q.E.D.

The nonmodular five-element lattice N_5 of Figure 1b is complemented but not relatively complemented.

THEOREM 15. *In a relatively complemented lattice L of finite length, every element a is the join of those points which it contains.*

PROOF. Given $a > O$, either a is a point or $a > b > O$ for some $b \in L$; let c be the relative complement of b in a. By induction on the length of $[O, a]$, we can assume that b and c are each joins of points; hence so is $a = b \vee c$.

COROLLARY. *In a complemented modular lattice of finite length, every element is the join of those points which it contains.*

Example 11. Let M be the lattice of all subspaces of Euclidean n-space E_n. Then M is modular (by Theorem 11), and complemented since the orthogonal complement $S^\perp$ of any subspace S satisfies $S \cap S^\perp = 0$, $S + S^\perp = E_n$.

10. Boolean Lattices; Boolean Algebras

By definition, a *Boolean lattice* is a complemented distributive lattice. We recall that, by Theorem 10, complements are unique in any distributive lattice. There follows

THEOREM 16. *In any Boolean lattice, each element x has one and only one complement x'. Moreover*

L8. $$x \wedge x' = O, \qquad x \vee x' = I,$$

L9. $$(x')' = x,$$

L10. $$(x \wedge y)' = x' \vee y', \qquad (x \vee y)' = x' \wedge y'.$$

PROOF. We have seen that the correspondence $x \to x'$ is single-valued. But by the symmetry of the definition of complement, x is a complement of x'; hence $x = (x')'$ by unicity, proving L9. Therefore the correspondence $x \to x'$ is one-one. We next prove

(7) $$x \wedge a = O \quad \text{if and only if } x \leqq a'.$$

This follows since: (i) if $x \leqq a'$, then $x \wedge a \leqq a' \wedge a = O$, and (ii) if $x \wedge a = O$, then

$$x = x \wedge I = x \wedge (a \vee a') = (x \wedge a) \vee (x \wedge a') = O \vee (x \wedge a') = x \wedge a'.$$

From (7), it follows that $a \leqq b$, which clearly implies $b' \wedge a \leqq b' \wedge b = O$, also implies $b' \leqq a'$: the one-one correspondence $x \to x'$ is antitone (order-inverting). Since $x' \to (x')' = x$ is also, it follows that $x \to x'$ is dual isomorphism, proving L10.

It follows that any Boolean lattice is dually isomorphic with itself (self-dual). Since complements are *unique* in Boolean lattices, the latter can also be regarded as algebras with two binary operations ($\wedge$, $\vee$) *and* one unary operation $'$. When so considered, they are called Boolean *algebras*.

DEFINITION. A *Boolean algebra* is an algebra with operations $\wedge$, $\vee$, $'$ satisfying L1–L10. A (Boolean) *subalgebra* of a Boolean algebra A is a nonvoid subset of A which contains with any a, b also $a \wedge b$, $a \vee b$, and a'.

Thus a proper interval sublattice $[a, b]$ of A, though a Boolean sublattice, is not a Boolean subalgebra.

We now show that any distributive lattice with O and I has a largest "Boolean subalgebra".

THEOREM 17. *The complemented elements of any distributive lattice with universal bounds form a sublattice.*

PROOF. If x and y are complemented, then

$$(x \wedge y) \wedge (x' \vee y) = (x \wedge y \wedge x') \vee (x \wedge y \wedge y') = O \vee O = O$$

and dually. Hence $x \wedge y$ has the complement $x' \vee y'$ as in L8. Dually $x \vee y$ has the complement $x' \wedge y'$, completing the proof.

Any field of sets is a Boolean algebra, and in particular the field of all subsets of a given set is one. Again, any subalgebra of a Boolean algebra is also a Boolean algebra. So is any direct product or interval sublattice of Boolean algebras.

Example 12. The class of all binary relations between elements of two classes I and J is a Boolean algebra, because this class is isomorphic with the class of all subsets of $I \times J$, under the correspondence mapping each relation ρ onto its graph: the set of all (x, y) with $x\rho y$.

Exercises for §§9–10:

1. Show that in any complemented modular lattice, all join-irreducible elements are atoms.
2. Show that if z is a relative complement of x in $x \vee y$ or of $x \wedge y$ in y, then it is a complement of $x \vee (x \vee y)'$.
3. Show that all relatively complemented lattices of eight or fewer elements are modular. (H. Rubin)
4. (a) Show that the lattice of Figure 1c is dually isomorphic to a sublattice of the lattice of all subgroups of the octic group.

   (b) Show that in Figure 1c, the middle element has no complement.
5. (a) Show that any lattice L is a sublattice of a complemented lattice, containing only three additional elements.

   (b) Show that if L is finite, only one additional element is needed.
6. Prove that every interval $[a, b]$ of a Boolean lattice L is a Boolean lattice, in which the meanings of $\leqq$, $\wedge$, $\vee$ are the same as in L, but complements in $[a, b]$ are "relative complements" in L.
7. (a) Show that in any Boolean lattice of finite length, every element is a join of points.

   (b) Conversely, show that if L is a distributive lattice of finite length in which I is a join of points, then L is a Boolean lattice.
8. Show that a nonvoid subset of a Boolean algebra is a subalgebra if it is closed under $\wedge$ and $'$.
9. Find a modular lattice of seven elements in which the complemented elements do not form a sublattice.

10. Show that each of the following properties holds in PQ if and only if it holds in the posets P and Q separately:

(a) PQ is a meet-semilattice,

(b) PQ is a modular lattice,

(c) PQ is complemented,

(d) PQ is a distributive lattice.

11. (a) Construct a relatively complemented lattice of order 9 not satisfying the Jordan–Dedekind chain condition.

(b) Show that every relatively complemented lattice of order 8 or less satisfies this condition.

(c) Construct a six-element complemented lattice which is not relatively complemented satisfying the Jordan–Dedekind chain condition.

12. Construct a relatively complemented lattice of order 14 which satisfies the Jordan–Dedekind chain condition but is not semimodular.

13. (a) Show that Π_3 is a relatively complemented semimodular lattice of order 12 which is not modular.

(b) Show that every relatively complemented semimodular lattice of order 11 or less is modular.

*14. Let L be a modular lattice with universal bounds. Prove that: (i) if $a \wedge b$, $a \vee b$ have complements in L, then a, b have complements in L, and (ii) if a, b have *unique* complements a', b' in L, then $a' \vee b'$ and $a' \wedge b'$ are complements of $a \wedge b$, $a \vee b$. (Bumcrot)

Problems

1. Given n, what is the smallest integer $\psi(n)$ such that every lattice with order $r \geqq \psi(n)$ elements contains a sublattice of exactly n elements?

2. Compute for small n, and find asymptotic estimates and bounds for the rates of growth of the functions $G(n)$ and $G^*(n)$ of Example 5 after §2.†

3. Same question for the numbers $H(n)$ and $H^*(n)$ of nonisomorphic resp. different lattice-orderings of a set of n elements.

4. (a) Compute the number $\gamma(n, \lambda)$ of connected posets of length λ having n elements. (b) How many of these are lattices? (c) Derive asymptotic formulas for this function as $n \to \infty$, for small fixed λ resp. $n - \lambda$.

5. Enumerate all finite lattices which are uniquely determined (up to isomorphism) by their diagram, considered purely as a graph. (Cf. Example 7 after §3.)

6. Determine all finite lattices in which every graph-automorphism is a lattice-automorphism. (A. G. Waterman)

† See A. P. Hillman, Proc. AMS **6** (1955), 542–8; V. Krishnamurthy, Amer. Math. Monthly **73** (1966), 154–7.

CHAPTER II

POSTULATES FOR LATTICES

1. Quasi-orderings

Most types of algebraic systems can be characterized by many sets of postulates. Thus, as we saw in §I.1, the usual postulates P1–P3 for posets are equivalent to the following two postulates on strict inclusion:

P1′. For no x is $x < x$. (Antireflexive)
P3. If $x < y$ and $y < z$, then $x < z$. (Transitive)

Other postulate systems for posets can be devised, describing for example properties of the ternary *betweenness* relation $(axb)\beta$, which is defined to mean that $a \leqq x \leqq b$ *or* $b \leqq x \leqq a$; see §9 below.

Such postulate systems, though intriguing, may be neither suggestive nor useful. In this chapter, we shall study a few sets of postulates for various types of lattices which are especially suggestive and useful. We shall also consider the effect of omitting one or more postulates from these sets.

In this spirit, we first define a *quasi-ordering* of a set S as a relation $\prec$ satisfying P1 and P3, but not necessarily P2. The couple $(S, \prec)$ is then called a *quasi-ordered set*.

Quasi-ordered sets can be constructed from *directed graphs*. These are sets of points connected by directed line segments. Figure 4a depicts such an oriented graph.

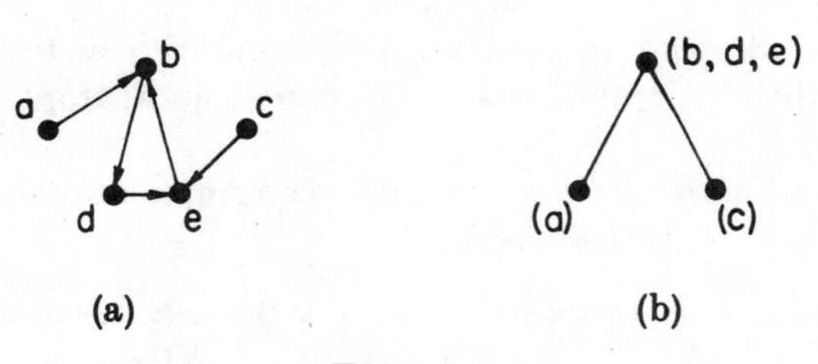

FIGURE 4

Given an oriented graph with vertices $x, y, \cdots, x \prec y$ is defined to mean that either $x = y$ or there exists a path from x to y in the direction of the arrows. Thus $b \prec e$ in Figure 4a because of the path $b \to d$, $d \to e$. This relation is clearly transitive. On the other hand, both $b \prec e$ and $e \prec b$ in Figure 4a, so the antisymmetric law fails to hold.

We will now show how to construct a *poset* from any given quasi-ordering.

LEMMA 1. *In any quasi-ordered set* $Q = (S, \prec)$, *define* $x \sim y$ *when* $x \prec y$ *and* $y \prec x$. *Then*:

(i) $\sim$ *is an equivalence relation on* S;

(ii) *if* E *and* F *are two equivalence classes for* $\sim$, *then* $x \prec y$ *either for no* $x \in E$, $y \in F$ *or for all* $x \in E$, $y \in F$;

(iii) *the quotient-set* $S/\sim$ *is a poset if* $E \leqq F$ *is defined to mean that* $x \prec y$ *for some (hence all)* $x \in E$, $y \in F$.

PROOF OF (i). Since $x \prec x$ for all $x \in S$, $\sim$ is reflexive.

Again, $x \sim y$ and $y \sim z$ imply $x \prec y$ and $y \prec z$ (by definition), hence $x \prec z$ by P3. Likewise, $z \prec x$, and so $x \sim z$, whence $\sim$ is transitive. It is symmetric by definition.

PROOF OF (ii). If $x \prec y$ for some $x \in E$, $y \in F$, and then $x_1 \prec x \prec y \prec y_1$ for all $x_1 \in E$, $y_1 \in F$, whence $x_1 \prec y_1$ by transitivity.

PROOF OF (iii). Clearly $E \sim E$ (since $x \sim x$) for all E. Again, $E \leqq F$ and $F \leqq G$ implies $x \prec y \prec z$ for all $x \in E$, $y \in F$, $z \in G$, whence $x \prec z$ by P3 on $\prec$, and $\leqq$ is transitive. Finally, $E \leqq F$ and $F \leqq E$ implies for all $x \in E$, $y \in F$ that $x \prec y$ and $y \prec x$, whence $x \sim y$ and $E = F$. Q.E.D.

In the oriented graph of Figure 4a, the equivalence classes are the sets $\{a\}$, $\{b, d, e\}$, $\{c\}$ and the corresponding poset has the diagram sketched in Figure 4b.

Because of Lemma 1, a quasi-ordering is often called a "pre-order"; Lemma 1 has many applications.

Example 1. In a commutative semigroup S with unity, let $a|b$ mean that $ax = b$ for some $x \in S$. The relation $|$ quasi-orders S; elements of S are "equivalent" in the sense of Lemma 1 if and only if they are "associates" in the sense of number theory.

2. Lattice Postulates; Semilattices

Postulates L1–L4 for a lattice are not independent: the idempotent laws follow from L4. Thus

$$x \vee x = x \vee [x \wedge (x \vee x)] = x,$$

where the first equality follows from the first contraction law $x = x \wedge (x \vee y)$ with $y = x$, and the second from the dual law with $y = x \vee x$. Dually, one can prove $x \wedge x = x$.

The remaining six identities of L2–L4 are independent. That is, no one of them can be proved from the other five. To show this, it suffices by duality to exhibit appropriate sets and operations defined on them which satisfy five of the six identities of L2–L4. This we now do.

Example 2. Let $\mathbf{Z}^+$ be the set of positive integers. Define $x \vee y = \max(x, y)$ and $x \wedge y = x$. Then all identities of L2–L4 except $x \wedge y = y \wedge x$ hold.

Example 3. Consider the diagram of Figure 5. Let joins be defined as usual, and meets also, except that $a \wedge b$ is the bottom element and not c. The resulting algebraic system satisfies L2, L4, and $x \vee (y \vee z) = (x \vee y) \vee z$, but not $x \wedge (y \wedge z) = (x \wedge y) \wedge z$. This fails for the triple a, b, c.

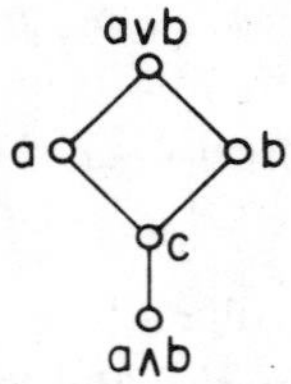

$a \vee b$

a b

c

$a \wedge b$

FIGURE 5. Nonassociative "latticoid"

Example 4. Let $\mathbf{Z}^+$ be the set of positive integers. Set $x \vee y = \max(x, y)$, $x \wedge y = 1$. Then the contraction law $x \wedge (x \vee y) = x$ fails to hold, but the other five identities of L2–L4 are satisfied. This completes the proof of

THEOREM 1. *Postulates* L2–L4 *for lattices imply* L1, *but the six identities of* L2–I4 *are independent.*

Semilattices. One gets interesting generalizations of the lattice concept (to skew-lattices and latticoids; see below) by dropping one or more of the six identities of L2–L4. But the most important generalization of the algebraic lattice concept is provided by semilattices.† As in §I.5, we define a *semilattice* as a set S with a binary operator $\circ$ which is idempotent, commutative, and associative.

LEMMA 2. *Any semilattice S is a poset under the divisibility relation $a|b$, meaning $a \circ x = b$ for some x. Moreover, in this poset, $c \circ d = c \vee d$ for any c, d.*

PROOF. Much as in Example 1, S is quasi-ordered by the relation $a|b$; $a|a$ since $\circ$ is idempotent. Moreover $a \circ x = b$ implies

$$a \circ b = a \circ (a \circ x) = (a \circ a) \circ x = a \circ x = b,$$

while the converse is trivial; hence $a|b$ if and only if $a \circ b = b$. It follows that $a|b$ and $b|a$ imply together $a = b \circ a = a \circ b = b$; the quasi-ordering is a partial ordering. Finally, $c|c \circ d$ trivially (since $c \circ d = c \circ d$), $d|c \circ d$ similarly (by commutativity), while $c|x$ and $d|x$ imply $x = c \circ x = c \circ d \circ x$, whence $(c \circ d)|x$; hence $c \circ d = c \vee d$.

DEFINITION. The poset of Lemma 2 is the *join*-semilattice defined by S; its dual is the *meet*-semilattice defined by S.

Clearly any lattice L is both a join-semilattice under $\vee$ and a meet-semilattice under $\wedge$; hence the terminology defined above is appropriate.

† Semilattices were already studied by Huntington [1, p. 294]; see *also* Fr. Klein, Math. Z. **48** (1943), 275–88 and 715–34.

Many semilattices are also lattices. For example, this is true if a meet-semilattice L has a universal upper bound I, and its length is finite. To see this, let $U = U(a, b)$ be the subset of L consisting of all upper bounds of two given elements a and b. The set U contains I, and it contains with any two elements x and y also $x \wedge y$—since $x \geqq a$ and $y \geqq a$ implies $x \wedge y \geqq a$ by definition, and similarly for b. We now construct a chain in U recursively, as follows.

Let $x_0 = I$. If x_n is not a (the) least element of U, take any $y \in U$ not satisfying $y \geqq x_n$, and let $x_{n+1} = x_n \wedge y < x_n$. As shown above, the chain $x_0 > x_1 > x_2 > \cdots$ lies in U. Since every chain in L is finite, it must have a *last* element $x_n \in U$. This x_n is the least element of U; hence $x_n = a \vee b$ exists and L is a lattice. Q.E.D.

Example 5. Let Q_n be the set of all quasi-orderings of a set S of n elements; then Q_n is a complete lattice. The subset P_n of all partial orderings of S is a meet-semilattice but not a lattice.

The lattice Q_n contains the symmetric partition lattice Π_n. Though not semimodular, it would seem to merit further study.

Bands, skew-lattices and latticoids. Between the concept of a semigroup and that of a semilattice lies the concept of a band.† A *band* is a semigroup whose elements are all idempotent (L1, L3 hold). It is known that every band is a semilattice of "rectangular" bands, each defined on a Cartesian product $S = X \times Y$, with

$$(x, y) \circ (x', y') = (x, y') \quad \text{for all } x, x' \in U \text{ and } y, y' \in V.$$

If one assumes L3 and $(a \wedge b) \vee a = a \vee (b \wedge a) = a$, one gets a "dual band" or skew-lattice, in which L1 holds. Such skew-lattices have been studied by P. Jordan;‡ they will not be discussed further below.

Concepts of "latticoid" in which one or both associative laws L3 are weakened, have also been studied by various authors.¶

Exercises for §§1–2:

1. In any semilattice, define $x_1 \circ \cdots \circ x_n$ by induction as $x_1 \circ (x_2 \circ \cdots \circ x_n)$.

(a) Prove from L3 by induction the *generalized associative* law: If $y_i = x_{s_{i-1}} \circ \cdots \circ x_{s_i}$ $[O = s_0 < x_1 < \cdots < x_m = n]$, then $y_1 \circ \cdots \circ y_m = x_1 \circ \cdots \circ x_n$.

(b) Prove that $x_1 \circ \cdots \circ x_n$ is invariant under every permutation of the x_i, using only L2 and L3.

2. Let $\wedge$ be the class of topological linear spaces, and let $S \rho T$ mean than T is topologically isomorphic with a subset of S. Define a poset of "linear dimensions", as in Theorem 3. Cf. S. Banach [1, Chapter XII].

3. (a) Let Φ be the class of groups, and let $G \rho H$ mean that H is isomorphic with a subgroup of G. Show that ρ is a quasi-ordering of groups, and a partial ordering of finite groups.

(b) Generalize to other classes of algebraic systems.

† This concept, due to Fr. Klein, was studied later by D. McLean, Amer. Math. Monthly **61** (1954), 110–13. For the theory of bands, consult Clifford-Preston [1]

‡ P. Jordan, Abh. Math. Sem. Univ. Hamburg **21** (1957), 127–38, and Crelle's J. **211** (1962), 136–61. See also J. A. Kalman, Math. Ann. **137** (1959), 362–70 and M. D. Gerhardt, ibid. **161** (1965), 231–240.

¶ N. Kimura, J. Sci. Tokushima Univ., 1950; W. Felscher, Arch. Math. **8** (1957), 171–4; G. Szasz, Publ. Math. Debrecen **10** (1963), 108–15.

4. (a) Show that in the case of finite groups, $G \sim H$ means that G is isomorphic with H in Exercise 3. Generalize.

(b) State analogous results for finite-dimensional vector spaces and for arbitrary cardinal numbers.

5. For positive continuous functions with domain $O \leqq x < \infty$, define $f = O(g)$ to mean that for some constants, $K, N, f(x) \leqq Kg(x)$ for all $x > N$. Show that this is a quasi-ordering, and define the associated equivalence relation $f \sim g$.

6. Show that from *any* relation ρ, a transitive relation $\tau = \tau(\rho)$ can be obtained by letting $x\tau y$ mean that for some finite set $a_0, \cdots, a_n$, $a_0 = x$, $a_n = y$, and $a_{i-1}\rho a_i$ $[i = 1, \cdots, n]$.

7. Show that the direct product PQ of two posets P and Q is a join-semilattice if and only if both factors are join-semilattices.

3. Morphisms and Ideals

The general "morphism" concept has four different (though related) interpretations in lattices, each of which has important applications. These will now be described.

DEFINITION. Let $\theta: L \to M$ be a function from a lattice L to a lattice M. Then θ is *isotone* when $x \leqq y$ implies $\theta(x) \leqq \theta(y)$; a *join*-morphism when

$$\theta(x \vee y) = \theta(x) \vee \theta(y) \quad \text{for all } x, y \in L; \tag{1}$$

a *meet*-morphism when the dual holds:

$$\theta(x \wedge y) = \theta(x) \wedge \theta(y) \quad \text{for all } x, y \in L; \tag{1'}$$

and a *morphism* (or "lattice-morphism") when (1)–(1′) both hold.

As always (see §VI.3), a morphism is called: (i) an *iso*morphism if a bijection, (ii) an *epi*morphism if onto, (iii) a *mono*morphism if one-one, (iv) an *endo*morphism if $L = M$, (v) an *auto*morphism if $L = M$ and it is an isomorphism.

Clearly, the concepts of join- and meet-morphism apply more generally to join-semilattices and meet-semilattices, respectively, while that of an isotone function applies to any poset! The following results are also obvious.

LEMMA 1. *Any join-morphism between join-semilattices is isotone; so is any meet-morphism between meet-semilattices.*

LEMMA 2. *Any isotone bijection with isotone inverse is a lattice-isomorphism.*

That is, for bijections the distinctions made above are unnecessary. However, they are needed for surjections *and* for injections. (For example, one can often "strengthen" the order in a lattice so as to obtain a join-monomorphic image, and closure operations are join-epimorphisms which are not usually meet-epimorphisms.)

The concept of the kernel of a morphism, familiar in the theory of groups, has more subtle properties when applied to lattices. The relevant concept is the following.†

† The concept of ideal was introduced by Stone, Proc. Nat. Acad. Sci. U.S.A. **20** (1934), 197–202, and **21** (1935), 103–5. See also A. Tarski [**1**] and [**2**]; Gr. C. Moisil, Ann. Sci. Jassy **22** (1936), 1–118. The distinction between join-morphisms and meet-morphisms was made by Ore [**1**, Chapter II]. The final results presented in §4 were due to many authors.

Definition. An *ideal* is a nonvoid subset J of a lattice (or join-semilattice)† L with the properties

$$a \in J, \quad x \in L, \quad x \leqq a \quad \text{imply} \quad x \in J, \tag{2}$$

$$a \in J, \quad b \in J \quad \text{imply} \quad a \vee b \in J. \tag{3}$$

The dual concept (in a lattice or meet-semilattice) is called a *dual ideal* (or meet-ideal).

It is easy to show that J is an ideal when $a \vee b \in J$ holds if and only if $a \in J$ and $b \in J$ (Caste property).

Example 6. In the "power set" $P(E)$ of all subsets of a set E, a dual ideal (not $P(E)$ itself) is called a *filter* of sets.

Theorem 2. *Under any join-morphism θ of a join-semilattice L onto a join-semilattice M with O, the set* Ker θ *of antecedents of O (the kernel of θ) is an ideal in L.*

Proof. If $a\theta = O$ and $b\theta = O$, then $(a \vee b)\theta = a\theta \vee b\theta = O$. Conversely, if $(a \vee b)\theta = O$, then $a\theta \vee b\theta = O$; but this implies $a\theta = b\theta = O$. Thus the set is an ideal by the definition given above.

There are both advantages and disadvantages in requiring ideals to be nonvoid. The advantage of requiring J to be nonvoid is that it makes the converse of Theorem 2 true; see Theorem 4. The main disadvantage is that the intersection of an infinite family of nonvoid ideals in a lattice without least element O may be void. If the lattice has O, however, all ideals contain this O and this difficulty does not arise.

Given an element a in any lattice L, the set $L(a)$ of all elements $x \leqq a$ is evidently an ideal; it is called a *principal ideal* of L. In any lattice of finite length, every (nonvoid) ideal is principal. More generally, this is true if every ascending chain in L is finite; see Chapter VIII.

If $J = L(a)$ is a principal ideal in L generated by a, the mapping $x \to \theta(x) = x \vee a$ is a join-endomorphism with kernel J since

$$\theta(x \vee y) = (x \vee y) \vee a = (x \vee a) \vee (y \vee a) = \theta(x) \vee \theta(y).$$

Also, if $z \in J$, $z \leqq a$ and $z \vee a = a$, whence $\theta(z) = a$. For every $x \in L$, $\theta(x) = x \vee a \geqq a$; thus $z \in J$ implies that $\theta(z) \leqq \theta(x)$. That is, a is the O of Im θ and $J =$ Ker θ is the kernel of the join-endomorphism.

Theorem 3. *The set of all ideals of any lattice L, ordered by inclusion, itself forms a lattice $\hat{L}$. The set of all principal ideals in L forms a sublattice of this lattice, which is isomorphic with L.*

Proof. Given any two ideals J and K of L, they have a common element since if $a \in J$ and $b \in K$, then $a \wedge b \in J \wedge K$. Thus we can take $J \wedge K$ as the set-intersection of J and K; this is clearly an ideal.

† For ideals in posets, see O. Frink, Amer. Math. Monthly **61** (1954), 223–34; E. S. Wolk, Proc. AMS **7** (1956), 589–94; D. R. Fulkerson, ibid., 701–2.

Again, any ideal which contains both J and K must contain the set M of all elements x such that $x \leqq a \vee b$ for some $a \in J$, $b \in K$. But the set M is an ideal: if $x \in M$ and $y \leqq x \leqq a \vee b$, then $y \leqq a \vee b$ by P3; and if $\{x, y\} \subset M$, then since $x \leqq a \vee b$ and $y \leqq a_1 \vee b_1$ for some $a, a_1 \in J$ and $b, b_1 \in K$,

$$x \vee y \leqq (a \vee b) \vee (a_1 \vee b_1) = (a \vee a_1) \vee (b \vee b_1),$$

where $a \vee a_1 \in J$ and $b \vee b_1 \in K$ since J and K are ideals. Hence $M = \sup(J, K)$ in the set of all ideals of L.

If J and K are principal ideals of L with generators a and b, then $J \vee K$ and $J \wedge K$ are the principal ideals generated by $a \vee b$ and $a \wedge b$, respectively. The principal ideals thus form a sublattice of $\hat{L}$ which is isomorphic with L.

Remark. When L is finite, $\hat{L}$ is isomorphic with L; the construction of $\hat{L}$ is important primarily for infinite lattices; cf. Chapter V.

4. Congruence Relations

A *congruence relation* on an algebraic system A is an equivalence relation θ on A which has the substitution property for its operations (cf. §VI.4). If A is a join-semilattice, this means that

$$(4) \qquad a \equiv b \pmod{\theta} \quad \text{implies} \quad a \vee x \equiv b \vee x \quad \text{for all } x \in A.$$

In a lattice, it means that the dual of (4) holds also.

LEMMA 3. *Let J be any ideal in a given join-semilattice S. Then the relation*

$$(5) \qquad a \equiv b \pmod{J} \quad \textit{when } a \vee d = b \vee d \textit{ for some } d \in J$$

is a congruence relation on S.

PROOF. One easily shows that (5) is an equivalence relation, and it is a congruence relation with respect to join since $(a \vee c) \vee d = (b \vee c) \vee d$ when $a \vee d = b \vee d$, whence

$$a \equiv b \pmod{J} \quad \text{implies} \quad a \vee c \equiv b \vee c \pmod{J}.$$

Lemma 3 yields as a corollary the following converse† of Theorem 2.

THEOREM 4. *If J is an ideal in a join-semilattice S with O, there is a join-epimorphism θ of L onto a join-semilattice T such that* Ker $\theta = J$.

PROOF. Consider the equivalence classes set up by the join congruence $a \equiv b \pmod{J}$. The mapping which takes an element of S into the equivalence class of which it is an element is a join-homomorphism because of the substitution property of the congruence relation. The set T of equivalence classes, is a join-semilattice. Since the set J forms one of the equivalence classes and since the O of S is in J, Ker $\theta = J$.

† This and many other related results are due to V. S. Krishnan, Proc. Indian Acad. Sci. Sect. A **22** (1945), 1–19.

THEOREM 5. *Given an ideal J in a distributive lattice L, the congruence relation $a \equiv b \pmod J$ defines an epimorphism θ from L onto a lattice M with O, such that $J = \text{Ker } \theta$.*

PROOF. We have just proved that $a \equiv b \pmod J$ implies $a \vee c \equiv b \vee c \pmod J$ for any $c \in J$. The proof is completed by proving that $a \equiv b \pmod J$ implies $a \wedge c \equiv b \wedge c \pmod J$ for any $c \in J$. This follows since, if $a \vee d = b \vee d$ for some $d \in J$, then

$$(a \wedge c) \vee d = (a \vee d) \wedge (c \vee d) = (b \vee d) \wedge (c \vee d) = (b \wedge c) \vee d.$$

It is important to notice that, contrary to the situation in groups, the kernel of a lattice-morphism does not in general uniquely determine the associated congruence relation. For instance, Figure 6 depicts two morphisms between chains

FIGURE 6. Distinct morphisms, same kernel

having the same kernel, but different congruence relations. In other examples, the epimorphic images may even be nonisomorphic.

However, the preceding ambiguity cannot arise in any relatively complemented lattice. To prove this is easy.

LEMMA. *If $u \equiv v \pmod \theta$ in a lattice L, then $x \equiv y \pmod \theta$ for all x, y in the interval $[u \wedge v, u \vee v]$.*

PROOF. By assumption, $x = x \vee (u \wedge v) \equiv x \vee (u \wedge u) \equiv x \vee u \pmod \theta$, and dually $x \equiv x \wedge u \pmod \theta$. Hence

$$u = u \wedge (u \wedge x) \equiv u \wedge x \equiv x \pmod \theta.$$

Similarly, $u \equiv y \pmod \theta$, whence $x \equiv y \pmod \theta$ by transitivity.

Standard ideals. In relatively complemented lattices, one can indeed construct a satisfactory bijection between congruence relations and a subclass of ideals, in the following way. In Theorem 5, it is easy to see that $a \equiv b \pmod J$ if and only if $(a \wedge b) \vee c = a \vee b$ for some $c \in J$. In a general lattice L, an ideal J is called *standard*† when this equivalence relation has the substitution property for meets as well as joins (cf. Lemma 3); that is, when it is a congruence relation. Theorem 5 asserts that every ideal of a distributive lattice is standard; the converse is also true: if every ideal of a lattice L is standard, the L is distributive.

† This notion is due to G. Gratzer, Magyar Tud. Akad. Mat. Fiz. Oszt. Közl. **9** (1959), 81–97; see also G. Gratzer and E. T. Schmidt, Acta Math. Acad. Sci. Hungar **12** (1960), 17–86.

We now define a lattice to be *sectionally complemented* when it has a O and every interval $[O, a]$ is complemented. Clearly, any relatively complemented lattice with O is sectionally complemented.

THEOREM 6. *Let θ be any congruence relation on a sectionally complemented lattice L. Then the $x \equiv O$ form a standard ideal $J(\theta)$ in L, and $x \equiv y\ (\theta)$ if and only if $(x \wedge y) \vee a = x \vee y$ for some $a \in J(\theta)$. Conversely, every standard ideal J of L defines a congruence relation on L in this way.*

PROOF. In any lattice, $x \vee a = y \vee a$ for some $a \equiv O$ implies that $x = x \vee O \equiv x \vee a = y \vee a \equiv y \vee O = y\ (\theta)$. Conversely, suppose that $x \equiv y\ (\theta)$ in a sectionally complemented lattice, and let a be a complement of $x \wedge y$ in $[O, x \vee y]$, so that $x \wedge y \wedge a = O$ and $(x \wedge y) \vee a = x \vee y$. Then $a \leqq x \vee y$; hence

$$a = (x \vee y) \wedge a \equiv x \wedge y \wedge a = O\ (\theta),$$

while $x \vee y \geqq x \vee a \geqq (x \wedge y) \vee a = x \vee y$. This makes $x \vee a = x \vee y$; similarly, $y \vee a = x \vee y = x \vee a$, completing the proof.

COROLLARY. *In any relatively complemented lattice of finite length, to every congruence relation θ there corresponds an element a such that $x \equiv y\ (\theta)$ if and only if $x \vee a = y \vee a$.*

Namely, a is the largest element of the (principal) ideal J of Theorem 6.

Prime and maximal ideals. An ideal of a lattice L is called *prime* when its set-complement is a dual ideal. This is clearly equivalent to the condition that $a \wedge b \in P$ implies $a \in P$ or $b \in P$; hence, in a distributive lattice, a principal ideal (c) is a prime ideal if and only if c is meet-irreducible. Likewise, a proper ideal is called *maximal* when L contains no larger proper ideal (i.e., when it is a dual atom of $\hat{L}$). From Theorems 2 and 4 we derive immediately the following important

COROLLARY. *The prime ideals of a given lattice L are the kernels of the lattice epimorphisms $\theta: L \to \mathbf{2}$.*

Unlike the situation in rings, maximal ideals need not in general be prime—neither is the converse true. But we note

THEOREM 7 (STONE [1]). *An ideal of a nontrivial Boolean lattice A is prime if and only if it is maximal.*

PROOF. If P is a prime ideal of A, then for any $a \notin P$ we have $a' \in P$ since $a \wedge a' = O \in P$; hence any ideal $J > P$ contains some $a \notin P$ and a', and so $I = a \vee a'$. Conversely, let M be maximal. Suppose $x \wedge y \in M$ and $x \notin M$; then $x \vee M > M$ must contain I. Hence, for some $z \in M$, $x \vee z = I$ and so

$$y = y \wedge I = y \wedge (x \vee z) = (y \wedge x) \vee (y \wedge z) \in M \vee M = M.$$

Hence M is prime. Q.E.D.

Exercises for §§3–4:

1. Prove Lemma 1.
2. Prove Lemma 2.

3. Show that the ideals of any lattice form a complete lattice.

4. (a) Show that the congruence relations on a finite chain correspond one-one to its subdivisions into intervals.

(b) Show that every order-morphism of a chain is a lattice-morphism.

(c) Show that, if every order-morphism of a lattice L is a lattice-morphism, then L is a chain.

5. Show that in the case of lattices of finite length, the antecedents of any element under any lattice-morphism form an interval $[a, b]$.

6. (a) Show that a meet-epimorphism of a lattice L onto a lattice M is necessarily an isomorphism if $x < x'$ implies $\theta(x) < \theta(x')$.

(b) Show that this is not true of all isotone functions.

7. Show that a lattice is a chain if and only if all its ideals are prime.

8. (a) Show that the correspondence from each subset S of a group G to the subroup S which it generates is a join-epimorphism of the lattice of all subsets of G onto the lattice of all subgroups of G.

(b) Obtain an isotone map of 2^2 onto the ordinal **4** which is neither join- nor meet-morphic.

9. Find a standard ideal in N_5 which is not neutral. (See also M. Janowitz and E. T. Schmidt,) Acta Math. Acad. Sci. Hungar. **16** (1965), 289–301, 435.

10. Show that any isotone surjection $\theta\colon P \to Q$ of posets can be uniquely represented as the effect of "strengthening" the order in P to a stronger quasi-order, and then applying Lemma 1 of §1.

11. Let S, T be any two join-semilattices with O.

(a) Show that, given an ideal J of S, there is a ("regular") *least* congruence relation $\theta(J)$ on S with Ker $\theta = J$.

(b) Show that any join-epimorphism $\phi\colon S \to T$ can be uniquely represented as the composite $\alpha\beta$ of a "regular" join-morphism $\alpha = \theta(J)$ with an "irreducible" join-morphism with kernel O. (V. S. Krishnan)

12. Discuss the lattice-epimorphisms $\theta\colon LM \to L$ from any product of two lattices onto the first factor.

13. Show that any finite lattice is a join-epimorphic image of a finite Boolean lattice. (*Hint.* Consider the subsets of its "join-irreducible" elements.)

14. (a) Show that a prime ideal of a lattice is a nonvoid subset P such that $a \wedge b \in P$ if and only if $a \in P$ or $b \in P$.

(b) Show that a principal ideal (c) of a lattice is prime if and only if $x \wedge y \leqq c$ implies $x \leqq c$ or $y \leqq c$.

15. Show that any distributive lattice of order $n > 2$ has two congruence relations θ and θ_1 such that $\theta \wedge \theta_1 = O$, the equality relation. (*Hint.* Consider $x \to x \wedge a$, $x \to x \vee a$.)

16. Show that any *non*distributive lattice L has a (principal) ideal which is not the kernel of any lattice-epimorphism.

17. (G. D. Crown) Let E be an equivalence relation on a lattice L. Show that E is a congruence relation if and only if:

(i) aEb and $a \wedge b \leqq x, y \leqq a \vee b$ imply xEy, and

(ii) $(a \wedge b)Ea$ holds if and only if $bE(a \vee b)$.

5. Lattice Polynomials

Expressions involving the symbols $\wedge$, $\vee$ and letters are called *lattice polynomials*. More precisely, we define individual letters $x, y, z, \cdots$ as polynomials of *weight* one. Recursively, if p and q are lattice polynomials of weights w and w', respectively, then $p \wedge q$ and $p \vee q$ will be called lattice polynomials of *weight* $w + w'$.

In this definition, the forms $x \wedge x$ and $x \vee (y \wedge x)$ are considered as different polynomials (of weights two and three, respectively), even though they are *equivalent* in the sense that they represent the same *function* $p: L \to L$ in any lattice. Indeed, the different lattice polynomial *functions* of x and y are easily enumerated. This is because of

LEMMA 1. *In any lattice L, the sublattice S generated by two elements x and y consists of x, y, $x \vee y = u$, and $x \wedge y = v$, with $\vee$ and $\wedge$ as in Figure 7.*

PROOF. By L4, $x \wedge u = x$; by L3, L1, $x \vee u = x \vee (x \vee y) = (x \vee x) \vee y = x \vee y = u$. The other cases are analogous, using symmetry in x and y and duality. (By L4, $u \vee v = x \vee y \vee (x \wedge y) = x \vee y = u$.)

COROLLARY. *Let $F_2 = \mathbf{2}^2$ be the lattice of Figure 7, and let $a, b \in L$ (L any*

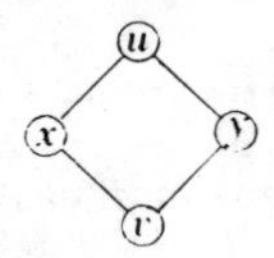

FIGURE 7

lattice) be given. Then the mapping $x \to a$, $y \to b$ can be extended to a morphism $\theta: F_2 \to L$.

The preceding results are usually summarized in the statement that F_2 is the *free lattice* with generators x, y; it has just four elements, and is distributive (in fact, a Boolean lattice).

Lattice polynomials in three or more variables can be extremely complicated (see §VI.8). However, they have a few simple properties.

LEMMA 2. *In any join-semilattice, every polynomial in r symbols $x_1, \cdots, x_r$ is equivalent to the join $\bigvee_S x_i$ of a nonempty subset of the x_i.*

PROOF. By L2–L3, every such polynomial is equivalent to a join of repeated occurrences of $x_1, x_2, \cdots, x_r$ *in that* order. And by L1, repeated occurrences of the same symbol can be replaced by a single occurrence of that symbol.

In other words, the free semilattice with r generators has $2^r - 1$ elements. (Cf. L. Martic, C.R. Acad Sci. Paris **244** (1957), 1953–55.)

LEMMA 3. *In any distributive lattice, every polynomial is equivalent to a join of meets, and dually*:

$$p(x_1, \cdots, x_r) = \bigvee_{\alpha \in A} \left\{ \bigwedge_{S_\alpha} x_i \right\} = \bigwedge_{\delta \in D} \left\{ \bigvee_{T_\delta} x_j \right\}, \tag{6}$$

where the S_α, T_δ are nonempty sets of indices.

PROOF. Each x_i can be so written, where A (or D) is the family of sets consisting of the single one-element set x_i. On the other hand, just as in Lemma 1 we have by L1–L3

$$\bigvee_{\alpha \in A} \left\{ \bigwedge_{S_\alpha} x_i \right\} \vee \bigvee_{\beta \in B} \left\{ \bigwedge_{S_\beta} x_i \right\} = \bigvee_{A \cup B} \left\{ \bigwedge_{S_\gamma} x_i \right\}. \tag{7}$$

Using the distributive law, we have similarly

$$(7') \qquad \bigvee_{\alpha \in A} \left\{ \bigwedge_{S_\alpha} x_i \right\} \wedge \bigvee_{\beta \in B} \left\{ \bigwedge_{S_\beta} x_i \right\} = \bigvee_{A \times B} \left\{ \bigwedge_{S \cup S_{\alpha\beta}} x_i \right\},$$

where $\bigwedge$ is the symbol for the meet and $\bigvee$ for the join of any finite number of terms, just as $\sum$ and $\prod$ represent the sum and product of any finite number of terms. This is a further generalization of the lattice analog of the familiar general distributive law of ordinary algebra

$$\left(\sum_{\alpha \in A} x_\alpha\right)\left(\sum_{\beta \in B} y_\beta\right) = \sum_{A \times B} x_i y_j.$$

It follows from L2 and L6′, combined with the relation $\{\bigwedge_S x_i\} \wedge \{\bigwedge_T x_i\} = \bigwedge_{S \cup T} x_i$, which follows from L1–L3 as in Lemma 2.

The isotonicity principle of §I.5, Lemma 3 can be extended by induction on weight to the following result.

Lemma 4. *If* $y_i \leqq z_i$ $(i = 1, \cdots, r)$, *then* $p(y_1, \cdots, y_r) \leqq p(z_1, \cdots, z_r)$ *for any lattice polynomial* p.

More generally, we have the *minimax principle*:

$$(8) \qquad \bigwedge_{\gamma \in C} \left(\bigvee_{B(\gamma)} x_{\gamma,\beta} \right) \geqq \bigvee_{P} \left(\bigwedge_{\gamma} x_{\gamma,\beta(\gamma)} \right).$$

Here each $\gamma \in C$ determines an index-set $B(\gamma)$ of β, so that the subscripts γ, β range through the Cartesian product $\prod_C B(\gamma)$. Inside the first parentheses, γ is fixed while β ranges over $\beta(\gamma)$; on the right side, $\beta(\gamma)$ designates any function assigning to each $\gamma \in C$ a $\beta(\gamma) \in B(\gamma)$; P is the set of all such functions. For example, when $C = \{1, 2\}$, $B(1) = \{1, 2\}$, and $B(2) = \{1, 2, 3\}$, the minimax inequality (8) is

$$(x_{11} \vee x_{12}) \wedge (x_{21} \vee x_{22} \vee x_{22}) \geqq \left[\bigvee_{j=1}^{3} (x_{11} \wedge x_{2j}) \right] \vee \left[\bigvee_{k=1}^{3} (x_{12} \wedge x_{2k}) \right].$$

Proof. Each term $\bigvee_{B(\gamma)} x_{\gamma,\beta}$ is an upper bound to each term $\bigwedge_\gamma x_{\gamma,\beta(\gamma)}$ on the right of (8), since they have some $x_{\gamma,\beta(\gamma)}$ in common. Hence the infimum of the terms on the left is also such an upper bound, and it is therefore also an upper bound to the supremum of the terms on the right. Q.E.D.

Exercises:

1. Show that each pair of congruence relations θ on L and θ' on M induces a congruence relation on LM (L, M lattices).
2. (a) Prove that $(a \vee b) \wedge (c \vee d) \geqq (a \wedge c) \vee (b \wedge d)$ identically.
   (b) Prove that $[a \wedge b \wedge (c \vee d)] \vee (c \wedge d) \leqq c \vee [b \wedge (a \vee d)] \vee (a \wedge d)$.
   (c) Show that L5 is equivalent in any lattice to $(a \vee b) \wedge (a \vee c) = a \vee [b \wedge (c \vee a)]$. (R. Dedekind)
3. Prove (4) directly from L1–L4.
4. An element a of a lattice is called *join-irreducible* if $x \vee y = a$ implies $x = a$ or $y = a$. Show that if all chains in a lattice L are finite, then every $a \in L$ can be represented as a join $a = x_1 \vee \cdots \vee x_n$ of a finite number of join-irreducible elements.

5. Let J be the set of join-irreducible elements of a finite lattice L. Associate with each $a \in L$ the set $S(a)$ of $x \leqq a$ in J. Show that this represents L order-isomorphically by subsets of J, and that meets in L correspond to intersections in the representation.†

Define the *breadth* of a finite lattice L, as the smallest integer b such that any meet $x_1 \wedge \cdots \wedge x_n$ $[n > b]$ is a meet of a subset of b of the x_i.

6. (a) Show that if $b(L) = n$, and S is a sublattice or lattice-epimorphic image of L, then $b(S) \leqq n$.

(b) Show that $b(LM) = b(L) + b(M)$.

(c) Show that the smallest lattice with $b(L) = n$ is $\mathbf{2}^n$.

7. (a) Show that a finite poset P whose diagram can be embedded in a plane is a lattice, if and only if P has universal bounds.

(b) Show that any lattice L whose graph is planar contains a join-irreducible element not O or I.

(*c) Show that a finite lattice L has a plane diagram if and only if there exists a "complementary" partial ordering $<'$ among its elements such that $a, b \in L$ are $<$-comparable if and only if they are incomparable under $<'$. (J. Zilber)

8. Show that in a modular lattice, any congruence relation which identifies two adjacent vertices of a quadrangle of elements linked by a covering relation identifies the other two vertices. (*Hint.* If $x \equiv x \wedge y\ (\theta)$, then $x \vee y \equiv y\ (\theta)$.)

9. Is Lemma 1 still true if L3 is not assumed (i.e., in "latticoids")?

6. Distributivity

As in §I.6, a lattice is distributive when it satisfies either (hence both) of the equivalent distributive laws L6′–L6″. We now prove a further result in the same direction.

THEOREM 8. *A lattice is distributive if and only if it satisfies the self-dual* median *law*

L6. $$(x \wedge y) \vee (y \wedge z) \vee (z \wedge x) = (x \vee y) \wedge (y \vee z) \wedge (z \vee x).$$

PROOF. Setting $x \geqq z$ in L6, the left side reduces to $(x \wedge y) \vee [(y \wedge z) \vee z] = (x \wedge y) \vee z$, by L4; dually, the right side reduces to $(x \vee y) \wedge (y \vee z) \wedge x = [x \wedge (x \vee y)] \wedge (y \vee z) = x \wedge (y \vee z)$. Hence L6 implies the modular law L5.

Again, abbreviating L6 to the form $u = v$, we get the identity $x \wedge u = x \wedge v$, where (by L2–L4):

$$x \wedge u = x \wedge ([y \wedge z) \vee [x \wedge y) \vee (x \wedge z]) = (x \wedge y \wedge z) \vee (x \wedge y) \vee (x \wedge z),$$

$$x \wedge v = [x \wedge (x \vee y)] \wedge (y \vee z) \wedge (z \vee x) = x \wedge (z \vee x) \wedge (y \vee z) = x \wedge (y \vee z)$$

the last step being justified by L5. Applying L4 to the first two terms of the last expression, we get $x \wedge u = (x \wedge y) \vee (x \wedge z)$. Substituting back into the identity $x \wedge u = x \wedge v$, we get L6′.

Conversely, applying L6′ to the right side of L6, we get

$$[(x \vee y) \wedge (y \vee z) \wedge z] \vee [(x \vee y) \wedge (y \vee z) \wedge x] = [z \wedge (x \vee y)] \vee [x \wedge (y \vee z)],$$

where L4 and L2 are used freely in the last step. Applying L6′ again to each term

† Cf. A. D. Campbell, Bull. AMS **49** (1943), 395–8. For any finite lattice L, it is interesting to find the "representation" in the above sense which minimizes the number of points needed.

of the last expression, we get

$$(z \wedge x) \vee (z \wedge y) \vee (x \wedge y) \vee (x \wedge z) = (x \wedge y) \vee (y \wedge z) \vee (z \wedge x),$$

using L2–L4 freely. Hence L6′ implies L6, completing the proof.

LEMMA 1. *In any distributive lattice, the sublattice generated by three given elements* x_1, x_2, x_3 *consists of the* x_j, $i = x_1 \vee x_2 \vee x_3$, $o = x_1 \wedge x_2 \wedge x_3$, $u_1 = x_2 \vee x_3$ *etc.*, $v_1 = x_2 \wedge x_3$ *etc.*, $c_1 = u_2 \wedge u_3$ *etc.*, $d_1 = v_2 \vee v_3$ *etc., and*

$$e = (x_1 \wedge x_2) \vee (x_2 \wedge x_3) \vee (x_3 \wedge x_1) = (x_1 \vee x_2) \wedge (x_2 \vee x_3) \wedge (x_3 \vee x_1). \tag{9}$$

Explanation. By "etc." is meant two analogous elements (e.g., $u_2 = x_3 \vee x_1$, $u_3 = x_1 \vee x_2$) obtained by cyclically permuting the subscripts of the same case specified.

PROOF. One can explicitly reduce the join and meet of each pair of the expressions specified to one of these expressions, by using L1–L6. Keeping in mind the six permutations of subscripts and duality, each such reduction yields eleven others, which simplifies the computation. Moreover for any $j \neq k$, we have

$$o \leqq v_j \leqq d_k \leqq x_k, \qquad e \leqq c_k \leqq u_j \leqq i.$$

Since joins and meets of comparable elements are trivial, one need therefore only verify the following relations (for $j \neq l$, $k \neq l$):

$$\begin{aligned} &x_j \vee u_j = u_j \vee u_k = i, \qquad u_j \wedge c_j = e, \qquad x_j \vee e = c_j, \\ &x_h \wedge c_k = d_j, \qquad x_j \vee c_k = u_l, \end{aligned} \tag{10}$$

and their duals. This completes the proof.

LEMMA 2. *There is a distributive lattice* D_{18} *with three generators in which all of the* 18 *expressions of Lemma* 1 *represent different elements.*

PROOF. We will construct D_{18} as a ring of sets. Let X_j ($j = 1, 2, 3$) be the set of four functions $\mathbf{f}: \{1, 2, 3\} \to \{0, 1\}$ with $f_j = 1$. One can identify each $\mathbf{f} = (f_1, f_2, f_3)$ as a three-vector with entries in $\mathbf{2} = \{0, 1\}$. Then $X_1 \cup X_2$ is the set U_3 of all $\mathbf{f}$ with $f_1 \vee f_2 = 1$; $X_1 \cap X_2$ is the set of all $\mathbf{f}$ with $f_1 \wedge f_2 = 1$, and cyclically. Therefore, $C_1 = U_2 \cap U_3$ is the set of all $\mathbf{f}$ with $f_1 \wedge (f_2 \vee f_3) = 1$, and cyclically. Finally, $E = U_1 \cap U_2 \cap U_3 = V_1 \cup V_2 \cup V_3$ is the set of all $\mathbf{f}$ with at least two components 1. These subsets of $\mathbf{2}^3$ are all different from each

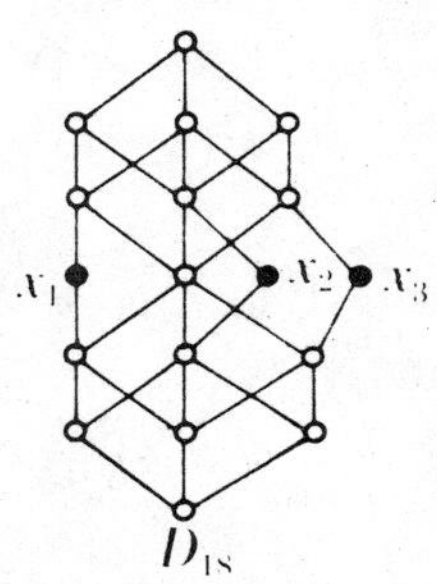

FIGURE 8

other, and from $O = \{(1, 1, 1)\}$ and $I = X_1 \cup X_2 \cup X_3$, the complement of $\{(0, 0, 0)\}$.

The diagram of D_{18} is shown in Figure 8; it has a very remarkable property.

THEOREM 9. *Let D_{18} be the lattice of Figure 8, with shaded elements x_1, x_2, x_3, and let a_1, a_2, a_3 be any three elements of any distributive lattice L. Then there is a morphism $\phi: D_{18} \to L$ with $\phi(x_i) = a_i$ $(i = 1, 2, 3)$.*

PROOF. By Lemma 2, we can identify each element $p_j(X_1, X_2, X_3)$ of the distributive lattice D_{18} with precisely one of the 18 polynomials $p_j(x_i, x_2, x_3)$ constructed in the proof of Lemma 1. We now set

(11) $$\phi(p_j(x_1, x_2, x_3)) = p_j(a_1, a_2, a_3) = p_j(\phi(a_1), \phi(a_2), \phi(a_3));$$

we shall show that this is a morphism. Indeed, by Lemma 1, each of the relevant 648 equations $p_j \vee p_k = p_l$ and $p_j \wedge p_k = p_m$ deduced in proving Lemma 1, from L1–L6, and defining the "multiplication tables" for $\vee$ and $\wedge$ in D_{18}, is true in any distributive lattice, and hence in L. This implies $\phi(p_j) \vee \phi(p_k) = \phi(p_j \vee p_k)$ in L, and dually.

Because of the property described in Theorem 9, D_{18} is called the *free* distributive lattice with three generators.

We next prove a result which shows that, for a distributive lattice with I, postulates L1–L4 and L6 are highly redundant. This result, in view of the independence of L2–L4 shown in §2, illustrates the power of the distributive law.

THEOREM 10. *The following four postulates on a system with two binary operations and a special element I imply that it is a distributive lattice:*

D1. $a \wedge a = a$ *for all* a,

D2. $a \vee I = I \vee a = I$ *for all* a,

D3. $a \wedge I = I \wedge a = a$ *for all* a,

D4. $a \wedge (b \vee c) = (a \wedge b) \vee (a \wedge c)$ *and* $(b \vee c) \wedge a = (b \wedge a) \vee (c \wedge a)$.

PROOF. We first show that, for all a, b,

D5. $$a = a \wedge I = a \wedge (a \vee I) = (a \wedge a) \vee (a \wedge I) = a \vee a,$$

D6. $$(a \wedge b) \vee a = (a \wedge b) \vee (a \wedge I) = a \wedge (b \vee I) = a \wedge I = a,$$

and similarly

D6′. $$a \vee (a \wedge b) = a \vee (b \wedge a) = (b \wedge a) \vee a = a.$$

We are now able to prove, using D4, D1, and D6′,

D7. $$a \wedge (a \vee b) = (a \wedge a) \vee (a \wedge b) = a \vee (a \wedge b) = a,$$

and similarly

D7′. $$a \wedge (b \vee a) = (a \vee b) \wedge a = (b \vee a) \wedge a = a.$$

Now we can prove the commutative law

D8. $$a \vee b = [a \wedge (b \vee a)] \vee [b \wedge (b \vee a)] = (a \vee b) \wedge (b \vee a)$$ by D7–D7′, D4,

$$= [(a \vee b) \wedge b] \vee [(a \vee b) \wedge a] = b \vee a$$ by D4, D7–D7′.

Preparatory to proving the associative law for joins, we show

$$a \wedge [(a \vee b) \vee c] = [a \wedge (a \vee b)] \vee [a \wedge c] = a \vee (a \wedge c) = a$$
by D4, D7, D6′,

D9. $$b \wedge [(a \vee b) \vee c] = [b \wedge (a \vee b)] \vee [b \wedge c] = b \vee (b \wedge c) = b$$
similarly,

$$c \wedge [(a \vee b) \vee c] = [c \wedge (a \vee b)] \vee [c \wedge c] = [c \wedge (a \vee b)] \vee c = c$$

by D4, D1, D6. Now we prove the associative law for joins,

$$a \vee (b \vee c) = \{a \wedge [(a \vee b) \vee c]\} \vee (\{b \wedge [(a \vee b) \vee c]\} \vee \{c \wedge [(a \vee b) \vee c]\})$$
by D9,

D10. $= [a \vee (b \vee c)] \wedge [(a \vee b) \vee c]$ by D4 used twice,

$= (a \vee b) \vee c$ similarly, by left-right symmetry.

We now prove the *dual* of D4, namely

$$(a \vee b) \wedge (a \vee c) = [a \wedge (a \vee c)] \vee [b \wedge (a \vee c)]$$

D11. $= a \vee [(b \wedge a) \vee (b \wedge c)]$ by D4,

$= [a \vee (b \wedge a)] \vee (b \wedge c) = a \vee (b \wedge c)$
by D10, D6′;

D11′. $(a \wedge b) \vee c = (a \vee c) \wedge (b \vee c)$ by left-right symmetry.

We have already proved the dual D5 of D1, while D6 and D7 are dual. But these were the only laws used in proving D8, D10; hence exact duals of the proofs of D8, D10 yield the commutative and associative laws for meets.

D12. $a \wedge b = b \wedge a$ and $a \wedge (b \wedge c) = (a \wedge b) \wedge c$.

This completes the proof of L1–L4; L6 was assumed as D4.

An ingenious related result is the following.

THEOREM 10′ (SHOLANDER†). *Any system with two binary operations* $\wedge$ *and* $\vee$ *which satisfy* $a \wedge (a \vee b) = a$ *and* $a \wedge (b \vee c) = (c \wedge a) \vee (b \wedge a)$ *is a distributive lattice.*

Medians. A glance at Figure 8 shows that the symmetric and self-dual ternary *median* operation

(12) $$(a, b, c) = (a \wedge b) \vee (b \wedge c) \vee (c \wedge a) = (a \vee b) \wedge (b \vee c) \wedge (c \vee a),$$

plays a unique role in distributive lattices. Its properties can also be used to define the latter. For instance, [**LT2**, pp. 137–8], one can prove‡

† M. Sholander, Canad. J. Math. **3** (1951), 28–30. See also R. Croisot, ibid., 24–7.

‡ Theorem 11 is due to S. A. Kiss and the author, Bull. AMS **53** (1947), 749–52. See also [**LT2**, pp. 137–8], and A. A. Grau, Bull. AMS **53** (1947), 567–72.

THEOREM 11. *Any system M with elements O, I and a ternary operation which satisfies*

$$(13) \qquad \begin{gathered} (O, a, I) = a, \qquad (a, b, a) = a, \\ (a, b, c) = (b, a, c) = (b, c, a), \\ ((a, b, c), d, e) = ((a, d, e), b, (c, d, e)), \end{gathered}$$

is a distributive lattice relative to the binary operations $a \wedge b = (a, O, b)$ and $a \vee b = (a, I, b)$.

Sholander, op. cit., has also shown that the postulates

$$(14) \qquad (O, a, (I, b, I)) = a \quad \text{and} \quad (a, (b, c, d), e) = ((a, c, e), d, (b, a, e))$$

are sufficient. The median operation is also related to various concepts of "betweenness"; see §9 below.

Exercises:

1. In any distributive lattice L, for given $a, b \in L$, let $J(a, b)$ denote the set of x such that $a \vee x = b \vee x$. Show that $J(a, b)$ is a dual ideal.
2. Show that any lattice in which
$$(x \vee y) \wedge [z \vee (x \wedge y)] = (x \wedge y) \vee (y \wedge z) \vee (z \wedge x),$$
for all x, y, z, is distributive.
3. Show that a lattice is distributive if and only if $x \vee (y \wedge z) \geqq (x \vee y) \wedge z$ for all x, y, z. (J. Bowden, 1936)
4. Show that if O, I exist with $a \vee O = a \wedge I = a$ for all a, then L2, L3, L6′–L6″ imply L1–L4. (E. V. Huntington [1, pp. 292–5])

*5. Show that the following postulates characterize distributive lattices: $a \wedge a = a$, $a \wedge b = b \wedge a$, $a \vee b = b \vee a$, $a \wedge (b \wedge c) = (a \wedge b) \wedge c$, $a \wedge (a \vee b) = a$, $a \wedge (b \vee c) = (a \wedge b) \vee (a \wedge c)$. (*Hint.* Define $a \geqq b$ to mean $a \wedge b = b$.)

6. Show that, in any distributive lattice L with O and I, if (a, b, c) is defined by (12), then (13) and (14) hold.
7. (a) Prove the identity $(ab(cde)) = ((abc)d(abe))$ in distributive lattices.
   (b) Show that this contains L6′ and L6″ as special cases.

*8. Prove that a given lattice L is distributive if and only if every ideal of L is standard.

9. Show that there is no way to define a median operation (a, b, c) in M_5, which self-dual and invariant under automorphisms.

7. Modularity

We now turn our attention from distributivity to modularity. As was pointed out in §I.7, the modular law L5 is a special case of the distributive law L6. Remarkable sets of postulates for modular lattices, somewhat analogous to those of Theorems 10–10′ have been constructed by M. Kolibiar and J. Riecan.†

The implications of modularity are more subtle than those of distributivity, as we shall see. We first prove a simple lemma.

LEMMA 1. *In a modular lattice, $x = y$ is implied by the conditions $a \wedge x = a \wedge y$ and $a \vee x = a \vee y$, provided that x and y are comparable (i.e., that $x \geqq y$ or $y \geqq x$).*

† Czechoslovak Math. J. **6** (1956), 381–6; Acta Fac. Natur. Univ. Comenian **2** (1958), 257–62; (MR **21** (1960), #1278, #1279).

Proof. If $x \geqq y$, then the contrary conclusion $x \neq y$ would imply $x > y$, and so the elements a, x, y would generate the five-element nonmodular lattice N_5 of Figure 1b, §I.3, contrary to hypothesis. The case $x \leqq y$ can be treated similarly, either by interchanging x and y or by duality.

Theorem I.12 shows that the conclusion of Lemma 1 cannot hold in any nonmodular lattice.

Distributive triples. In modular lattices, many triples $\{x, y, z\}$ of elements generate distributive sublattices. When this is the case, we write $(x, y, z)D$ and call $\{x, y, z\}$ a *distributive triple.*

Theorem 12. *Given a, b, c in a modular lattice M, either of the two equations*

$$a \wedge (b \vee c) = (a \wedge b) \vee (a \wedge c) \quad \text{or} \quad a \vee (b \wedge c) = (a \vee b) \wedge (a \vee c)$$

implies that $\{a, b, c\}$ is a distributive triple.

Proof. We first assume $a \wedge (b \vee c) = (a \wedge b) \vee (a \wedge c)$, and apply the preceding lemma to the three elements a, $x = (a \wedge b) \vee c, y = (a \vee c) \wedge (b \vee c)$. These elements satisfy $x \leqq y$ by the distributive inequality. Also,

$$a \vee x = a \vee (a \wedge b) \vee c = a \vee c,$$

$$a \vee y = y \vee a = [(a \vee c) \wedge (b \vee c)] \vee a$$

$$= (a \vee c) \wedge [(b \vee c) \vee a] \qquad \text{by L5}$$

$$= a \vee c \qquad \text{by L4.}$$

Likewise,

$$a \wedge x = x \wedge a = [(a \wedge b) \vee c] \wedge a = (a \wedge b) \vee (c \wedge a) \qquad \text{by L5}$$

$$= (a \wedge b) \vee (a \wedge c) = a \wedge (b \vee c) \qquad \text{by hypothesis.}$$

$$a \wedge y = a \wedge [(a \vee c) \wedge (b \vee c)] = [a \wedge (a \vee c)] \wedge (b \vee c) = a \wedge (b \vee c).$$

Hence $x = y$, by Lemma 1. In summary, $a \wedge (b \vee c) = (a \wedge b) \vee (a \wedge c)$ implies that $c \vee (a \wedge b) = (c \vee a) \wedge (c \vee b)$.

The result that $c \vee (a \wedge b) = (c \vee a) \wedge (c \vee b)$ implies $b \wedge (c \vee a) = (b \wedge c) \vee (b \wedge a)$ follows by the Duality Principle. Combining with the result proved above, we see that $a \wedge (b \vee c) = (a \wedge b) \vee (a \wedge c)$ implies $b \wedge (c \vee a) = (b \wedge c) \vee (b \wedge a)$.

That is, in a modular lattice, the distributive law for one arrangement of $\{a, b, c\}$ implies its validity for the cyclically permuted arrangement $\{b, c, a\}$. Repeating and using the argument above, we obtain all six distributive laws on $\{a, b, c\}$. That $\{a, b, c\}$ generates a distributive lattice follows immediately.

Again, Theorem I.12 shows that the hypothesis of modularity is necessary as well as sufficient in Theorem 12.

Theorem 16 below also shows that the hypothesis of modularity is essential in Theorem 12. However, we have

THEOREM 12′. *For any three elements a, b, c of any lattice L, $(a \wedge b, b \wedge c, c \wedge a)D$ and $(a \vee b, b \vee c, c \vee a)D$.*

PROOF. Let $p = a \wedge b$, $q = b \wedge c$, $r = c \wedge a$, and $o = a \wedge b \wedge c$; and let $u = p \vee q$, $v = p \vee r$, $w = q \vee r$, $i = p \vee q \vee r$. We will show that $\{o, p, q, r, u, v, w, i\}$ form a sublattice, in which joins and meets are formed as in the Boolean algebra $\mathbf{2}^3$ of all subsets of $\{p, q, r\}$. Indeed:

$$p \leqq u \wedge v = [(a \wedge b) \vee (b \wedge c)] \wedge [(a \wedge b) \vee (c \wedge a)]$$
$$\leqq [b \vee b] \wedge [a \vee a] = b \wedge a = p,$$

whence $u \wedge v = p$. By symmetry, $u \wedge w = q$ and $v \wedge w = r$. Finally, trivially, $p \wedge q = q \wedge r = r \wedge p = o$ and

$$u \vee v = u \vee r = u \vee w = p \vee w = v \vee w = v \vee q = i.$$

LEMMA 2. *In any modular lattice, for all x, y, z,*

$$(15) \qquad [x \wedge (y \vee z)] \vee [y \wedge (z \vee x)] = (x \vee y) \wedge (y \vee z) \wedge (z \vee x).$$

PROOF. Since $x \wedge (y \vee z) \leqq x \leqq z \vee x$ and $y \leqq y \vee z$, repeated use of L5 gives

$$[x \wedge (y \vee z)] \vee [y \wedge (z \vee x)] = \{[x \wedge (y \vee z)] \vee y\} \wedge (z \vee x)$$
$$= [(y \vee z) \wedge (x \vee y)] \wedge (z \vee x).$$

Formula (15) now follows by L2–L3.

Applying the Duality Principle to (15), we get the

COROLLARY. *In any modular lattice, we have also*

$$(16) \qquad [x \vee (y \wedge z)] \wedge [y \vee (z \wedge x)] = (x \wedge y) \vee (y \wedge z) \vee (z \wedge x).$$

LEMMA 3. *In a modular lattice, let*

$$e = (y \wedge z) \vee [x \wedge (y \vee z)] = [(y \wedge z) \vee x] \wedge (y \vee z) \qquad \text{(by L5)},$$

and let f and g be defined similarly by permuting x, y, z cyclically (thus $f = (z \wedge x) \vee [y \wedge (z \vee x)]$, etc.) Then

$$(17) \qquad e \wedge f = f \wedge g = g \wedge e = (x \wedge y) \vee (y \wedge z) \vee (z \wedge x),$$

and

$$(17') \qquad e \vee f = f \vee g = g \vee e = (x \vee y) \wedge (y \vee z) \wedge (z \vee x).$$

PROOF. By definition and L2–L3, using Lemma 2:

$$e \vee f = (y \wedge z) \vee [x \wedge (y \vee z)] \vee [y \wedge (z \vee x)] \vee (z \wedge x)$$
$$= (y \wedge z) \vee [(x \vee y) \wedge (y \vee z) \wedge (z \vee x)] \vee (z \wedge x).$$

But $y \wedge z$ and $z \wedge x$ are both contained in the expression in square brackets, since they are lower bounds to $x \vee y$, $y \vee z$, and $z \vee x$ individually. From this, (17′) follows, whence (17) follows by duality.

LEMMA 4. *In Lemma* 3, *if any two of the elements* e, f, g *are equal, then* $(x, y, z)D$.

PROOF. If any two are equal, say e and f, then $e \wedge f = e \vee f$ and so, by (17)–(17′)

$$(x \wedge y) \vee (y \wedge z) \vee (z \wedge x) = (x \vee y) \wedge (y \vee z) \wedge (z \vee x). \tag{18}$$

But, just as in the proof of Theorem 6 (second paragraph):

$$x \wedge [(x \wedge y) \vee (y \wedge z) \vee (z \wedge x)] = (x \wedge y) \vee (x \wedge z),$$

$$x \wedge [(x \vee y) \wedge (y \vee z) \wedge (z \vee x)] = x \wedge (y \vee z),$$

in any modular lattice. Hence, by Theorem 12, we have $(x, y, z)D$ if (18) holds.

THEOREM 13. *Any modular, nondistributive lattice* M *contains a sublattice isomorphic with the lattice* M_5 *of Figure* 1a, *Chapter* I.

PROOF. Unless M is distributive, it contains a nondistributive triple $\{x, y, z\}$. By Lemma 4, the elements e, f, g will be distinct; by Lemma 3, they will generate a sublattice isomorphic with M_5.

COROLLARY. *For a lattice to be distributive, the following condition is necessary and sufficient*:

$$a \wedge x = a \wedge y \quad \text{and} \quad a \vee x = a \vee y \quad \text{imply} \quad x = y. \tag{19}$$

That is, it is necessary and sufficient that relative complements be unique.

Exercises:

1. Prove that $x \wedge y = [(x \wedge y) \vee (x \wedge z)] \wedge [(x \wedge y) \vee (y \wedge z)]$ in any modular lattice. (Cf. H. Löwig, Ann. of Math. **44** (1943), 573–9.)
2. Prove that, in any lattice, L5 is equivalent to each of the identities

$$x \wedge (y \vee z) = x \wedge \{[y \wedge (x \vee z)] \vee z\}, \qquad [(x \wedge z) \vee y] \wedge z = [(y \wedge z) \vee x] \wedge z.$$

3. Prove that any sublattice or epimorphic image of a modular lattice is modular.
4. Prove that any direct product of modular lattices is modular.
5. Assuming Theorem 12, give a short proof of Theorem 8.
6. Show that Theorem 12 fails in any nonmodular lattice.

*7. Show that every nonmodular complemented lattice L of finite length contains a complemented nonmodular sublattice of five elements which includes O and I. (R. P. Dilworth, Tôhoku Math. J. **47** (1940), 18–23)

8. Show that, for any fixed a in a modular lattice L, the binary operation $[x, y]_a = [x \wedge (a \vee y)] \vee (a \wedge y)$ is idempotent and associative (defines a "band"). (S. A. Kiss)

8. Semimodularity and Length

Let P be any poset of finite length with O. We shall call P (upper) *semimodular* when it satisfies:

(σ) If $a \neq b$ both cover c, then there exists $d \in P$ which covers both a and b.†

† O. Ore, Bull. AMS **49** (1943), 558–66.

Lower semimodular posets of finite lengths are defined dually. A poset of finite length which is both upper and lower semimodular is called *modular*.

Condition (σ) is closely related to condition (ξ) of §I.7. Indeed, we first prove

LEMMA 2. *If a poset P is a lattice, then (σ) is equivalent to condition (ξ).*

For, in (σ), $d \geqq a \vee b$ since d is an upper bound to the set $\{a, b\}$. But $d > a \vee b$ is impossible if $a \neq b$, since otherwise $d > a \vee b > a$, and d could not cover a. Hence (σ) implies (ξ) in a lattice; the converse is trivial. Dual results and proofs also hold.

THEOREM 14. *The Jordan-Dedekind chain condition holds in any (upper or lower) semimodular poset of finite length.*

PROOF. By duality, it suffices to prove that (σ) implies the Jordan-Dedekind chain condition in a poset of finite length with O. This we prove by induction, as follows (see Figure 9).

For each positive integer m, let $P(m)$ be the assertion that if one connected chain $\gamma: a = x_0 < x_1 < \cdots < x_m = b$ has length m, then every connected chain between a and b has length m. Now $P(1)$ is trivial. We shall prove that $P(m - 1)$ implies $P(m)$. Let $\gamma': a = y_0 < y_1 < \cdots < y_m = b$ be any other (finite) connected chain connecting a and b. Then by (σ), there will exist an element u which covers x_1 and y_1, except in the trivial case that $x_1 = y_1 = I$. Form any connected chain γ'' connecting u and b. By $P(m - 1)$, the connected chain (x_1, γ'') has length $m - 1$; since the connected chain $x_1 < x_2 < \cdots < x_m = b$ does. Hence γ'' has length $m - 2$; hence (y_1, γ'') has length $m - 1$; hence by $P(m - 1)$ the connected chain $y_1 < y_2 < \cdots < y_n = b$ has length $m - 1$ and so $m = n$.

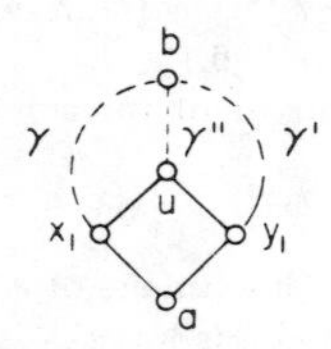

FIGURE 9

Referring to §I.3, we obtain the

COROLLARY. *Any modular or semimodular poset of finite length is graded by its height function $h[x]$.*

A more important further consequence is

THEOREM 15. *A graded lattice of finite length is semimodular if and only if*

$$h[x] + h[y] \geqq h[x \vee y] + h[x \wedge y]. \tag{20}$$

Dually, it is lower semimodular if and only if

$$h[x] + h[y] \leqq h[x \vee y] + h[x \wedge y]. \tag{20'}$$

PROOF. It is immediate that (20) implies condition (ξ) in a graded lattice of finite length. Conversely, let L be a semimodular lattice of finite length. Then there exist two finite connected chains

$$x \wedge y = x_0 < x_1 < \cdots < x_m = x,$$

$$x \wedge y = y_0 < y_1 < \cdots < y_n = y,$$

one between $x \wedge y$ and x, the other between $x \wedge y$ and y. If we assume by induction on $i + j - 1$ that $x_{i-1} \vee y_j$ and $x_i \vee y_{j-1}$ at most cover $x_{i-1} \vee y_{j-1}$ (that is, each either covers or equals $x_{i-1} \vee y_{j-1}$), then, since L is upper semimodular we know that

$$x_i \vee y_j = (x_{i-1} \vee y_j) \vee (x_i \vee y_{j-1})$$

at most covers $x_{i-1} \vee y_j$ and $x_i \vee y_{j-1}$. Thus, by induction,

$$h[x \vee y_j] - h[x \vee y_{j-1}] \leqq 1,$$

and in particular $d[x \vee y_j]$ is finite for all j. Summing over j, this gives

$$h[x \vee y] - h[x] \leqq n = h[y] - h[x \wedge y]$$

which proves that semimodularity implies (20). The second statement follows by duality.

COROLLARY. *In any modular lattice of finite length:*

$$h[x] + h[y] = h[x \vee y] + h[x \wedge y]. \tag{21}$$

THEOREM 16. *Let L be a lattice of finite length. Then the following conditions are equivalent:*

(i) *the modular identity* L5,
(ii) L *is both upper and lower semimodular,*
(iii) *the Jordan–Dedekind chain condition, and* (21).

PROOF. The implication (i) → (ii) has already been proved in §I.8, while (ii)–(iii) follows from Theorem 14 and the preceding Lemma. Thus it remains to prove that (iii) → (i). The Jordan–Dedekind chain condition makes $d[x]$ well defined. If L were nonmodular, then by Theorem I.12 it would contain a sublattice N_5; with elements $x < z$ and y such that $x \wedge y = z \wedge y$ and $x \vee y = z \vee y$. But if (iii) holds, then by (21)

$$h[x] + h[y] = h[x \wedge y] + h[x \vee y] = h[z \wedge y] + h[z \vee y] = h[z] + h[y].$$

contradicting $x < z$. Hence (iii) implies (i). Q.E.D.

Complemented modular and semimodular lattices will be considered in Chapter IV, where a notion of semimodularity for lattices of infinite length will be defined in §IV.2.

9*. Betweenness

The properties of the intuitive concept of betweenness in chains are quite easy to describe.† We first observe

LEMMA 1. *For triples of elements in any chain, the following conditions are equivalent:*

(i) $a \leqq b \leqq c$ *or* $c \leqq b \leqq a$,
(ii) $b \in [a \wedge c, a \vee c]$,
(iii) $(a, b, c) = b$.

We omit the proof. If one (hence all) of the conditions of Lemma 1 is fulfilled, then b is said to be *between* a and c, a ternary relation written $(abc)\beta$. It evidently possesses the following properties:

(B1) $(abc)\beta$ implies $(cba)\beta$,
(B2) $(abc)\beta$ and $(acb)\beta$ imply $b = c$,
(B3) $(abc)\beta$ and $(axb)\beta$ imply $(axc)\beta$,
(B4) $(abc)\beta$, $(bcd)\beta$, and $b \neq c$ imply $(abd)\beta$,
(B5) $(abc)\beta$ and $(acd)\beta$ imply $(bcd)\beta$.

Unfortunately, Lemma 1 is no longer true in lattices generally, though (ii) and (iii) are equivalent in any distributive lattice. Hence there are three *inequivalent* concepts of "betweenness" in lattices, each sharing with betweenness in chains many of properties (B1)–(B5). What fails is

(B6) Given $a, b, c \in S$, one of the following is true: $(abc)\beta$, $(bca)\beta$, or $(cab)\beta$.

Curiously, none of the three conditions of Lemma 1 is commonly used to define betweenness in lattices. Instead, use is made of the following fourth condition, due to Glivenko.‡

DEFINITION. In a lattice, we will write $(abc)\beta$ when

$$(a \wedge b) \vee (b \wedge c) = b = (a \vee b) \wedge (b \vee c). \tag{22}$$

It may be shown that (22) is equivalent to the condition that b be between a and c in the metric space defined by the *graph* of any modular lattice M, each segment being understood to have length one. (Here, "between" means $\partial(a, b) + \partial(b, c) = \partial(a, c)$.)

The "betweenness" (22) has many interesting quasi-geometric properties. For example, in a distributive lattice, the "median" (a, b, c) is the unique element

† See for example D. Hilbert, *Grundlagen der Geometrie*, 7th ed., p. 5 ff.; E. V. Huntington and J. R. Kline, Trans. AMS **18** (1917), 301–25; E. V. Huntington, ibid., **38** (1935), 1–19.

‡ V. Glivenko, Amer J. Math. **58** (1936), 799–828, and **59** (1937), 941–56. See also E. Pitcher and M. F. Smiley, Trans. AMS **52** (1942), 95–114; and M. F. Smiley and R. Transue, Bull. AMS **49** (1943), 280–7.

which is "between" any two of the three elements a, b, c. For the study of these properties, the reader is referred to the periodical literature.†

Exercises for §§8–9:

1. If one defines $(a, x, b)\beta$ by condition (i) of Lemma 1, which of conditions (B1)–(B5) holds in every poset? (See M. Altwegg, Comm. Math. Helv. **24** (1950), 149–155.)

2. Show that if $(a, x, b)\beta$ is defined to mean $x \in [a \wedge b, a \vee b]$, then (B2) fails in some lattices. What about (B3)–(B5)?

3. Which of conditions (B1)–(B5) hold for the ternary relation $(abc)\gamma$, defined to mean that $a < x < b$ or $a > x > b$?

4. Let L be any distributive lattice of finite length, and define $\partial(a, b) = h[a \vee b] - h[a \wedge b]$.

   (a) Show that $(a, x, b)\beta$ if and only if $\partial(a, x) + \partial(x, b) = \partial(a, b)$.

   (b) Infer that (a, b, c) gives $\partial(a, x) + \partial(b, x) + \partial(c, x)$ its minimum value

$$\{\partial(a, b) + \partial(b, c) + \partial(c, a)\}/2.$$

   (c) Conclude that any homeomorphism of the *graph* of L is an automorphism of L with respect to the ternary operation (a, b, c).

   (d) Infer that $\mathbf{2}^n$ has $2^n(n!)$ automorphisms with respect to the ternary operation (a, b, c)—as contrasted with $n!$ ordinary automorphisms.

5. Show that in neither nondistributive lattice of five elements, does $\partial(a, x) + \partial(b, x) + \partial(c, x)$ assume a minimum at a unique x, for all a, b, c.

*6. Let P be any poset in which every bounded chain is finite. Show that any maximal chain between two elements can be deformed into any other maximal chain between the same elements, by repeatedly replacing one side of a "simple cycle" by the other.‡

7. Show that the following axioms characterize the ternary betweenness relation in chains (M. Altwegg):

   (i) $(x, y, x)\beta$ implies $x = y$,

   (ii) $(x, y, z)\beta$ implies $(z, y, x)\beta$,

   (iii) $(x, y, z)\beta$, $(y, z, u)\beta$ and $y \neq z$ imply $(x, y, u)\beta$,

   (iv) either $(x, y, z)\beta$, $(y, z, x)\beta$, or $(z, x, y)\beta$.

8. Show that, for any congruence relation θ on a complemented modular lattice, the following conditions are equivalent:

   (i) $x \equiv y\ (\theta)$,

   (ii) $x \wedge y \equiv x \vee y\ (\theta)$,

   (iii) $(x \wedge y)' \wedge (x \vee y) \equiv O\ (\theta)$ for *some* complement of $x \wedge y$,

   (iv) $(x \wedge y)' \wedge (x \vee y) \equiv O\ (\theta)$ for *every* complement of $x \wedge y$.

*9. (Grätzer–Schmidt) Let $\theta(L)$ be the lattice of join-endomorphisms of a finite lattice L.

   (a) Show that $\theta(L)$ is distributive if L is distributive.

   (b) Show that if L is not distributive, then $\theta(L)$ is not semimodular (and not every principal ideal is the kernel of a lattice-morphism).

10. Boolean Algebras

A *Boolean lattice* was defined in §I.10 as a complemented distributive lattice; by definition, such a lattice must contain universal bounds O and I. As was shown (loc. cit.), complements in any Boolean lattice are *unique*, and the function $a \to a'$

† L. M. Kelly, Duke Math. J. **19** (1952), 661–9; M. Sholander, Proc. AMS **3** (1952), 369–81 and **5** (1954), 801–7 and 808–12. See also J. Hashimoto, Osaka Math. J. **10** (1958), 147–58, and the references given there.

‡ S. Mac Lane, Bull. AMS **49** (1943), 567–8.

is a dual automorphism of period two (an "involution"):

$$(a \wedge b)' = a' \vee b', \qquad (a \vee b)' = a' \wedge b', \tag{23}$$

$$(a')' = a \quad \text{for all } a \in L. \tag{24}$$

These facts were well known by 1900; see Schröder [1].

Alternative systems of postulates for Boolean algebra were intensively studied during the decades 1900–1940; E. V. Huntington wrote an influential early paper [1] on this subject. No attempt will be made here to survey the extensive literature on such postulate systems; a typical result is the following.†

HUNTINGTON'S THEOREM. *Let A have one binary operation $\vee$ and one unary operation $'$, and define $a \wedge b = (a' \vee b')'$. Suppose*

$$a \vee b = b \vee a, \tag{25a}$$

$$a \vee (b \vee c) = (a \vee b) \vee c, \tag{25b}$$

$$(a \wedge b) \vee (a \wedge b') = a. \tag{25c}$$

Then A is a Boolean algebra.

Although it is fairly obvious from (23)–(24) that one can eliminate $\wedge$ from the set of undefined (primitive) operations, it is surprising that L1–L4, L6, and L8–L10 all follow from the three identities (25a)–(25c)—i.e., half of L2–L3 and one identity on complements!

It was shown in Chapter I that complements are unique in any distributive lattice; we now consider the converse question. We shall see in Chapter VIII that the existence of a unique complement to every element a implies the distributive law in any lattice where every element is a join of points, though not (Theorem VI.15) in lattices generally. We here prove a weaker result.

THEOREM 17. *If every element in a lattice L has a unique complement a', and if* (23) *holds, then L is a Boolean lattice.*

PROOF. By commutativity, $a \wedge a' = O$ and $a \vee a' = I$ imply $a' \wedge a = O$ and $a' \vee a = I$; hence (24) holds. We next show that

$$b \geqq a \quad \text{implies} \quad (b \wedge a') \vee a = b. \tag{26}$$

Indeed, setting $c = b \wedge a'$ $(b \geqq a)$, $c \wedge a \leqq a' \wedge a = O$ is immediate. Moreover, by (23)–(24),

$$O = (b' \vee a) \wedge (a' \wedge b) = [(b' \vee a) \wedge a'] \wedge b,$$

and $(b' \vee a) \wedge a' = (b \wedge a')' \wedge a' = [(b \wedge a') \vee a]' = (c \vee a)'$. Substituting back, we get $O = (c \vee a)' \wedge b$. On the other hand, $c \vee a = (b \wedge a') \vee a \leqq b \vee a = b$, so $(c \vee a)' \vee b \geqq b' \vee b = I$. Combining, $(c \vee a)'$ is *the* complement of b, and so

$$b = (b')' = ((c \vee a)')' = c \vee a = (b \wedge a') \vee a.$$

† E. V. Huntington, Trans. AMS **35** (1933), 274–304, 557, 977. The theorem quoted below, which presents A as a "complemented semilattice", refers to Huntington's Fourth Set of postulates.

This proves (26); since the hypotheses are self-dual, the dual of (26) also holds.

To prove distributivity, we now use the Corollary of Theorem 13. Suppose $x \wedge y = a$, $x \vee y = b$; let $e = b' \vee (x \wedge a')$. Since $y = a \vee y = b \vee y$, we have by L4, (26), and its dual:

$$e \vee y = b' \vee (x \wedge a') \vee a \vee y = b' \vee x \vee y = b' \vee b = I,$$
$$e \wedge y = [b' \vee (x \wedge a')] \wedge b \wedge y = x \wedge a' \wedge y = a' \wedge a = O.$$

Hence y is the (unique) complement e' of e. Likewise, $x = e'$, so $x = y$: *relative complements are unique*. Therefore (19) holds, and L is distributive. Q.E.D.

Sheffer† showed in 1913 that all Boolean functions could be obtained from the single binary "rejection" operation

(27) $$a|b = a' \wedge b'.$$

(The bar | is called Sheffer's stroke symbol.) Namely:

(28) $$a' = a|a, \qquad a \vee b = (a|b)|(a|b), \qquad a \wedge b = (a|a)(b|b).$$

One then obtains Boolean algebra by assuming the following two ingenious postulates:

I. $$(b|a)|(b'|a) = a,$$

and

II. $$a|(b|c) = [(c'|a)|(b'|a)]'.$$

Unfortunately, these somewhat *ad hoc* postulates are not very easy to work with.

11*. Brouwerian Lattices

In any Boolean algebra A, a' is the *largest* element x such that $a \wedge x = O$ (i.e., such that a and x are "disjoint"). More generally, $a \wedge x \leqq b$ if and only if $a \wedge x \wedge b' = O$, that is $(a \wedge b') \wedge x = O$ or $x \leqq (a \wedge b')' = b \vee a'$. Hence, given $a, b \in A$, there exists a *largest* element $c = b \vee a'$ such that $a \wedge c \leqq b$.

In connection with the foundations of logic (see Chapter XII), Brouwer and Heyting‡ characterized an important generalization of Boolean algebra through an extension of the preceding property.

DEFINITION. A *Brouwerian lattice*¶ is a lattice L in which, for any given elements a and b, the set of all $x \in L$ such that $a \wedge x \leqq b$ contains a greatest element $b:a$, the *relative pseudo-complement* of a in b.

THEOREM 18. *Any Browerian lattice is distributive.*

---

† H. M. Sheffer, Trans. AMS **14** (1913), 481–8. The adequacy of Postulates I and II below was shown by B. A. Bernstein, Bull. AMS **39** (1933), 783–7.

‡ A. Heyting, S. B. Preuss, Akad. Wiss. (1930), 42–56. The relation to lattice theory was pointed out by the author [**LT1**, §§161–2].

¶ For Brouwerian *semilattices*, see O. Frink, Duke Math. J. **29** (1962), 505–14; also J. R. Büchi, Portugal. Math. **7** (1948), 119–78.

PROOF. Given a, b, c form $d = (a \wedge b) \vee (a \wedge c)$, and consider $d:a$. Since $a \wedge b \leqq d$ and $a \wedge c \leqq d$, we have $b \leqq d:a$ and $c \leqq d:a$. Hence $b \wedge c \leqq d:a$, and so $a \wedge (b \vee c) \leqq a \wedge (d:a) \leqq d = (a \wedge b) \vee (a \wedge c)$. But this implies distributivity, by Theorem I.9 and the distributive inequality.

One easily verifies that any Boolean algebra is a Brouwerian lattice, in which $b:a = a' \vee b$ is the relative complement of a in $[a \wedge b, I]$. Likewise, any finite distributive lattice L is Brouwerian, since the join $u = \bigvee x_\alpha$ of the x_α such that $a \wedge x_\alpha \leqq b$ satisfies $a \wedge u = a \wedge \bigvee x_\alpha = \bigvee (a \wedge x_\alpha) \leqq b$. Any chain is also a Brouwerian lattice.

The complete distributive lattice of all *open* subsets of any topological space is also Brouwerian. However, the complete distributive lattice of all *closed* subsets of the line is not Brouwerian: there is no greatest closed set satisfying $p \wedge x = \varnothing$. Hence not all distributive lattices are Brouwerian.

In a Brouwerian lattice with O, the element $O:a$ is called the *pseudo-complement* of a, and is denoted by a^*. Pseudo-complements in Brouwerian lattices have various interesting formal properties.

Exercises for §§10–11:

1. Show that any lattice-epimorphism $\theta: A \to B$ between Boolean algebras preserves complements.
2. Show that the following postulates define Boolean algebra:
 (a) $x \wedge y = y \wedge x$,
 (b) $x \wedge (y \wedge z) = (x \wedge y) \wedge z$,
 (c) $x \wedge y' = z \wedge z'$ if and only if $x \wedge y = y$. (Lee Byrne)
3. Let L be any complemented lattice in which $a \wedge x = O$ implies $x \leqq a'$ for *any* complement a' of a. Prove:
 (a) Complementation is unique.
 (b) If $a \leqq b$, then $a \wedge b' = O$ and $b' \leqq a'$.
 (c) L is a Boolean algebra. (E. V. Huntington [1])
4. Prove that $(a, b, c)' = (a', b', c')$ in any Boolean algebra.
5. Show that, in a Boolean lattice A, $b \leqq c$ if and only if $a \wedge c = O$ $(a \in A)$ implies $a \wedge b = O$.
6. Show that for pseudo-complements in a Brouwerian lattice:
 (i) $a \leqq b$ implies $b^* \leqq a^*$, (ii) $a \leqq a^{**}$,
 (iii) $a^* = a^{***}$, (iv) $(a \vee b)^* = a^* \wedge b^*$,
 and, in the lattice of closed elements, (v) $(a \wedge b)^* = a^* \vee b^*$.

*7. Prove that postulates I, II on Sheffer's stroke symbol imply all the laws of Boolean algebra.

*8. (J. Abbott–P. R. Kleindorfer) In a Boolean algebra A, define $a*b = a' \wedge b$. Show that:

(i) $(a*b)*a = a$, $(a*b)*b = (b*a)*a$, $a*(b*c) = b*(a*c)$,
(ii) $a*a = O$, $O*a = a$, $a*O = O$, (iii) $\exists$ 1 such that $1*a = O$.

Show that (i)–(iii) are a system of postulates for Boolean algebras as subtractive "implication algebras". (Cf. B. A. Bernstein, Trans. AMS **36** (1934), 876–84.)

*9. Show that the ideals of any distributive lattice with O form a complete Brouwerian lattice. (Ribenboim)

10. In each of the two nondistributive five-element lattices, M_5 and N_5, find a, b such that $a*b$ does not exist.
11. Show that N_5 is pseudo-complemented.
12. Show that every chain is relatively pseudo-complemented.

13. Show that the elements of any finite distributive lattice form a Brouwerian lattice.

14. Show that a Brouwerian lattice is a Boolean lattice if it satisfies either (a) $(x^*) = x$, or (b) $x \wedge x^* = O$.

15. Find a, b in a five-element Brouwerian lattice such that $(a \wedge b)^* > a^* \vee b^*$.

*16. Show that the following postulates for Brouwerian lattices form an independent set:

$$a{:}a = I, \qquad a \wedge O = O,$$
$$(a{:}b) \wedge a = a, \qquad (a{:}b) \wedge b = a \wedge b,$$
$$(a \wedge b){:}c = (a{:}c) \wedge (b{:}c), \qquad a{:}(b \vee c) = (a{:}b) \wedge (a{:}c).$$

(A. A. Monteiro, Rev. Union Mat. Argentina **17** (1955), 148–60)

12*. Boolean Rings

The analogy between the algebra of logic and ordinary algebra has been emphasized by many mathematicians, from Boole [**1**] on. However, it was M. H. Stone [**1**] who first made clear around 1935 the precise connection between Boolean algebra and rings.†

A ring may be formed from the elements of any Boolean algebra by defining multiplication as meet and addition as the "symmetric difference":

$$ab = a \wedge b \quad \text{and} \quad a + b = (a \wedge b') \vee (a' \wedge b). \tag{29}$$

It is obvious that both operations are commutative. We next check the associative law for addition:

$$(a + b) + c = \{[(a \wedge b') \vee (a' \wedge b)] \wedge c'\} \vee \{[(a \wedge b') \vee (a' \wedge b)]' \wedge c\}.$$

After simplification by repeated use of distributivity, this becomes

$$(a + b) + c = (a' \wedge b \wedge c') \vee (a \wedge b' \wedge c') \vee (b' \wedge c \wedge a') \vee (a \wedge b \wedge c)$$

which is symmetric in a, b, and c. Because of this symmetry, we have

$$(a + b) + c = a + (b + c). \tag{30}$$

The element O of the Boolean algebra acts as the additive zero in the ring:

$$a + O = (a \wedge O') \vee (a' \wedge O) = (a \wedge I) \vee (a' \wedge O) = a \vee O = a.$$

The additive inverse of a exists and is a since, by (29),

$$a + a = O \vee O = O. \tag{31}$$

It remains to establish the distributive laws. By the definition

$$a(b + c) = a \wedge [(b \wedge c') \vee (b' \wedge c)] = (a \wedge b \wedge c') \vee (a \wedge b' \wedge c),$$

and

$$\begin{aligned} ab + ac &= (a \wedge b \wedge (a \wedge c)') \vee ((a \wedge b)' \wedge a \wedge c) \\ &= (a \wedge b \wedge (a' \vee c')) \vee ((a' \vee b') \wedge a \wedge c) \\ &= (a \wedge b \wedge a') \vee (a \wedge b \wedge c') \vee (a' \wedge a \wedge c) \vee (b' \wedge a \wedge c) \\ &= (a \wedge b \wedge c') \vee (b' \wedge a \wedge c). \end{aligned}$$

Thus $a(b + c) = ab + ac$, completing the proof.

† For related earlier work, see the references of [**LT2**, p. 153, footnote 3].

Trivially the ring just constructed has a multiplicative unity, namely I, since $I \wedge a = a \wedge I = a$ for all a. Also trivially, $aa = a$ for all a in the ring defined by (29). Hence, if we define a *Boolean ring* as an (associative) ring in which

$$aa = a \quad \text{for all } a \tag{32}$$

(i.e., multiplication is idempotent), we can summarize the results proved above as follows.

THEOREM 19. *Under the operations defined by* (29), *every Boolean algebra is a Boolean ring with unity.*

From any Boolean ring with unity, we can conversely construct a Boolean algebra by defining:

$$a \wedge b = ab \quad \text{and} \quad a \vee b = a + b + ab. \tag{33}$$

If one defines $x \geqq y$ to mean $xy = y$, then clearly

(i) $1 \geqq x \geqq 0$ for all x,
(ii) $x \geqq x$ by idempotence,
(iii) if $x \geqq y$ and $y \geqq x$, then $x = yx = xy = y$ by hypothesis and commutativity,
(iv) if $x \geqq y$ and $y \geqq x$, then $x = xy = x(yz) = (xy)z = xz$, and so $x \geqq z$,
(v) $x \geqq xy$, since $x(xy) = (xx)y = xy$,
(v′) similarly, using commutativity, $y \geqq xy$,
(vi) if $x \geqq z$ and $y \geqq z$, then $xyz = xz = z$ and so $xy \geqq z$,
(vii) the correspondence $x \to 1 - x$ is obviously one-one.

Moreover since $xy = y$ implies $(1 - y)(1 - x) = 1 - y - x + xy = (1 - x)$, it inverts inclusion; hence it is a dual automorphism.

Hence our definition makes R into a poset with O and I by (i)–(iv), in which $x \wedge y$ exists and is xy by (v)–(vi), whence by (vii) $x \vee y$ exists and is

$$1 - (1 - x)(1 - y) = x + y - xy,$$

as in (33). Also, if we define $x' = 1 - x$, then $x \wedge x' = x(1 - x) = 0$ and $x \vee x' = x + (1 - x) + x(1 - x) = 1$; hence our definition makes R into a complemented lattice. Finally,

$$\begin{aligned} x \wedge (y \vee z) &= x(y + z - yz) = xy + xz - xyz \\ &= xy + xz - xyxz = (x \wedge y) \vee (x \wedge z), \end{aligned}$$

and so R is a Boolean lattice.

For any Boolean ring $R(B)$ with unity obtained from a Boolean algebra B by the definitions (29), the Boolean lattice $B(R(B))$ obtained from the ring $R(B)$ by (33) is isomorphic with B. We omit the proof.

Stone [**1**] has defined a *generalized Boolean algebra* as a *relatively* complemented distributive lattice with a O (but not necessarily a I), such as the following.

Example 8. The lattice $\mathbf{2}^{(\omega)}$ of all *finite* subsets of the set $\mathbf{Z}^+ = \omega$ of positive integers is a generalized Boolean algebra. The corresponding characteristic functions (with values in $\mathbf{Z}_2$) form a Boolean ring, which is the restricted direct sum (product) of countably many copies of $\mathbf{Z}_2$ (see Exercise 6 below). Note that $\mathbf{2}^{(\omega)}$ is an *ideal* in $\mathbf{2}^{\omega}$, the set of *all* subsets of ω; the quotient-lattice $\mathbf{2}^{\omega}/\mathbf{2}^{(\omega)}$ has many interesting properties (see Exercise 3 below; also Exercise 1 after Chapter VIII). In a generalized Boolean algebra we still define $a + b$ by (33), where b' and a' are *relative* complements in $[O, c]$ for any $c \geqq a \vee b$.

There also exist nonassociative Boolean rings, as the following example shows.

Example 9. Consider the *nonassociative* linear algebra R over the field $\mathbf{Z}_2$ of integers mod 2, with basis element 1 (unity), a, b, c, and the multiplication rules

$$(34) \qquad a^2 = a,\ ab = ba = c, \qquad \text{and cylically.}$$

Then $(a + b)^2 = a^2 + 0 + b^2 = a + b$, etc.

13*. Newman Algebras

A remarkable synthesis of Boolean algebras and Boolean rings was made in 1941–2 by Newman.† Define a *Newman algebra* as an algebra A with two binary operations such that

N1. $\quad a(b + c) = ab + ac,$
N1′. $\quad (a + b)c = ac + bc,$
N2. $\quad \exists 1$ such that $a1 = a$ for all a,
N3. $\quad \exists 0$ so that $a + 0 = 0 + a = a$ for all a,
N4. $\quad$ To each a corresponds at least one a' with $aa' = 0$ and $a + a' = 1$.

Observe that neither idempotence, commutativity, nor associativity is assumed for either operation. Neither is the unicity of a'.

It is easily verified that every Boolean algebra *and* every Boolean ring with unity is a Newman algebra; the Boolean ring need *not* be associative. Hence so is any direct product of a Boolean algebra with a Boolean ring. We now establish the converse of this result, through a sequence of lemmas (T1–T7). First,

T1. $\quad$ *For all* a, $aa = a$.

PROOF. $aa = aa + 0 = aa + aa' = a(a + a') = a1 = a$.

Next, we show that *any* complement $(a')'$ of *any* complement a' of a must equal a:

T2. $\quad (a')' = a$.

PROOF.
$$\begin{aligned}(a')' &= 0 + (a')'(a')' + (a')'(a')' \\ &= [a' + (a')'](a')' = 1(a')' = (a + a')(a')' \\ &= a(a') + 0 = 0 + a(a')' = aa' + a(a')' \\ &= a(a' + (a')') = a1 = a. \qquad \text{Q.E.D.}\end{aligned}$$

† M. H. A. Newman, J. London Math. Soc. **16** (1941), 256–72, and **17** (1942), 34–47. See also G. D. Birkhoff and G. Birkhoff, Trans. AMS **60** (1946), 3–11.

Hence, if a' and a^* are any two complements of a, then $(a^*)' = a$ and so $((a^*)')' = a^*$ by the above; yet $((a^*)')' = a'$ also; hence $a^* = a'$ and *complements are unique.* As a corollary of T2 and N4, we also have

N4′. $a'a = 0$ and $a' + a = 1$.

Further, by N4, N3, N1′, N4, and N3,

$$0 = aa' = a(a' + 0) = aa' + a0 = 0 + a0 = a0,$$

and likewise, using N4′, N3, N1′, N4′, and N3,

$$0 = a'a = (a' + 0)a = a'a + 0a = 0 + 0a = 0a.$$

Thus we have proved

T3. $a0 = 0 = 0a$, *for all* $a \in A$.

We note that if $1 = 0$ then $0 = a0 = a1 = a$ for all a. Hence, except in this trivial one-element case, $0 \neq 1$. We shall assume $0 \neq 1$ from now on. We next prove

N2′. $1a = (a + a')a = aa + a'a = a + 0 = a$.

Therefore, there is complete left-right symmetry in the properties of addition and multiplication.

We now define $1 + 1 = 2$, and call the left-multiplies $x2$ of 2 *even* elements. Note that $2 + 2 = 2(1 + 1) = 2 \cdot 2 = 2$ (by T1). Using these definitions, we prove

T4. *x is even if and only if $x + x = x$.*

For clearly $y2 + y2 = y(2 + 2) = y2$; and conversely, if $x = x + x$ then $x = x1 + x1 = x(1 + 1) = x2$.

T5. *Any multiple xt or ux of an even element is even. For if*

$x = x + x$ then $xt = (x + x)t = xt + xt$ and $ux = u(x + x) = ux + ux$.

T6. *The function $x \to x2$ is an idempotent endomorphism*:

$$(x + y)2 = x2 + y2, \qquad (xy)2 = (x2)(y2), \qquad (x2)2 = x2.$$

Proof. $(x + y)2 = x2 + y2$

$$(x2)2 = x2 + x2 = x(2 + 2) = x2$$

and

$$\begin{aligned}(x2)(y2) &= (x + x)(y + y) = (x + x)y + (x + y)y + (x + y)y \\ &= (xy + xy) + (xy + xy) \\ &= (xy)2 + (xy)2 = (xy)(2 + 2) = (xy)2.\end{aligned}$$

It is a corollary that the even elements form a subalgebra B in which addition is idempotent. In this subalgebra we have

$$a + 1 = (a + 1)(a + a') = (aa + a) = (aa' + 1a')$$
$$= (a + a) + (0 + a') = a + a' = 1$$

so $a + 2 = a2 + 2 = (a + 1)2 = 2$, and $2 \in B$. Thus the elements of B satisfy the postulates given previously for a distributive lattice in which 2 acts as the universal upper bound I. Moreover, if $a \in B$ then $a'2 \in B$ is its complement in B, since

$$a + a'2 = a2 + a'2 = (a + a')2 = 1 \cdot 2 = 2$$

and

$$a(a'2) = a(a' + a') = aa' + aa' = 0 + 0 = 0.$$

Thus, by Theorem 10, B is a Boolean algebra; in particular, the multiplication of "even" elements is associative.

We now define all left-multiples of $2'$ to be *odd*. Then $2' + 2' = 2' \cdot 1 + 2' \cdot 1 = 2'(1 + 1) = 2'2 = 0$ and we have

T7. *x is odd if and only if $x + x = 0$.*

Clearly if x is odd then $x = y2'$ and $x + x = y2' + y2' = y(2' + 2') = y0 = 0$. On the other hand, if $x + x = 0$ then $x = x \cdot 1 = x(2' + 2) = x2' + x2 = x2'$ since $x2 = x + x = 0$.

Arguments similar to those used to prove T5 and T6 show that the set B' of odd elements form a Boolean ring with unity $2'$, which need not be associative.

Finally, for any x of our Newman algebra

$$x = x \cdot 1 = x(2 + 2') = x2 + x2'.$$

In conclusion, we have sketched a proof of

THEOREM 20. *Any Newman algebra is the direct product of a Boolean algebra and a (possibly nonassociative) Boolean ring with unity.*

Exercises for §§12–13:

1. Show that $x + y = (x \wedge y) + (x \vee y)$ in any Boolean algebra.
2. Show that every group-translation $x \to x + a$ of a Boolean algebra is an automorphism for the median operation.
3. Show that the finite sets of integers form a generalized Boolean algebra without unity under $\wedge$ and $\vee$.
4. (a) Show that the idempotent elements of any commutative ring of characteristic two form a Boolean subring.

   (b) Show that any commutative ring with one is also a ring under the "dual operations"

$$x \circ y = x + y - xy \quad \text{and} \quad x \oplus y = x + y - 1.$$

5. Show that, under the correspondence of Theorem 19, automorphisms correspond to automorphisms, Boolean subalgebras to subrings, ideals to ideals, and lattice-theoretic prime ideals to ring-theoretic prime ideals.

6. Generalize Theorem 19 to generalized Boolean algebras and (associative) Boolean rings not assumed to have a 1.

7. Let R be a commutative ring with 1 in which $1 + 1 = 0$ and $x(1 - x)y(1 - y) = 0$. Show that $a = (a + aa) + aa$ decomposes every $a \in R$ *uniquely* into the sum of a nilpotent and an idempotent component.

8. Using only N1–N4 and their consequences, prove

(a) $a + b = (a + b)(b + b') = (a + 1)b + ab' = \cdots = b + a$.

(b) $1 + (1 + c) = [1 + (1 + 1)]c + (c' + c') = \cdots = (1 + 1) + c$, by right-multiplying by $c + c'$, expanding, and using (a) to infer $1 + (1 + 1) = (1 + 1) + 1$.

(c) $1 + (b + c) = (1 + b) + c$, using (b) and a similar right-multiplication by $b + b'$.

(d) $a + (b + c) = (a + b) + c$, using (c).

*9. Show that, in N1–N4, condition $0 + a = a$ is redundant.

10. Let R be a commutative ring with 1 of characteristic two, in which $ab(a + b + ab) = ab$ for all a, b. Prove that:

(i) $a^4 = a^2$,

(ii) R is the direct sum of a Boolean ring and a zero-ring (its radical), and conversely.

(Cf. A. L. Foster, Trans. AMS **59** (1946), 166–87.)

14*. Ortholattices

We finally consider *nondistributive* analogs of Boolean algebras (as contrasted with Boolean lattices; cf. §I.10), in which a unary *complementation* operation $a \to a^\perp$ is given as well as a partial ordering in which any two elements have a g.l.b. and l.u.b.

DEFINITION. An *ortholattice* is a lattice with universal bounds, and a unary operation $a \to a$ satisfying:

L8. $a \wedge a^\perp = O, \quad a \vee a^\perp = I, \quad$ for all a.

L9. $(a \wedge b)^\perp = a^\perp \vee b^\perp, \quad (a \vee b)^\perp = a^\perp \wedge b^\perp$.

L10. $(a^\perp)^\perp = a$.

Trivially, any ortholattice is complemented, and a distributive ortholattice is a Boolean algebra. The most familiar nondistributive ortholattice is that of subspaces of a finite-dimensional Euclidean space (§I.9, Example 11); it is modular. We now describe an extremely important *nonmodular* ortholattice.

Example 10. Let $L(\mathfrak{H})$ be the lattice of all closed subspaces of the (separable) Hilbert space $\mathfrak{H} = L^2(0, 1)$. For each such closed subspace S, let $S^\perp$ be its orthocomplement. Then L is an ortholattice.

DEFINITION. In an ortholattice we write aCb (in words, a commutes with b) when $a = (a \wedge b) \vee (a \wedge b^\perp)$.

LEMMA 1. *In any ortholattice, $a \leqq b$ implies aCb.*

PROOF. If $a \leqq b$, then $(a \wedge b) \vee (a \wedge b^\perp) = a \vee (a \wedge b^\perp) = a$ by the contraction law L4.

In Example 10, SCT means that the projections E_S and E_T onto S and T commute: $E_S E_T = E_T E_S$. In this example, therefore, aCb implies bCa. Moreover, in this example, $a \leqq b$ implies aCb. Nakamura† has proved

† M. Nakamura, Kodai Math. Sem. Rep. **9** (1957), 158–60.

THEOREM 21. *In any ortholattice L, the following two conditions are equivalent*:

(35) $x \leqq y$ *implies* $x \vee (x^{\perp} \wedge y) = y$ (*i.e.*, yCx),

(35′) xCy *implies* yCx (*commutativity is symmetric*).

The proof can be found in S. Holland [**1**], the exposition of whose Theorems 1–3 we summarize here.

An ortholattice which satisfies either (hence both) of the equivalent conditions (35)–(35′) is called an *orthomodular lattice*. By (35), any modular ortholattice is orthomodular. Conversely, in any orthomodular lattice, every "orthogonal" pair is a "modular" pair (cf. §IV.2), though the modular law need *not* hold.

COROLLARY. *Any orthomodular lattice is relatively complemented.*

THEOREM 22 (FOULIS–HOLLAND). *In any orthomodular lattice L, for any $a \in L$, the set $C(a)$ of all elements x such that aCx is a sub-ortholattice.*†

We note also (Holland [**1**, Theorem 3], and D. J. Foulis, op. cit., Theorem 5):

THEOREM 23. *In any orthomodular lattice L, let aCb and aCc. Then $\{a, b, c\}$ generate a distributive sublattice of L.*

COROLLARY 1. *Let a_iCa_j for all $a_i \in S$ in an orthomodular lattice L. Then the $a_i \in S$ generate a Boolean algebra under the operations $\wedge$, $\vee$, $a^{\perp}$ of L.*

COROLLARY 2. *In any orthomodular lattice L, any chain generates a Boolean subalgebra of L.*

COROLLARY 3. *In any orthomodular lattice, each interval sublattice $[a, b]$ is orthomodular and closed under $\wedge$, $\vee$, and the relative complementation operation $c^* = (a \vee c^{\perp}) \wedge b = a \vee (c^{\perp} \wedge b)$.*

Exercises (see also Exercises after §IV.15):

1. (a) Show that any distributive ortholattice is a Boolean algebra.

(b) Let LM be the direct product of two ortholattices, regarded as posets. Make LM into an ortholattice, which is orthomodular when L and M are.

2. Show that, in any ortholattice, $a \leqq b$ if and only if $a^{\perp} \geqq b^{\perp}$.

3. (a) In a symmetric ortholattice, show that $a \leqq b$ if and only if $b \wedge c = O$ implies $a \wedge c = O$.

(b) Show that no five-element lattice can be made into an ortholattice.

4. Show that the following conditions on elements a, b of an orthomodular lattice are equivalent:

(i) aCb,

(ii) $(a \vee b^{\perp}) \wedge b = a \wedge b$,

(iii) $a = (a \vee b) \wedge (a \vee b^{\perp})$,

(iv) $a = e \vee g$ and $b = f \vee g$ for pairwise orthogonal elements e, f, $g \in L$. (Foulis-Holland)

5. Show that any lattice of finite order $n > 2$ can be embedded as a sublattice in:

(i) a complemented lattice of order $n + 1$, and

(ii) an ortholattice of order $2n - 2$.

† See D. J. Foulis, Portugal. Math. **21** (1962), 65–72, Lemma 3; Holland [**1**].

6. Prove that any modular ortholattice is orthomodular.

7. Prove that each of the following conditions is necessary and sufficient for an ortholattice to be orthomodular:

(i) $a \leqq b$ and $a^{\perp} \wedge b = O$ imply $a = b$.

(ii) If $a \leqq b \leqq c$, then $a \vee (b^{\perp} \wedge c) = (a \vee b^{\perp}) \wedge c$.

8. Show that, in Example 10, the elements of finite height form an ideal F which is also a modular sublattice.

*9. Describe the quotient-lattice $L(\mathfrak{H})/F$, where F is as in Example 8. (See Chapter VIII.)

*10. Find a lattice having 18 elements and length 5, which has a dual automorphism but no involutory dual automorphism.

*11. Let R be a ring with unity 1, and a mapping $a \to a^*$ with $a^{**} = a$, $(a + b)^* = a^* + b^*$, and $(ab)^* = b^*a^*$. Let L be the set of $x \in R$ with $x = x^* = x^2$. Prove that:

(a) The relation "$x \leqq y$ means $x = xy$" is a partial ordering.

(b) If L is a lattice under this order, then it is an orthomodular lattice for $x^{\perp} = 1 - x$.

(c) In L, $xy = yx$ if and only if xCy.

(d) The bounded linear transformations of $\mathfrak{H}$ satisfy the above hypotheses, for a^* the adjoint of a.

CHAPTER III

STRUCTURE AND REPRESENTATION THEORY

1. Cardinal Arithmetic

The fundamental *structure problem* of algebra is that of *analyzing* a given algebraic system into simpler *components*, from which the given system can be reconstructed by *synthesis*. Thus, it is classic that any finite-dimensional vector space can be factored into one-dimensional subspaces (copies of the base field), and that any finite Abelian group can be factored into cyclic components (of prime-power order).

Such decomposition theorems reveal the structure of a given algebraic system. The relevant components usually form *lattices*, considered either as distinguished subsets (e.g., the normal subgroups of a group) or as partitions (the congruence relations) or both. This gives two natural connections between lattice theory and the study of the structure of algebraic systems (cf. Examples 5 and 6 of Chapter I). These connections will be investigated in Chapters VI–VIII.

The present chapter will begin with a study of the structure of lattices, and especially with the problem of synthesizing a lattice from smaller components. It turns out that, in this synthesis, an essential role is played by generalizations of the familiar arithmetic binary operations of addition, multiplication, and exponentiation to posets.

Although cardinal multiplication and addition can be defined for sets with arbitrary relations,† we shall only be concerned with posets.

DEFINITION. Let X, Y be posets. The *cardinal sum* $X + Y$ of X and Y is the set of all elements in X or Y, considered as disjoint. Here $\geqq$ keeps its meaning in X and in Y, while neither $x \geqq y$ nor $x \leqq y$ for any $x \in X$, $y \in Y$. The *cardinal product* XY is the direct product already defined in §I.4. The *cardinal power* Y^X with base Y and exponent X is the set of all isotone functions $y = f(x)$ from X to Y, partly ordered by letting $f \geqq g$ mean that $f(x) \geqq g(x)$ for all $x \in X$.

For finite posets, the diagram of $X + Y$ consists of the diagrams of X and Y laid side by side; thus $\substack{\circ\\|\\\circ}\ \substack{\circ\\|\\\circ}$ is the diagram of $\mathbf{2} + \mathbf{2}$.

Likewise, if X is a finite chain and Y has a plane diagram, then the Cartesian product of a linear diagram for X and the plane diagram for Y is a space diagram for XY. More generally, if X and Y have space diagrams in which no vector $\mathbf{x}_h - \mathbf{x}_k$ is equal to a vector $\mathbf{y}_i - \mathbf{y}_j$, then the $\mathbf{x}_h + \mathbf{y}_i$ form the vertices of a space

† Whitehead and Russell [**1**, §§162, 172]. Applications to posets were given by F. Hausdorff, *Grundzüge der Mengenlehre*, Leipzig, 1914. Application to lattice theory, and a definition of cardinal exponentiation, were given by the author ([**5**] and Duke J. Math. **3** (1937), 311–6).

diagram for XY, in which $\mathbf{x}_h + \mathbf{y}_i$ is covered by the $\mathbf{x}_k + \mathbf{y}_i$ for which $\mathbf{x}_k$ covers $\mathbf{x}_h$, and by the $\mathbf{x}_h + \mathbf{y}_j$ in which $\mathbf{y}_j$ covers $\mathbf{y}_i$.

It is trivial to show that the cardinal sum, product, and power of any two posets is again a poset. Moreover by Theorem I.7, we know that the product of any two lattices is a lattice. We now prove a less obvious result.

THEOREM 1. *If M is a lattice and X a poset, then M^X is a lattice. If M is modular or distributive, then so is M^X.*

PROOF. Let $y = f(x)$ and $y = g(x)$ be any two isotone single-valued functions from X to M. Then if for all x, we define $h(x) = f(x) \vee g(x) \in M$, we get a function h which is isotone, single-valued, and a least upper bound of f and g in M^X. Dually $h^*(x) = f(x) \wedge g(x)$ is a greatest lower bound, completing the proof of the first statement. To prove the second, one need only verify the relevant identity at each x.

An important application of cardinal powers is to the construction of $\mathbf{2}^X$, where $\mathbf{2}$ denotes the ordinal two and X is a general poset. Call a subset A of X *J-closed*† (or a "dual semi-ideal") when it contains with any a all $x \geqq a$, and a semi-ideal or *M-closed* when it contains with any a all $x \leqq a$. (Thus, if X is a lattice, every ideal of X is M-closed, and every dual ideal is J-closed.) We prove

LEMMA 1. *The distributive lattice $\mathbf{2}^X$ is isomorphic to the ring of all J-closed subsets of X, ordered by inclusion.*

Specifically, let any such J-closed subset A correspond to its *characteristic function* Q_A: $X \to \mathbf{2}$, defined by the conditions that

$$Q_A(x) = 1 \quad \text{if } x \in A$$
$$= 0 \quad \text{if } x \notin A.$$

This correspondence is obviously one-one; and is also onto $\mathbf{2}^X$ because any $f \in \mathbf{2}^X$ corresponds to the set J_f of $x \in X$ such that $f(x) = 1$. In other words, $Q_{J_f} = f$.

Furthermore, J_f is J-closed in X by the isotonicity of f; for if $a \in J_f$ and $x \geqq a$, then $f(x) \geqq f(a) = 1$ and hence $x \in J_f$ by definition of J_f. Finally, the correspondence $A \to Q_A$ is an isomorphism because $B \subset A$ implies $Q_B(x) \leqq Q_A(x)$ for all $x \in X$.

COROLLARY. *Dually, $\mathbf{2}^{\check{X}}$ is dually isomorphic to the ring of all M-closed subsets of X.*

2. Formal Identities

If we use the equality sign to signify "is isomorphic with", we can generalize most laws of arithmetic to posets generally, and prove some further laws for dualization.‡

† Generally, as will appear in Chapter IX, the T_0-space associated through "J-closed subsets" with XY is the product of those associated with X and Y, respectively.

‡ Many other laws may be found in G. Birkhoff [5] and M. M. Day [1]. Some of these are stated in the Exercises below.

THEOREM 2. *The following laws are identities for the cardinal arithmetic of posets:*

$$(1)\quad X + Y = Y + X, \qquad X + (Y + Z) = (X + Y) + Z,$$

$$(2)\quad XY = YX, \qquad X(YZ) = (XY)Z,$$

$$(3)\quad X(Y + Z) = XY + XZ, \qquad (X + Y)Z = XZ + YZ,$$

$$(4)\quad X^{Y+Z} = X^Y X^Z, \qquad (XY)^Z = X^Z Y^Z, \qquad (X^Y)^Z = X^{YZ},$$

$$(5)\quad \widecheck{X + Y} = \check{X} + \check{Y}, \qquad \widecheck{XY} = \check{X}\check{Y}, \qquad \widecheck{Y^X} = \check{Y}^{\check{X}}.$$

PROOF. The proof of (1) is trivial. Again, $XY = YX$ follows since the mapping $(x, y)\theta = (y, x)$, defined for all $x \in X$ and $y \in Y$, is one-one and order-preserving from XY onto YX—that is, it is an isomorphism. Likewise, the mapping ϕ: $(x, (y, z))\phi = ((x, y), z)$ defines a trivial isomorphism from $X(YZ)$ onto $(XY)Z$. The proofs of the distributive laws (3) are also trivial. So are those of the dualization laws (5); they are applications of the Duality Principle.

The proofs of the three "laws of exponents" (4) are less trivial; we present them in turn.

If $f \in X^{Y+Z}$, it is a function on the union of two disjoint sets Y and Z; let f correspond to the couple $(f_Y, f_Z) \in X^Y X^Z$, where f_Y and f_Z are defined as f restricted to Y and Z, respectively. Clearly f_Y and f_Z are isotone if f is; the correspondence is one-one because Y and Z together make up all of $Y + Z$. Conversely, each member (f_1, f_2) of $X^Y X^Z$ corresponds to the function defined as equal to f_1 on Y and f_2 on Z. This is well defined since Y and Z are disjoint, and isotone because the ordering on $Y + Z$ agrees with the original ones on Y and Z. Finally, the correspondence is an isomorphism: if $f \leqq g$, $f(x) \leqq g(x)$ for all $x \in Y$ and all $x \in Z$, which implies $f_1 \leqq g_1$ and $f_2 \leqq g_2$—and conversely.

Suppose $f \in (XY)^Z$ and let it correspond to the couple (f_X, f_Y) where $f_X(z)$, $z \in Z$, is the X-component of f and similarly for f_Y; then this defines an isomorphism of $(XY)^Z$ and $X^Z Y^Z$. The isotonicity of (f_X, f_Y) is immediate from consideration of coordinates. The correspondence is one-one because f is defined in terms of its components. And if $(f_1, f_2) \in X^Z Y^Z$ it corresponds to the function $f(z) = (f_1(z), f_2(z)) \in (XY)^Z$, so that it is also onto. Further, $f \leqq g$ in $(XY)^Z$ means that $f_X(z) \leqq g_X(z)$ and $f_Y(z) \leqq g_Y(z)$ for all $z \in Z$, so that it is isotone.

If $f \in (X^Y)^Z$, then f is an isotone function from Z into the set of isotone functions from Y to X. Denote the mapping of $z \in Z$ by f_z. Then define ϕ from $(X^Y)^Z$ into X^{YZ} by $f_\phi(y, z) = f_z(y)$. Since f is isotone so is f_ϕ, because if $z \leqq z_1$ and $y \leqq y_1$, $f_z \leqq f_{z_1}$, which in turn implies that $f_\phi(y, z) = f_z(y) \leqq f_z(y_1) \leqq f_{z_1}(y_1) = f_\phi(y_1, z_1)$. ϕ is trivially one-one, and it is onto since if $g(X, Y) \in X^{YZ}$, the function defined by $g_z(y) = g(z, y)$ is taken by ϕ into g. If $f \leqq g$, $f, g \in (XY)^Z$, then $f_z \leqq g_z$ for all $z \in Z$ which means that $f_z(y) \leqq g_z(y)$ for all $z \in Z$ and $y \in Y$.

Exercises for §§1–2:

1. Show that if X and Y are posets, then so are $X + Y$, XY, and Y^X.
2. Prove (a) formulas (1)–(3), (b) formula (5).

3. Prove that $X + Y$ cannot be a lattice unless X or Y is void.

4. Show that, if X is a poset, then M^X is a lattice if and only if M is a lattice.

5. Show that if we define $0^0 = 1$, then $X^0 = 1$ for all X. Show that $0^X = 0$ for all nonvoid X.

6. Show that the lattice of all J-closed subsets of $X + Y$ is the product of the lattices of all J closed subsets of X and Y separately.

7. Show that the poset of Figure 1a is $\mathbf{2}^3$. (G. Birkhoff, [1, Theorem 5.1])

8. Prove that if $n(X)$ signifies the order of X, then $n(X + Y) = n(X) + n(Y)$ and $n(XY) = n(X)n(Y)$, but that $n(Y^X) < n(Y)^{n(X)}$ is possible.

9. (a) Prove that $X^1 = X$, while $X^m X^n = X^{m+n}$ for any cardinal numbers (unordered sets) m and n.

(b) Let P, Q, R be finite posets, and let P have a least element. Show that $P^Q \cong P^R$ implies $Q \cong R$.

10. (a) Show that the set of intervals of a partly ordered set P, partly ordered by set-inclusion, is isomorphic to a subset of $\check{P}P$.

(b) Show that it is in one-one correspondence with $P^{\mathbf{2}}$, where $\mathbf{2}$ represents the lattice of two elements.

(c) Show that if (with Ore [1, p. 425]) $[x, y] \geqq [x_1, y_1]$ is defined to mean $x \geqq x_1$ and $y \geqq y_1$, we get $P^{\mathbf{2}}$.

3. Representation of Distributive Lattices

We will now show that any distributive lattice of finite length has an isomorphic representation as a ring of sets. Specifically, $L \cong \mathbf{2}^X$, where X is the poset of join-irreducible elements of L. In Chapter VIII, we will prove the first result (but not the second) without assuming finite length!

DEFINITION. A *representation* of a distributive lattice L is a lattice-epimorphism from L to a ring R of subsets $S, T, \cdots$ of a set X. An element $a \neq O$ of a lattice L is *join-irreducible* when $b \vee c = a$ implies $b = a$ or $c = a$; meet-irreducibility is defined dually.

LEMMA 1. *If p is join-irreducible in a distributive lattice L, then $p \leqq \bigvee_{i=1}^k x_i$ implies $p \leqq x_i$ for some i.*

PROOF. By $p \leqq \bigvee_{i=1}^k x_i$, we have $p = p \wedge \bigvee_{i=1}^k x_i = \bigvee_{i=1}^k (p \wedge x_i)$ by distributivity. Since p is join-irreducible, $p = p \wedge x_i$ for some i. Hence $p \leqq x_i$ for some i.

COROLLARY. *In a distributive lattice of finite length, each element a has a strictly unique representation as the join of a join-irredundant set of join-irreducible elements.*

PROOF. Namely, a is the join of the maximal members of the set of join-irreducibles contained in it. (By join-irredundant, we mean that the join of any proper subset is a smaller element.)

LEMMA 2. *If a distributive lattice L contains n join-irreducible elements $p_1, \cdots, p_n$, then $d[L] = n$.*

PROOF. Renumber the p_i so that $p_i < p_j$ implies $i < j$; this is possible since partial ordering is anti-circular. Then the chain $O < p_1 < p_1 \vee p_2 < \cdots < \bigvee_{i=1}^n p_i$ has length n, because $p_1 \vee p_2 \cdots \vee p_j = p_1 \vee p_2 \cdots \vee p_j \vee p_{j+1}$

would imply $p_{j+1} \leqq p_1 \vee p_2 \cdots \vee p_j$, which by Lemma 1 implies $p_{j+1} \leqq p_i$ for some i, $1 \leqq i \leqq j$, giving a contradiction.

THEOREM 3. *Let L be any distributive lattice of length n. Then the poset X of meet-irreducible elements $p_i > O$ has order n and $L = \mathbf{2}^{\breve{X}}$.*

PROOF. By finite induction, every a in L is the join $\bigvee_A p_i$ of the set A of the join-irreducible elements $p_i > O$ which it contains. Also, if $p_i \leqq a$ and $p_j \leqq p_i$, then $p_j \leqq a$; that is, every set A is M-closed in X. But conversely, by Lemma 1, if A is M-closed, then $\bigvee_a p_i$ contains no p_k not in A. Hence the two mappings $a \leftrightarrow A$, which are obviously isotone, are one-one, and so isomorphisms of L with the ring of J-closed subsets of $\breve{X}$ or, by Lemma 1 of §1, with $\mathbf{2}^{\breve{X}}$. But it is immediate that the length of $\mathbf{2}^X$ is the order of X, completing the proof.

COROLLARY 1. *The number of (nonisomorphic) distributive lattices of length n is equal to the number of posets of n elements.*

Thus there are 2 nonisomorphic distributive lattices of length two, 5 of length three, 16 of length four, and 63 of length five.

COROLLARY 2. *Any distributive lattice of length n is isomorphic with a ring of subsets of a set X of n elements.*

Since every Boolean lattice L is relatively complemented (Theorem I.13), the join-irreducible elements $P > O$ of L are its atoms. Hence, if L is complemented, the X of Theorem 3 is totally unordered. We conclude

THEOREM 4. *Every Boolean algebra of finite length n is isomorphic with the field of all subsets of a set of n elements. Thus in particular there is just one Boolean algebra of length n, namely $\mathbf{2}^n$.*

Exercises for §3:

1. Show that, for any finite poset P, Aut $P \cong$ Aut $(\mathbf{2}^P)$.
2. (a) Show that a principal ideal $a \wedge L$ of a distributive lattice L is prime, if and only if a is meet-irreducible.

   (b) Show that if L has finite length n, then L has exactly n prime ideals, apart from the void set and L itself.
3. Show that in a distributive lattice L, the representation $a = x_1 \wedge \cdots \wedge x_r$ of a as a meet of meet-irreducible elements is irredundant unless $x_i > x_j$ for some i, j.
4. Show that the join-irreducible elements in Example 2 of §I.1 are the powers of primes.
5. Show that the order $f(m, n)$ of $\mathbf{m}^\mathbf{n}$ is defined by $f(1, n) = 1$, $f(m, 1) = m$, and the recurrence relation $f(m, n) = f(m - 1, n) + f(m, n - 1)$.

4. Free Distributive Lattices

The notions of lattice polynomial and "free" lattice have already been introduced informally in §II.5, where it was shown that the free distributive lattice D_{18} with three elements had exactly 18 elements. By inspecting Figure 8 there, we see that in fact $D_{18} \cong \mathbf{2}^P$, where P is the poset consisting of $\mathbf{2}^3$ with O and I deleted. If we adjoin O and I to D_{18}, we therefore get $\mathbf{2}^{\mathbf{2}^3}$. We will now generalize this result, using similar considerations.

DEFINITION. A lattice polynomial of the form

$$(6) \qquad f_J(x_1, \cdots, x_n) = f_J(\mathbf{x}) = \bigvee_{S \in J} \left\{ \bigwedge_{k \in S} x_k \right\},$$

where J is a J-closed family of subsets S of the indices $k = 1, \cdots, n$, is called J-normal.

LEMMA 1. *In any lattice, the J-normal polynomial functions f_J satisfy*

$$(7) \qquad f_J(\mathbf{x}) \vee f_K(\mathbf{x}) = f_{J \cup K}(\mathbf{x}).$$

PROOF. The identity (7) follows trivially from L1–L3 and (6), after eliminating repeated occurrences of each S.

LEMMA 2. *In any distributive lattice L,*

$$(8) \qquad f_J(\mathbf{x}) \wedge f_K(\mathbf{x}) = f_{J \cap K}(\mathbf{x}).$$

PROOF. By distributivity, writing $y_S = \bigwedge_{i \in S} x_i$, etc.,

$$(9) \qquad \left\{ \bigvee_{S \in J} y_S \right\} \cap \left\{ \bigvee_{T \in K} y_T \right\} = \bigvee_{J \times K} \{ y_S \wedge y_T \} = \bigvee_{J \times K} y_{S \cup T},$$

where the last inequality follows from L1–L3 on meets. Since J and K are J-closed, $S \cup T = U \in J \cap K$ for any $S \in J$, $T \in K$—and conversely, given $u \in J \cap K$, defining $S = T = U$, $y_U = y_S \wedge y_T$ appears as a term in the right member of (9), we obtain

$$f_J \wedge f_K = \bigvee_{J \cap K} y_U = \bigvee_{J \cap K} \left\{ \bigwedge_{k \in U} x_k \right\} = f_{J \cup K}. \qquad \text{Q.E.D.}$$

Since the one-element set $S_i = \{i\}$, consisting of i alone, is trivially J-closed, each $x_i = \{\bigvee_{S_i}\} = \{\bigwedge_{k \in S_i} x_k\}$ can be written in J-normal form. From this fact, (7), and (8), we conclude

LEMMA 3. *Let $a_1, \cdots, a_n$ be any elements of a distributive lattice L. Then the sublattice of L generated by the a_i is the set of all $f_J(a_1, \cdots, a_n)$ where the f_J are the (formally) different J-normal polynomials* (6).

We will now construct an example in which different J-normal polynomials (6) give different functions.

Example 1. In the Boolean algebra $\mathbf{2}^n$, with atoms $p_1, \cdots, p_n$, let X_i denote the principal ideal of all $x \leqq p_i'$. By Lemmas 1–2, the ring of sets generated by the X_i consists of the $f_J(X_1, \cdots, X_n)$, where $\bigwedge_S X_i$ is the (principal) ideal of all $x \leqq \bigwedge_S p_i' = (\bigvee_S p_i)'$. Every $f_J(X_i, \cdots, X_n)$ will be a J-closed subset, therefore, which includes O and excludes I in $\mathbf{2}^n$. Conversely, since any such J-closed subset of $\mathbf{2}^n$ is a set-union of the principal ideals generated by its maximal elements, we get them all and they are all different!

From the preceding results, it follows by Theorem VI.13 that, just as in Theorem II.9, the J-normal polynomials in $x_1, \cdots, x_n$ define the *free distributive lattice with n generators*. Moreover (7) and (8) show that this is isomorphic with the ring of

J-closed sets of subsets of $\{1, \cdots, n\}$. Moreover, the sets of J of such subsets which can be obtained in (6) are nonvoid themselves and exclude the void set $\varnothing$ of indices. Otherwise, they include all subsets of $\mathbf{2}^S$, just as in Example 1. If we adjoin O (to represent $J = \varnothing \cdot \subset \mathbf{2}^S$) and I (to represent the set of all S including the void set $\varnothing$), we get:

THEOREM 5. *The free distributive lattice generated by n symbols, with O and I adjoined, is isomorphic with the ring of all J-closed subsets of the lattice $\mathbf{2}^n$ of all subsets of n points.*

By Theorem 3, since $\mathbf{2}^n$ is self-dual, we infer immediately the following.

COROLLARY. *The free distributive lattice generated by n symbols, with O and I adjoined, is $\mathbf{2}^{\mathbf{2}^n}$.*

Free distributive lattices with n generators have been extensively studied.†

In conclusion, we state without proof the general finite distributive law

$$\bigwedge_{i=1}^{r} \left\{ \bigvee_{j=1}^{s(i)} x_{i,j} \right\} = \bigvee_{F} \left\{ \bigwedge_{i=1}^{r} x_{i,f(i)} \right\}, \tag{10}$$

where F is the set of all *functions* which assign to each i an $f(i)$ from the set $1, \cdots, s(i)$.

5. Free Boolean Algebras

A *Boolean polynomial* of n variables $x_1, \cdots, x_n$ is built up similarly from the three basic operations: join, meet, and complement, operating on $x_1, \cdots, x_n$. For instance, the Boolean polynomials of one variable x are just x, x', $x \vee x'$, $x \wedge x'$, as already observed in §II.5.

By systematically applying the laws of Boolean algebra, a canonical *disjunctive normal form* for any Boolean polynomial at n variables $y_1, \cdots, y_n$ can be computed as follows:

(i) If any prime appears outside any parenthesis in the polynomial, move it inside by applying the dualization laws

$$(p \wedge q)' = p' \vee q' \quad \text{and} \quad (p \vee q)' = p' \wedge q'. \tag{11}$$

This makes the polynomial an expression involving primed and unprimed letters combined by joins and meets.

(ii) By repeated application of the distributive law (combined with L1–L4), reduce the resulting expression to the form (6), where now S can contain both $y_i = x_i$ and $y_i' = x_{n+i}$, $i = 1, \cdots, n$.

(iii) If any S contains both y_i and y_i', then $\bigwedge_S x_k = O$ and the term can be dropped.

† K. Yamamoto, J. Math. Soc. Japan **6** (1954), 343–53. For related questions (expressed in the language of relay circuits), see E. F. Moore and C. Shannon, J. Franklin Inst. **262** (1956), 191–208 and 281–97.

(iv) If any S contains neither y_i nor y_i', then write

$$\bigwedge_S x_k = \left(\bigwedge_S x_k\right) \wedge (y_i \vee y_i') = \left(\bigwedge_{S_1} x_k\right) \vee \left(\bigwedge_{S_2} x_k\right),$$

where $S_1 = y_i \cup S$ and $S_2 = y_i' \cup S$. By repeating this expansion, we can make each S in (6) contain *exactly one* of each pair y_i, y_i'.

For example, consider Boolean polynomials in three variables, such as

$$\begin{aligned} F &= [(x \wedge y')' \vee z'] \wedge (z \vee x')' \\ &= [x \vee y \vee z'] \wedge (z \wedge x) \qquad \text{by (i)} \\ &= (x' \wedge z' \wedge x) \vee (y \wedge z' \wedge x) \vee (z' \wedge z' \wedge x) \qquad \text{by (ii)} \\ &= (y \wedge z' \wedge x) \vee (z' \wedge x) \qquad \text{by (iii)} \\ &= (y \wedge z' \wedge x) \vee (y \wedge z' \wedge x) \vee (y' \wedge z' \wedge x) \qquad \text{by (iv).} \end{aligned}$$

Steps (i)–(iv) reduce any such polynomial either to O or a join of some selection of the terms

$$\begin{array}{cccc} x \wedge y \wedge z, & x' \wedge y \wedge z, & x \wedge y' \wedge z, & x \wedge y \wedge z', \\ x \wedge y' \wedge x', & x' \wedge y \wedge z', & x' \wedge y' \wedge z, & x' \wedge y' \wedge z'. \end{array}$$

These 2^3 expressions are called *minimal* Boolean polynomials.

In the same way any Boolean polynomial of n variables $x_1, \cdots, x_n$ can be reduced to a join at some subset of the 2^n minimal Boolean polynomials

$$f_i = x_{i1} \wedge x_{i2} \wedge x_{i3} \cdots \wedge x_{in},$$

where x_{ij} is either x_j or x_j' depending on i.

THEOREM 6. *There is one and only one way to write a given Boolean polynomial as a join of minimal polynomials.*

PROOF. The assertion that there is at least one such representation was proved by the reduction process described above. To prove that there is at most one representation—in other words, that joins of distinct sets of minimal polynomials represent distinct Boolean functions—we reconsider Example 1 of §4, in which X_i is the set of all elements $t \in \mathbf{2}^n$ with $t \wedge p_i = O$, i.e., omitting p_i. Therefore X_i' is the set of all $x \in \mathbf{2}^n$ which contain p_i. Given *any* set T of p_i, we can therefore represent it as a meet Z_i, where Z_i is X_i or X_i' according as $p_i \in T$ or not. Hence, in Example 1, the 2^{2^n} joins of minimal polynomials referred to in Theorem 6 all represent distinct functions. This proves that the family of *all* sets of subsets of the set $\{1, \cdots, n\}$ provides an *isomorphic* representation of the free Boolean algebra with n generators. In summary:

THEOREM 7. *The free Boolean algebra with n generators is* $\mathbf{2}^{2^n}$.

Exercises for §§4–5:

1. Show that the free lattice with two generators is a Boolean lattice, but not the free Boolean algebra with two generators.
2. For what X is $\mathbf{2}^X$ the free distributive lattice with n generators without O and I adjoined?

3. Show that the sublattice generated by a finite subset of n elements of a distributive lattice, contains at most 2^{2^n} elements, and hence is finite.

4. Let $f(n)$ denote the number of elements of $FD(n)$. Show that $f(1) = 3$, $f(2) = 6$, $f(3) = 20$, $f(4) = 168$, $f(5) = 7581$, $f(6) = 7{,}828{,}354$, $f(7) = 2{,}414{,}682{,}040{,}998$.†

5. Prove that a Boolean polynomial is equivalent to a lattice polynomial if and only if it is isotone.

6. Show that, for even n, the number of elements in $FD(n)$ is even.

7. Consider algebras with two binary idempotent, commutative, associative, and mutually distributive operations, and two elements O and I satisfying $O \vee a = I \wedge a = a$. Show that the "free" such algebra with one generator has exactly five elements.‡

6. Free Modular Lattice M_{28}

The algebraic implications of the modular law are far more subtle than those of the distributive law. However, one can establish the following remarkable result.

THEOREM 8 (DEDEKIND). *The free modular lattice with three generators has* 28 *elements; its diagram is shown in Figure 10a.*

PROOF. The first statement is a corollary of the second. It therefore remains to show that Figure 10a is the diagram of a modular lattice M_{28} with generators x, y, z and that the equations satisfied in M_{28} are truly consequences of L1–L5. This we now do, grouping the 21 elements of M_{28} other than $x, y, z, O = x \wedge y \wedge z$, $I = x \vee y \vee z$, and

$$o = (x \wedge y) \vee (y \wedge z) \vee (z \wedge x), \qquad i = (x \vee y) \wedge (y \vee z) \wedge (z \vee x)$$

into triples which are equivalent under permutations of the generators x, y, z. We thus define

$$p = x \wedge y, \quad q = x \wedge z, \quad r = y \wedge z; \qquad p^* = x \vee y, \quad q^* = x \vee z, \quad r^* = y \vee z;$$
$$u = p \vee q, \quad v = p \vee r, \quad w = q \vee r; \qquad \text{and } u^*, v^*, w^* \text{ dually};$$
$$a = x \wedge (y \vee z), \quad b = y \wedge (x \vee z), \quad c = z \wedge (x \vee y);$$
$$a^* = x \vee (y \wedge z), \quad b^* = y \vee (x \wedge z), \quad c^* = z \vee (x \wedge y),$$
$$e = a \vee o, \quad f = b \vee o, \quad g = c \vee o.$$

Since $(x \wedge y) \vee (y \wedge z) \vee (z \wedge x) \leqq y \vee z$, by L2,

$$\begin{aligned} e = a \vee o = o \vee a &= [(x \wedge y) \vee (y \wedge z) \vee (z \wedge x)] \vee [x \wedge (y \vee z)] \\ &= [(x \wedge y) \vee (y \wedge z) \vee (z \wedge x) \vee x] \wedge (y \vee z) \qquad \text{by L5} \\ &= [(y \wedge z) \vee x] \wedge (y \vee z) = (y \wedge z) \vee [x \wedge (y \vee z)] \qquad \text{by L2–L4, L5.} \end{aligned}$$

Hence $e = a \vee o = a^* \wedge i = e^*$; moreover similar equations hold for $f = f^*$ and $g = g^*$.

The eight-element sublattices of elements $t \leqq o$ and $t^* \leqq i$ in Figure 10a can be obtained using Theorem II.13. This leaves only a limited number of relations to

† Randolph Church, Abstract 65T–447, Notices AMS **12** (1965), 724.

‡ See J. A. Kalman, Math. Chronicle 1 (1971), 147-50. For other generalizations of distributive lattices, see Gr. C. Moisil [1, pp. 1-5]; M. Smiley, Trans. AMS **56** (1944), 435-47.

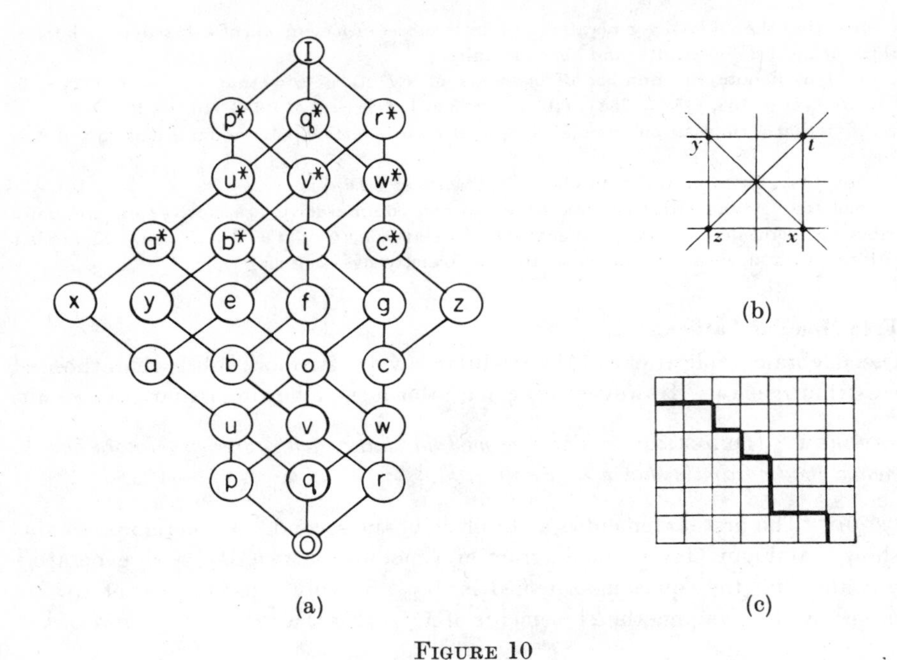

(a) (b) (c)

FIGURE 10

verify, after advantage is taken of symmetry and duality. As in Lemma 2 of §II.7, we have by L5, since $x \wedge r^* \leqq x \leqq x \vee z = q^*$,

$$a \vee b = (x \wedge r^*) \vee (y \wedge q^*) = [(x \wedge r^*) \vee y] \wedge q^*$$
$$= [r^* \wedge (x \vee y)] \wedge q^* = (y \vee z) \wedge (x \vee y) \wedge (x \vee z) = i.$$

From this $b \vee c = c \vee a = i$ follows by symmetry, whence by duality $a \wedge b = b \wedge c = c \wedge a = o$. As in Lemma 3 of §II.7, we have also $e \wedge f = f \wedge g = g \wedge e = o$, and dually $e \vee f = f \vee g = g \vee e = i$.

The other entries in the table of joins in M_{28} can be obtained from the exercises below—and the table of meets in M_{28} deduced from this by duality.

REMARK. The free modular lattice with four generators is infinite. To verify this, consider the sublattice of the modular (ortho)lattice of subspaces of a three-dimensional real vector space generated by the one-dimensional subspaces of $(\lambda x, 0, \lambda x)$, $(0, \lambda y, \lambda y)$, $(0, 0, \lambda z)$, and $(\lambda t, \lambda t, \lambda t)$. Considered as points in the projective plane, these generate a sublattice S of projective points and lines which contains four points x, y, z, t, no three of which are collinear. The sublattice S also contains with any two points the line through them, and with any two lines their intersection. Figure 10b shows the projection of some points and lines of S in the "finite" (x, y)-plane of vectors $(x, y, 1)$; it is geometrically evident (since $(x \vee t) \wedge (y \vee z)$ and $(y \vee t) \wedge (z \vee x)$ are on the "line at infinity" of vectors $(x, y, 0)$) that an infinite subset of S is obtained by successive bisections.

7. Free Modular Lattices Generated by Two Chains

We next consider the sublattice of a general modular lattice generated by two chains. Let L be any lattice, and let $O = x_0 < x_1 < \cdots < x_m = I$ and $O = y_0 < y_1 < \cdots < y_n = I$ be any two chains in L between O and I. Clearly the set of $u^i_j = x_i \wedge y_j$ includes all x_i and y_j (for $x_i \wedge y_n = x_i$ and $x_m \wedge y_j = y_j$); dually, the set of $v^i_j = x_i \vee y_j$ includes them. Hence so does the set of joins of the u^i_j, and that of the meets of the v^i_j.

LEMMA 1. *Any join of the u^i_j can be written in the form*

$$(x_{i(1)} \wedge y_{j(1)}) \vee \cdots \vee (x_{i(r)} \wedge y_{j(r)}),$$

where $i(1) > \cdots > i(r)$ and $j(1) < \cdots < j(r)$.

PROOF. If two u^i_j have the same superscript, then since the y_j are a chain, one u^i_j must be contained in, and hence by L4 can be absorbed by the other. Thus we can make all the $i(k)$, and similarly all the $j(k)$, distinct. Moreover if $i > i'$ and $j \geqq j'$, then $(x_i \wedge y_j)$ will absorb $(x_{i'} \wedge y_{j'})$, since it includes it. Hence after we have absorbed as many elements as possible, and utilized L2 to arrange the $i(k)$ in descending order, we will have $j(1) < \cdots < j(r)$ also.

LEMMA 2. *If $a_i \geqq a_{i+1}$ and $b_i \leqq b_{i+1}$ for all i in a modular lattice, then*

$$(a_1 \wedge b_1) \vee \cdots \vee (a_r \wedge b_r) = a_1 \wedge (b_1 \vee a_2) \wedge \cdots \wedge (b_{r-1} \vee a_r) \wedge b_r,$$
$$(b_1 \vee a_1) \wedge \cdots \wedge (b_r \vee a_r) = b_1 \vee (a_1 \wedge b_2) \vee \cdots \vee (a_{r-1} \wedge b_r) \vee a_r.$$

PROOF. By duality and induction on r, we need only prove the first identity on the assumption that the second holds when there are fewer than r summands. But by L5 applied twice, $(a_1 \wedge b_1) \vee \cdots \vee (a_r \wedge b_r)$ can be rewritten in the form

$$a_1 \wedge [b_1 \vee (a_2 \wedge b_2) \vee \cdots \vee (a_{r-1} \wedge b_{r-1}) \vee a_r] \wedge b_r.$$

And by the second identity for the case $(r - 1)$, the two expressions

$$(b_1 \vee a_2) \wedge (b_2 \vee a_3) \wedge \cdots \wedge (b_{r-1} \vee a_r),$$
$$b_1 \vee (a_2 \wedge b_2) \vee (a_3 \wedge b_3) \vee \cdots \vee (a_{r-1} \wedge b_{r-1}) \vee a_r$$

are equal. And if we substitute the former for the latter in the square brackets above, we get the right-hand side of the first identity. Q.E.D.

LEMMA 3. *The joins of the u^i_j are a sublattice.*

PROOF. Evidently any join of joins of u^i_j is a join of u^i_j; but any meet of joins of u^i_j is by Lemmas 1–2 a meet of meets of v^i_j, hence a meet of v^i_j, and hence by Lemmas 1–2 a join of u^i_j.

Now observe that if X_i denotes the set of points (x, y) of the rectangle $O \leqq x \leqq m$, $O \leqq y \leqq n$ satisfying $x \leqq i$, and if Y_j denotes the set of points satisfying $y \leqq j$, and if joins and meets are interpreted as set-unions and set-products, then all of the expressions admitted in Lemma 1 describe different sets (of saw-tooth shape). For example, Figure 10c represents $(x_2 \wedge y_5) \vee (x_3 \wedge y_4) \vee (x_4 \wedge y_3) \vee (x_6 \wedge y_1)$

$[m = 6, y = 5]$. By Lemma 3, these sets describe a ring of sets (i.e., a sublattice of the lattice of all subsets of the square), which thus represents isomorphically the free modular lattice generated by the two chains. We infer that this lattice is distributive and finite. In summary, we have proved

THEOREM 9. *The free modular lattice generated by two finite chains is a finite distributive lattice.*†

COROLLARY. *Any two finite chains between the same end points have refinements whose subintervals are projective* (*see* §11) *in pairs.*

Exercises for §§6–7:

1. Show that, in a modular lattice, each of the following conditions implies $(x, y, z)D$ (notation of §II.7):
 (i) $x \wedge (y \vee z) = (x \wedge y) \vee (x \wedge z)$,
 (ii) $e = f$.
2. Show that, in a nonmodular lattice, neither condition of Exercise 1 implies $(x, y, z)D$.
*3. (a) Represent M_{28} as a sublattice of $2^6 M_5$.
 (b) Infer that Figure 10a is the diagram of a modular lattice.
4. Show that M_{28} has breadth 3.
5. Let A be the commutative group with 256 elements, and generators $e_1, \cdots, e_8$ of order two. Let X_1, X_2, X_3, be the subgroups generated by $\{e_1, e_2, e_4, e_7\}$, $\{e_2, e_3, e_5, e_8\}$, $\{e_1, e_3, e_6, e_7 + e_8\}$. Show that all 28 subgroups of Figure 10a are distinct.
6. (a) Show that the free modular lattice generated by $a > b$ and $c > d$ has 18 elements. (This is not a special case of Theorem 9 unless O and I are adjoined.)
 (b) Show that if $m = 1$, $n = 3$, and O, I are not adjoined, the diagram of the lattice of Theorem 9 is that shown in Figure 10b.
7. (a) Show that if $m = 2$, the lattice of Theorem 9 is planar.
 (b) What is its breadth in general?
8. (a) Let the two chains in Theorem 9 have $m - 1$ and $n - 1$ elements respectively. Show that if O, I are added to the resulting distributive lattice, it becomes $2^{\mathbf{mn}}$.
 (b) Show that it has $(m + n)!/m!n!$ elements. (*Hint.* Identify the heavy line in Figure 10c with a representation $xxyxyxyyxxy$ of x^6y^5.)

Exercises 9–11 *are results of B. Jonsson, Proc. AMS* **6** (1955), 682–8.

9. Let $X_1, \cdots, X_p$ be chains in a modular lattice. Show that $X_1 \cup \cdots \cup X_p$ generates a distributive lattice if and only if $(x_1, x_2, \cdots, x_p)D$ for all $x_i \in X_i$.
10. Let S, T be distributive sublattices of a modular lattice. Show that $S \cup T$ generates a distributive sublattice if and only if $(x, y, z)D$ for all $x, y, z \in S \cup T$.
11. Let X be a subset of a modular lattice. Show that X generates a distributive sublattice if and only if, for all $y_i, z_j \in X$, $(\bigvee y_i) \wedge \bigwedge z_j = \bigvee [y_i \wedge \bigwedge z_j]$.
*12. Extend Theorem 9 to arbitrary chains.

8. Center

The factorization of a poset with universal bounds O and I are best analyzed by considering its *center*, defined as follows.‡

† This result is due to the author [**LT1**, p. 51]. The proof is practically that of O. Schreier, *Über den J-H'schen Satz*, Abh. Math. Sem. Univ. Hamburg **6** (1928), 300–2, and H. Zassenhaus, *Zum Satz von Jordan–Hölder–Schreier*, ibid. **10** (1934), 106–8.

‡ For complemented modular lattices, by J. von Neumann; for posets, by the author [**LT1**, p. 24].

DEFINITION. The *center* of a poset P with O and I is the set of elements $e \in P$ which have one component I and the other O, under some direct factorization of P.

Since cardinal multiplication of lattices is commutative and associative, we see that for e to be in the center of P it is necessary and sufficient that $e = (I, O)$ under some representation $P = XY$ of P as a product of two factors.

From now on, let P be any poset with universal bounds O and I, and let C be the center of P.

LEMMA 1. *Each element $e \in C$ has a unique complement e', also in C.*

PROOF. Clearly $(I, O) \wedge (x, y) = (x, O)$ and $(I, O) \vee (x, y) = I(y)$; hence (x, y) is a complement of $e = (I, O)$ if and only if $(x, y) = (O, I) = e'$, which is unique and in the center. Again, P is the product of the $x \leqq (I, O)$ and $y \leqq (O, I)$ if and only if it is the product of the $s \geqq (I, O)$ and the $t \geqq (O, I)$.

LEMMA 2. *Given e in C, then $z \wedge e$ and $z \vee e$ exist for any $z \in P$, and the mapping $z \to (z \wedge e, z \vee e)$ is an isomorphism from P onto EE^*, where E is the ideal $[O, e]$ and E^* is the dual ideal $[e, I]$ in P.*

PROOF. If $P = XY$ and $e = (I, O)$, then the mapping $z \to (z \wedge e, z \vee e)$ is

$$(x, y) \to ((x, y) \wedge (I, O), (x, y) \vee (I, O)) = ((x, O), (I, y)).$$

From this the assertion follows since $[O, e]$ is the principal ideal (isomorphic with X) of elements $(x, O) \leqq (I, O)$ and $[e, I]$ is the dual ideal of $(I, y) \geqq (I, O)$.

For any $e \in C$, let ϕ_e and ψ_e denote the *projections* $z \to z \vee e = \phi_e(z)$ and $z \to z \wedge e = \psi_e(x)$, respectively. One easily verifies that, for any complementary pair $e, e' \in C$, ϕ_e is an *isomorphism* from $[e', I]$ to $[O, e]$ with inverse $\psi_{e'}$. Moreover, for any $e, f \in C$, $\psi_{e \wedge f} = \psi_e \psi_f = \psi_f \psi_e$, since $z \wedge (e \wedge f) = (z \wedge e) \wedge f$. etc. Dually, $\phi_{e \vee f} = \phi_e \phi_f$, and (on $[O, e \wedge f]$)

$$\psi_{e \wedge f}^{-1} = (\psi_e \psi_f)^{-1} = \psi_f^{-1} \phi_e^{-1} = \phi_{f'} \phi_{e'} = \phi_{e' \vee f'}.$$

This and other similar formulas prove

LEMMA 3. *If e, f are in C, then $e \wedge f$ and $e \vee f$ (which exist in P by Lemma 2) are in C.*

THEOREM 10. *The center C of any poset P with O and I is a Boolean lattice in which joins and meets represent l.u.b. and g.l.b. in P.*

PROOF. By Lemma 3, C is a sublattice of P. Moreover $e \wedge f = e \wedge g$ and $e \vee f = e \vee g$ imply $f = g$ in this sublattice (by Lemma 2). Hence, by the Corollary of Theorem II.13, it is a *distributive* sublattice. Finally, by Lemma 1, C is complemented—which proves Theorem 10.

Since the definition of a direct product is invariant under isomorphism and dual isomorphism, it is evident that the center of any poset P with universal bounds is a subset mapped onto itself by all automorphisms and dual automorphisms of P.

Unique factorization. A unique factorization theorem for cardinal products†

† The original proof of this theorem was different. See G. Birkhoff, Bull. AMS **40** (1934), 613–9.

follows rather easily from the preceding results. We first prove

LEMMA 4. *Given a direct factorization $P = \prod_{i=1}^{m} X_i$ of a poset P with universal bounds, let $e_i \in P$ have X_i-component I and all other components O. Then the e_i are disjoint elements of the center with join I. Conversely, any such subset of the center corresponds to a direct factorization of P.*

PROOF. The first statement is obvious, when one represents the e_i as vectors with appropriate components. (By *disjoint*, we mean that $e_i \wedge e_j = O$ if $i = j$.)

As for the converse, consider the mapping

$$z \to (z \wedge e_1, z \wedge e_2, \cdots, z \wedge e_m), \tag{12}$$

defined by Lemma 2. Moreover $e_1' = e_2 \vee \cdots \vee e_m$ is the unique complement of e_1 mentioned in Lemma 1, and $z \in [e_1, I]$ if and only if $z \geqq e_1$. Hence, as in the proof of Lemma 1, $P \cong E_1 E_1^*$, where E_1^* is the ideal $[e_1, I] \cong [O, e_1']$. By induction on m,

$$E_1^* \cong E_2 E_3 \cdots E_m,$$

completing the proof.

THEOREM 11. *With any two factorizations of a poset P with O and I into factors X_i and Y_j respectively, there is associated a factorization into Z_j^i such that the product of Z_j^i with fixed j is X_i and the product of the Z_j^i with fixed i is Y_j.*

PROOF. Let there be given factorizations $P = \prod X_i$ and $P = \prod Y_j$ into factors associated with elements e_i and f_j of the center of P, respectively, so that $X_i \cong [O, e_i]$ and $Y_j \cong [O, f_j]$. Then the ideals $Z_j^i = [O, e_i \wedge f_j]$, some of which may reduce to the trivial ideal $[O]$, define a decomposition of I by elementary Boolean algebra. Hence, again by Lemma 4, $P = \prod Z_j^i$. Moreover, since

$$\bigvee_j (e_i \wedge f_j) = e_i \wedge \bigvee_j f_j = e_i \wedge I = e_i,$$

$X_i = \prod_j Z_j^i$ as claimed. Similarly, $Y_j = \prod_i Z_j^i$, completing the proof.

COROLLARY 1. *If P can be factored into indecomposable factors, this factorization is unique is the strict sense that any irredundant*† *factorization of P is obtainable by grouping these indecomposable factors into subfamilies.*

COROLLARY 2. *If P contains a finite maximal chain, then P can be factored into nontrivial indecomposable posets in one and only one way.*

PROOF. The uniqueness of the factorization follows from Corollary 1; its existence by induction on the length of the shortest maximal chain. We omit the details.

Further results. Nakayama‡ has shown that factorization into indecomposables

† Factorizations which contain the trivial factor O are called redundant, and are thus excluded.

‡ T. Nakayama, Math. Japon. **1** (1948), 49–50. See also J. Hashimoto, Math. Japon. **1** (1948), 120–3, and Ann. of Math. **54** (1951), 315–8; C. C. Chang, [**SLT**, p. 124].

is unique more generally in any bidirected poset. It is not unique in all finite disconnected posets; for example

$$(1 + \mathbf{2}^3)(1 + \mathbf{2} + \mathbf{2}^2) \cong (1 + \mathbf{2}^2 + \mathbf{2}^4)(1 + \mathbf{2}),$$

much as for ordinary polynomials with *positive* integral coefficients. However, the semiring $\mathscr{P}$ of finite posets can be *embedded* in $\mathbf{Z}[x_1, x_2, x_3, \cdots]$ which is a unique factorization domain (and a commutative l-ring, see Chapter XIII).

9. Distributive and Standard Elements

We now show that the center of any *lattice* L consists of those elements $a \in L$ which are complemented and *distributive* (or "neutral"†) in the following sense.

DEFINITION. An element a of a lattice L is *neutral* (alias "distributive") if and only if $(a, x, y)D$ for all $x, y \in L$—i.e., any triple including a generates a distributive sublattice of L.

LEMMA 1. *An element a is distributive (neutral), if and only if the functions $\phi_a: x \to x \wedge a$ and $\psi_a: x \to x \vee a$ are (lattice-) endomorphisms of L; moreover the correspondence $x \to (x\phi_a, x\psi_a)$ is a monomorphism of L with a sublattice of the product of the ideal A of $y \leqq a$ and the dual ideal A_d of $z \geqq a$.*

PROOF. The first two statements are obvious, while the last follows just as in the proof of Chapter II, (19). The preceding conditions display L as a *subdirect product* of A and A_d; the monomorphism maps a onto (I, O) in $A \times A_d$. From this the converse is obvious.

It is also clear that a' is a *complement* of a if and only if $(a'\phi_a, a'\psi_a) = (O, I)$ in $A \times A_d$; but a' must therefore be neutral and unique, as well as in the center of L. We have proved

THEOREM 12. *A neutral element can have at most one complement, and any such complement is neutral. The center of a lattice consists of its complemented, neutral elements.*

Following Grätzer and Schmidt (loc. cit. supra in §II.4), we now define an element s of a lattice L to be *standard* when:

(i) the function $\psi_s: x \to x \vee s$ is an endomorphism of L, and
(ii) $x \vee s = y \vee s$ and $x \wedge s = y \wedge s$ imply $x = y$.

Trivially, an element is neutral in L if and only if it is standard in both L and its dual, since (i), its dual (i'), and our self-dual (ii) are precisely the conditions of Lemma 1. Moreover if L is modular, then (i) implies (i') since any one distributivity relation between s, x, y implies all six. Finally, by (ii) a standard element can have at most one complement. We have proved the following results.

† Ore [**1**, pp. 419–21] first studied "neutral" elements in modular lattices; the concept was extended to general lattices by the author, Bull. AMS **46** (1940), 702–5. Though the synonym "distributive" is more suggestive, the phrase "distributive ideal" would be ambiguous. Hence both "neutral" and "distributive" are used below.

LEMMA 2. *In a modular lattice, every standard element is dually standard (neutral); in a general lattice, an element is neutral if and only if it is standard and dually. Complements of standard elements are unique (if they exist).*

THEOREM 13. *The set of neutral elements of a lattice L is the intersection of its maximal distributive sublattices.*

PROOF. First, if a is not neutral, then some triple $\{a, x, y\}$ is not distributive. Hence *no* maximal distributive sublattice obtained by enlarging the distributive sublattice generated by $\{x, y\}$ can contain a. We conclude that the intersection of the maximal distributive sublattices of L contains no nonneutral elements.

Conversely, let S be any maximal distributive sublattice of L; let a be neutral; it is easily shown that the sublattice $\{a, S\}$ of $L \subset AA_d$ generated by $a = [I, O]$ and S is distributive—whence, S being maximal, $a \in S$. (Detailed proof: It is sufficient to prove that the A-components of $\{a, S\}$ and the A_d-components of $\{a, S\}$ form distributive lattices—and for these, we only adjoin I resp. O.)

It is a corollary that the neutral elements of any lattice form a distributive lattice. Grätzer has proved more: the *standard* elements of any lattice form a distributive sublattice.

10. Sectionally Complemented Lattices

In this section, L will denote an arbitrary sectionally complemented lattice (§II.4). For example (Theorem I.13), L might be any complemented modular lattice. Such lattices have an especially simple structure.

For example (Theorem II.6), the congruence relations θ on L correspond one-one to its standard ideals J, through the condition that $x \equiv y\ (\theta)$ if and only if $(x \wedge y) \vee a = x \vee y \vee a$ for some $a \in J$.

Now let L be of finite length, so that $J(\theta) = (s)$ is a *principal* ideal with chief element s. Then $(x \vee y) \vee a = x \vee y \vee a$ for some $a \in (s)$ if and only if $(x \wedge y) \vee s = x \vee y \vee s$—i.e., if and only if s is standard. This proves

LEMMA 1. *In any sectionally complemented lattice of finite length, let a be the largest element of the kernel $J(\theta)$ of a congruence relation θ. Then $x \equiv y\ (\theta)$ if and only if $(x \wedge y) \vee s = x \vee y \vee s$ for some standard element s.*

COROLLARY 1. *The quotient-lattice L/θ is isomorphic with the dual ideal of $x \geqq s$, under the projective endomorphism ψ_s: $x \to x \vee s$ of L.*

By duality, we have also

COROLLARY 2. *In a relatively complemented lattice of finite length, let d be the least element of the "dual kernel" of θ: the dual ideal $D(\theta)$ of all $x \equiv I\ (\theta)$ in L. Then $x \equiv y\ (\theta)$ if and only if $x \wedge d = y \wedge d$, and ϕ_d: $x \to x \wedge d$ is a projective endomorphism, mapping L/θ isomorphically onto $[O, d]$.*

Note that, for each x, $x \vee c$ is the largest element and $x \wedge d$ the smallest element of the residue class containing x. Hence $d \vee c = I$ and $c \wedge d = O$:

c and d are complements. Actually, c and d are in the center of L, as we now show.

Indeed, if $x \wedge d = O$, then

$$c = O \vee c = (x \wedge d) \vee c = (x \vee c) \wedge (d \vee c) = x \vee c,$$

and so $x \leqq c$. In other words, d is a *pseudo-complement* of c (§II.11). Also, for *any* $x \in L$, $c \wedge x$ has a relative complement t in $[O, x]$, such that $(c \wedge x) \wedge t = O$ and $(c \wedge x) \vee t = x$. Since d is a pseudo-complement of c, we infer that $t = x \wedge t \leqq d$ as well as $t \leqq x$, whence $t \leqq d \wedge x$ and

$$x = (c \wedge x) \vee t \leqq (c \wedge x) \vee (d \wedge x) \leqq x,$$

and the correspondence $x \rightarrow (x \wedge c, x \wedge d)$ is an isomorphism between L and the product $[O, c][O, d]$. We infer

THEOREM 14 (DILWORTH†). *A relatively complemented lattice of finite length is either simple (i.e., without proper congruence relations) or directly decomposable.*

COROLLARY 1. *Any relatively complemented lattice of finite length is a product of simple lattices.*

Dilworth's Theorem can be proved independently in a modular lattice M as follows. Since M has finite length, Theorem II.5 applies, and every congruence relation on M is associated with an endomorphism $x \rightarrow x \vee a$. Since $(x \wedge y) \vee a = (x \vee a) \wedge (y \vee a)$, we have $(a, x, y)D$ by Theorem II.12; hence a is neutral. But, in a *complemented* lattice, every neutral element is in the center. There follows

COROLLARY 2. *Any complemented modular lattice of finite length is a product of simple complemented modular lattices of finite length.*

We conclude this section with an unpublished lemma of M. F. Janowitz. Following F. Maeda [**KG**, p. 20], we define $a \bigtriangledown b$ in a lattice L with O to mean that $a \wedge b = O$ and $(a \vee x) \wedge b = x \wedge b$ for all $x \in L$. In a modular lattice, $a \bigtriangledown b$ if and only if $a \wedge b = O$ and $(a, b)D$. The interesting result is

LEMMA 2. *Let L be a relatively complemented lattice with O and I. Then the following conditions are equivalent:* (i) $a \bigtriangledown b$, (ii) *b is contained in every complement of a,* (iii) *$a_1 \leqq a$, $b_1 \leqq b$, and‡ $a_1 \sim b_1$ imply $a_1 = b_1 = O$,* (iv) *$x = (x \vee a) \wedge (x \vee b)$ for all $x \in L$.*

PROOF. (i) implies (ii). If c is a complement of a, then $b = (a \vee c) \wedge b = c \vee b$, whence $b \leqq c$.

(ii) implies (iii). Let c be a common complement of a_1 and b_1; then $c \vee a \geqq c \vee a_1 = I$. Let d be a relative complement of $a \wedge c$ in $[O, c] : d \wedge a \leqq c \wedge a = O$

† R. P. Dilworth, Ann. of Math. **51** (1950), 348–59. This paper contains many other interesting related results. See also J. E. McLaughlin, Pacific J. Math. **3** (1953), 197–208.

‡ By $a \sim b$ in a lattice L, we mean that a and b have a common complement c (cf. §XV.9).

and $d \vee (a \wedge c) = c$. Hence $d \vee a \geqq d \vee (c \wedge a) = c$, so $d \vee a \geqq c \vee a = I$. Therefore d is a complement of a and so $b \leqq d$ by (ii). A fortiori, $b_1 \leqq d$, so $b_1 = b_1 \wedge d \leqq b_1 \wedge c = O$, whence $c = b_1 \vee c = I$, and $a_1 = a_1 \wedge I = a_1 \wedge c = O$.

(iii) implies (iv). For any $x \in L$, $(x \vee a) \wedge (x \vee b) \geqq x$; let y be a relative complement of $(x \vee a) \wedge (x \vee b)$ in $[x, I]$. Then $y \vee a = y \vee x \vee a = y \vee ((x \vee a) \wedge (x \vee b)) = I$ and, similarly, $y \vee b = I$. Moreover, $y \wedge a$ has a relative complement a_1 in $[O, a]$, and $y \wedge b$ has a relative complement b_1 in $[O, b]$. Thus $y \wedge a_1 \geqq (y \wedge a) \vee a_1 = a$, which implies $y \vee a_1 \leqq (y \wedge a) \vee a_1 = a$, which implies $y \vee a_1 \geqq y \wedge a = I$. Also, y is a complement of b; (but by (iii), $a_1 = b_1 = O$, so $y = I$ and $x = (x \vee a) \wedge (x \vee b)$.

(iv) implies (i). If $x = (x \vee a) \wedge (x \vee b)$ for all $x \in L$, then $x \wedge b = (x \vee a)$, which completes the proof.

COROLLARY 1. *In any relatively complemented lattice L, the relation $a \nabla b$ is symmetric; moreover $a \nabla b$ and $a_1 \leqq a$, $b_1 \leqq b$ imply $a_1 \nabla b_1$.*

Exercises for §§8–10:

1. (a) Show that N_5 contains a standard element which is not neutral.

(b) Show that the lattice L_5 of congruence relations on N_5 is $\mathbf{2}^P$, where P = [figure]; hence it is not a Boolean algebra.

2. (a) Prove that, in a relatively complemented lattice, every prime ideal is maximal (cf. Theorem II.7).

(b) Prove that a distributive lattice L is complemented if and only if every prime ideal of L is maximal. (Nachbin–Monteiro; cf. also J. Hashimoto [**1**, p. 162])

3. Show that, in a lattice L, the principal ideal $[O, a]$ is standard if and only if a is a standard element.

4. (a) Show that the lattice of Figure 11a contains a nonstandard element which has a unique complement.

(b) Show that there exist principal ideals $[O, a]$ which are the inverse images of O under a lattice-epimorphism, for which a is not standard. (See Figure 11b.)

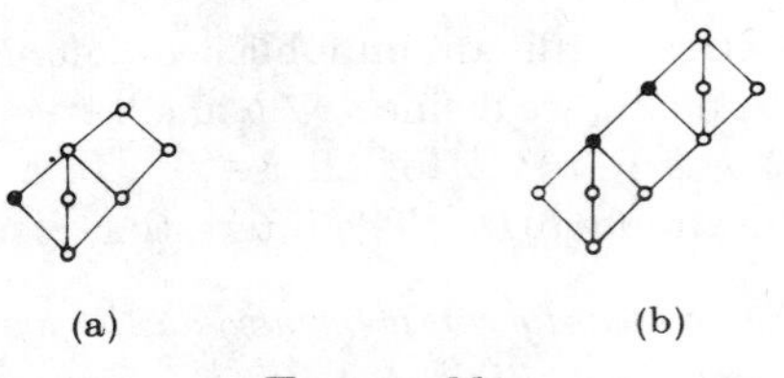

FIGURE 11

5. Show that any lattice-epimorphism carries neutral elements into neutral elements, and standard elements into standard elements. What about dual epimorphisms?

6. (a) Show that an element n of a lattice L is neutral if and only if, for all $x, y \in L$,

$$(n \wedge x) \vee (n \wedge y) \vee (x \wedge y) = (n \vee x) \wedge (n \vee y) \wedge (x \vee y).$$

(b) In a relatively complemented lattice with O and I, show that n is neutral if and only if it has a unique complement.†

† Exercise 6 is due to Hashimoto and Kinugawa, Proc. Japan Acad. Sci. **39** (1963), 162–3. Exercises 7–8 are due to G. Grätzer and E. T. Schmidt, Acta. Math. Acad. Sci. Hungar. **12** (1961), 17–86, and M. F. Janowitz, ibid. **16** (1965), 289–301.

*7. Show that an element s of a lattice L is standard if and only if $x \wedge (s \vee y) = (x \wedge s) \vee (x \wedge y)$ for all $x, y \in L$.

*8. Show that the standard elements of any lattice L form a distributive sublattice.

11. Schwan's Lemma; Independence

For given a, b in a lattice L, the assignments $\phi_a: x \to x \wedge a$ *and* $\psi_b: y \to y \vee b$ are trivally *isotone* from $[b, a \vee b]$ to $[a \wedge b, a]$ and vice-versa. In a modular, lattice, they are inverse isomorphisms (§I.7). In a general lattice, by the modular inequality of Chapter I, (6), moreover, $x\phi_a\psi_b \leqq x$ for all $x \in [b, a \vee b]$ and $y \leqq y\psi_b\phi_a$ for all $y \in [a \wedge b, a]$. Since ϕ_a maps $[b, a \vee b]$ into $[a \wedge b, a]$ and ψ_b maps $[a \wedge b, a]$ into $[b, a \vee b]$, it follows that

$$\phi_a\psi_b\phi_a = \phi_a \quad \text{on} \quad [b, a \vee b], \tag{14}$$

and

$$\psi_b\phi_a\psi_b = \psi_b \quad \text{on} \quad [a \wedge b, a]. \tag{14'}$$

This proves a simple but fundamental result.†

SCHWAN'S LEMMA. *In any lattice, the istone maps* $\phi_a: x \to x \wedge a$ *and* $\psi_b: y \to y \vee b$ *are istone from* $[b, a \vee b]$ *to* $[a \wedge b, a]a$ *and vice-versa; they satisfy* (14)–(14′).

In a *modular* lattice, the inequalities proved above become identities, and we obtain as a special case the classic transposition principle of Dedekind (Theorem I.13).

We now study more deeply the consequences of Dedekind's transposition principle, in arbitrary modular lattices.

THEOREM 15. *The sublattice of a modular lattice generated by* $[u \wedge v, u] = U$ *and* $[u \wedge v, v] = V$ *is the cardinal product* UV.

PROOF. Let (x, y) denote $x \vee y$, for any $x \in X$, $y \in Y$. Then by L2–L3, for any $x, x' \in U$ and $y, y' \in V$,

$$(x, y) \vee (x', y') = x \vee y \vee x' \vee y' = (x \vee x') \vee (y \vee y') = (x \vee x', y \vee y').$$

Also, by Theorem I.13 and condition L5, respectively, we have

$$(x, y) = x \vee y = [(v \vee x)] \wedge y = (v \vee x) \wedge (u \vee y).$$

Hence $(x, y) \wedge (x', y') = [(v \vee x) \wedge (v \vee x')] \wedge [(u \vee y) \wedge (u \vee y')]$ by L2–L3. Again by Theorem I.13, $(v \vee x) \wedge (v \vee x') = v \vee (x \wedge x')$ and $(u \vee y) \wedge (u \vee y') = u \vee (y \wedge y')$. Hence

$$(x, y) \wedge (x', y') = [v \vee (x \wedge x')] \wedge [u \vee (y \wedge y')] = (x \wedge x', y \wedge y')$$

as before. This completes the proof.

In particular, setting $u = a \vee c$ and $v = b \vee d$, and observing that two-element subsets such as $\{a, c\}$ and $\{b, d\}$ generate distributive sublattices in any lattice, we have

† W. Schwan, Math. Z. **51** (1948), 126–34 and 346–54.

COROLLARY 1. *In any modular lattice, if* $(a \vee c) \wedge (b \vee d) = O$, *then*
$$(a \vee b) \wedge (c \vee d) = (a \wedge c) \vee (b \wedge d).$$

We leave the details to the reader. By induction, Theorem 15 also yields

COROLLARY 2. *If* $(x_1 \vee \cdots \vee x_k) \wedge x_{k+1} = a$ *for* $k = 1, \cdots, n-1$, *in a modular lattice, then the sublattice generated by the intervals* $X_k = [a, x_k]$ *is* $X_1X_2 \cdots X_n$.

But the definition of $X_1X_2 \cdots X_n$ is symmetric in the subscripts. Hence the condition of the preceding corollary is invariant under all permutations of the subscripts, and we can legitimately state the

DEFINITION. Under the hypothesis of the preceding corollary, $x_1, \cdots, x_n$ are said to be *independent* over a.

Using the concept of independence, we now prove another elementary lemma which is useful in the theory of von Neumann lattices (Chapter XI).

VON NEUMANN–HALPERIN LEMMA. *Let* x *and* $a \leqq b$ *be given in a complemented modular lattice* M. *Then there exists a relative complement* y *of* a *in* b *such that* $x = (x \vee y) \wedge (x \vee a)$.

PROOF. Let $x_1 = x \wedge a$; let x_2 be a relative complement of $x \wedge a$ in $x \wedge b$; let x_3 be a relative complement of $a \wedge x$ in a; let y_2 be a relative complement of $a \vee x_2$ in b. Then x_1, x_2, x_3, y_1, y_2 are independent, and so $y = x_2 \vee y_2$ has the desired property.

12. Perspectivity; Kurosh-Ore Theorem

In *complemented* modular lattices, the definition of transposed quotients can be replaced by that of perspective elements. To pave the way for proving this, we first establish a lemma.

LEMMA 1. *The "differences"* $d_k = x'_{k-1} \wedge x_k$ *between successive terms of any chain* $O = x_0 < x_1 < \cdots < x_s$ *are independent elements (over* O*) whose join is* x_s. *(Note that* x'_{k-1} *need not be unique.)*

PROOF. Evidently $d_1 = x'_0 \wedge x_1 = I \wedge x_1 = x_1$. We now proceed by induction on k. Assume $d_1 \vee \cdots \vee d_k = x_k$; then
$$\bigvee_{i=1}^{k+1} d_i = x_k \vee d_{k+1} = x_k \vee (x'_k \wedge x_{k+1})$$
$$= (x_k \vee x'_k) \wedge x_{k+1} = I \wedge x_{k+1} = x_{k+1}.$$
The d_k are independent over O since for $k = 1, 2, \cdots, s-1$,
$$(d_1 \vee d_2 \vee \cdots \vee d_k) \wedge d_{k+1} = x_k \wedge d_{k+1}$$
$$= x_k \wedge x'_k \wedge x_{k+1} = O \wedge x_{k+1} = O.$$

DEFINITION. Two elements a and b of a complemented modular lattice are *perspective* if and only if they have a common complement c, called an "axis of perspectivity" for a and b.

In any finite-dimensional complemented modular lattice, as will be shown in Chapter IV, perspectivity is an equivalence relation. It is obviously reflexive and symmetric in *any* complemented modular lattice; a basic question concerns the *transitivity of perspectivity* (cf. §IV.6). Since, in a modular lattice, $a \leqq b, a \wedge c = b \wedge c$, and $a \vee c = b \vee c$ together imply $a = b$ (otherwise they would generate N_5), clearly no element in a complemented modular lattice can be perspective to a proper part of itself.

On the other hand, if a and b are perspective, then $[O, a]$ and $[O, b]$ are projective via $[d, I]$, which is a transpose of both. Hence, if perspectivity is transitive, a and b are perspective if and only if $[O, a]$ and $[O, b]$ are projective. We now give an example in which both conditions fail.

Example 3. Let M be the complemented modular lattice of all subgroups of the Abelian group $A = Z_2{}^{\omega}$ of countably many copies of the cyclic group of order two. Then *any* two subgroups $S \subset A$ and $T \subset A$ of infinite order and index are projective in M (related by a sequence of perspectivities), whereas there exist two such subgroups with $S < T$ (**LT1**, §72]. More generally, we have

LEMMA 2. *Let M be the lattice of all subspaces of a countable-dimensional vector space V over any field F. In M, any two ideals $[O, a]$ and $[O, b]$ having infinite dimensions and infinite-dimensional complements are projective.*

PROOF. As a first step, observe that if A and B are any two such subspaces which satisfy $A \wedge B = O$, then there is an isomorphism $\theta: A \rightarrow B$, and the "diagonal" subspace D of sums $x + \theta(x)$ $[x \in A]$ is a common relative complement of A and B in $A \vee B$. Hence, if C is any complement of $A \vee B$, $C \vee D$ is a common complement of A and B, which are thus perspective with "axis of perspectivity" $C \vee D$.

The proof can be easily extended to the case that $A/A \wedge B$ and $B/A \wedge B$ are both infinite-dimensional, by extending the perspectivity from any *relative* complement A' of $A \wedge B$ in A to a relative complement B' of $A \wedge B$ in B. Finally, if $A/A \wedge B$ or $B/A \wedge B$ is finite-dimensional, then $A \vee B$ has an infinite-dimensional complement C_1 which is disjoint from both A and B—and hence perspective to both.

Kurosh–Ore Theorem. We shall now show that the strictly unique irredundant representation of an element a in a distributive lattice as a join of join-irreducible elements, whose existence was proved in §3, has as analog in modular lattices a weak uniqueness theorem concerning the *number* of such components. The proof again utilizes Dedekind's transposition principle (Theorem I.13).

THEOREM 16. *Let $a = x_1 \wedge \cdots \wedge x_r = x_1^* \wedge \cdots \wedge x_s^*$ be any two irredundant reductions of a into irreducible components. Then one can substitute for any x_i a suitable x_j^*, and get a new reduction of a.*

PROOF. Set $y_i = x_1 \wedge \cdots \wedge x_{i-1} \wedge x_{i+1} \wedge \cdots \wedge x_r$. Then by irredundancy, $y_i > a$, yet $x_i \wedge y_i = a$. Now form $z_j = y_i \wedge x_j^*$; clearly $y_i \geqq z_j \geqq a$, and since $z_j \leqq x_j^*$,

$$a \leqq z_1 \wedge \cdots \wedge z_s \leqq x_1^* \wedge \cdots \wedge x_s^* \leqq a.$$

But by Theorem I.13 the sublattice between $a = x_i \wedge y_i$ and y_i is isomorphic to the sublattice between x_i and $x_i \vee y_i$, and since x_i is irreducible in the latter, so is a in the former. Hence some z_j is a, and

$$x_1 \wedge \cdots \wedge x_{i-1} \wedge x_j^* \wedge x_{i+1} \wedge \cdots \wedge x_r = a.$$

THEOREM 17. *The number of components in irredundant reductions of any element is independent of the reduction: in Theorem* 16, $r = s$.

PROOF. Choose r minimal, and replace the x_i by x_j^* one at a time. We get in the end $a = x^*_{j(1)} \wedge \cdots \wedge x^*_{j(r)}$, where by the irredundancy of the x_j^* there are at least s $j(i)$, and $s \leqq r$. Hence by minimality, $s = r$.

Exercises for §§11–12:

1. (a) In the diagram of a modular lattice of finite length, show that the opposite sides of a quadrilateral represent transposed prime intervals.

(*b) Show that two prime intervals are projective if and only if one can pass from one to another by a finite sequence of substitutions of one side of a quadrilateral in the diagram for the opposite side.

2. Show that, in a modular lattice, c is a relative complement of a in b (where $a \leqq b$), if and only if $c = a' \wedge b$ for some complement a' of a.

3. Show that a and b are perspective in a modular lattice if d exists with $a \wedge d = b \wedge d = O$ and $a \vee d = b \vee d$.

4. Show that any product of endomorphisms of a distributive lattice of the form $x \rightarrow x \vee c$ or $x \rightarrow x \wedge c$ can be written in the form $x \rightarrow (x \vee a) \wedge b$, for suitable a, b.

5. (a) Using Exercise 4, show that if intervals $[x, y]$ and $[x_1, y_1]$ are projective, in a distributive lattice, then $x_1 = (x \vee a) \wedge b$ and $y_1 = (y \vee a) \wedge b$, for some a, b.

(b) Infer that in a distributive lattice, *no interval can be projective to a proper subinterval of itself*.

6. (a) Show that, for any a, b of a lattice L, the correspondence $x \rightarrow [(a \wedge b) \vee x] \wedge [a \vee b]$ is an isotone and idempotent map of L onto $[a \wedge b, a \vee b]$.

(b) Show that if θ and ϕ are both isotone transformations of a poset P, then the set of fixpoints of $\theta\phi$ is isomorphic with the set of fixpoints of $\phi\theta$. (W. Schwan)

* 7. For any elements a, b of any lattice L, let $a/a \curlywedge b$ denote the set of all $x = (x \vee b) \wedge a$, and let $a \curlyvee b/b$ denote the dual set of all $y = (y \wedge a) \vee b$. Show that $a/a \curlywedge b$ and $a \curlyvee b/b$ are always isomorphic lattices, though not in general sublattices of L. Show that L5 is equivalent to the identity $a/a \curlywedge b = a/a \wedge b$. (W. Schwan)

8. Show that in the proof of Theorem 16, the y_i are in fact independent over a.

9. Show that in Theorem 16, we can renumber the x_j^* so that every x_i^* can be substituted for x_i, $i = 1, \cdots, r$.

*10. (a) Show that in Theorem 16, for each x_i, an x_j^* exists which can be substituted for x_i and such that x_i can also be substituted for x_j^*.

(b) Show that in one case, no renumbering can be found such that, for all i, x_i^* can be substituted for x_i and vice-versa.

11. (a) Show that the conclusion of the Kurosh–Ore Theorem is not true in the semimodular lattice of Figure 12.

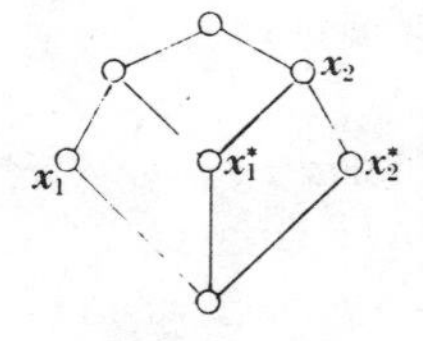

FIGURE 12

(*b) Show that the conclusion of the Kurosh–Ore Theorem holds in a finite semimodular lattice M if and only if the sublattice $L(a)$, generated by the elements p_i covering any $a \in M$, is modular.†

*12. (a) Given $y \wedge z \leqq x \leqq z$ in a semimodular lattice L, show that t exists in L such that $y \wedge z < t \leqq y$ and $x = (x \vee y) \wedge z$.

(b) Given a meet-irreducible element a in a semimodular lattice L, show that $a \wedge c = b \wedge c$ and $a > b$ imply $b \geqq c$. (Lesieur [1], p. 179, Prob. 7.1).

13. Show that the conclusion of Theorem 15 remains valid if uCv in an orthomodular lattice (§II.14).

13. Neutral Elements in Modular Lattices

Neutral (i.e., distributive) elements in complemented modular lattices have many special properties. First, Lemma 2 of §9 can be sharpened as follows.

LEMMA 1. *In a modular lattice, an element a is neutral if $x \to x \vee a$ or $x \to x \wedge a$ is an endomorphism.*

For, any one distributivity relation between a, x, y implies all six; hence Lemma 1 of §9 applies.

COROLLARY. *In a complemented modular lattice, an element a is in the center if $x \to x \vee a$ or $x \to x \wedge a$ is an endomorphism.*

THEOREM 18 (VON NEUMANN‡). *An element is in the center of a complemented modular lattice L if and only if it has a unique complement.*

PROOF. In any lattice $L = M \times N$, $[I, O]$ can have only $[O, I]$ for complement; hence no element of the center can have more than one complement. Conversely, suppose a has a unique complement a', and that $u \wedge a = O$. Then by hypothesis u, a, and any complement $(u \vee a)'$ of $u \vee a$ are independent elements with join I, hence $u \vee (u \vee a)' = a'$ and $u \leqq a'$. Thus a' contains every u with $u \wedge a = O$. In particular, since $[(a \wedge x)' \wedge x] = O$ irrespective of x, $(a \wedge x)' \wedge x$ is included in a' as well as x, and

$$x = (a \wedge x) \vee [(a \wedge x)' \wedge x] \leqq (a \wedge x) \vee (a' \wedge x) \leqq x \vee x = x.$$

Hence $x = (a \wedge x) \vee (a' \wedge x)$, and every $x \in L$ can be written in the form $y \vee z$ $[y \leqq a, z \leqq a']$. It follows by Theorem 15 that $L = [O, a] \times [O, a']$. Q.E.D.

THEOREM 19. *In a complemented modular lattice L, two elements a and b are connected by a sequence of perspectivities if and only if the ideals $[O, a]$ and $[O, b]$ are projective.*

PROOF. If a and b are perspective by c, then $[O, a]$, $[c, I]$, and $[O, b]$ are transposes in that order; hence $[O, a]$ and $[O, b]$ are projective. Since projectivity is a transitive relation, $[O, a]$ and $[O, b]$ are projective (in the sense of §I.7) if they are connected by a sequence of perspectivities. Conversely, let $[O, a]$ and

† See R. P. Dilworth, *The arithmetical theory of Birkhoff lattices*, Duke Math. J. **8** (1941), 286–99; also Trans. AMS **49** (1941), 325–53.

‡ [CG], Part I, Theorems 5.3–5.4. (Cf. Excercise 6b after §10.)

$[O, b]$ be projective, and let $[u \wedge v, v]$ and $[u, u \vee v]$ be any adjacent intervals in the sequence of transpositions connecting (by definition) $[O, a]$ and $[O, b]$. Then any relative complements w of $u \wedge v$ in v and w_1 of u in $u \vee v$ are perspective by any $(u \vee v)' \vee u$, where t is any relative complement of $u \wedge v$ in u. (Proof. Consider the chains $O \leqq u \wedge v \leqq u \leqq u \vee v \leqq I$ and $O \leqq u \wedge v \leqq v \leqq u \vee v \leqq I$.) Hence, by induction, any relative complements of O in a and O in b are connected by a sequence of perspectivities. But a and b are the only such relative complements. Q.E.D.

THEOREM 20. *In a complemented modular lattice L, the congruence relations correspond one-one with the neutral ideals N of L, or ideals containing with any element a all perspective elements.*

PROOF. Suppose that N defines a congruence relation on L, as in Theorem II.3, and that $a \in N$ and b are perspective, with the axis of perspectivity c. Then

$$b = b \wedge I = b \wedge (a \vee c) \equiv b \wedge c = O \ (\text{mod } N).$$

Hence $b \in N$, and N is neutral. Conversely, let N be neutral, and let $x = y$ mean $x \vee a = y \vee a$ for some $a \in N$. This is an equivalence relation and even a join-homomorphism, for any ideal in any lattice (§II.6, Example 3). We wish to show that $x \equiv y$ implies $x \wedge u \equiv y \wedge u$ in a complemented modular lattice, if N is neutral. Since we are dealing with an equivalence relation, and $x \equiv x \vee a = y \vee a \equiv y$, we can reduce to the case $y = x \vee a$; hence it is sufficient to show that

$$(x \wedge u) \vee a \equiv [(x \vee a) \wedge u] \vee a = (x \vee a) \wedge (u \vee a),$$

in any modular lattice. But, setting $u = y$ and $a = z$, this is equivalent to showing that $w^* = c^*$ in the free modular lattice with three generators (Figure 10a). By inspection, however, $[c^*, w^*]$ is projective to part of $z = a$. Hence, by Theorem 19, z and some element d which contains $w^* \wedge c^{*\prime}$ ($c^{*\prime}$ any complement of c^* in L) are perspective. Therefore $w^* \wedge c^{*\prime} \in N$, and $c^* = c^* \vee (c^{*\prime} \wedge w^*) = w^*$, completing the proof.

Exercises:

1. Show that an ideal in a complemented modular lattice is neutral if and only if it satisfies the condition of Theorem 20.
2. Construct a modular lattice of ten elements, containing $\mathbf{3}^2$ as a sublattice, one of whose elements has a unique complement without being neutral. (M. Hall)

PROBLEMS

7. Define natural extensions of the notions of neutral element and standard element to posets.

8. The ϕ-sublattice of a lattice L is defined as the intersection $\phi(L)$ of its maximal proper sublattices. Find necessary and sufficient conditions on L for $\phi(L)$ to be void.

9. Is every true identity on the operations of cardinal addition, multiplication and exponentiation implied by the identities (1)–(4) of Theorem 2? If dualization is

included, are (1)–(5) a complete set of identities (i) for posets, (ii) for positive reals, (iii) for positive integers?

10. Describe a systematic algorithm for reducing an arbitrary Boolean polynomial to shortest form—i.e., one involving the fewest occurrences of letters.†

11. What is the order $f(n)$ of $\mathbf{2}^{2^n}$? Describe the asymptotic behavior of $\log \log f(n)$ for large n.‡

† For a related problem, see T. Bartee, IRE Trans. Electronic Computers, EC-10, (1961), 21–6.

‡ See V. K. Korobkov, Doklady Acad. Nauk SSSR **150** (1963), 744–7.

CHAPTER IV

GEOMETRIC LATTICES

1. Introduction

The idea of considering various collections of points, lines, planes, etc., as geometrical "configurations" partially ordered by set-inclusion is a very old one. Many important such configurations are geometric (alias matroid) lattices, in the sense of the following definition.†

DEFINITION. A *point lattice* (or "atomic lattice") is a lattice in which every element is a join of points. A *geometric lattice* (or "matroid lattice") is a finite-dimensional (upper) semimodular point lattice.

Example 1. The affine subspaces of the n-dimensional vector space $D^n \cong V_n(D)$ over any field or division ring D form a nonmodular geometric lattice $AG_n(D)$, if the void set is included.

Example 2. In the "affine" geometric lattice $AG_n(D)$, those "linear" subspaces which contain the origin form a *modular* geometric lattice $PG_{n-1}(D)$, called the $(n - 1)$-dimensional *projective geometry* over D (see §7 below). When $D = \mathbf{Z}_p$, the field of integers mod p, $PG_{n-1}(\mathbf{Z}_p)$ is also isomorphic with the lattice of all subgroups of the additive group $\mathbf{Z}_p{}^n$.

Note that, in the lattice $AG_n(D)$, the height $h[x]$ (§I.3) of any element exceeds the geometric dimension of the corresponding affine subspace by one: "points" have height one, "lines" height two, and so on. Whereas in $PG_{n-1}(D)$, the lattice-theoretic height of an element equals the geometric dimension of the corresponding subspace. If one renames the lines of $PG_{n-1}(D)$ "points", the planes of $PG_{n-1}(D)$ "lines", etc., as in projective geometry, then the height again exceeds the (projective) geometric dimension by one.

Another interesting geometric lattice, which does not arise directly from geometry, is the symmetric partition lattice Π_n of degree n (§I.8, Example 9). This will be studied below in §9.

The preceding geometric lattices are highly symmetric. Thus, the automorphisms of $PG_{n-1}(D)$, considered as permutation groups, are transitive on the elements of any given dimension, and triply transitive on the points of any "line". Those of $AG_n(D)$ are transitive on the elements of any given dimension, and doubly

† G. Birkhoff, Amer. J. Math. **57** (1935), 800–4. This work was stimulated by the "matroid" concept of H. Whitney (ibid., 509–33), which the author recognized as a natural generalization of the concept of a finite-dimensional complemented modular lattice. K. Menger, F. Alt, and O. Schreiber developed closely related ideas independently during the years 1928–35; see K. Menger [**1**].

transitive on the points of any line. Those of Π_n constitute the symmetric group of degree n.

In §II.8, Theorems 14–15, it was shown that a lattice of finite length was semimodular if and only if it satisfied the Jordan–Dedekind chain condition (was *graded*) and also

$$(1) \qquad h[x] + h[y] \geqq h[x \wedge y] + h[x \vee y].$$

This basic inequality has several useful implications.

Thus, since $h[x \wedge y] \geqq 0$, (1) implies $h[x \vee y] \leqq h[x] + h[y]$, and hence (by induction)

$$(2) \qquad h[x_1 \vee \cdots \vee x_n] \leqq h[x_1] + \cdots + h[x_n].$$

Likewise, it implies condition (ξ) of §I.7.

COROLLARY 1. *A lattice of finite length is semimodular if and only if, for all elements x, y,*

$$(3) \qquad x \text{ covers } x \wedge y \text{ implies that } x \vee y \text{ covers } y.$$

COROLLARY 2. *If $x_1 < x_2 < \cdots < x_n$ is a connected chain in a semimodular lattice, then the range of the mapping ψ_a: $x_i \to x_i \vee a$ is a connected chain.* (*It may, however, be a shorter chain.*)

COROLLARY 3. *In a semimodular lattice of finite length,*

$$(4) \qquad \text{if } p \text{ is a point, then either } p \leqq a \text{ or } a \vee p \text{ covers } a.$$

PROOF. $p \wedge a \leqq p$. If $p \wedge a = p, p \leqq a$. If $p \wedge a < p, p \wedge a = O$. Then $h[p \vee a] \leqq h[p] + h[a] = 1 + h[a]$, by (2). Since $p \nleqq a$, $h[p \vee a] \neq h[a]$, hence $h[p \vee a] = h[a] + 1$.

COROLLARY 4. *In a semimodular lattice of finite length,*

$$(5) \qquad \text{if } p, q \text{ are points and } a < a \vee q \leqq a \vee p, \text{ then } a \vee p = a \vee q.$$

PROOF. By Corollary 3. The relation (5) is called the Steinitz–Mac Lane exchange axiom.

2. Modular Pairs

We now introduce the basic concept of a modular pair,† which involves the considerations of W. Schwan already introduced in §III.11. As observed there, each ordered pair of elements a, b of a lattice L defines four *isotone* functions: ψ_a: $x \to x \vee a$ from $[a \wedge b, b]$ to $[a, a \vee b]$, ψ_b: $x \to x \vee b$ from $[a \wedge b, a]$ to $[b, a \vee b]$. ϕ_a: $x \to x \wedge a$ from $[b, a \vee b]$ to $[a \wedge b, b]$ and ϕ_b: $x \to x \wedge b$ from $[a, a \vee b]$ to $[a \wedge b, b]$. Moreover $\phi_a\psi_b\phi_a = \phi_a$ and $\psi_b\phi_a\psi_b = \psi_b$. In modular

† This notion is due to L. R. Wilcox, Ann. of Math. **40** (1939), 490–505, who used a terminology dual to ours.

lattices, these functions are all bijections by Dedekind's Theorem I.13; moreover $\psi_b = \phi_a^{-1}$ and $\psi_a = \phi_b^{-1}$ (two-sided inverses).

DEFINITION. We say (a, b) is a *modular pair* in a lattice L, and write aMb or or $(a, b)M$, when

(6) $$x = (x \vee a) \wedge b = x\psi_a\phi_b \quad \text{for all } x \in [a \wedge b, b].$$

LEMMA 1. *We have aMb if and only if*

(7) $$t \leqq b \quad \textit{implies} \quad t \vee (a \wedge b) = (t \vee a) \wedge b.$$

PROOF. For $t \in [a \wedge b, b], t \vee (a \wedge b) = t$, and so (7) implies (6). Conversely, if (6) holds and $t \leqq b$, then setting $x = t \vee (a \wedge b) \in [a \wedge b, b]$, we have

$$x = t \vee (a \wedge b) = [t \vee (a \wedge b) \vee a] \wedge b = [t \vee a) \wedge b,$$

where the second equality follows from (6). This proves (7).

COROLLARY. *A lattice L is modular if and only if aMb for all $a, b \in L$.*

The *dual aM^*b* of the relation aMb is

(7′) $$y = (y \wedge a) \vee b = y\phi_a\psi_b \quad \text{for all } y \in [b, a \vee b].$$

Evidently, ψ_a and ϕ_b are *inverse isomorphisms* if and only if aMb and bM^*a. Hence we have

LEMMA 2. *If aMb and bM^*a in a lattice L, then $[a \wedge b, b]$ and $[a, a \vee b]$ are isomorphic.*

More generally, aMb alone implies that ψ_a is one-one (an order-*mono*morphism). In particular, ψ_a carries chains of length n in $[a \wedge b, b]$ into chains of the same length in $[a, a \vee b]$. Hence, if $[a \wedge b, a \vee b]$ is an interval of *finite length* in a lattice satisfying the Jordan–Dedekind chain condition, aMb implies $h[b] - h[a \wedge b] \leqq h[a \vee b] - h[a]$. Transposing, we get

THEOREM 1. *In any lattice L of finite length which satisfies the Jordan–Dedekind chain condition, aMb implies that*

(8) $$h[a] + h[b] \leqq h[a \wedge b] + h[a \vee b].$$

The preceding theorem applies in particular to (upper) *semimodular* lattices of finite length. But in these, by (1), the reverse of inequality (8) holds. Hence aMb implies $h[a] + h[b] = h[a \wedge b] + h[a \vee b]$ in any such (upper) semimodular lattice. We will now show that the converse is also true. First, we prove

LEMMA 3. *In any lattice, the following statements are equivalent:* (i) aMb, (ii) ψ_a *is one-one,* (iii) ϕ_b *is onto.*

PROOF. For $\psi_a\phi_b$ to be the identity on $[a \wedge b, b]$, in any event ψ_a must be one-one and ϕ_b must be onto: (i) implies (ii) and (iii) almost trivially. The sharper converses follow from Schwan's Lemma (§III.11):

(9) $$\psi_a\phi_b\psi_a = \psi_a \quad \text{and} \quad \phi_b\psi_a\phi_b = \phi_b.$$

Therefore, unless aMb, we have $x < (x \vee a) \wedge b = x'$ for some $x \in [a \wedge b, b]$, although $x'\psi_a = x\psi_a\phi_b\psi_a = x\psi_a$, hence ψ_a is not one-one and (ii) implies (i). Finally, if ϕ_b is onto, then $x = y\phi_b = y\phi_b\psi_a\phi_b = x\psi_a\phi_b$ for all $x \in [a \wedge b, b]$ and so aMb; hence (iii) implies (i).

THEOREM 2. *In an (upper) semimodular lattice L of finite length, aMb if and only if*

$$h[a] + h[b] = h[a \wedge b] + h[a \vee b]. \tag{10}$$

PROOF. The implication $aMb \Rightarrow (10)$ was established above. But conversely, suppose aMb fails. Then the join-homomorphism ψ_a is not one-one, by Lemma 3: it must map two distinct elements x and $x \vee z > x$ of some connected chain in $[a \wedge b, b]$ onto the same image $y = x\psi_a$ in $[a, a \vee b]$. But, by Corollary 2 of §1, ψ_a carries connected chains in $[a \wedge b, b]$ into connected chains in $[a, a \vee b]$. Hence, if aMb fails, connected chains joining $a \wedge b$ to b through x and $x \vee z$ are shortened by ϕ_a, and so

$$h[b] - h[a \wedge b] > h[a \vee b] - h[a].$$

COROLLARY 1. *In any (upper) semimodular lattice L of finite length, aMb implies bMa.*

COROLLARY 2. *In any lower semimodular lattice L of finite length, aM^*b implies bM^*a.*

The preceding corollaries enable one to extend the notion of semimodularity to lattices of infinite length, as follows.

DEFINITION. A lattice L is *semimodular* if aMb implies bMa for all $a, b \in L$. It is *lower* (or "dually") semimodular if aM^*b implies bM^*a for all $a, b \in L$.

3. Examples

Although every nonmodular sublattice contains a five-element sublattice which is not semimodular (Theorem I.12), one has trivially

LEMMA 1. *Any convex sublattice of a semimodular lattice is semimodular.*

A less obvious result is

LEMMA 2. *Let $L = L_1L_2$ be the cardinal product of lattices L_1 and L_2. Then $[a_1, a_2]M[b_1, b_2]$ in L if and only if a_iMb_i in L_i for $i = 1, 2$.*

PROOF. Since for any fixed c_1, the set of all $[c_1, x_2]$ is a convex sublattice isomorphic with L_2, and similarly for L_1, the "only if" follows from Lemma 1. To prove the "if", recall that $[a_1, a_2]M[b_1, b_2]$ if and only if $a_i \wedge b_i \leqq a_i \vee b_i$ for $i = 1, 2$ implies that $\psi_a: x \to x \vee a$ is one-one from $[a \wedge b, b]$ to $[a, a \vee b]$ in L_1L_2—i.e., if and only if $\psi_{a_i}: x_i \to x_i \vee a_i$ is one-one from $[a_i \wedge b_i, b_i]$ to $[a_i, a_i \vee b_i]$ in each L_i. But this is the condition that a_iMb_i for $i = 1, 2$.

COROLLARY. *The cardinal product of two lattices is semimodular if and only if both factors are semimodular.*

We now give a far-reaching generalization of Examples 1–2. The genesis of the construction may be found in a paper of Tutte;† the present formulation is due to G.-C. Rota and A. G. Waterman (personal communications).

LEMMA 3. *Let G be any geometric lattice, and let P be any set of atoms ("points") of G. Then the set L of all joins of $p_i \in P$ is a geometric lattice (if $O = \bigvee_{\phi} p_i$ is included).*

PROOF. Clearly b covers a in L if and only if $b = a \vee p$ for some $p \in P$. Hence, if b and c both cover a in L, then $b \vee c = (a \vee p) \vee (a \vee q) = a \vee p \vee q$ covers b and c in G, and a fortiori in L. That is, L satisfies condition (3). But it is trivial that L is a point lattice of finite length, which completes the proof.

Now let Φ be any n-dimensional subspace of the vector space D^X of all functions $f: X \to D$, where D is a division ring and X any set. For each $x \in X$, let $F_x \subset \Phi$ be the subspace of all $f \in \Phi$ with $f(x) = 0$. Trivially, either $F_x = \Phi$ or F_x is a hyperplane of Φ: in other words, a *dual* point in the lattice $L(\Phi) \cong PG_{n-1}(D)$ of all linear subspaces of Φ. Call Φ *discriminating* when, for each $x \in X$, there is a nonzero $f \in \Phi$ with $f(x) = 0$—i.e., when F_x is *always* a dual point. Then the set of all $\bigcap_S F_x$ for subsets $S \subset X$, partly ordered by inclusion, is isomorphic with the set of all meets of the set of dual points $F_x \in L(\Phi)$. Hence, by Lemma 3, it is the dual of a geometric lattice. The conclusion can be summarized as follows.

LEMMA 4. *Let Φ be any discriminating finite-dimensional linear space of functions on a set X. For each $x \in X$, let F_x be the set of all $f \in \Phi$ with $f(x) = 0$. Then the set $P(\Phi)$ of all intersections $\bigcap F_x$, partly ordered by inclusion, is dual to a geometric lattice.*

Finally, the dual of $P(\Phi)$ has a direct interpretation. For any subset $S \subset X$, define the *closure* $\bar{S}$ of S by

$$(11) \qquad p \in \bar{S} \quad \text{if and only if } f(p) = 0 \quad \text{for all } f \in \bigcap_{x \in S} F_x.$$

Then the "closed" subsets of X form a geometric lattice dual to $P(\Phi)$; this is a special instance of the concept of "polarity", to be treated systematically in Chapter V. These considerations prove

THEOREM 3. *Let Φ be a discriminating finite-dimensional linear space of functions on a set X. Call a subset $S \subset X$ Φ-closed when, given p not in S, there exists $f \in \Phi$ such that $f(p) \neq 0$ yet $f(x) = 0$ for all $x \in X$. Then the Φ-closed subsets of X form a geometric lattice.*

In Example 1, Φ was the set of all affine functions on $V_n(D)$; one gets $PG_{n-1}(D)$ by letting Φ be the set of all linear functions on $V_n(D)$ (the dual space $V_n^*(D)$). By letting Φ be the set of all real functions of the form $f = a + \sum b_i x_i + c \sum x_i^2$, one gets *spherical* geometry. Many other intriguing examples can be constructed similarly.

† W. T. Tutte, Trans. AMS 88 (1958), 144–74.

"Geometric" lattices of infinite length. The hypothesis of finite length will play an essential role in this chapter; if it is not made, the concept of "semimodularity" ramifies in a most unpleasant way.† Some idea of the facts can be had by considering the following important example.

Example 3. Let $\mathcal{B}$ be any Banach space, and let $L(\mathcal{B})$ be the lattice of all closed subspaces of $\mathcal{B}$.

Mackey has proved that aMb in $L(\mathcal{B})$ if and only if the linear sum of the closed subspaces a and b is itself closed,‡ and that aM^*b is also equivalent to this condition. But this condition is symmetric in a and b. Hence we have

THEOREM 4. *For any Banach space $\mathcal{B}$, $L(\mathcal{B})$ is an upper and lower semimodular lattice.*

The case $\mathfrak{H}$ of Hilbert space is especially interesting, as defining a semimodular orthomodular lattice (§II.14). Combining Theorem II.16 with Theorem 4 above, we obtain the

COROLLARY. *In any semimodular ortholattice, all interval sublattices of finite length are modular.*

In addition to this property, $L(\mathfrak{H})$ can be shown to satisfy also

$$aMa^{\perp} \quad \text{for all } a \in L(\mathfrak{H}). \tag{12}$$

Orthomodular, semimodular lattices which satisfy (12) may be called *strongly* orthomodular. Thus $L(\mathfrak{H})$ is a strongly orthomodular lattice, in which every element is a join of points (i.e., it is atomic).

Kakutani and Mackey have considered the possibility of defining an orthocomplementation operation in $L(\mathcal{B})$ for other Banach spaces $\mathcal{B}$. They have shown§ and this is possible if and only if X is topologically and linearly isomorphic to some Hilbert space.

Exercises for §§1–3:

1. Show that, in any lattice L, for any point p, pMx for all $x \in L$.
2. Show that unless F consists of 0, 1 alone, every "line" in $PG(F; n)$ contains at least four points.
3. Show that if p is any point of $AG_n(D)$, the dual ideal $[p, I]$ of $x \geqq p$ forms a modular lattice isomorphic with $PG_{n-1}(D)$.
4. Show that if, in any semimodular lattice L, we equate to I all elements with $d[x] \geqq n$ (n any fixed integer), we get a semimodular lattice. Show that this is a join-epimorphic image of L. (S. Mac Lane)
5. Show that if $\mathbf{R}$ is the real number system, the portions of affine subspaces contained in any open convex subset S of $V(\mathbf{R}; n)$ form a semimodular lattice. (This gives n-dimensional hyperbolic geometry, if S is an n-sphere.)

---

† R. Croisot, Ann. Sci. École Norm. Sup. **68** (1951), 203–65, discusses the interdependence of 23 such ramifications! See also §VIII.9.

‡ G. Mackey, Trans. AMS **57** (1945), 155–207.

§ S. Kakutani and G. Mackey, Ann. of Math. **45** (1944), 50–8 (real $\mathcal{B}$) and Bull. AMS **52** (1946), 727–33 (complex $\mathcal{B}$).

6. Show that the lattice of all subgroups of any p-group is dually semimodular.

*7. Show that the subspaces of any finite-dimensional real vector space which are left invariant by any finite group of linear operators form a modular ortholattice.

*8. Construct a 16-element nonmodular ortholattice of length 3 in which every sublattice is closed under $a \wedge b'$ and $b \vee (a \wedge b')$. (Dilworth, Tôhoku Math. J. **47** (1940), 18–23.)

*9. Show that the free modular ortholattice with two generators is $2^4 \times PG(Z_3, 1)$, and has 96 elements.

10. Show that the lattice with diagram sketched below is not monomorphic to a sublattice of a geometric lattice of the same length. (A. G. Waterman)

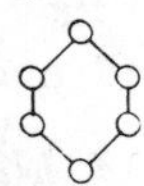

11. Let $L(G)$ be the lattice of all subgroups of a group G.

(a) Show that if $ST = TS$ in G, then (S, T) is a dually modular pair in $L(G)$.

(b) Show that if N is a normal subgroup of G, then (N, S) is a modular pair in $L(G)$ for any subgroup S of G. (*Hint.* Use the remark following Theorem I.11.)

*12. Let L be a lattice in which any $a < b$ can be joined by at least one finite chain. Show that (3) implies semimodularity, and generalize (8) to this case. (I. Kaplansky)

*13. In Theorem 3, show that $\bar{S}$ is the set of all $p \in X$ such that $f(x) = 0$ for all $x \in S$ implies $f(p) = 0$.

4. Dependence and Rank

We shall consider only lattices of finite length for the rest of this chapter. We begin by defining a natural extension of the notion of linear independence† to geometric lattices.

DEFINITION. A sequence $x_1, \cdots, x_r$ of elements of a semimodular lattice of finite length is called *independent* if and only if it satisfies the symmetric condition

$$h[x_1 \vee \cdots \vee x_r] = h[x_1] + \cdots + h[x_r]. \tag{13}$$

The significance of (13) becomes apparent when we compare it with (2). For any $x_1, \cdots, x_r$ in a semimodular lattice L of finite length, by (1):

$$h[x_1 \vee \cdots \vee x_r] = h[x_1 \vee \cdots \vee x_{r-1}] + h[x_r] - h[(x_1 \vee \cdots \vee x_{r-1}) \wedge x_r].$$

Moreover, by (10), equality holds if and only if

(i) $(x_1 \vee \cdots \vee x_{r-1}) \wedge x_r = O$,
(ii) $x_1 \vee \cdots \vee x_{r-1}$ and x_r form a modular pair, and
(iii) $x_1, \cdots, x_{r-1}$ are independent.

From (iii) and the symmetry of (13), using induction, it follows that *any subset of an independent set is independent.*

The most interesting case occurs when the x_i are points; then condition (ii) above holds automatically,

LEMMA. *A sequence $x_1, \cdots, x_r$ of points is independent if and only if*

$$(x_1 \vee \cdots \vee x_k) \wedge x_{k+1} = O \quad \textit{for } k = 1, \cdots, r-1. \tag{14}$$

† See H. Whitney, Amer. J. Math. **57** (1935), 507–33; also S. Mac Lane [1].

PROOF. Since by Corollary 1 of Theorem 1, $x_1 \vee \cdots \vee x_{k+1}$ at most covers $x_1 \vee \cdots \vee x_k$, we have $h[x_1 \vee \cdots \vee x_{k+1}] = h[x_1 \vee \cdots \vee x_k] + 1$ or $h[x_1 \vee \cdots \vee x_k]$ accordingly as $(x_1 \vee \cdots \vee x_k) \wedge x_{k+1} = O$ or not. More generally, we define a sequence of elements $x_1, \cdots, x_r$ in a lattice with O to be *sequentially independent* if and only if Equation (14) holds:

$$(x_1 \vee \cdots \vee x_{k-1}) \wedge x_k = O \quad \text{for } k = 2, \cdots, r.$$

We have just seen that the relation of sequential independence for any sequence of *points* is *symmetric* (i.e., invariant under all permutations), in any semimodular (hence in any geometric) lattice.

The following definition presents a concept which is closely related to that of independence.

DEFINITION. Let X be a set of points; let their least upper bound be denoted sup X. Then $r(X) = h[\sup X]$ is called the *rank* of X. (Note that sup X is the *closure* $\bar{X}$ of X in the notation of §3.)

LEMMA 2. *In a semimodular lattice, any set of points contains an independent subset of the same rank (i.e., with the same join).*

PROOF. Let $x_1, \cdots, x_r$ be any sequence of points, independent or not. We can construct term-by-term a subsequence, by deleting x_k if $x_k \leqq x_1 \vee \cdots \vee x_{k-1}$, none of whose members is contained in the join of the preceding, yet whose join is $x_1 \vee \cdots \vee x_r$. This will be independent.

THEOREM 5. *Let X be a set of independent elements $x_i > O$ in a semimodular lattice L of finite length. Then the x_i generate a sublattice isomorphic with the field of all subsets of X.*

PROOF. Associate with each subset S of X, sup S. The set of all sup S is clearly closed under join. By the minimax inequality (Chapter II, §5, (8)) sup $(S \cap T) \leqq \sup S \wedge \sup T$, while by L1–L3, $\sup (S \cup T) = \sup S \vee \sup T$. By this and (2),

$$\begin{aligned} h[\sup S \wedge \sup T] &\leqq h[\sup S] + h[\sup T] - h[\sup S \cup T] \\ &= \sum_S h[x_i] + \sum_T h[x_i] - \sum_{S \cup T} h[x_i] \\ &= \sum_{S \cap T} h[x_i] = h[\sup S \cap T]. \end{aligned}$$

Hence the minimax inequality is an equality, and meets correspond to set-products as well as joins to set-unions. The condition $x_i > O$ guarantees that we have not merely an epimorphism, but an isomorphism.

5. Postulates for Geometric Lattices

The definition of a geometric lattice given in §1 conforms to intuitive geometrical ideas, especially about affine and projective geometry. But it does not conform to the classification of lattices given in Chapters I–II. To connect it with the latter, we will now characterize geometric lattices as (relatively) *complemented* semimodular lattices of finite length.

As in §I.9, a lattice is said to be *complemented* if it satisfies

L7. Given x, there exists x' such that $x \wedge x' = O$, $x \vee x' = I$.

It is said to be *relatively complemented* if it satisfies

L7R. Given $a \leqq x \leqq b$, there exists y such that $x \wedge y = a$, $x \vee y = b$.

Although any relatively complemented lattice with O and I is clearly complemented, the converse is not necessarily true, even though any complemented *modular* lattice is relatively complemented (cf. Theorem I.14). In particular, a complemented (upper) *semimodular* lattice need not be relatively complemented, as the diagram of Figure 13a shows.

(a) (b)

FIGURE 13

THEOREM 6. *A semimodular lattice of finite length is complemented if and only if it satisfies*

L7′. *I is the join of points.*

It is relatively complemented if and only if

L7R′. *Every element $x > O$ is the join of points.*

PROOF. In any lattice L of finite length, L7 implies L7′. For let a be the join of all points of L; $a' = O$ since it cannot contain a point. Hence $I = a \vee a' = a \vee O = a$ is a join of points.

Conversely, let L be a semimodular lattice of finite length satisfying L7′; let $a \in L$ be given. There exist in succession points $p_1 \nleqq a$, $p_2 \nleqq a \vee p_1, \cdots$, until $a \vee p_1 \vee \cdots \vee p_r = I$. As in Lemma 1 of §4,

$$h[a \vee (p_1 \vee \cdots \vee p_r)] = h[a] + r;$$

hence

$$\begin{aligned} h[a \wedge (p_1 \vee \cdots \vee p_r)] &\leqq h[a] + h[p_1 \vee \cdots \vee p_r] - h[a \vee p_1 \vee \cdots \vee p_r] \\ &= h[a] + r - (h[a] + r) = 0. \end{aligned}$$

It follows that $a' = p_1 \vee \cdots \vee p_r$ is a complement of a, and even that a, a' form a modular pair.

Again, in any lattice of finite length, L7R implies L7R′ by Theorem I.14. Conversely, let L be a semimodular lattice satisfying L7R′; and let $a \leqq x \leqq b$ be given. There exist in succession points $p_1, \cdots, p_r \leqq b$ such that $(x \vee p_1 \vee \cdots \vee p_k) \wedge p_{k+1} = O$, $x \vee p_1 \vee \cdots \vee p_r = b$. As in Lemma 1 of §4 and in the preceding paragraph, if we let $z = p_1 \vee \cdots \vee p_r$, then $h[a \vee z] = d[a] + r$ and $h[x \vee z] = h[x] + r = d[b]$. Let $y = a \vee z$. Then $x \vee y = x \vee a \vee z = x \vee z = b$; while $x \wedge y \geqq a$ and $h[x \wedge y] \leqq h[x] + h[a \vee z] - h[x \vee y] = h[x] + h[a] + r -$

$(h[x] + r) = h[a]$. Hence $x \wedge y = a$, and L7R and L7R′ are equivalent in any semimodular lattice of finite length.

Thus, a geometric lattice is simply a relatively complemented semimodular lattice of finite length. Hence a modular geometric lattice is a complemented modular lattice of finite length.

A lattice in which L7′ or L7R′ holds need not be complemented unless it is semimodular, however. Figure 13b gives an example of this; the shaded element has no complement.

We will now prove a strong converse of the preceding results.†

THEOREM 7. *A relatively complemented lattice L of finite length is geometric if any one of the following conditions holds*: (i) *L satisfies* (ξ)—i.e., *is semimodular*, (ii) *L satisfies* (4), (iii) *L satisfies* (5), *or* (iv) *sequential independence is symmetric in L.*

PROOF. Since L7R implies L7R′ in any lattice of finite length, it suffices to show that if L7R′ holds, then conditions (ξ), (iv), (4), and (5) imply each other cyclically. For then any of these conditions with L7R′ will imply (ξ) and L7R′, hence (ξ) and L7R, and hence that L is a geometric lattice.

First, (ξ) implies (iv) as in Lemma 1 of §4. Next, (iv) implies (5). Using L7R′, one easily obtains independent points $p_1, \cdots, p_r$ with $p_1 \vee \cdots \vee p_r = a$. Moreover, since $a \vee q \leqq a \vee p$, the sequence $p_1, \cdots, p_r, p, q$ is not independent; hence by (13) neither is $p_1, \cdots, p_r, q, p$; hence $p \geqq (a \vee q) \wedge p > O$; hence $p \leqq a \vee q$, and $a \vee p \leqq a \vee q$, giving $a \vee p = a \vee q$ by hypothesis and P2.

Then (5) implies (4). If $a < b \leqq a \vee p$, there exists by L7R′ a point q contained in b but not in a; clearly $a < a \vee q \leqq b$. By (5), $a \vee p = b$, and $a \vee p$ covers a.

Finally, (4) implies (ξ). If x, y covers a, then by L7R′ there exist points p, q with $a < a \vee p \leqq x$, $a < a \vee q \leqq y$, whence $a \vee p = x$ and $a \vee q = y$. Hence $x \vee y = a \vee p \vee q$, which by (4) covers $x = a \vee p$ and $y = a \vee q$.

THEOREM 8. *A point-lattice of finite length is (upper) semimodular if and only if aMp for any $a \in L$ and any point $p \in L$.*

PROOF. Suppose L is upper semimodular. It is to be shown that for any element a and point p, $b \leqq a$ implies $a \wedge (p \vee b) = (a \wedge p) \vee b$. By Corollary 3 of §1, either $p \leqq a$ or $a \vee p$ covers a. Suppose $p \leqq a$; then $p \wedge a = p$, hence $(a \wedge p) \vee b = p \vee b$. But since $b \leqq a$, $b \vee p \leqq a$, hence $a \wedge (p \vee b) = b \vee p$. If $p \nleqq a$, then $p \nleqq b$, hence $b \vee p$ covers b. Therefore, $a \wedge (b \vee p) = b = (a \wedge p) \vee b$; for $p \nleqq a$ implies $p \wedge a = O$.

Conversely, suppose $a \wedge (b \vee p) = (a \wedge p) \vee b$ for all $b \leqq a$, and points p. Since every element of the lattice is assumed to be a join of points, it will suffice to show that (4) holds; for by Theorem 7 this implies that the lattice is geometric, hence certainly semimodular. Assume that $p \nleqq b$ and that there exists c such that $b \leqq b \leqq b \vee p$. If $p \leqq c$, then $b \vee p \leqq c \leqq b \vee p$. Hence $c = b \vee p$. If p is

† See [**LT2**, p. 106, Theorem 7], and [**Tr**, p. 99, Theorem 4].

not contained in c, then $c = c \wedge (b \vee p) = b \vee (c \wedge p) = b$. So $b \vee p$ covers b and (4) holds.

Exercises:

1. (a) Show that a point-lattice L is geometric if and only if, for any point p, pMa for all $a \in L$.

(b) Prove that this is equivalent to the condition that $b \leqq a$ implies $a \wedge (b \vee p) = b \vee (a \wedge p)$.

2. In a lattice of finite length satisfying L7′, show that every interval $[a, I]$ is complemented.

3. Let b and $c \leqq b$ be complements of a in a geometric lattice. Show that if aMb and aMc, then $b = c$.

4. Show that, in a semimodular lattice, if points $p_1, \cdots, p_s$ are independent, and if $q_1, \cdots, q_{s+1}$ are independent, then some set $p_1, \cdots, p_s, q_t$ is independent. (H. Whitney)

5. Show that, even though $x_1, \cdots, x_r$ are independent in a geometric lattice, then the intervals $[O, x_i]$ need not generate a sublattice isomorphic with their cardinal product. (*Hint*. Consider the semimodular lattice obtained by deleting a point from $\mathbf{2}^4$.) (Phillip Chase)

6. (a) Show that, in the lattice of Figure 13b, aMb fails to imply bMa.

(b) Show that the conclusion of the Kurosh–Ore Theorem fails to hold in this lattice.

(*c) Show that the conclusion of the Kurosh–Ore Theorem holds in a finite semimodular lattice M if and only if the sublattice $L(a)$, generated by the elements p_i covering any $a \in M$, is modular.†

7. Correlate Fr. Klein's observation that independence is symmetric in a semimodular lattice, with Wilcox's condition that modularity of pairs is symmetric.

8. Show that any lattice of finite length possesses an isotone function assuming the real values $0, \frac{1}{2}, \frac{3}{4}, \frac{7}{8}, \cdots$, such that (1) holds.

9. Construct an eleven-element relatively complemented semimodular lattice of length 3, which contains a six-element relatively complemented sublattice not satisfying the Jordan–Dedekind chain condition. (Dilworth)

10. The Exchange Axiom holds for the R-submodules of an R-module if and only if R is "left-regular".

6. Modular Geometric Lattices

By the results of §5, the concept of a modular geometric lattice is the same as that of a complemented modular lattice of finite length. Moreover, by Dilworth's Theorem (Theorem III.14), we know that any geometric lattice G (being relatively complemented) is a direct product of *simple* (geometric) lattices. We will now show that, if G is *modular*, these factors are projective geometries. To this end, we first prove

LEMMA 1. *Let p and q be two distinct points of a modular geometric lattice M. Then $p \vee q$ contains a third point s if and only if p and q have a common complement.*

PROOF. Any third point s on $p \vee q$ is a common *relative* complement of p and q in $[O, p \vee q]$. Hence, as in Theorem I.14, $s \vee (p \vee q)' = c$ is a common complement of p and q, for any complement $(p \vee q)'$ of $p \vee q$.

Conversely, let p and q have a common complement c. Clearly $I = p \vee c = q \vee c$ covers c; define $s = (p \vee q) \wedge c$. Then

$$d[s] = d[p \vee q] + (d[c] - d[p \vee q \vee c]) = 2 - 1 = 1,$$

† See R. P. Dilworth, *The arithmetical theory of Birkhoff lattices*, Duke Math. J. **8** (1941), 286–99; also Trans. AMS **49** (1941), 325–353.

since $p \vee q \vee c = I$ covers c. Hence $r \leqq p \vee q$ is a point. Finally, since $r \leqq c$ whereas $p \wedge c = q \wedge c = O$, it is not p or q: it is a third point on $p \vee q$.

DEFINITION. Two elements a and b of a complemented lattice L which have a common complement are *perspective* (in symbols, $a \sim b$).

LEMMA 2. *In a modular geometric lattice M, perspectivity is an equivalence relation.*

PROOF. In any complemented lattice, perspectivity is reflexive and symmetric; it remains to prove that *perspectivity of points is transitive* in any modular geometric lattice. The proof will utilize the condition of Lemma 1, which states that if $p \sim q$ and $q \sim r$, then (except in the trivial case that $p = q$ or $q = r$) there exist points s on $p \vee q$ and t on $q \vee r$ as in Figure 14 (see next page).

Now construct the "line" $s \vee t$; clearly

$$d[(s \vee t) \wedge (p \vee r)] = d[s \vee t] + d[p \vee r] - d[s \vee t \vee p \vee r]$$
$$\geqq 2 + 2 - 3 = 1,$$

since $s \vee t \vee p \vee r \leqq p \vee q \vee r$. This means that $s \vee t$ and $p \vee r$ have a point u in common. But clearly $s \vee t$ cannot contain p; otherwise it would contain $p \vee s$ and hence q; this would make it contain $q \vee t$ and hence r, which would imply $d[p \vee q \vee r] = d[s \vee t] = 2$. Likewise $s \vee t$ cannot contain r. Hence u is a third point on $p \vee r$ and so, again by Lemma 1, $p \sim r$. Q.E.D.

It follows that, in any modular geometric lattice M (i.e., in any complemented modular lattice of finite length), the relation of perspectivity partitions the points into equivalence classes E_i, such that for $p \in E_i$ and $q \in E_j$, $p \sim q$ if and only if $i = j$.

Now let $E_1, \cdots, E_s$ be the equivalence classes of mutually perspective points, and let $e_1, \cdots, e_s$ be the joins of the sets of points in these equivalence classes. It has not yet been proved that s is finite, and it is not generally true that the individual equivalence classes are finite. We now prove

LEMMA 3. *In a finite-dimensional complemented modular lattice, the elements e_i just defined are independent.*

PROOF. To show that $(e_1 \vee \cdots \vee e_k) \wedge e_{k+1} = O$, it suffices to observe that no point in e_{k+1} can be in $e_1 \vee \cdots \vee e_k$, since the E_i are disjoint, and that any $x > O$ contains a point.

THEOREM 9. *Any complemented modular lattice of finite length is the product of the ideals $[O, e_i]$, where the e_i are the joins of the different equivalence classes E_i of perspective points.*

PROOF. Any element a is the join of the points which it contains, hence $a = a_i$, where $a_i \in [O, e_i]$ is the join of the points in E_i contained in a. Hence, by Corollary 2 to Theorem III.15, it suffices to show that the e_i are independent. But this has just been shown.

It follows that the e_i belong to the center of the given lattice. The converse is however an immediate consequence of the fact that an element of the center of M must have a *unique* complement (be perspective to itself alone).

7. Projective Geometries

The factors of length one in Theorem 9 are (trivially) just copies of **2**; their product is a Boolean algebra. We now consider the factors of length $d[L] > 1$ in Theorem 9; they are *simple* (modular) *geometric lattices*. We call the atoms of any such factor "points", and the elements λ with $d[\lambda] = 2$ "lines". We say p is "on" λ, and write $p \in \lambda$, when $p \leqq \lambda$.

LEMMA 1. *The lattices $PG_{n-1}(D)$ of Example* 1, §1, *are all simple geometric lattices.*

SKETCH OF PROOF. The dependence conditions of §4 are familiar properties of linear dependence over a field or division ring. The fact that any "line" contains three "points" is also trivial. Modularity follows by Theorem I.11.

LEMMA 2. *In any simple modular geometric lattice of length $d > 1$:*

PG1. *Two distinct points are on one and only one line.*

PG2. *If a line λ intersects two sides of a triangle (not at their intersection), then it also intersects the third side.*

PG3. *Every line contains at least three points.*

PG4. *The set of all points is spanned by d points, but not by fewer than d points.*

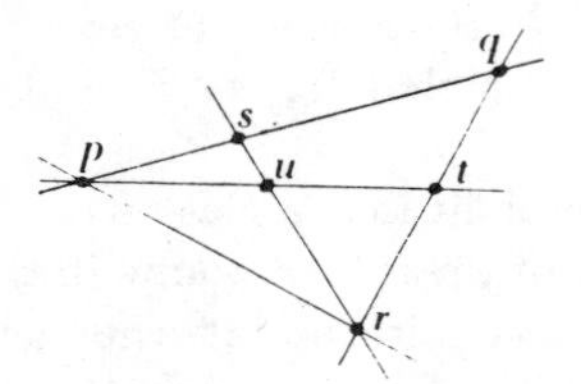

FIGURE 14

PROOF. First, if $p \neq q$ are distinct points, then any line λ containing both must contain $p \vee q$, and hence be $p \vee q$. Second, the hypotheses of PG2 for the triangle with vertices p, q, r imply $\lambda \wedge (p \vee q) = s$ and $\lambda \wedge (p \vee r) = t\ (s, t \neq p)$, and so (see Figure 14)

$$d[\lambda \wedge (q \vee r)] = d[\lambda] + d[q \vee r] + d[\lambda \vee q \vee r] = 2 + 2 - 3 = 1;$$

hence $\lambda \wedge (q \vee r)$ is a point, proving PG2. Finally, PG3 follows as in §6 from the hypothesis of simplicity.

But conditions PG1–PG4 constitute the classic† definition of a *projective geometry* of dimension $d - 1$. We have thus proved

† O. Veblen and J. W. Young, *Projective geometry*, vol. 1. The results of §§6–7 are due to the author [2].

Theorem 10. *Any modular geometric lattice is a product of a Boolean algebra with projective geometries.*

Conversely, one can construct from any projective geometry Λ a simple geometric lattice, consisting of the flats of Λ in the following sense.

Definition. A set of points in a projective geometry is called a *flat* when it contains, with any two distinct points p and q, all other points on $p \vee q$.

It is obvious that the flats of any projective geometry form a *complete atomic lattice* (every flat is the join of points). Using some care, one can prove further the Steinitz–Mac Lane Exchange Condition for finite subsets. If the space is spanned by a finite set of points, then it follows that the lattice is geometric. Moreover, one can prove that $d[x] + d[y] = d[x \wedge y] + d[x \vee y]$: the lattice is *modular*. By PG3, it is simple.

Corollary. *Each of the following conditions on a modular geometric lattice L is necessary and sufficient that it be a projective geometry*:

(i) *L is simple*;

(ii) *L is directly indecomposable*,

(iii) *all points of L are perspective.*

Slightly less trivial is the following result.

Theorem 11. *In a finite-dimensional projective geometry, x and y are perspective if and only if $d[x] = d[y]$.*

Proof. If x and y are perspective with common complement c, then $d[x] = d[y] = d[I] - d[c]$. Conversely, suppose $d[x] = d[y]$, and let $p_1, \cdots, p_r$ and $q_1, \cdots, q_r$ be bases for $x \wedge (x \wedge y)'$ and $y \wedge (x \wedge y)'$ respectively. Let $t_1, \cdots, t_r$ be third points on $p_1 \vee q_1, \cdots, p_r \vee q_r$. Then evidently

$$\begin{aligned} x \vee t_1 \vee \cdots \vee t_r &= (x \wedge y) \vee p_1 \vee t_1 \vee \cdots \vee p_r \vee t_r \\ &= (x \wedge y) \vee p_1 \vee q_1 \vee \cdots \vee p_r \vee q_r = x \vee y, \end{aligned}$$

and so by dimensionality, $x \wedge y$, the p_i, the t_j, and $(x \vee y)'$ are independent elements whose join is I. Hence $(x \vee y)' \vee t_1 \vee \cdots \vee t_r$ and $x = (x \wedge y) \vee p_1 \vee \cdots \vee p_r$ are complements; likewise, it and y are complements, whence x and y are perspective. This completes the proof.

Exercises for §§6–7:

1. Prove in detail the Corollary of Theorem 10.
2. Show that, if perspectivity is transitive between points of a geometric lattice G, then G is modular.
3. (a) Show that if a and a_1 are perspective elements in a lattice A, then $[a, b]$ and $[a_1, b]$ are perspective elements in any cardinal product AB, for any $b \in B$.

   (b) Show that the relation of perspectivity between elements is transitive in any modular geometric lattice M.
4. Show that, in a simple modular geometric lattice (projective geometry), all lines contain the same number of points. (*Hint.* They are perspective, by Theorem 11.)
5. Show that, if all lines of a modular geometric lattice contain the same number of points, then it is a projective geometry or a finite Boolean algebra.
6. Prove that every epimorphic image of a geometric lattice is a geometric lattice.

7. Show that a modular geometric lattice is a projective geometry, if and only if O, I are the only elements with unique complements.

8. (a) Show that the modular lattice of all finite-dimensional subspaces of an infinite-dimensional vector space is relatively complemented, but not complemented.

(b) Show that it also satisfies L7R′.

(c) Construct modular lattices satisfying L7R, but not L7R′, and L7R′ but not L7R. (Cf. §III.12, Example 3.)

9. Let $\mathbf{Z}_p$ be any prime field. Show that $PG_2(\mathbf{Z}_p)$ is a modular lattice with four generators.

8. Nonmodular Geometric Lattices

Many kinds of nonmodular geometric lattices arise in geometry. We will begin by considering in more detail Example 1 of §1: the geometric lattice $AG_n(D)$ of all *affine* subspaces of an n-dimensional linear space $V_n(D)$ over a division ring.

Such "affine geometries" were first studied by Menger.† In any $AG_n(D)$, for any $a > O$, the dual ideal $[a, I]$ is the projective geometry $PG_{n-1}(D)$. A second, related condition is the following. Given any element $a > O$ and any point $p \nleq a$, there exists a unique "parallel b to a through p", such that: $a \wedge b = O$, $p \leqq b$, and $d[b] = d[a]$, and $a \vee b = a \vee p$. Finally, for any $a > O$, the ideal $[O, a]$ is also an affine geometry, and each line contains at least two points. The converse of these results has been considered by Sasaki,‡ in two important papers. He has shown that one can characterize those geometric lattices which are "affine geometries" by the following *parallel postulate*

PP. Given independent points p, q, r, there is a unique line λ such that $p < \lambda < p \vee q \vee r$ and $\lambda \wedge (q \vee r) = O$.

Sasaki and Fujiwara have also generalized the author's decomposition theorem for modular geometric lattices to nonmodular geometric lattices, as follows.

Definition. In a geometric lattice G, two points p and r are *pseudo-perspective* when $p = r$ or, for some x with $q \wedge x = O$, $r \leqq p \vee x$.

Theorem of Sasaki–Fujiwara. *The relation of pseudo-perspectivity is an equivalence relation. The center of any geometric lattice consists of the joins e_i of the equivalence classes E_i of pseudo-perspective elements.*

The difference between "perspectivity" in the original sense of von Neumann, and "pseudo-perspectivity", is well illustrated by the geometric lattice of Figure 15. In this lattice, q is perspective to p and r, but p is not perspective to r—although it is pseudo-perspective to r.

Another interesting geometric lattice is provided by real n-dimensional *spherical* geometry. Here we deal with the geometric lattice of all those subsets of the real n-sphere (*not* n-ball) which either: (i) contain at most two points, or (ii) contain with any three points, the unique circle passing through them. The structure of

† K. Menger, F. Alt, and O. Schreiber [**1**].

‡ U. Sasaki and S. Fujiwara, J. Sci. Hiroshima Univ. Ser. A. **15** (1951), 183–8; U. Sasaki, ibid. **A16** (1952), 223–8 and 409–16. Sasaki speaks of "perspective" elements, instead of "pseudo-perspective" elements as we do.

this geometric lattice is harder to analyze;† it is the "polar" in $V_n(\mathbf{R})$ of the set of all functions of the form $f = a + \sum b_i x_i + c \sum x_i^2$ (cf. §3).

Still another kind of geometric lattice is obtained, for any field F, from the "polar" in the same sense of *all* quadratic functions $f = a + \sum b_i x_i + \sum c_{ij} x_i x_j$, where $\|c_{ij}\|$ is an arbitrary geometric lattice. For F real, classic theorems of Brianchon and Pascal describe some of the basic properties of this lattice.

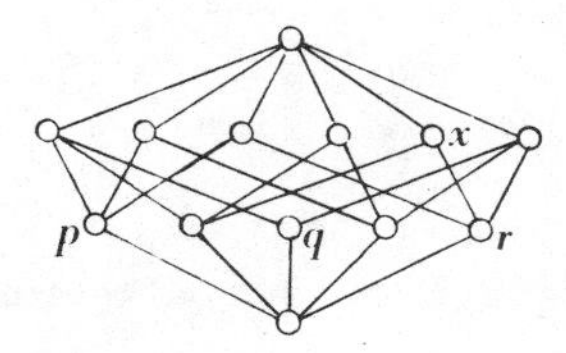

FIGURE 15

Such geometric lattices are characterized by a high degree of symmetry: the group of automorphisms of each is transitive on the elements of any given dimension height. The same is true of *hyperbolic* geometry, defined as the lattice of all intersections of flats with the interior of a given quadratic. However, of all the preceding geometries, only projective geometry (which can be also defined as the geometry of geodesics in *elliptic* space) defines a modular lattice.

9. Partition Lattices; Algebraically Closed Subfields

In recent years, the theory of *nonmodular* "geometric" lattices has undergone important developments. It has been increasingly apparent that many more geometric lattices arise in *combinatory analysis* than in geometry or abstract algebra, the two branches of mathematics (besides logic and set theory) to which lattice theory was first applied.

The first such lattice to be studied‡ was the symmetric partition lattice Π_n of all partitions of an aggregate of n objects, already considered in §I.8. Here the first result is

THEOREM 12. *The lattice Π_n of all partitions of any finite set of n objects is a geometric lattice of length n.*

PROOF. The "points" of Π_n are the equivalence relations which identify two elements. The covering relation (ξ) is easily verified, as is the fact that Π_n is a point-lattice.

THEOREM 13. *Let θ and θ' be permutable equivalence relations on a set S. Then $\theta M^* \theta'$: θ and θ' are a modular pair in the dual of $\Pi(S)$.*

† See A. J. Hoffman, Trans. AMS **71** (1951), 218–42; also W. Prenowitz, Canad. J. Math. **2** (1950), 100–19.

‡ G. Birkhoff, [**3**, §§16–22]; also Abstract 40-11-323, Bull. AMS **40** (1934), 797.

Proof. By definition, the problem is to show that if $\theta \leqq \theta''$, then $(\theta \vee \theta') \wedge \theta'' = \theta \vee (\theta \wedge \theta'')$. By the modular inequality (Chapter I, (6)), it suffices to show that $(\theta \vee \theta') \wedge \theta'' \leqq \theta \vee (\theta' \wedge \theta'')$; this we now do.

If $\theta\theta' = \theta'\theta$ and $\theta \leqq \theta''$, then $(\theta \vee \theta') \wedge \theta'' \leqq \theta \vee (\theta' \wedge \theta'')$. But if $\theta\theta' = \theta'\theta$, then $\theta \vee \theta' = \theta\theta' = \theta'\theta$; hence $x \equiv y\ (\theta \vee \theta')$ implies that t exists such that $x \equiv t\ (\theta)$ and $t \equiv y\ (\theta')$. If also $\theta \leqq \theta''$ and $x \equiv y\ (\theta'')$, then $x \equiv t\ (\theta'')$ and so (since θ'' is an equivalence relation) $t \equiv y\ (\theta' \wedge \theta'')$. In conclusion, since $x \equiv t\ (\theta)$ and $t \equiv y\ (\theta' \wedge \theta'')$, we obtain $x \equiv y\ (\theta(\theta' \wedge \theta''))$. That is, if $\theta\theta' = \theta'\theta$ and $\theta \leqq \theta''$, then

$$(\theta \vee \theta') \wedge \theta'' \leqq \theta(\theta \wedge \theta'') \leqq \theta \vee (\theta \wedge \theta''),$$

which completes the proof.†

Corollary 1. *Any sublattice of permutable equivalence relations is a modular sublattice.*

Corollary 2. *Let θ, θ' be any two permutable equivalence relations on a set S. Then*

$$(14) \qquad \theta \wedge \theta' = O \quad \textit{and} \quad \theta \vee \theta' = I \quad \textit{imply} \quad S = (S/\theta) \times (S/\theta').$$

Proof. Since $\theta \wedge \theta' = O$, each intersection $X_i \cap Y_j$ of a π-subset X_i and a π'-subset Y_j contains at most one element: $X_i \cap Y_j = z_{ij}$ or is empty. But since $\theta\theta' = \theta \vee \theta' = I$, for any $x \in X_i$ and $y \in Y_j$, there exists $s \in S$ such that $x\theta s$ and $s\theta' y$; this $s \in X_i \cap Y_j$; hence $X_i \cap Y_j$ contains exactly one $s = z_{ij} \in S$. Finally, $z_{ij}\theta z_{kl}$ if and only if $j = l$, completing the proof.

A kind of geometric lattice very different from Π_n arises in algebraic number theory. We have

Theorem 14. *Let $I = F(t_1, \cdots, t_r)$ be any finite transcendental extension of a field F. Then the algebraically closed‡ fields S, $F \subset S \subset I$, form a geometric lattice of length r.*

Proof. Let S be any algebraically closed subfield of the field I, and let H and K cover S. Choose x in H, y in K, neither in S. Then H is the set of numbers in I algebraic over the ring $\{S, x\}$, K of those algebraic over $\{S, y\}$, and $H \vee K$ of those algebraic over $\{S, x, y\}$. Now if z is in $H \vee K$, then some polynomial $p(x, y, z) = 0$; if z is not in H then this polynomial must involve positive powers of y; hence y is algebraic over $\{S, x, z\}$. It follows that if $H < Z \leqq H \vee K$, then Z contains y, and so $Z = H \vee K$, which proves (ξ) and the Steinitz-Mac Lane "exchange axiom" (5). Finally, the lattice satisfies L7R′: if $G > F$, choose a maximal set of numbers x_α of G which are algebraically independent over F; then the algebraic closures of the $\{F, x_\alpha\}$ will be points whose join is G.

We can define the transcendence degree of G over F as the lattice-theoretic dimension of G over F. Then the above theorems on dependence, rank, etc.,

† For interesting characterizations of Π_n, see O. Ore, Duke Math. J. **6** (1942), 573–627, esp. Chapter IV; D. Sachs, Pacific J. Math. **11** (1961), 325–45.

‡ By "algebraically closed" we mean that if $x \in K$ and if $p(x) = 0$, where p is a polynomial with coefficients in S, then $x \in S$. Theorem 14 is due to S. Mac Lane [1].

include as corollaries most of Steinitz' theory of algebraic dependence; for example, the number of x_α is $d[G/F]$.

Both Π_n and the geometric lattices described in Theorem 14 are *symmetric*, in the sense that the group of their automorphisms is transitive on points (atoms). In Π_n, all dual ideals $[\pi, I]$ of length l are also isomorphic to Π_l, and hence with each other. In Theorem 14, moreover, the group of automorphisms is even transitive on elements of *any* given height.

The main differences between Π_n and the lattices of Theorem 14 relate to the *finiteness* of Π_n. In Π_n, *each "line" contains at most three points.* Specifically, if $l[\pi_1, \pi_2] = h[\pi_2] - h[\pi_1] = 2$, then the interval $[\pi_1, \pi_2]$ contains three elements of height $h[\pi_1] + 1$ if π_2 is obtained from π_1 by uniting three π_1-subsets, and 2 such elements if π_2 is obtained by uniting two pairs of π_1-subsets.

An unsolved problem. A very interesting combinatorial problem concerns the question of whether every *finite* lattice is isomorphic with a sublattice of some *finite* Π_n. (A deep result of Whitman† asserts that *any* lattice is isomorphic with a sublattice of some infinite Π_∞.) An important related result is

DILWORTH'S EMBEDDING THEOREM.‡ *Every finite lattice is isomorphic with some sublattice of a finite semimodular lattice.*

The proof begins by constructing, for *any* lattice L of finite length, an integer-valued *pseudo-rank* $r[a]$ satisfying $r[O] = 0$, $a > b$ implies $r[a] > r[b]$, and (5).

Exercises for §§8–9:

1. Show that unless $F = \mathbf{Z}_2$ every "line" in $PG_n(F)$ contains at least four points.
2. Show that if p is any point of $AG_n(F)$, the $x \geqq p$ form a modular lattice isomorphic with $PG_{n-1}(F)$. (Menger)
3. (a) Show that the portions of affine subspaces contained in any open convex subset S of $\mathbf{R}^n$ form a geometric lattice. (This gives real n-dimensional *hyperbolic* geometry, if S is an n-sphere.)
   (*b) Generalize to other ordered fields, so far as possible.
4. Show that the lattice of subalgebras of the Boolean algebra $\mathbf{2}^3$ is modular, whereas that of $\mathbf{2}^4$ is not.
5. (a) Show that Π_n is modular if and only if $n \leqq 3$.
   (b) Show that the number $\pi(n)$ of partitions of n objects satisfies the recursion formula

$$\pi(n+1) = \sum_{i=0}^{n} \binom{n}{i} \pi(i).$$

   (c) Show that $(n!)\pi(n)$ has the generating function $e^{e^x - 1}$.
   (d) Compute $\pi(n)$ for $n = 1, \cdots, 10$. ($\pi(10) = 115{,}975$.)
6. Show that there is a join-morphism which maps the geometric lattice $\mathbf{2}^2$ onto the chain $\mathbf{3}$.
7. Show that any interval sublattice or cardinal product of geometric lattices is a geometric lattice.
8. (a) Show that Π_n is isomorphic to a sublattice of the subgroup-lattice of the symmetric group of degree n. (*Hint.* Associate with π the permutations having the π-subsets as imprimitive sets.)

† P. Whitman, Bull. AMS **52** (1946), 507–22. The proof makes extensive use of the theory of free lattices (Chapter VI). Cf. J. Hartmanis, [**Symp**, pp. 22–30] for more recent results.

‡ [**LT2**, p. 105]. For the proof, see D. T. Finkbeiner, Canad. J. Math. **12** (1960), 582–91.

(b) Extend to the infinite case. (H. Löwig) (*Hint.* Do the same, but take only permutations leaving all but a finite number of objects fixed.)

9. Show that Π_n has no proper congruence relations (is "simple"), for finite n.

10. (a) Show that, in Π_n, any dual ideal $[a, I]$ of length k is isomorphic with Π_k.

(b) Show that, in Π_n, any *interval* $[a, b]$ of length k is isomorphic with a product of $s = d[I] + 1 - d[b]$ symmetric partition lattices $\Pi_{m(i)}$, where $m(i) = d[I] - d[a]$.

11. Show that $\theta M^*\theta'$ in Π_n if and only if $\theta \vee \theta' = \theta\theta'\theta$. (A. Waterman)

12. Show that, in real spherical geometry, for any point p, the dual ideal $[p, I]$ forms a real affine geometry. What about the dual ideal $[p \vee q, I]$, if $p \neq q$ are two points?

13. (a) Show that the k-spheres and k-flats $[k = 0, 1, \cdots, n - 1]$ in Euclidean n-space I form, together with I and the void set O, a geometric lattice L_s of length $n + 2$. (*Note.* A 0-sphere is a point pair.)

(b) Show that this is also true in spherical n-space and elliptic n-space, giving lattices L_s' and L_s''.

(c) Show that every "inversion" generates a lattice-automorphism of L_s'.

14. (a) Using Ptolemaic projection, show that (for given n), L_s is isomorphic to a sublattice of L_s' obtained by deleting a single point.

(b) Show further that, if p is any point of L_s', the dual ideal of $x \geqq p$ in L_s is isomorphic with $AG_{n-1}(\mathbf{R})$ where $\mathbf{R}$ is the real field.

(c) State a corresponding result about L_s.

10. Graphs; Width and Breadth

Almost by definition, a *graph* is a graded point-lattice of length 3 in which each "line" (*link*) contains exactly two points (*nodes*).† It would be impractical to enter here in any detail into the vast subject of graphs.‡ But since any finite lattice has a *directed* graph, and is determined by this, we will mention two connections between the theory of graphs and the theory of geometric (or "matroid") lattices, having important applications to lattice theory. In the first place, we mention

KÖNIG'S THEOREM.§ *Let G be a graph, and let π partition the set P of "points" of G into complementary subsets S and T. Define an (S, T)-cut as a minimal subset $C \subset P$ such that every link joining a point of S to a point of T has one end in C. Define an (S, T)-join as a set J of links joining S to T, without a common end point. The* max $|J| =$ min $|C|$, *where the absolute value signs refer to numbers of elements.*

Width and breadth. Now define the *width* $w[P]$ of a poset P as the maximum number of elements in a set of incomparable elements. Evidently, P cannot be represented as the set-union of fewer than $w[P]$ chains. The following converse to this observation, first established by Dilworth¶, is a fairly direct corollary of König's Theorem.

† In graph theory, as in treating topological complexes of higher dimension generally, the void set and I are usually excluded.

‡ For a study of self-dual graphs having a high degree of symmetry, see H. S. M. Coxeter, Bull. AMS **56** (1950), 413–55.

§ J. König, *Theorie der Graphen*, Akad. Verlag, Leipzig, 1936, p. 232.

¶ R. P. Dilworth, Ann. of Math. **51** (1950), 161–6. The connection with König's Theorem was pointed out by D. R. Fulkerson, Proc. AMS **7** (1956), 701–2.

THEOREM 15. *If $w[P]$ is finite, then P can be expressed as the set-union of $w[P]$ chains.*

To determine $w[P]$ is not always easy to practice. However, for the case $P = \mathbf{2}^n$ we have

SPERNER'S LEMMA. *The number $w[\mathbf{2}^n]$ is equal to the number $\binom{n}{k}$ of subsets containing $k = [n/2]$ elements in a set of n elements.*†

(It is obvious that any two such subsets are incomparable; hence that

$$w[\mathbf{2}^n] \leqq \binom{n}{k}.)$$

Closely related to $w[P]$ is the notion of the *dimension* $\delta[P]$ of a poset, not to be confused with its *length*. This has been defined‡ as the smallest number m of *chains* $P_1, \cdots, P_m$, such that P is isomorphic with a subset of the product $P_1 P_2 \cdots P_m$.

The preceding numbers are related to the concept of the *breadth* of a lattice L, already defined (§II.5, Exercise 6) as the least positive integer $b = b[L]$ such that any meet $x_1 \wedge \cdots \wedge x_n$ $[n > b]$ is always a meet of a subset of b of the x_i. It is easy to show that $b[L] \leqq \delta[L]$ in any lattice. In a finite distributive lattice $L = \mathbf{2}^P$, one easily verifies that

$$b[\mathbf{2}^P] = w(P). \tag{15}$$

COROLLARY. *The breadth of $\mathbf{2}^{\mathbf{2}^n}$, the free distributive lattice with n generators, is $\binom{n}{k}$ where $k = [n/2]$.*

11*. Polyhedral Complexes

A very important further generalization of the concept of a geometric lattice is provided by the simplicial and polyhedral *complexes* studied in combinatorial topology. These are finite *graded posets*, whose minimal elements are called "points" or "0-cells". It is required that:

(i) Every ideal is a "point-semilattice"—i.e., every set having an upper bound has a l.u.b. which is a join of points,

(ii) Every "line" (or "1-cell") contains exactly two points. A *graph* is thus simply a complex of height one.

Figures 16a–16c depict the complexes corresponding to a segment, a triangle, and a square, respectively. An *n-simplex* is a complex consisting of $\mathbf{2}^{n+1}$ with 0

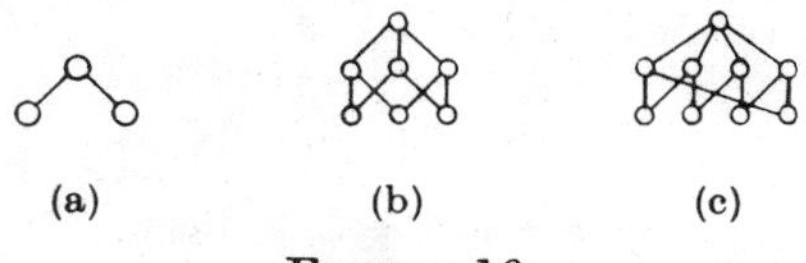

(a) (b) (c)

FIGURE 16

† A. Sperner, Hamb. Abh. **7** (1930), 149–63. See also A. Learner, MR **21** 4930; P. Erdös and R. Rado, Quart. J. Math. **12** (1961), 313–20.

‡ B. Dushnik and E. W. Miller, Amer. J. Math. **63** (1941), 600–10.

deleted; Figures 16a–16b represent n-simplexes with $n = 1, 2$. A complex whose ideals are all simplexes is called a *simplicial* complex, for obvious geometric reasons.

Surprisingly, no simple characterization seems to be known of those complexes (semilattices) which represent solid n-dimensional *polyhedra* for $n > 3$; to find such a characterization is a basic problem proposed by van Kampen.† However, if we define the *boundary* ∂a_i of a k-cell a_i as the set of all $(k - 1)$-cells *covered* by a_i, and the boundary ∂R of a set R of k-cells $a_1, \cdots, a_m$, as the Boolean sum (in the sense of Boolean rings, mod 2) $\sum_R \partial a_i$, then a classic result is

$$\partial(\partial R) = \varnothing \quad \text{for all } R. \tag{17}$$

We therefore postulate that

(iii) Condition (17) must hold in any complex.

Interesting abstract theories of combinatorial topology, based on such axioms, have been proposed by W. Mayer, P. Alexandroff, and others.‡ One can also impose the self-dual Orientability Condition:

(iv) One can assign a value ± 1 to each covering relation so that, if ∂a_i has values $\pm 1 \in \mathbf{Z}$, (17) still holds.

We will confine our attention here to a few obvious facts about the preceding concepts.

First, the cardinal *sum* of two complexes C_1 and C_2 represents the topological sum of the corresponding manifolds. Likewise, the cardinal *product* C_1C_2 represents the Cartesian product $M(C_1) \times M(C_2)$ of the corresponding manifolds. (Thus, Figure 16c represents $S_1{}^2$, where S_1 is the simplex of Figure 16a.) Finally, the *dual* of a subdivision of a manifold without boundary is simply its lattice dual.

In a polyhedral complex C, a set S of h-cells with $\partial S = 0$ is called an *h-cycle*. The hth *Betti number* β_h of a complex C is the maximum number of (linearly) independent *h-cycles* of C, minus the maximum number of independent§ *boundary* h-cycles $B = \partial A$, where A is an arbitrary set of $(h + 1)$-cycles. Conversely, any geometric lattice has its Euler–Poincaré characteristic and homology groups.¶

Exercises for §§10–11:

In Exercises 1–8 below, P denotes an arbitrary poset; most of the exercises are due to Kirby Baker.

1. Show that $b(P) = 1$, $\delta(P) = 1$, and P a chain are equivalent.
2. Show that, if P is finite and has universal bounds, then it has a plane diagram if and only if it is a lattice and of dimension two or less.
3. Show that $b[L] = b[\breve{L}]$. (*Hint.* See Theorem II.13.)
4. Construct a nine-element lattice L with $b[L] = 2$, $\delta[L] = 3$.

---

† E. R. van Kampen, *Die Kombinatorische Topologie*, Hague, 1929.

‡ See W. Mayer, Monatsh. Math. Phys. **36** (1929), 1–42 and 219–58; P. Alexandroff, Mat. Sb. **44** (1937), 501–19.

§ Above, consider the Boolean algebra $\mathbf{2}^{n(h)}$ of h-chains as a Boolean ring $V_{n(h)}(\mathbf{Z}_2)$, for "linear dependence".

¶ J. Folkman, J. Math. Mech. **15** (1966), 631–6.

5. Show that $b[L] \leqq \delta[L]$ for any lattice L.

6. Show that $b[L] = \delta[L]$ in any finite distributive lattice. (*Hint.* Use Dilworth's Theorem.)

*7. (a) Show that if P and Q are posets with universal bounds, then

$$\delta[PQ] = \delta[P] + \delta[Q], \qquad b[PQ] = b[P] + b[Q].$$

(b) Generalize to arbitrary cardinal products.

*8. Show that, if P is finite and $o[P] \geqq 3$, then $\delta[P] \leqq o[P]/2$. (Hiraguchi)

9. Show that any complex is a lattice if O is adjoined.

10. Draw the diagram representing a triangular prism (product of a 1-simplex and 2-simplex).

11. Draw the dual of the diagram of the complex of a cube, and explain how (with I adjoined and O deleted) it represents an octahedron.

12. (a) Characterize the abstract configurations which represent polygonal subdivisions of a 1-sphere (circle).

(b) Same for polyhedral subdivisions of the 2-sphere. (*Hint.* Each edge has two vertices, and the faces meeting at a point are cyclically ordered.)

*13. (a) Show that π and π' are a modular pair if and only if the topological graph of π and π' (with π-subsets A_i and π'-subsets A_j as vertices, and an edge joining A_i and A_j if and only if $A_i \wedge A'_j > 0$) is without cycles.†

(b) Infer that Wilcox's definition of semimodularity is satisfied, even for partitions of infinite sets.

*14. (a) Show that the partitions of any measure space into a finite number of disjoint measurable subsets form a semimodular lattice.

(b) In what sense do the partitions into countable disjoint measurable subsets form a semimodular lattice? Suppose sets of measure zero are ignored?

15. (a) Call a complex "h-symmetric" when its automorphism group is transitive on h-cells. Show that this implies the condition (E. H. Moore) that the number (h, k) of h-cells incident with each k-cell depends only on h and k.

(b) Show that this holds in an n-simplex, with

$$(h, k) = \binom{h+1}{k+1} \quad \text{if } h > k.$$

12. Moebius Functions

Let P be any poset. Define the *incidence* function $n(x, y)$ of P by the formula (classic for complexes):

$$n(x, y) = \begin{cases} 1 & \text{if } x < y \\ 0 & \text{otherwise.} \end{cases} \tag{17}$$

Thus $n(x, y)$ is the characteristic function of the relation $<$, and $\xi(x, y) = \delta_{xy} + n(x, y)$ that of the inclusion relation $\leqq$ on P.

If P is *finite*, then its natural order can be strengthened to a simple order, which lists its elements in a finite sequence $x_1, \cdots, x_n$ so that $x_i < x_j$ implies $i < j$. For any such "compatible" simple ordering of P, the numbers $n_{ij} = n(x_i, x_j)$ then form a strictly triangular matrix. $N = \|n_{ij}\|$, the *incidence* matrix of P (relative to the compatible ordering).

Hence $I + N$ (the zeta matrix) is invertible. The *inverse* $(I + N)^{-1}$ is a matrix which has many important applications to *inversion formulas* of probability,

† O. Ore, op. cit., p. 583; also P. Dubreil and M.-L. Dubreil-Jacotin, J. Pures Appl. Math. **18** (1939), 63–96.

number theory, combinatory analysis, etc. For a full account of these, we refer to the basic paper of G.-C. Rota [**1**];† only a very brief sketch of results of lattice-theoretic interest will be given here.

DEFINITION. The *Moebius function*‡ $\mu(x, y)$ of a finite poset is defined by: (i) $\mu(x, x) = 1$, (ii) $\mu(x, y) = 0$ unless $x \leqq y$, and

$$(18) \qquad \mu(x, y) = -\sum_{x \leqq z < y} \mu(x, z) \qquad \text{if } x < y.$$

LEMMA 1 (ROTA). *The Moebius function satisfies*

$$(19) \qquad \|\mu(x, y)\| = \|I + N\|^{-1} = I - N + N^2 - N^3 + \cdots \pm N^{l[p]}.$$

The proof by induction is straightforward. As a virtual corollary, we obtain the following theorem by Philip Hall (op. cit):

HALL'S FIRST THEOREM. *If $\lambda(x, y; n)$ denotes the number of chains of length n which can be interpolated between x and y, then $\mu(x, y) = 1 + \sum_{k=1}^{n} (-)^k \lambda(x, y; k)$.*

LEMMA 2. *Let $\theta: x \to x^*$ be a dual isomorphism $P \to P^*$; then the Moebius function μ^* in P^* satisfies*

$$(20) \qquad \mu^*(x, y) = \mu(x^*, y^*).$$

We omit the proof, and also that of another important theorem of Philip Hall, and one of Rota (op. cit.).

HALL'S SECOND THEOREM. *If P is a lattice, then $\mu(x, y) = 0$ unless y is the join of elements covering x.*

ROTA'S THEOREM. *In a finite geometric lattice, let $x \leqq y$. Then $\mu(x, y) \neq 0$; moreover $\mu(x, y)$ is positive if $h[y] - h[x]$ is even, and negative if $h[y] - h[x]$ is odd.*

We next define the *characteristic polynomial* of a finite *graded* poset (§I.3) as

$$(21) \qquad p_x(\lambda) = \lambda^{h(x)+1} - \sum_{y < x} p_y(\lambda).$$

This is related to the Moebius function by the formula

$$(21') \qquad p_x(\lambda) = \sum_{y \leqq x} \mu(x, y)\lambda^{h(y)+1}.$$

This formula yields as a special case the number of ways of coloring a graph (map) Ω in λ colors.

Consider the poset P in which a "submap" of Ω is formed by obliterating boundaries, then there are λ^n ways of coloring n regions in λ colors, each of which colors either Ω or a unique submap of Ω so that no two adjacent regions have the

---

† The author wishes to acknowledge many stimulating conversations with Professor Rota.

‡ A. F. Möbius, J. Reine Angew. Math. **9** (1832), 105–23 (cf. Example 1 below). The generalization to posets was due to L. Weisner [**1**] and P. Hall, Quart. J. Math. **7** (1936), 134–51; see also [**LT1**, §§39–40].

same color. Hence $p_\Omega(\lambda)$ is the number of ways of coloring Ω in λ colors, so that no two regions have the same color.† A related application to flows in networks has been given by Tutte.

The Moebius function has been applied by Dilworth‡ to prove the following remarkable result.

DILWORTH'S THEOREM. *In a finite modular lattice M, let V_k be the set of elements covered by precisely k elements, and let W_k be the set of elements covering precisely k elements. Then V_k and W_k have the same cardinal number.*

Setting $k = 1$, we get the

COROLLARY. *In any finite modular lattice, the number of meet-irreducibles equals the number of join-irreducibles.*

(In a finite distributive lattice $L = \mathbf{2}^P$, both numbers are just $o(P) = l[L]$, the length of L by Theorem III.3.)

Exercises:

1. Let L be the distributive lattice of positive integers under $m|n$, and let $\mu[n] = \mu(1, n)$.
   (a) Show that $\mu[n] = 0$ unless n is square-free, while $\mu[n] = (-1)^s$ if n is the product of s distinct prime factors.
   (b) Show that $\mu[mn] = \mu[m]\mu[n]$, if $(m, n) = 1$.
   (c) Show that, if $n = p_1 \cdots p_s$ (p_i distinct primes), then $p_n[\lambda] = \lambda(\lambda - 1)^s$.
2. (a) Prove Lemma 1.
   (b) Prove Lemma 2.
3. Prove that $\mu(x, y)$ has the same meaning on $[x, y]$ as on P, for any $x \leqq y$ in a poset P.
4. Show that on $P = \mathbf{2}^n$, $\mu(x, y) = (-1)^{h(y)-h(x)}$ if $x \leqq y$.
5. Let $P = PG_n(F)$, where F is a Galois field with $q = p^r$ elements. Show that for $x \leqq y$:

$$(*) \qquad \mu(x, y) = (-)^h q^{\binom{h}{2}}, \quad \text{where } h = h(y) - h(x).$$

6: (a) Let μ_P denote $\mu(O, I)$ in P. Show that $\mu_{PQ} = \mu_P\mu_Q$.
   (b) Let μ_1, μ_2, μ denote the Moebius functions on finite posets P, Q and PQ, respectively. Prove that $\mu((x, y), (u, v)) = \mu_1(x, u)\mu_2(y, v)$. (Rota)
7. Let $\mu_n = \mu(O, I)$ for the symmetric partition lattice Π_n.
   (a) Show that $\mu_n = (-1)^{n-1}(n - 1)!$
   (b) Compute $\mu(x, y)$ in Π_n. (*Hint.* Use Exercise 6, and Exercise 10b after §9.)
8. Show that the "submaps" of a finite map are a dual geometric lattice.
9. Show that, in any finite distributive lattice,

$$\mu(O, x \wedge y)\mu(O, x \vee y) = \mu(O, x)\mu(O, y).$$

10. Show that the "contractions" of any (linear) graph form a geometric lattice. (A "contraction" consists of a sequence of replacements of edges by vertices.)
11. What is the number of ways of coloring the map of n triangles in λ colors, whose sides are the edges of a regular n-gon and radii to its center?

---

† G. D. Birkhoff, Ann. of Math. **14** (1913), 42–6; G. D. Birkhoff and D. C. Lewis, Trans. AMS **60** (1946), 355–451, and references given there; W. T. Tutte, Canad. J. Math. **6** (1953), 80–91. The connection with Moebius functions was first noted by the author.

‡ R. P. Dilworth, Ann. of Math. **60** (1954), 359–64; see also S. P. Avann, Proc. AMS **11** (1960), 9–16.

12. Any selection of λ elements of a group G of finite order g generates either G or a subgroup S of G. Infer that the number of ways of generating G by λ elements is $P_G[\lambda]$, determined recursively by $P_G[\lambda] = \lambda^g - \sum_{S<G} p_S[\lambda]$.

13. Show that the partitions of the *number n* (§I.8, Example 10) form a modular poset, in which $\mu(0, x) = (-1)^d$ for the two elements of height $d = 0$, 1 and 0 for all others. (D. B. Wales)

13*. Projectivities and Collineations

The study of the group Aut G of all automorphisms ("symmetries") of a geometric lattice is a fascinating one, with a long history. Consider first the case $G = PG_n(D)$ of an n-dimensional projective geometry over a division ring D. The case $n = 1$ is trivial: Aut G is the *symmetric* group of all permutations of the points of G.

For any n, the group Proj G of all *projectivities*

$$w_i = \frac{z_1 a_{1i} + \cdots + z_n a_{ni} + a_{n+1,i}}{z_1 b_1 + \cdots + z_n b_n + b_{n+1}} \qquad (b_j = a_{j,n+1}, \quad i = 1, \cdots, n), \tag{22}$$

where $\|a_{ij}\|$ is a nonsingular $(n + 1) \times (n + 1)$ matrix, is certainly a *subgroup* of Aut G, provided matrices which differ by a scalar factor are identified. For $n = 1$, this subgroup is exactly *triply transitive*; for $n > 1$, it is doubly transitive, and "almost" triply transitive (it carries any three collinear *or* noncollinear points into any three others with the same property). If D is the real field $\mathbf{R}$, then for $n > 1$, Proj G = Proj $PG_n(\mathbf{R})$ equals Aut G, which thus has the same properties.

In general (for $n > 1$) Aut G is the *collineation group*; for any automorphism ϕ of D,

$$w_i = \frac{z_1^\phi a_{1i} + \cdots + z_n^\phi a_{ni} + a_{n+1,i}}{z_1^\phi b_1 + \cdots + z_n^\phi b_n + b_{n+1}} \tag{23}$$

is also a collineation, *and conversely*. Thus Aut G is the semidirect product of Proj G with the Galois group of D (over its prime subfield). Cf. also **[LT2]**, Chapter VIII, §7.

Finally, it is classic that any projective plane which satisfies Desargues' Theorem† (is "Desarguesian") is a $PG_2(D)$; this result will be discussed in §14. Moreover, for $h > 2$, any projective geometry of length $h + 1$ is Desarguesian, and a $PG_h(D)$. However, there exist *non-Desarguesian projective planes*.

The preceding results reduce the study of *modular* geometric lattices and their automorphisms to (i) the study of division rings and their Galois groups, and (ii) the study of non-Desarguesian projective planes. Moreover in the *finite* case, which is most interesting from the combinatorial standpoints, all division rings *and* their automorphism groups are easily enumerated: there is just one‡ (a "Galois field" with cyclic automorphism group of order r) of each prime-power order p^r. Hence,

† This theorem states that two triangles which are perspective from a point are perspective from a line, and conversely.

‡ Every finite division ring is commutative; for the Galois groups of finite fields, see **[B-M]**, Chapter XV, §6.

at least so far as *finite* modular geometric lattices are concerned, the main mystery concerns non-Desarguesian projective planes.

The theory of finite non-Desarguesian projective planes has received enormous attention in the mathematical literature.† A remarkable Theorem of Ostrom and Wagner, for example, states that a finite projective plane P is Desarguesian if and only if its collineation group Aut P is doubly transitive on its points. Other theorems concern the integers n for which there exist projective planes with $n + 1$ points on each line (hence $n^2 + n + 1$ points in all). These fail to exist for $n = 6$ (M. G. Tarry) and more generally for any $n \equiv 1$ or 2 (mod 4) which is not the sum of two squares (Bruck–Ryser Theorem).

14*. Coordinatization Problem

We now turn to the coordinatization problem: the problem of representing a given geometric lattice by "closed" subsets of an algebraic system—i.e., the subspaces or "flats" of vector space or affine space. The possibility of doing this in projective geometries was recognized over a century ago by von Staudt,‡ who invented an "algebra of throws". Technically, as was first pointed out by E. Artin and M. Hall,§ this problem is easier in *affine* planes (in which parallelism is defined) than in projective planes.

In principle, its solution goes back to the "geometric algebra" of the Greeks, who used configurations as substitutes for algebraic identities (presumably because they did not have a good algebraic notation). First fix (arbitrarily) an origin (0, 0), and three sets of parallel lines which may be thought of as the lines $x = a$, $y = b$, and $x + y = c$ for arbitrary constants a, b, c. The x-axis and y-axis are the lines $y = 0$ and $x = 0$ through the "origin"; one must also choose a *unit* 1, by fixing (1, 0) on the x-axis; the line $x + y = 1$ will pass through this point, and intersect $x = 0$ in (0, 1). Each point in the plane will then have coordinates (a, b).

To construct $a + b$, draw a parallel to the line through (0, 0) and $(a, 1)$ through $(b, 0)$; this will intersect $(x, 1)$ in $(a + b, 1)$. To construct ab, draw through $(0, b)$ a parallel to the line through (0, 1) and $(a, 0)$; it will intersect the x-axis in $(ab, 0)$. The commutativity of multiplication is then equivalent to the validity of Pappus' Theorem: if the six vertices of a hexagon lie alternately on two lines, then the three points of intersection of pairs of opposite sides are collinear. The other identities which define fields (i.e., those which define division rings) are equivalent to the validity of other geometric theorems ("configurations").¶

† For an excellent, now standard exposition, see M. Hall, *The Theory of Groups*, Macmillan, 1959, Chapter XX. For the theorem of Ostrom and Wagner, see Math. Z. **71** (1959), 186–99. See also K. D. Fryer, [**Symp**, pp. 71–8].

‡ For historical remarks on the subject see O. Veblen and J. W. Young, *Projective Geometry*, Ginn, 1910, vol. 1, p. 141 and §55.

§ E. Artin, *Coordinates in affine geometry*, Reps. Math. Colloq. Notre Dame, 1940; *Geometric Algebra*, Interscience, 1957, Chapter II; M. Hall, Trans. AMS **54** (1943), 229–77, §5.

¶ For a classic discussion, see O. Veblen and J. W. Young, *Projective Geometry*, Boston, 1910, 2 vols.; for modern ones, see G. Pickert, *Projektive Ebene*, Springer, 1955; R. Baer, *Linear algebra and projective geometry*, Chapter VII; or the books of Artin and Hall cited above.

Conversely, let there be given any algebra H with a single ternary operation $a + sx$ which satisfies:

(i) $a + sx = a' + sx$ implies $a = a'$,

(ii) if $s \neq s'$, then $a + sx = a' + s'x$ for just one x,

(iii) given x, y, $x \neq x'$, and y', the simultaneous equations $a+sx=y$, $a+sx'=y'$ have just one solution a, s.

We shall call such an algebra a *planar ternary ring*.†

Given a planar ternary ring A, we define "points" to be number-pairs $(x, y) \in A^2$, and "lines" to be the sets defined by equations of the form $x = c$ and $y = a + sx$, for any given c, s, a in A. Then the following axioms for an *affine plane* are satisfied:

AP1. Any two distinct points p, q are on one and only one line $p \vee q$.

AP2. Given a line $p \vee q$ and $r \nleq p \vee q$, there exists just one line $r \vee s$ such that $(p \vee q) \wedge (r \vee s) = \varnothing$.

AP3. There are three points not on a line.

We omit the (almost trivial) verifications. There results

THEOREM 16. *One can coordinatize any affine plane by a planar ternary ring, and every planar ternary ring defines an affine plane.*

However, the same affine plane can give rise to many planar ternary rings depending on the selection of origin and sets ("pencils") of parallel lines. In the Desarguesian case, these are all isomorphic, but not generally.

Coordinates in rings. An ingenious and very different "coordinatization" for modular geometric lattices which are products of projective geometries was proposed by von Neumann [**CG**, Part II]. The basic idea stems from the analogy between the fundamental Theorem 10 about modular geometric lattices.

Namely, any such "Desarguesian" geometric lattice G is the product $\mathbf{2}^r \prod_{i=1}^{s} P_i$ of a Boolean algebra $\mathbf{2}^r$ and Desarguesian projective geometries P_i. Each $P_i = PG_{n(i)}(D_i)$ is isomorphic with the lattice of all right-ideals in the "full matrix ring" $M_i = M_{n(i)}(D_i)$ of all $n(i) \times n(i)$ matrices over a division ring D_i, while $\mathbf{2}$ is isomorphic with the lattice of all right-ideals of any field F. Hence G is isomorphic with the lattice of all right-ideals of the (semi-simple) direct sum $\bigoplus_{i=1}^{s} M_i \oplus F^r$ of the s rings M_i, and r copies of F. This proves

THEOREM 17 (VON NEUMANN). *Any Desarguesian geometric lattice G is isomorphic with the lattice of all right-ideals of a suitable semi-simple ring.*

Note that, though F is arbitrary, the set of D_i and exponents $n(i)$ in the above representation are unique. Conversely, if R is any semi-simple ring, and if the lattice L of right-ideals of R has finite length, then L is a modular geometric lattice.

† Planar ternary rings are related to loops. See D. R. Hughes, Proc. AMS **6** (1955), 973–80. As in [**LT2**, p. 111], we do not assume a 0 or a 1; cf. J. R. Wesson, Amer. Math. Monthly **73** (1966), 36–40.

Binary geometric lattices. Another, very original approach to the coordinatization problem has been made by Tutte,† who interprets geometric lattices as "matroids" following the original approach of Whitney. Call a geometric lattice G *binary* when no interval sub-lattice of length two in G contains more than three points; the partition lattices Π_n are "binary". Tutte shows that the matroid of all "polygons" (simple circuits) of any graph Γ is also binary and "regular", and that this regularity is equivalent to the exclusion of two specific finite configurations ("Fano" and "heptahedral"). Such regular binary matroids can be (uniquely) coordinatized by means of "chain-groups" over $\mathbf{Z}_2$.

15. Orthocomplementation in $PG_{n-1}(D)$

It is relatively easy to describe the operations of orthocomplementation which can be constructed in *Desarguesian* modular geometric lattices; we now turn our attention to this question.‡ By the results of §14, since the center is a Boolean algebra which has precisely one orthocomplementation, it suffices to describe the possible orthocomplementation operations in an arbitrary "concrete" projective geometry $PG_{n-1}(D)$—i.e., the lattice of all *subspaces* of the n-dimensional right vector space $V_n(D) = D^n$ over an arbitrary division ring.

The *dual* space V_n^* consists of all *left*-linear functionals $f(x)$ on D^n, satisfying

$$f(x+y) = f(x) + f(y) \quad \text{and} \quad f(x\lambda) = f(x)\lambda \quad \text{for all } \lambda \in D. \tag{24}$$

These form a *left* n-dimensional vector space over D—an isomorphic copy of $D^{\#n}$, where $D^{\#}$ is formed from D by *inverting* the order of multiplication. Moreover, by the theory of simultaneous linear equations over division rings, for any k-dimensional subspace $S_k \subset V_n$, the set of all $f \in V_n^*$ such that $f(x) = 0$ for all $x \in S_k$ is an $(n-1)$-dimensional subspace $T_{n-k} = S_k^*$ of V_n^*. This proves the following

Duality Theorem for Linear Equations. *The mapping $\delta: S_k \to S_k^*$ is a dual isomorphism from $PG_{n-1}(D)$ onto $PG_{n-1}(D^{\#})$; if $\delta^*: T_k \to T_k^{\dagger}$ is the analogous correspondence from $PG_{n-1}(D^{\#})$ to $PG_{n-1}(D)$, then $\delta\delta^*$ and $\delta^*\delta$ are the identity mappings of their respective domains* (*i.e., mutually inverse*).

It follows trivially that, if $P = PG_{n-1}(D)$, then PP^* can always be made into an ortholattice. However, to make P itself into an ortholattice (i.e., to construct an orthocomplementation on P) is less easy, and not possible unless P is *self-dual*!

If P is self-dual, then the composite $\phi\delta^*$ of *any* dual automorphism of P with the canonical duality δ^* defined above is an *automorphism* of P, and hence (for $n > 2$) a *collineation*. That is, it is defined by a transformation of the form (23) of P, corresponding to a *semilinear* transformation $x_i \to \sum a_{ij}x_j^{\alpha}$ (α an *automorphism* of D, $A = \|a_{ij}\|$ a nonsingular matrix) of V_n. From this result we obtain as a corollary

† W. R. Tutte, Trans. AMS **88** (1958), 144–74; **90** (1959), 527–52; Proc. AMS **11** (1960), 905–17; J. Res. Nat. Bureau Standards Sect. **69B** (1965), 1–47.

‡ See G. Birkhoff and J. von Neumann [**1**, Appendix].

THEOREM 18. *The most general dual isomorphism* $PG_{n-1}(D) \to PG_{n-1}(D^{\#})$ *associates with each* $S_k \in PG_{n-1}(D)$ *the set* T_{n-k} *of all* $f \in D^{\#n}$ *such that*

$$\sum f_j a_{jk} x_k^{\alpha} = 0 \quad \textit{for all } \mathbf{x} = (x_1, \cdots, x_n) \in S_k, \tag{25}$$

for some nonsingular $n \times n$ *matrix* $A = \|a_{jk}\|$ *and some automorphism* α *of* D.

COROLLARY 1. *The most general dual automorphism of* $PG_{n-1}(D)$ *is obtained by choosing* $A = \|a_{ij}\|$ *and an anti-automorphism* $x \to x^{\#}$ *of* D, *and associating with each* $S \in PG_{n-1}(D)$ *the set* $S^{\perp}$ *of all* $y \in D^n$ *such that*

$$\sum x_j^{\#} a_{jk} y_k = 0 \quad \textit{for all } \mathbf{x} \in S. \tag{26}$$

COROLLARY 2. *A Desarguesian geometry* $PG_{n-1}(D)$, $n > 2$, *has a dual automorphism if and only if* D *has an anti-automorphism.*

For a given anti-automorphism $x \to x^{\#}$ and matrix a_{jk}, we always have

$$(S \vee T)^{\perp} = S^{\perp} \wedge T^{\perp} \quad \text{and} \quad (S \wedge T)^{\perp} = S^{\perp} \vee T^{\perp}. \tag{27}$$

In order that $(S^{\perp})^{\perp} = S$, it is necessary and sufficient that $a_{kj} = a_{jk}^{\#}$ (up to a scalar factor, which must be one unless all $a_{jj} = O$). In order that $S \wedge S^{\perp} = O$ (which is equivalent to $S \vee S^{\perp} = I$ by (27)), it is necessary and sufficient that $A = \|a_{jk}\|$ be *definite* for the anti-automorphism $\#$. That is, the condition is

$$\sum x_j^{\#} a_{jk} x_k = 0 \quad \text{implies} \quad x_1 = \cdots = x_n = 0. \tag{28}$$

Since the real field has no proper automorphisms, we infer

THEOREM 19. *To make* $PG_{n-1}(\mathbf{R})$ *into an ortholattice for* $n > 2$, $S^{\perp}$ *must be the orthocomplement of* S *with respect to some positive definite quadratic form.*

COROLLARY. *For* $n > 2$, *any ortholattice which is isomorphic to* $PG_{n-1}(\mathbf{R})$ *as a lattice, is isomorphic as an ortholattice with the ortholattice of subspaces of Euclidean* n*-space under inclusion and orthocomplementation.*

Exercises for §§13–15:

1. Show that a "line" with n points can be coordinatized by a division ring if and only if $n = p^k + 1$ for some prime p and positive integer k.
2. Show directly that, under the operation $\vee$, the points of $PG_n(D)$ generate a semilattice.
3. (a) Show that, in $PG_n(\mathbf{R})$, $n > 1$, every orthocomplementation is similar to that defined by the inner product $\sum x_i y_i$.

   (b) Show that, in $PG_n(\mathbf{C})$, $n > 1$, every orthocomplementation is isomorphic to that defined by the unitary product $\sum x_i y_i^*$.
4. Let F be a "formally real" field, in which $\sum x_i^2 = 0$ implies that all $x_i = 0$. In $V_n(F,$ define $S^{\perp}$ to be the set of all $\mathbf{y}$ such that $\sum x_i y_i = 0$.

   Show that this makes $PG_{n-1}(F)$ into a modular ortholattice.
5. (a) Show that, if $\mathbf{R}$ is the real field and $\mathbf{Q}$ the quaternion ring, then $PG_2(\mathbf{Q})$ is isomorphic with a sublattice of $PG_{11}(\mathbf{R})$.

   (*b) Infer that no lattice-theoretic identity can be equivalent to Pascal's Theorem. (*Hint.* See Chapter VI.)

*6. In a projective plane, for given a_i ($i = 1, 2, 3, 4$), define

$$b_{12} = (a_1 \vee a_2) \wedge (a_3 \vee a_4) \text{ etc.,} \qquad c_{12} = (a_1 \vee a_2) \wedge (b_{13} \vee b_{14}), \text{ etc.}$$

Show† that Desargues Theorem is equivalent to $c_{12} \leqq c_{23} \vee c_{31}$.

*7. In a projective plane, for all a_i, b_i ($i = 1, 2, 3$), set

$$y = (a_1 \vee a_2) \wedge (b_1 \vee b_2) \wedge [\{(a_1 \vee a_3) \wedge (b_1 \vee b_3)\} \vee \{(a_2 \vee a_3) \wedge (b_2 \vee b_3)\}.$$

Show that Desargues' Theorem is equivalent to

$$\bigwedge_{i=1}^{3} (a_1 \vee b_i) \leqq [a_1 \wedge (a_2 \vee y)] \vee [b_1 \wedge (b_2 \vee y)].$$

*8. (a) Show that no non-Desarguesian projective geometry is isomorphic with a lattice of normal subgroups (under meet and join).

(b) Show that the same result holds for the free modular lattice with four generators.

9. (a) Show that if M is a sublattice of a modular geometric lattice whose prime quotients are all projective, and if $[a, b]$ and $[c, d]$ are intervals of M of length 3, then both or neither are Desarguesian planes.

(b) Show that not every modular lattice is isomorphic with a sublattice of a complemented modular lattice.‡

A "null system"§ is a projective geometry P with a polarity $S \to S^*$ such that $p \leqq p^*$ for every point $p \in P$.

10. (a) Show that, in a "null system", $p \leqq q^*$ implies $p^* \geqq q$.

(b) Prove that no plane projective geometry can be made into a null system.

*11. (a) Show that any "null system" contains a chain $0 < x_1 < x_2 < \cdots < x_{n-1} < I$ in which x_i covers x_{i-1} and $x_i^* = x_{n-i}$.

(b) Show that, given a null system in $PG_n(D)$, if we take the x_i as coordinate subspaces, then we can make the first two rows and columns of the "polarity" matrix a_{ij} consist of zeros, except for $a_{01} = 1$, $a_{10} = -1$.

(c) Show that n is odd, and that we can make all coefficients zero except $a_{2i,2i+1} = 1$, $a_{2i+1,2i} = 0$.

(d) Infer that D must be commutative.

Problems

12. How should one define "modular pairs" (i.e., the relation aMb) in a general poset?

13. Find necessary and sufficient condition on a lattice, for all its epimorphs to be semimodular. Is every epimorph of a semimodular lattice semimodular? (See [**Tr**, p. 95, Theorem 2].)

14. For which m, n, is $\breve{\Pi}_m$ monomorphic to Π_n?

15. Is every finite lattice an epimorphic image of a sublattice of the symmetric partition lattice Π_n, for some finite n?

16. Is every epimorphic image of a sublattice of Π_n monomorphic to some Π_m (m finite)?

17. Is every lattice of finite length isomorphic with a sublattice of a geometric lattice of finite length?

---

† Exercise 6 is due to M. P. Schuetzenberger, C. R. Acad Sci. Paris **221** (1945), 218–20; Exercises 7–8 to B. Jonsson, Math. Scand. **1** (1953), 193–206.

‡ M. Hall and R. P. Dilworth, Ann. of Math. **45** (1944), 450–6.

§ See R. Brauer, Bull. AMS **42** (1936), 247–54, and R. Baer, ibid. **51** (1945), 903–6.

18. Which modular lattices M can be represented as sublattices of complemented modular lattices M? When this is possible, can one make the length of L equal to the length of M? (Dilworth–Hall)

19. Determine all symmetric configurations having small length and few elements.†

20. For given m, n, which (m, n)-systems admit a group of automorphisms which is transitive on points?‡

21. For which n do there exist plane projective geometries having $n + 1$ points on each line?

22. Enumerate all finite plane projective geometries. Which are self-dual? Which admit orthocomplements?

23. Which finite plane geometries have groups of automorphisms which are transitive (a) on points, (b) on lines? Do all these conditions together imply Desargues' Theorem? Does any have no nontrivial automorphism?

24. Is there a projective plane which has a dual automorphism of period two, permutable with every collineation (lattice-automorphism)?

25. Characterize combinatorially all the posets which represent polyhedral subdivisions of the n-sphere.§

26. Obtain an easily applied test for deciding whether or not a given configuration is orientable.

27. Which finite posets with I satisfying the Jordan–Dedekind condition are characterized up to isomorphism by $p_I[\lambda]$? By the set of all $p_x[\lambda]$?

28. Let L be the "configuration" defined by a convex polyhedron in n-space. Bound the "dimension" (Dushnik–Miller) of n. (Kurepa)

29. Characterize abstractly (up to isomorphism) the lattice Γ_n of all convex polyhedra in real n-space. What are its modular pairs?

30. Is every finite modular ortholattice Desarguesian? (D. Foulis)

---

† See E. H. Moore, Amer. J. Math. **18** (1896), 264; F. Levi, *Geometrische Konfigurationen*, Springer, Berlin, 1927.

‡ For problems 18–21, see M. Hall, *Combinatory Analysis*, Ginn Blaisdell, 1967.

§ See E. R. van Kampen, "*Die Kombinatorische Topologie*", Hague, 1929.

CHAPTER V

COMPLETE LATTICES

1. Closure Operations

In §I.4, a *complete lattice* was defined as a poset in which every subset has a least upper bound and a greatest lower bound. Clearly, any lattice of finite length is complete. The direct product of any two complete lattices is also a complete lattice. The cardinal power X^Y of two posets X, Y is a complete lattice whenever the base X is a complete lattice.

We also showed in Theorem I.6 that the subsets of a set I "closed" with respect to any closure property on I form a complete lattice, a property of subsets of I being called a *closure* property when (i) I has the property, and (ii) any intersection of subsets having the property itself has the property.

Trivially, the subsets "closed" under a given closure property form a Moore family in the sense of the following definition, and conversely.

DEFINITION. A *Moore family* of subsets of I is a family of subsets which contains I, and contains $\bigcap X_\alpha$ when it contains every X_α (is closed under intersection).

We now show that the concept of a "closure property" is also effectively equivalent to that of a "closure operation", as defined below, and give an alternative proof of Theorem I.6.

DEFINITION. A *closure operation* on a set I is an operator $X \to \bar{X}$ on the subsets of I such that:

C1. $X \subset \bar{X}$. (Extensive)
C2. $\bar{X} = \bar{\bar{X}}$. (Idempotent)
C3. If $X \subset Y$, then $\bar{X} \subset \bar{Y}$. (Isotone)

A subset X of I is *closed* under a given closure operation when it is equal to its "closure" $\bar{X}$.

THEOREM 1. *The subsets of I "closed" under any closure operation form a Moore family, and conversely.*

In other words, being "closed" is a "closure property", and conversely.†

PROOF. The intersection $D = \bigcap_\Phi X_\phi$ of any collection of closed sets is closed, since by C3 $\bar{D} \subset \bar{X}_\phi = X_\phi$ for all $X_\phi \in \Phi$, whence $\bar{D} \subset D$. By C1, this implies that D is closed. Conversely, if $\mathscr{F}$ is a Moore family of subsets of I, let $\bar{X}$ be the

† Theorem 1 is due to E. H. Moore, [GA, pp. 53–80]. Moore refers to "closure properties" as "extensionally attainable". Cohn [UA] calls a "Moore family" as a "closure system".

intersection of the sets $F_\alpha \in \mathscr{F}$ which contain X. Since $I \in \mathscr{F}$, there is at least one such set; moreover $\bar{X} = \bigcap F_\alpha$ contains X since every F_α contains X; this proves C1. Since every $F_\alpha \in \mathscr{F}$ which contains X contains $\bar{X}$ (by definition of $\bar{X}$), and conversely, C2 holds. Finally, C3 is trivial, completing the proof.

THEOREM 2. *Any Moore family $\mathscr{F}$ of subsets of a set I forms a complete lattice under set-inclusion.*

PROOF. The existence in $\mathscr{F}$ of $\inf\{X_\alpha\} = \bigcap X_\alpha$ and of $\sup\{X_\alpha\} = \overline{\bigcup X_\alpha}$, for any $\{X_\alpha\} \subset \mathscr{F}$, is obvious.

COROLLARY. *The subsets "closed" with respect to any closure operation on sets form a complete lattice, in which g.l.b. means intersection.*

Example 1. Let G be any monoid of continuous transformations acting on a space S. Define $X \in \mathscr{F}$ to mean that $X \subset S$ and that $x \in X$ and $\gamma \in G$ imply $x\gamma \in X$. Then $\mathscr{F}$ is a complete lattice under set-inclusion.

The preceding results can be generalized as follows.

THEOREM 3. *If P is a poset and every subset of P (including the void subset) has a g.l.b. in P, then P is a complete lattice.*

PROOF. Let $X \subset P$ be given, and let U be the set of upper bounds for X. Let $a = \inf U$. If $x \in X$, then x is a lower bound for U so that $x \leqq a$. Hence a is an upper bound for X. If also b is an upper bound for X, then $b \in U$ so that $a \leqq b$. Hence $a = \sup X$ exists in P, which is therefore a complete lattice.

More generally, if S is any subset of a complete lattice L such that $\inf X \in S$ whenever $X \subset S$, then S is also a complete lattice.

DEFINITIONS. If $S \subset L$ contains both $\inf X$ and $\sup X$ whenever $X \subseteq S$, then S is called a *closed sublattice*. A *closure operation* on a lattice L is an operation $x \to \bar{x}$ on the elements of L which satisfies conditions C1–C3 above. The elements $x \in L$ such that $x = \bar{x}$ are called *closed*.

The following straightforward generalizations of Theorems 1–3 can be obtained by paraphrasing their proofs.

THEOREM 4. *Let $x \to \bar{x}$ be a closure operation on a complete lattice; and let S be the subset of closed elements of L. Then $x_\alpha \in S$ implies $\bigwedge x_\alpha \in S$.*

COROLLARY. *The closed elements of L form a complete lattice.*

The lattice of all closure operators on a given lattice has been studied by Morgan Ward and others.†

Exercises:

1. Show that the quotients a/b ($a \geqq b$) of any complete lattice form a complete lattice, either (i) if $a/b \leqq c/d$ is defined to mean $[b, a] \subset [d, c]$, or (ii) if $a/b \leqq c/d$ is defined to mean that $a \leqq c$ and $b \leqq d$.

† M. Ward, Ann. of Math. **43** (1942), 191–6; O. Ore. ibid. **44** (1943), 514–33 and Duke Math. J. **10** (1943), 761–85; A. Monteiro and H. Ribeiro, Portugaliae Math. **3** (1942), 191–284; J. Morgado, Math. Revs. **24** (1962), 3092–4.

2. Show that, if L is a complete lattice and P any poset, then L^P is a complete lattice.

3. (a) Show that the binary relations on any set A form a complete Boolean algebra, $R(A) = 2^{A^2}$.

(b) Show that the reflexive relations form a dual principal ideal which is also a sublattice.

(c) Show that the symmetric relations form a closed Boolean subalgebra.

(d) Show that the transitive relations form a complete lattice but not a sublattice.

4. Show that the identical relation $x \vee \bar{\bar{x}} \leqq \overline{x \vee y}$ characterizes closure operations in any complete lattice. (K. Iseki)

5. Show that C1–C3 are equivalent to the single condition $y \cup \bar{y} \cup \bar{\bar{x}} \subset \overline{x \cup y}$. (A. Monteiro)

6. A lattice-morphism is called "complete" if and only if it preserves sup and inf of arbitrary sets. Show that the inverse image of any element under any complete morphism between complete lattices is a closed interval.

7. Show that lattice-morphisms between complete lattices need not preserve infinite joins and meets.

8. Show that any epimorphic image of a complete chain is complete.

*9. Let $2^\omega = A$ be the Boolean lattice of all subsets of a countable set, and let F be the ideal of finite subsets. Show that A is complete, whereas its epimorphic image A/F is not.

10. In the set of all closure operators on a lattice L, define $\phi \leqq \psi$ to mean that $\phi(x) \leqq \psi(x)$ for all $x \in L$.

(a) Prove that, if L is complete, the preceding definition defines a complete lattice $C_L \subset L^L$.

(b) Prove that, if L is not complete, then neither is C_L.

11. Show that C_L is modular if and only if L is a (complete ?) chain.

2. Ideal Lattices

In §II.3, we have already discussed the lattice $\hat{L}$ of all ideals of a given lattice L, which contains L as a sublattice (the sublattice of principal ideals). Evidently, the ideals of any lattice L form a Moore family of subsets of L. If L is considered as an algebra with the binary join operation $x \vee y$ and, for each $a \in L$, a unary operation $f_a(x) = x \wedge a$, the ideals of L are just its subalgebras. As a corollary, we have

THEOREM 5. *Any lattice L can be embedded as a sublattice in the complete lattice $\hat{L}$ of all its ideals.*

Joins and meets can be constructed in $\hat{L}$ explicitly, as follows.

LEMMA 1. *Given two nonempty ideals A, B of L, the meet $A \wedge B = A \cap B$ is the set of all $a \wedge b$ ($a \in A, b \in B$); the join $A \vee B$ is the set of all $x \leqq a \vee b$ for some $a \in A$, $b \in B$.*

PROOF. Since $a \wedge b \leqq a, b$, clearly any such $a \wedge b$ is in A and in B; hence it is in their intersection $A \cap B$. Conversely, if $c \in A \cap B$, then $c = c \wedge c$ where $c \in A, c \in B$. Again any ideal which contains A and B must contain all $x \leqq a \vee b$ ($a \in A$, $b \in B$). Conversely, the set of $x \leqq a \vee b$, $a \in A$, $b \in B$ clearly contains both A and B and is contained in any ideal which contains A and B. Since for all a, $a' \in A$, and b, $b' \in B$, we have $(a \vee b) \vee (a' \vee b') = (a \vee a') \vee (b \vee b')$, this set is an ideal.

THEOREM 6. *The ideals of any modular lattice $\hat{M}$ form a modular lattice M. (Dilworth)*

PROOF. Suppose $X \subset Z$, Y are ideals, by the one-sided modular law, we need only show that every $t \in (X \vee Y) \wedge Z$ is in $X \vee (Y \wedge Z)$. But for ideals in general, $t \in (X \vee Y) \wedge Z$ means $t \leqq (x \vee y) \wedge z$ for some $x \in X$, $y \in Y$, $z \in Z$. In the present case, since $X \subset Z$, $x \in Z$, whence $w = x \vee z \in Z$; therefore every such $t \leqq (x \vee y) \wedge w$ with $x \leqq w$. By the modular law, this implies $t \leqq x \vee (y \wedge w)$; hence $t \in X \vee (Y \wedge Z)$. Q.E.D.

COROLLARY. *Any modular lattice can be embedded in a complete modular lattice.*

LEMMA 2. *If A and B are ideals in a distributive lattice, then*

$$A \vee B = \{a \vee b \mid a \in A, b \in B\}.$$

PROOF. We use Lemma 1. If $x \leqq a \vee b, a \in A, b \in B$, then $x = x \wedge (a \vee b) = (x \wedge a) \vee (x \wedge b) = a' \vee b'$, with $a' = x \wedge a \in A$ and $b' = x \wedge b \in B$.

THEOREM 7. *The ideals of any distributive lattice L form a distributive lattice $\hat{L}$.*

PROOF. If X, Y, Z are ideals, then by the one-sided distributive law

$$(X \wedge Y) \vee (X \wedge Z) \subset X \wedge (Y \vee Z).$$

If, conversely, $a \in X \wedge (Y \vee Z)$, then $a = x \wedge (y \vee z)$ for $x \in X$, $y \in Y$, $z \in Z$. Then $a = (x \wedge y) \vee (x \wedge z)$ by the distributive law and $a \in (X \wedge Y) \vee (X \wedge Z)$.

COROLLARY. *Any distributive lattice can be embedded in a complete distributive lattice.*

However, it is not true that the ideals of any Boolean lattice form a Boolean lattice. For example, let X be the ideal of all finite sets in the complete Boolean lattice $A = [\mathbf{2}^S]$ of all subsets of some infinite set S. If $X \wedge Y = O$ for some ideal Y of A, then $x \wedge y = \varnothing$ for all $x \in X$, $y \in Y$. Let $y \in Y$. Then for any $p \in S$, we have $\{p\} \in X$, $\{p\} \wedge y = \varnothing$, $p \notin y$, so that $y = \varnothing$ and $Y = O$. Then $X \vee Y = X \neq A$, and so Y is not a complement of X. Hence X has no complement in A, which is therefore not a Boolean lattice.

3. Conditional Completeness; Fixpoint Theorem

Many important lattices, though not complete, have the property that every nonempty *bounded* subset has a g.l.b. and l.u.b. Lattices (and posets) with this property are called *conditionally complete.* Thus the real number field $\mathbf{R}$ is conditionally complete; so is the set of all functions convex in a given closed interval and assuming given values at its end points.

We have the following partial analog of Theorem 3.

THEOREM 8. *For a lattice L to be conditionally complete, it is sufficient that every bounded nonempty subset of L have a g.l.b.*

PROOF. Let X be a bounded nonempty subset of L. Let U be the set of upper bounds of S. Since X is bounded, U is nonempty. Select $a \in U$. Let V be the set of all $a \wedge u$, $u \in U$. Select $x_0 \in X$. Then V is bounded below by x_0 and above by a, and is nonempty since $a = a \wedge a \in V$. Hence $b = \inf V$ exists in L. If $x \in X$, then x is a lower bound of V so that $x \leqq b$. Hence the b just constructed is an upper bound of X. If $u \in U$, then $a \wedge u \in V$ so that $b \leqq a \wedge u \leqq u$. Hence $b = \sup X$.

COROLLARY. *For a lattice L to be conditionally complete, it is sufficient that every bounded nonempty subset of L have a l.u.b.*

THEOREM 9. *In a conditionally complete lattice L, every nonempty subset which has a lower bound has a g.l.b.—and dually.*

PROOF. Let X be a nonempty subset of L with lower bound b. Select $c \in X$. Then the set of all $c \wedge x$ with $x \in X$ is bounded and nonempty, and hence has a g.l.b. which is then also a g.l.b. of X.

We shall now show that the only difference between conditionally complete posets and complete lattices is the absence of universal bounds O and I. This generalizes the fact that one can make $\mathbf{R}$ into a complete lattice by adjoining $-\infty$ or $+\infty$.

THEOREM 10. *Let P be any conditionally complete poset. If elements O, I are added to P, we get a complete lattice $\bar{P}$.*

PROOF. Let X be any subset of $\bar{P}$. We will show that X has a g.l.b. If $O \in X$, then $O = \inf X$, so we may assume $O \notin X$. Let $X' = X - \{I\}$ so that $X' \subset P$. If $X' = \varnothing$, then $X = \varnothing$ or $X = I$. In either case, $I = \inf X$, so we may assume $X' = \varnothing$. If X' has no lower bound in P, then $O = \inf X' = \inf X$. If X' has a lower bound in P, let $b = \inf_P X'$. Then $b = \inf X$. Hence X has a g.l.b. and we can apply Theorem 3.

Fixpoint theorem. For isotone operators satisfying C3 alone, the following striking result was discovered by Tarski in 1942, but not published by him until 1955.†

THEOREM 11. *Let L be a complete lattice and f an isotone function on L into L. Then $f(a) = a$ for some $a \in L$.*

PROOF. Let S be the set of all $x \in L$ such that $x \leqq f(x)$, and let $a = \sup S$. Trivially, $O \in S$. Moreover, by hypothesis,

$$x \leqq f(x) \leqq f(a) \qquad \text{for all } x \in S, \text{ by isotonicity.}$$

Hence $a = \sup S \leqq f(a)$, and so $a \in S$. By isotonicity again, $f(a) \leqq f(f(a))$ follows; hence $f(a) \in S$, and so $f(a) \leqq \sup S = a$. Combining with $a \leqq f(a)$, we have $f(a) = a$, as desired.

Exercises for §§2–3:

1. Show that the modular, nondistributive lattice M_5 of five elements has two ideals J, K whose join is not the set of all $x \vee y$, with $x \in J$, $y \in K$.

2. (a) Show that a principal ideal $a \wedge L$ of a distributive lattice L is prime if and only if a is meet-irreducible.

---

† See Pacific J. Math. **5** (1955), 285–309; also A. C. Davis, ibid., 311–9. A related result was obtained earlier by L. Kantorovich, Acta. Math. **71** (1939), 63–97.

(b) Show that if L has finite length n, then L has exactly n proper prime ideals, including O.

(c) Prove that every maximal ideal of a distributive lattice is prime.

3. Show that the interval [0, 1], though a complete chain, is not isomorphic to the complete lattice of its ideals.

4. Show that the ideals of a Boolean lattice A form a Boolean lattice A if and only if A is finite.

5. Prove that the subharmonic functions assuming given continuous values on the boundary of a plane region R form a conditionally complete lattice.

6. Show that if $f(x)$ is an isotone operator in a conditionally complete lattice, and if $a \leqq f(a) \leqq f(b) \leqq b$, then $c = f(c)$ for some c between a and b.

7. Show that in Theorem 11 there is a *least* fixpoint. Show that the fixpoints need not form a sublattice. Do they form a lattice?

8. Prove that Lemma 2 of §2 holds for any ideal A and standard ideal B in any lattice. (Grätzer-Schmidt)

9. Prove that if, in a lattice L, every isotone mapping has a fixpoint, then L is complete. (A. C. Davis)

4. Topological Closure

The most intensively studied closure operations arise in set theory, though the concept is also important in algebra. Define a *topological space*† as a set Q together with a family R of "closed" sets having the following three properties:

(α) The sum of any two closed sets is closed;

(β) Any intersection of closed sets is closed;

(γ) Q and the empty set $\varnothing$ are closed.

It follows from (β) and the first statement of (γ), by Theorem 1, that there is defined in any toplogical space a closure operation satisfying C1–C3. Because of (α), this operation also satisfies

C3*. $$\overline{X \cup Y} = \bar{X} \cup \bar{Y}.$$

Finally, the last statement of (γ) implies that $\bar{\varnothing} = \varnothing$.

In general, a family of sets which satisfies (α) and (β)—i.e., is closed under finite union and arbitrary intersection—will be called a $\bigcap$-*ring* of sets. The concept of a $\bigcup$-ring of sets is defined dually.

Thus, in any topological space Q, the closed sets form a $\bigcap$-ring whose members include $\varnothing$ and Q; the "open" sets (i.e., the complements of closed sets) form a dually isomorphic $\bigcup$-ring also including $\varnothing$ and α. The following result is immediate.

THEOREM 12. *The closed (and dually, the open) subsets of any topological space form a complete distributive lattice.*

More generally, if L is any complete distributive lattice with a "closure"

† Topological spaces will be studied systematically in Chapter IX. Conditions C1–C4 were originally formulated by F. Riesz (1909). See C. Kuratowski, *Topologie* I, Warsaw, 1948, p. 20.

operation $x \to \bar{x}$ which satisfies C1, C2, C3*, then the subset S of "closed" elements is a complete distributive sublattice of L.

A topological space is called a T_1-*space* when it satisfies

C4. If p is a point, then $\bar{p} = p$.

It is called a T_0-*space* when it satisfies the weaker condition

C4′. For two points p and q, $\bar{p} = \bar{q}$ implies $p = q$.

There is a natural one-one correspondence between finite T_0-spaces and finite posets. Define the *M-closure* of any subset S of a poset P, as the set $\bar{S}$ of all $t \in P$ such that $t \leqq s$ for some $s \in S$.

Then C1 follows from P1, C2 from P3; C4′ from C1, P2; and C3* is clear. Hence P is a T-space. Moreover, in this space, $q \in \bar{p}$ if and only if $q \leqq p$. But conversely, the definition

$$q \leqq p \quad \text{if and only if} \quad q \in \bar{p}$$

partly orders the points of any T_0-space. For a finite set $S = \bigvee_{i=1}^{n} p_i$, we have $\bar{S} = \bigvee_{i=1}^{n} \bar{p}_i$ for some $p \in S$, and so the correspondences between T_0-topologies and partial orderings are inverse.

THEOREM 13. *There is a one-one correspondence between finite posets P and finite T_0-spaces Q: $q \leqq p$ in P means $q \in \bar{p}$ in Q.*

A family of subsets of any set is called a "ring" if it contains with any two subsets S and T, their union $S \cup T$ and their intersection $S \cap T$.

THEOREM 14. *In the preceding theorem, $\mathbf{2}^P$ is isomorphic with the ring of all open subsets of Q, and hence anti-isomorphic with the ring of all closed subsets of Q.*

PROOF. Associate with each subset S of Q its "characteristic function" f_S: $f_S(p) = 1$ or 0, according as $p \in S$ or $p \notin S$. Then $S \subseteq T$ if and only if $f_S \leqq f_T$, and S is closed if and only if $f_S(q) \geqq f_S(p)$ whenever $q \leqq p$; and hence S is open if and only if f_S is isotone. This establishes the isomorphism asserted.

The preceding result is closely related to Theorem III.3. It shows that each finite distributive lattice is isomorphic with the lattice of all closed subsets of the finite T_0-space of its join-irreducible elements.

5. Infinite Distributivity

In any complete lattice L, the operations $\inf S = \bigwedge S$ and $\sup S = \bigvee S$ apply to infinite as well as finite subsets $S \subset L$. We will now consider some basic properties of these *infinitary* operations.†

The following "generalized associative laws" are immediate.

L*. If Φ is any family of subsets S_ϕ of L, then

$$\bigwedge_{\Phi} \left\{ \bigwedge_{S_\phi} x \right\} = \bigwedge_{\cup S_\phi} x \quad \text{and, dually} \quad \bigvee_{\Phi} \left\{ \bigvee_{S_\phi} x \right\} = \bigvee_{\cup S_\phi} x.$$

† Cf. Exercises 6–9 of §1. Also R. P. Dilworth and J. E. Mc Laughlin, Duke J. Math. **19** (1952), 683–93.

Conversely, assuming properties of the set-union operation $\cup$, they imply L1–L3: L*, with L4, characterizes complete lattices.

THEOREM 15. *Any system L with operations $S \to \bigvee S \in L$ and $S \to \bigwedge S \in L$ which satisfy* L4 *and* L*, *for arbitrary $S \subset L$, is a complete lattice.*

PROOF. We define $x \leqq y$ to mean that $\bigwedge \{x, y\} = x$ and $\bigvee \{x, y\} = y$; by L4, these conditions are equivalent. Since L4 implies L1, we have $x = x \wedge x = \bigwedge \{x, x\} = \bigwedge \{x\}$ for all $x \in L$. If $x, y, z \in L$, then

$$x \wedge (y \wedge z) = \bigwedge \{x, y \wedge z\} = \bigwedge \{\bigwedge \{x\}, \bigwedge \{y, z\}\} = \bigwedge \{x, y, z\}$$

by L*. Hence L is a lattice.

Now let $S \subset L$. Then

$$(\bigwedge S) \wedge x = \bigwedge \{\bigwedge S, x\} = \bigwedge \{\bigwedge S, \bigwedge \{x\}\} = \bigwedge (S \cup \{x\}) = \bigwedge S$$

for all $x \in S$ and, if $x \wedge t = t$ for all $x \in S$, then

$$\begin{aligned}(\bigwedge S) \wedge t &= \bigwedge \{\bigwedge S, t\} = \bigwedge \{\bigwedge S, \bigwedge \{t\}\} = \bigwedge (S \cup \{t\}) \\ &= \bigwedge (\bigcup \{\{x, t\} \mid x \in S\}) = \bigwedge \{\bigwedge \{x, t\} \mid x \in S\} \\ &= \bigwedge \{x \wedge t \mid x \in S\} = \bigwedge \{t\} = t,\end{aligned}$$

so that $\bigwedge S = \inf S$.

In any distributive lattice, we have by induction

$$a \wedge \bigvee_S x_\sigma = \bigvee_S (a \wedge x_\sigma) \tag{1}$$

and, dually,

$$a \vee \bigwedge_S x_\sigma = \bigwedge_S (a \vee x_\sigma), \tag{1'}$$

for any *finite* index-set S. Formulas (1)–(1') hold for arbitrary S in any complete Boolean lattice, but not in every complete distributive lattice.

For example, (1) does not hold in the complete distributive lattice of all closed subsets of the plane: if c denotes the circle $x^2 + y^2 = 1$, and s_k denotes the disc $x^2 + y^2 \leqq 1 - k^{-2}$, then $c \wedge \bigvee_{k=1}^{\infty} d_k = c$, yet $\bigvee_{k=1}^{\infty} (c \wedge d_k)$ is the empty set $\varnothing$. On the other hand, (1') holds in the preceding lattice, since $\vee$ and $\bigwedge$ coincide in it with the set-theoretic operations $\cup$ and $\bigcap$.

Complete lattices in which (1) holds will be studied in §§10–11, and various refinements of (1) will be considered in §X.11. But for the present, we confine ourselves to the case of Boolean lattices. We first prove

THEOREM 16. *The distributive laws* (1) *and* (1') *hold in any complete Boolean lattice A, for any index set S.*

PROOF. By isotonicity, $a \wedge \bigvee_S x_\alpha$ is an upper bound to every $a \wedge x_\alpha$; therefore $\bigvee_S (a \wedge x_\alpha) \leqq a \wedge \bigvee_S x_\alpha$. To prove the reverse inequality and hence (1), it suffices that $a \wedge \bigvee_S x_\sigma \leqq u$ for every upper bound u to the $a \wedge x_\sigma$. But, if $a \wedge x_\sigma \leqq u$ for all α, then

$$x_\sigma = (a \wedge x_\sigma) \vee (a' \wedge x_\sigma) \leqq u \vee a'$$

for all $x_\alpha \in S$, and so

$$a \wedge \bigvee_S x_\sigma \leqq a \wedge (u \vee a') = (a \wedge u) \vee (a \wedge a') = a \wedge u \leqq u,$$

which proves (1). By duality, we have (1′).

LEMMA. *In any complete lattice, the infinite distributive law* (1), *implies*

$$\left(\bigvee_S x_\sigma\right) \wedge \left(\bigvee_T y_\tau\right) = \bigvee_{ST} \left(x_\sigma \wedge y_\tau\right), \tag{2}$$

PROOF. By (1) and its equivalent under L2 (commutativity):

$$\left(\bigvee_S x_\sigma\right) \wedge \left(\bigvee_T y_\tau\right) = \bigvee_S \left(x_\sigma \wedge \left(\bigvee_T y_\tau\right)\right) = \bigvee_S \left(\bigvee_T (x_\sigma \wedge y_\tau)\right).$$

From this (2) follows by the generalized associative law.

COROLLARY. *In any complete Boolean algebra,* (2) *holds together with its dual*

$$\left(\bigwedge_S x_\sigma\right) \vee \left(\bigwedge_T y_\tau\right) = \bigwedge_{ST} (x_\sigma \vee y_\tau). \tag{2′}$$

For *finite* sets of elements (and index-sets) in any distributive lattice L, one has the general finite distributive law of Chapter III, (10):

$$\bigwedge_{h=1}^{r} \left[\bigvee_{i=1}^{s(h)} x_{h,i}\right] = \bigvee_F [x_{1,f(1)} \wedge \cdots \wedge x_{r,f(r)}] = \bigvee_F \left[\bigwedge_{h=1}^{r} x_{h,f(h)}\right], \tag{3}$$

where F is the set of all functions assigning to each $h = 1, \cdots, r$ some value $f(h)$ in the set $\{1, \cdots, s(h)\}$. In view of Theorem 16, it would be natural to guess that the infinite analog of (3) held in every complete Boolean algebra.

However, this is not true. Call a (complete) lattice *completely distributive* when it satisfies the dual *extended distributive laws:*

$$\bigwedge_C \left[\bigvee_{A_\gamma} x_{\gamma,\alpha}\right] = \bigvee_\Phi \left[\bigwedge_C x_{\gamma,\phi(\gamma)}\right], \tag{4}$$

$$\bigvee_C \left[\bigwedge_{A_\gamma} x_{\gamma,\alpha}\right] = \bigwedge_\Phi \left[\bigwedge_C x_{\gamma,\phi(\gamma)}\right], \tag{4′}$$

for *any* nonvoid family of index-sets A_γ, one for each $\gamma \in C$, provided Φ is the set of *all* functions ϕ with domain C and $\phi(\gamma) \in A_\gamma$. We then have

THEOREM 17 (TARSKI†). *If a complete Boolean algebra A is completely distributive, then it is isomorphic with the algebra $\mathbf{2}^\aleph$ of all subsets of some aggregate.*

Remark. Actually, it suffices to assume (4); see below.

PROOF. Let C be the set of complementary pairs (x_γ, x'_γ) of elements of A; let Φ be the set of functions ϕ which select from each complementary pair one

† A. Tarski, Fund. Math. **16** (1929), 195–7. See also O. Ore, Ann. of Math. **47** (1946), 56–72, especially Theorem 23, which refines Tarski's proof.

member $x_{\phi(\gamma)}$; and let $p_\phi = \bigwedge x_{\phi(\gamma)}$. Then, by (4):

$$I = \bigwedge_C (x_\gamma \vee x'_\gamma) = \bigvee_\Phi \left\{\bigwedge_C x_{\phi(\gamma)}\right\} = \bigvee_\Phi p_\phi.$$

We next show that each p_ϕ at most covers O (i.e., either $p_\phi = O$ or p_ϕ is an atom of A). Indeed, if $x_\gamma < p_\phi$, then $x_{\phi(\gamma)} = x'_\gamma$ since $x_{\phi(\gamma)} = x_\gamma$ would imply $p_\phi = \bigwedge x_{\phi(\gamma)} \leqq x_\gamma$; hence $x_\gamma < p_\phi$ implies

$$x_\gamma = x_\gamma \wedge p_\phi \leqq x_\gamma \wedge x_{\phi(\gamma)} = x_\gamma \wedge x'_\gamma = O.$$

Moreover, each $x_\alpha = x_\alpha \wedge I = x_\alpha \wedge \bigvee_\Phi p_\phi = \bigvee (x_\alpha \wedge p_\phi)$, where $x_\alpha \wedge p_\phi$ is the point p_ϕ if $p_\phi \leqq x_\alpha$, and O otherwise. Hence each $x_\alpha \in A$ is the join of those "points" which it contains. Finally, a join $g_S = \bigvee_S p_\phi$ of points p_ϕ contains no point p not in S, since

$$g_S \wedge p = \left(\bigvee_S p_\phi\right) \wedge p = \bigvee_S (p_\phi \wedge p) = \bigvee_S O,$$

unless $p = p_\phi$ for some $p_\phi \in S$, by (1). This establishes an isomorphism between the elements $a \in A$ and the subsets S of the set of all points p_ϕ, completing the proof.

Raney's Theorems. The class of all completely distributive (complete) lattices contains many non-Boolean lattices. For example:

LEMMA. *Any complete chain is a completely distributive lattice.*

PROOF. In any complete lattice,

$$\bigwedge_C \left[\bigvee_{A_\gamma} x_{\gamma,\alpha}\right] \geqq \bigvee_\Phi \left[\bigwedge_C x_{\gamma,\phi(\gamma)}\right] = b,$$

since the left side is an upper bound to $\bigwedge_C x_{\gamma,\phi(\gamma)}$ for every $\phi \in \Phi$. To prove the reverse inequality, observe that if $b < a$, then for every $\gamma \in C$, there exists $\alpha = \phi(\gamma)$ such that $x_{\gamma\alpha} > b$.

If a covers b, this implies $x_{\gamma\alpha} \geqq a$, and so $\bigwedge_C x_{\gamma,\phi(\gamma)} \geqq a$, whence the reverse inequality holds. Otherwise, a is the join of $y<a$; but for any $y<a$, $\bigwedge_C x_{\gamma,\phi(\gamma)} \geqq y$, for some $\phi = \phi_y \in \Phi$. as above. Hence $\bigvee_\Phi [\bigwedge_C x_{\gamma,\phi(\gamma)}]C \geqq \bigvee_{y<a} y = a$, completing the proof.

COROLLARY. *Any closed sublattice of a direct product of complete chains is a completely distributive* (*complete*) *lattice.*

Raney has proved some powerful converses to the preceding results, using the notion of "semi-ideal" (M-closed subset).† Thus he has proved that, in any complete lattice L, each of conditions (4)–(4′) implies the other (hence complete distributivity), and is equivalent to the condition that L be an epimorphic image

† G. N. Raney, Proc. AMS **3** (1952), 677–80 and **4** (1953), 518–22. See also G. Bruns, J. Reine Angew. Math. **209** (1962), 167–200 and **210** (1962), 1–23.

of a complete ring of sets, with respect to *arbitrary* joins and meets. Finally, he has proved a deep converse of the corollary above: he has shown that every completely distributive complete lattice is a closed sublattice of a direct product of complete chains.

Exercises for §§4–5:

1. In any semilattice, define $x_1 \circ \cdots \circ x_n$ by induction as $x_1 \circ (x_2 \circ \cdots \circ x_n)$.

(a) Prove from L3 by induction the *generalized associative* law: If $y_i = x_{s_i - 1} \circ \cdots \circ x_{s_i}$ $[O = s_0 < s_1 < \cdots < s_m = n]$, then $y_1 \circ \cdots \circ y_m = x_1 \circ \cdots \circ x_n$.

(b) Prove that $x_1 \circ \cdots \circ x_n$ is invariant under every permutation of the x_i, using only L2 and L3.

(c) Prove (L*) by induction, using L1–L3, in case Φ and all S_ϕ are finite.

2. Derive L1, L2, L3 are special cases of (L*).

3. State a weakened form of (L*) valid in any σ-lattice.

4. (a) Show that a point a is in the *closure* $\bar{X}$ of a set X, if and only if every open set U containing a contains a point of X.

(b) Show that this result requires only (1). (O. Ore)

5. Show that the proof of Theorem 17 only requires the "intermediate" distributive law

$$\bigwedge_{i \in I} (a_i^{(1)} \vee a_i^{(2)}) = \bigvee_{2^I} \{\bigwedge a_i^{f(i)}\}.$$

6. Show that, for any poset P, $\mathbf{2}^P$ is a completely distributive (complete) lattice. (*Hint.* It is a $\bigcap$-sublattice and $\bigcup$-sublattice of $\mathbf{2}^{o(P)}$).

7. Generalize Exercise 6 to D^P, where D is any completely distributive lattice.

8. Let $\phi: A \to B$ be an epimorphism between two complete Boolean algebras A, B. Show that ϕ is a $\bigwedge\bigvee$-morphism if and only if Ker ϕ is a principal ideal. (Ph. Dwinger)

6. Lattices with Unique Complements

A lattice in which each element has one and only one complement is called *uniquely complemented*; we have seen in §I.10 that Boolean lattices are uniquely complemented. We now prove a partial converse of this result.

LEMMA. *If p and q are distinct points (atoms) in a uniquely complemented lattice, then $p \leqq q'$.*

PROOF. Let $p \nleqq q'$. Then $p \wedge q' < p$ so that $p \wedge q' = O$. Define $x = p \vee q'$; clearly $x > q'$. Unless $x \geqq q$, $x \wedge q < q$; hence $x \wedge q = O$ and $x \vee q \geqq q' \vee q = I$. Therefore $x = q'$, contradicting $x > q'$. Hence $x \geqq q$. Then $x \geqq q \vee q' = I$, $x = I$, $p \vee q' = I$, so that $q' = p'$ and $q = p$.

THEOREM 18. *Let L be a complete, uniquely complemented lattice in which every element $a > O$ contains a point. Then L is isomorphic with the Boolean lattice A of all subsets of its points.*

PROOF. For each $a \in L$, let $S(a)$ denote the set of all points $p \leqq a$. Conversely, with each set S of points of L, associate the join $j(S)$ of the points $p \in S$. The correspondences $a \to S(a)$ and $S \to j(S)$ are clearly isotone from L to the power set $\mathbf{2}^{o(P)} = A$ of all subsets of the set P of all points of L, and vice-versa. Moreover since $j(P)$ contains every point, $[j(P)]'$ can contain no point. Hence $[j(P)]' = O$, and $j(P) = O' = I \in L$.

Again, by the preceding lemma, $q \notin S$ implies $q' \geqq p$ for all $p \in S$, whence it implies $q' \geqq j(S)$. Hence $p \leqq j(S)$ implies $p \in S$; but $p \in S$ trivially implies $p \leqq j(S)$. Consequently

(5) $$S(j(S)) = S \quad \text{for any set } S \text{ of points of } L.$$

Now let $A = S(a)$; trivially, $j(A) = j(S(a)) \leqq a$, and $j(A) \leqq a$ contains no point not $\leqq a$. Hence $a \wedge j(A')$ contains no point not in A and in A', which shows that $a \wedge j(A') = O$. But $a \vee j(A') \geqq j(A) \vee j(A') = j(A \vee A') = j(P) = I$. Hence

(6) $$j(A') = a' \quad \text{if } A = S(a).$$

Interchanging $a = (a')'$ and a', we get $a = j([S(a')]')$: a is the join of the points $p \in [S(a')]'$. This implies *a fortiori* that a is the join of the set A of *all* points $p \leqq a$. This proves $j(S(a)) = a$ for all $a \in L$. Hence, by (5), the correspondences $a \to S(a)$ and $S \to j(S)$ are inverse *isomorphisms* of L and $\mathbf{2}^{o(P)}$.

COROLLARY. *Any finite, uniquely complemented lattice is a Boolean lattice.*

However, using the "free algebra" concept (Chapter VI), Dilworth has proved that *every* lattice is a sublattice of a lattice with unique complements, in which every element has one and only one complement. It follows that lattices with unique complements need not be distributive, nor even modular.

Exercises:

1. Show that $(a')' = a$ in any lattice with unique complements.
2. Show that *any* modular lattice with unique complements is a Boolean lattice. (See §III.7.)
3. Show that a complete, atomic lattice L is a Boolean lattice if and only if every element of L has a unique complement.
4. Show that a complete Boolean lattice A is atomic if and only if it is completely distributive.

7. Polarities

We now describe a fundamental construction, which yields from any binary relation two inverse dual isomorphisms. These are called "polarities", because they generalize the dual isomorphism between "polars" in analytic geometry;† see Example 4 below.

Let ρ be any binary relation between the members of two classes I and J. For any subsets $X \subset I$ and $Y \subset J$, define $X^* \subset J$ (the "polar" of X) as the set of all $y \in J$ such that $x \rho y$ for all $x \in X$, and we define $Y^\dagger \subset I$ (the "polar" of Y) as the set of all $x \in I$ such that $x \rho y$ for all $y \in Y$. Clearly

LEMMA. *For any binary relation* ρ:

(7) $$X \subset X_1 \quad \textit{implies} \quad X^* \supset X_1^*,$$

(7′) $$Y \subset Y_1 \quad \textit{implies} \quad Y^\dagger \supset Y_1^\dagger,$$

(8) $$X \subset (X^*)^\dagger \quad \textit{and} \quad Y \subset (Y^\dagger)^*.$$

† The construction was apparently first described in general terms in [**LT1**, §32].

COROLLARY. *In the preceding lemma*

(9) $$((X^*)^\dagger)^* = X^* \quad and \quad ((Y^\dagger)^*)^\dagger = Y^\dagger.$$

PROOF. By (8), we have $X^* \subset ((X^*)^\dagger)^*$ and $X \subset (X^*)^\dagger$, whence, by (7), $X^* \supset ((X^*)^\dagger)^*$. The proof that $((Y^\dagger)^*)^\dagger = Y^\dagger$ is similar.

THEOREM 19. *The operations $X \to (X^*)^\dagger$ and $Y \to (Y^\dagger)^*$ are closure operations. Moreover, the correspondences $X \to X^*$ and $Y \to Y^\dagger$ define a dual isomorphism between the complete lattices of "closed" subsets of I and J.*

PROOF. Condition C1 follows from (8), C2 from (9), and C3 from (7)–(7′). By (9), the closed subsets of I, J are just those of the form $Y^\dagger$, X^*, respectively. Also by (9), the correspondences $Y^\dagger \to (Y^\dagger)^*$ and $X^* \to (X^*)^\dagger$ are inverse; hence one-one and onto. Finally, by (7)–(7′), they invert inclusion.

Example 2. Let $I = J$ be any ring, and let $x \,\rho\, y$ mean that $xy = O$. Then every X^* is a right-ideal, every $X^\dagger$ a left-ideal.

Example 3. Let I be any field or division ring, and let I' be any finite group of automorphisms α of I. If we define $x \,\rho\, \alpha$ to mean $\alpha(x) = x$, we get the well-known dual isomorphism between subgroups of J and certain subfields of I.

There are other important examples where $I = J$ and ρ is symmetric. The following case is typical.

Example 4. Let $A = \|a_{ij}\|$ be any symmetric nonsingular $n \times n$ matrix. For two vectors $\xi = (x_0, \cdots, x_n)$ and $\eta = (y_0, \cdots, y_n)$ of projective n-space I, define $\xi \,\rho\, \eta$ to mean that the sum $O(\xi, \eta) = \sum_{i,j} x_i a_{ij} y_j = O$. The "closed" subsets of I are then its points, lines, planes, and other subspaces; if X is any such subspace, X^* is its polar with respect to Q, in the sense of classical geometry.†

The preceding example suggests the following result.

COROLLARY. *If $I = J$ and the relation ρ is symmetric, then $X^\dagger = X^*$, and in the complete lattice of closed sets $X = (X^*)^*$: the correspondence $X \to X^*$ is an involution. In symbols, for closed sets X, Y, $\cdots$;*

(10) $$(X^*)^* = X,$$

(11) $$(X \wedge Y)^* = X^* \vee Y^*, \qquad (X \vee Y)^* = X^* \wedge Y^*.$$

If ρ is also anti-reflexive (if $x \,\rho\, x$ for no x), or if $x \,\rho\, x$ implies $x \,\rho\, y$ for all y, then

(12) $$X \wedge X^* = O \quad \text{and} \quad X \vee X^* = I.$$

PROOF. Since $X^* = X^\dagger$, (10) is implied by (9), and (11) by Theorem 19—a dual isomorphism interchanges joins and meets. We leave the proof of (12) to the reader.

Example 5. Let I be any group, and let $x \,\rho\, y$ mean that $xy = yx$. Then (10)–(11) hold; every closed set is a subgroup; and the correspondence $X \to X^*$ carries each subgroup into its "centralizer".

† W. C. Graustein, *Introduction to higher geometry*, New York, 1940, Chapter XIV, §§1–3.

Example 6. Let I be any class, and let $x \rho y$ mean $x \neq y$. Then (10)–(12) hold; every set is closed; and the involution $X \to X^*$ carries each set into its set-complement.

Example 7. Let I be Cartesian n-space, and let $x \rho y$ mean $x \perp y$ (x is orthogonal to y). Then (10)–(12) hold; the "closed" subsets are the linear subspaces; and the involution carries each subspace into its orthogonal complement. (This is the special case $A = I$ of Example 4.) Note that $X \wedge X^*$ contains only elements x such that $x \rho x$; hence it is the void set $\varnothing$ in Example 6 and the origin in Example 7.

Example 8. Let F be any set of functions $f(\mathbf{x}) = f(x_1, \cdots, x_n)$ from $\mathbf{R}^n$ to $\mathbf{R}$. Define $f \rho \mathbf{x}$ to mean that $f(\mathbf{x}) = O$. Then the "closed" sets in $\mathbf{R}^n$ defined by the resulting polarity include many families of sets fundamental for geometry. (See Exercise 7 below, and §IV.3.)

8. Galois Connections

The preceding results can be generalized to posets† as follows.

DEFINITION. Let P, Q be any posets, and let $x \to x^*$, $y \to y^\dagger$ be any correspondences $\phi: P \to Q$ and $\psi: Q \to P$, such that

$$x \leqq x_1 \quad \text{implies} \quad x^* \geqq x_1^*, \tag{13}$$

$$y \leqq y_1 \quad \text{implies} \quad y^\dagger \geqq y_1^\dagger, \tag{13'}$$

$$x \leqq (x^*)^\dagger \quad \text{and} \quad y \leqq (y^\dagger)^*. \tag{14}$$

The correspondences $x \to x^*$ and $y \to y^\dagger$ are said to define a *Galois connection* between P and Q.

The formal proofs in §7, of (9), and of the fact that the correspondences $x \to (x^*)^\dagger$ and $y \to (y^\dagger)^*$ are closure operations if P and Q are complete lattices, then go through without change. It is also known that Galois connections are defined by the condition that $b \leqq a^*$ if and only if $a \leqq b^\dagger$. (J. Schmidt)

Applying Theorems 1–2, we get

THEOREM 20. *Any Galois connection $x \to x^*$, $y \to y^\dagger$ between two complete lattices L and M gives a dual isomorphism between the complete lattices S and T of "closed" subsets of L and M.*

Other results of §7 also generalize to arbitrary Galois connections. Thus, if $L = M$ and $\phi = \psi$, then (11) holds.

Indeed, Everett‡ has shown that for $L = \mathbf{2}^S$ and $M = \mathbf{2}^T$ completely distributive (complete) Boolean lattices, all Galois connections between L and M can be

† Cf. O. Ore, Trans. AMS **55** (1944), 494–513; C. J. Everett, ibid, 514–25; G. Pickert, Archiv. Math. J. **29** (1962), 505–14.

‡ C. J. Everett, Trans. AMS **55** (1944), 514–25. For Raney's results, see ibid. **97** (1960), 418–26. For related results see G. Aumann, S.-B. Bayer. Akad. Wiss. (1955), 281–4; E. F. Wright, Pacific J. Math. **10** (1960), 723–30.

obtained by contraction from suitable polarities. Conditions under which Galois connections between other complete lattices have analogous properties have been determined by Raney.

Pseudocomplements in semilattices.† Although most Galois connections arise from polarities, and hence are between lattices of sets, others arise naturally in lattices. Thus, let S be a *meet-semilattice* with O in which each element a has a *pseudocomplement* a^* such that $a \wedge x = O$ if and only if $x \leqq a^*$. In other words, we assume that, for each $a \in S$, the set of elements disjoint from a has a largest element a^* (this will be unique). Any Boolean lattice is such a "pseudocomplemented" semilattice (cf. §II.9); other examples will be described in the next section. Thus, by assumption

$$a \wedge a^* = O. \tag{15}$$

Moreover, since $O \wedge x = O$ for all $x \in S$, $O^* \geqq x$ for all $x \in S$: any such S has the universal upper bound O^*.

Again, since $a \leqq b$ implies $a \wedge b^* \leqq b \wedge b^* = O$,

$$a \leqq b \quad \text{implies} \quad b^* \leqq a^*. \tag{16}$$

Further, since $a^* \wedge a = O, a \leqq (a^*)^*$. In summary, we have proved the following result.

LEMMA. *In any pseudocomplemented meet-semilattice, pseudocomplementation is a symmetric Galois connection.*

Exercises for §§7–8:

1. Show that, for any elements a, b in any lattice L, the correspondences $x \to x \vee a$ and $y \to y \wedge b$ define the dual of a Galois connection between $[a \wedge b, b]$ and $[a, a \vee b]$.
2. Let ρ be any binary relation. Amplify the notation of §5 by letting $X \cap X_1$ and $X \cup X_1$ denote set-product and set-union. Show that, for any closed sets $X = (X^*)^\dagger$, $X_1 = (X_1^*)^\dagger$, we have $X \wedge X_1 = X \cap X_1$ but $X \vee X_1 = ((X \cup X_1)^*)^\dagger = (X^* \cap X_1^*)^\dagger < X \cup X_1$ in general. (F. W. Levi)
3. Let V be a real vector space, and V^* its dual; define $x \rho f$ to mean $|f(\pm x)| \leqq 1$. Show that the polar of a symmetric convex "unit ball" in V is the dual unit ball in V^*.
4. Let I be any finite Abelian group; let J be the group of its characters, and let $a \rho \chi$ $[a \in I, \chi \in J]$ mean $\chi(a) = O$.
   (a) Show that the "closed" subsets of I and J are their subgroups.
   (b) Infer that the subgroup-lattice of I is dual to that of J.
   (c) Prove that the subgroup-lattice of I is self-dual.
5. (a) Let $x \rho y$ mean that $x \perp y$ in Hilbert space. Show that the complete lattice of closed subspaces of Hilbert space is self-dual.
   (b) Let I be any Banach space, and let J be the space of its linear functionals; define $x \rho \lambda [x \in I, \lambda \rho J]$ to mean $\lambda(x) = O$. Show that the lattices of weakly closed subspaces of I and J are dually isomorphic.

*6. Let M_n be the class of all complex $n \times n$ matrices: let $A \rho B [A, B \in M_n]$ mean $AB = BA$. Apply the Frobenius–Burnside–Schur Theorem to ascertain how Theorem 19 applies to this case of Example 2.

---

† See O. Frink, Duke Math. J. **29** (1962), 505–14; W. C. Nemitz, Trans. AMS **115** (1965), 128–42.

7. In Example 8, describe specifically the lattice of configurations when F consists of: (a) all linear functions, (b) all affine functions, (c) all homogeneous quadratic functions, (d) all quadratic functions, (e) all polynomial functions, (*f) all analytic functions.

8. Let Ω be the set of all isotone maps $x \to \omega(x)$ of a complete lattice L into itself. Define $x \rho \omega$ to mean $\omega(x) \leqq x$.

(a) Show that, if $X \subset L$, X^* is a semigroup in Ω; and if $\Sigma \subset \Omega$, $\Sigma^\dagger$ is a complete lattice.

(b) Show that, if $X_\alpha \in \Sigma^\dagger$ for all α, then $(\bigcap X_\alpha) \in \Sigma^\dagger$.

9. Let ρ be any relation between the elements of complete lattices L and M such that: (i) $x \rho y$ and $t \leqq x$, $u \leqq y$ imply $t \rho u$, and (ii) $x_\alpha \rho y_\beta$ for all α, β implies $(\bigvee x_\alpha) \rho (\bigvee y_\beta)$. For any $x \in L$ and $y \in M$, define $Y(x)$ and $X(y)$ as the sets of $y_\beta \in M$ such that $x \rho y_\beta$ and of $x_\alpha \in L$ such that $x_\alpha \rho y$. Show that the functions $x \to \sup Y(x)$ and $y \to \sup X(y)$ define a Galois correspondence.

9. Completion by Cuts

Let P be any poset, and let $x \rho y$ $(x, y \in P)$ mean that $x \leqq y$ in P. Then, in the notation of §7, X^* is the set of upper bounds, and $X^\dagger$ the set of lower bounds to X, hence $(X^*)^\dagger$ is the set of all lower bounds to the set of all upper bounds of X.

In particular, if x is any element of P, then x^* is the set of $u \geqq x$ and $(x^*)^\dagger$ is the principal ideal of elements $t \leqq x$. Hence if $x > y$, then $(x^*)^\dagger > (y^*)^\dagger$. Moreover, if $a = \inf X$, then $t \leqq a$ if and only if $t \leqq x$ for all $x \in X$; hence $(a^*)^\dagger$ is the intersection of the sets $(x^*)^\dagger$, $x \in X$. This proves the following representation theorem.

THEOREM 21. *Any poset P is isomorphic to a family $\phi(P)$ of subsets of P in such a way that infima in P (when they exist) go into intersections.*

COROLLARY. *The inclusion relation on sets is completely characterized by postulates* P1–P3. (See §I.2.)

A modification of the preceding proof shows that Dedekind's celebrated construction of irrational numbers by "cuts" actually works in any poset.

THEOREM 22 (MAC NEILLE†). *Any poset P can be embedded in a complete lattice L, so that inclusion is preserved, together with g.l.b. and l.u.b. existing in P.*

PROOF. First adjoin a O to P, unless P has a least element already. Then let L consist of all nonvoid "closed" subsets $X = (X^*)^\dagger$ of P; this L is a complete lattice by Theorem 19. By Theorem 21, the correspondence $a \to (a^*)^\dagger$ embeds P in L with preservation of inclusion and g.l.b.; $(a^*)^\dagger$ is the principal ideal of all $x \leqq a$ in L. Now suppose $a = \sup X$ in P. For $T \subset L$, $(T^*)^\dagger \supset (X^*)^\dagger$ in L if and only if $T^* \subset X^*$, by (9); but $X^* = a^*$ by definition of l.u.b. Hence $(T^*)^\dagger \supset (X^*)^\dagger$ if and only if $T^* \subset a^*$, or, by (7), if and only if $(T^*)^\dagger \supset (a^*)^\dagger$. Hence $(a^*)^\dagger$ is a l.u.b. of $(X^*)^\dagger$ in L, whence l.u.b. are preserved.

Note that if L is a lattice, and X is any subset of L, then $(X^*)^\dagger$ is an ideal. For if $a, b \in (X^*)^\dagger$, then since $a \leqq y$, $b \leqq y$ for any $y \in X^*$, $a \vee b \leqq y$ for all

† For the original proof, see H. Mac Neille, [**1**]. For some extensions, see G. Bruns, J. Reine Angew. Math. **209** (1962), 167–200.

$y \in X^*$, and $a \vee b \in (X^*)^\dagger$. Also, $t \leqq a \in (X^*)^\dagger$ trivially implies $t \in (X^*)^\dagger$. This makes the following terminology legitimate.

DEFINITION. A *closed ideal* of a poset P is a subset of P which contains (in fact, consists of) all lower bounds to the set of its upper bounds.

THEOREM 23. *A lattice L is complete if and only if every closed ideal of L is a principal ideal.*

PROOF. Let every closed ideal be principal and $X \subset L$. Then the ideal $(X^*)^\dagger$ is closed by (9), and hence is the principal ideal determined by some $a \in L$. By (8), $X \subset (X^*)^\dagger$ so that $x \leqq a$ for all $x \in X$, and $a \in (X^*)^\dagger$ implies $a \leqq b$ for every upper bound b of X, for then $b \in X^*$. Hence $a = \sup X$, and so L is complete.

Conversely, let L be complete. If J is a closed ideal, $J = (J^*)^\dagger$; let $a = \sup J$. Then $a = \inf J^*$. If $x \in J$, then $x \leqq a$ and if $x \leqq a$ then $x \in (J^*)^\dagger = J$. Hence J is the principal ideal determined by a.

Conditional completion. We now prove a refinement of Theorem 22, for use in Chapter XV.

COROLLARY. *The conditional completion $D^\#$, by nonvoid cuts, of any bidirected set D is a conditionally complete lattice.*

PROOF. Obviously $D^\#$ is obtained from $\bar{D}$, the completion by cuts of D, by deleting O and I unless these exist in D. Hence $D^\#$ is always a conditionally complete poset. If D is bidirected, and a and b are in $D^\#$, then one can find $u, v \in D$ such that the principal ideals of $x \leqq u$ and $y \leqq v$ in D both contain the lower parts of the "cuts" defining a and b. Hence $D^\#$ is also directed, and so (being conditionally complete) it is a join-semilattice. The proof can be finished by appealing to the Duality Principle.

Example 9. Consider the modular ortholattice $M(\mathfrak{H})$ of all closed subspaces of Hilbert space $\mathfrak{H}$ which have finite dimension or codimension. Then the completion of $M(\mathfrak{H})$ by cuts is isomorphic with the *nonmodular* (weakly orthomodular) lattice $L(\mathfrak{H})$ of all closed subspaces of $\mathfrak{H}$ (§II.14, Example 9).

Contrast with ideal completion. It is interesting to contrast completion by cuts (alias "normal" completion) with the ideal completion $L \to \hat{L}$ of §2. Whereas the latter preserves the modular and distributive laws (when they hold in L), Example 9 shows that this is not true of completion by cuts.

On the other hand, the notion of completion by cuts has the advantage of being *self-dual*—whereas ideal-completion is not. Also, as will be shown in §11, the completion by cuts of any Boolean algebra is again a Boolean algebra—while this is also not true of ideal-completion.

Still other methods of completion will be studied in Chapter X, ("metric completion") and Chapter XV.

Exercises:

1. A "segment" of a poset has been defined (W. D. Duthie) as an intersection of closed intervals. Show that an ideal of a lattice is a segment if and only if it is closed.
2. Show that the closed ideals of a lattice L are not necessarily a sublattice of $\hat{L}$, the lattice of all ideals of L.

3. Draw a diagram of the completion by cuts of the poset P_6 of Figure 1e of §I.3.

4. Prove that any conditionally complete bidirected set is a lattice.

5. (a) Show that if $=$ and $\leqq$ are interpreted in the sense of isomorphism and monomorphism, then the operation $P \rightarrow L(P)$ of "completion by cuts" has the following characteristic† properties: $P \leqq L(P)$; $P \leqq Q$ implies $L(P) = L(Q)$; $L(L(P)) = L$.

(b) Show that it is self-dual, in the sense that $L(\breve{P})$ is dually isomorphic to $L(P)$.

6. Let $\mathbf{R} = (-\infty, +\infty)$ be the real axis in its natural ordering. Show that the completion by cuts of $\mathbf{R}^2$ is *not* isomorphic to the square of the completion by cuts of R.

*7. Show that breadth and dimension are both preserved under completion by cuts.

*8. Show that the completion by cuts of the (complemented) modular ortholattice of all subspaces of Hilbert space which have finite dimension or codimension is nonmodular.

*9. Show that, if A is a complete Boolean lattice and J an ideal of A, then A/J is complete if and only if $\bar{J}/J$ is complete. (Ph. Dwinger)

*10. Construct a distributive lattice L which cannot be embedded "regularly" (i.e., preserving all $\bigwedge$ and $\bigvee$ which exist in L) in any complete distributive lattice.‡

10. Complete Brouwerian Lattices

Brouwerian lattices were defined in §II.11 as lattices in which any two elements a, b has a "relative pseudo-complement" $b : a$, the largest element x with $a \wedge x \leqq b$. We now give the class of *complete* Brouwerian lattices a simple alternative characterization.

THEOREM 24. *A complete lattice is Brouwerian if and only if the join operation is completely distributive on meets, so that*

L6*. $a \wedge \bigvee x_\alpha = \bigvee (a \wedge x_\alpha)$ *for any set* $\{x_\alpha\}$.

PROOF. If L is a complete Brouwerian lattice, for any set $\{x_\alpha\}$ let $b = \bigvee(a \wedge x_\alpha)$. By definition of sup, $a \wedge x_\alpha \leqq b$ for all α; hence every $x_\alpha \leqq b : a$, and so $\bigvee x_\alpha \leqq b : a$. Substituting in the identity $a \wedge (b : a) \leqq b$, we obtain $a \wedge \bigvee x_\alpha \leqq b = \bigvee (a \wedge x_\alpha)$. Combining with the distributive inequality, we get L6*.

Conversely, given elements a, b of any complete lattice B which satisfies L6*, let $X = X(a, b)$ be the set of all elements x_α of B such that $a \wedge x_\alpha \leqq b$, and let $b : a = \bigvee_X x_\alpha$. Then, by L6*,

$$a \wedge (b : a) = a \wedge \bigvee_X x_\alpha = \bigvee (a \wedge x_\alpha) \leqq b.$$

Hence $b : a \in X$, and so X contains a largest element. Clearly $a \wedge x \leqq b$ if and only if $x \leqq b : a$; hence $b : a$ is indeed the desired relative pseudocomplement. This proves that L is Brouwerian. Q.E.D.

It is a corollary that any $\bigcup$-ring of sets is a complete Brouwerian lattice, since the operations referred to in L6* are then just set-union and (binary) set-intersection. Hence the open sets of any topological space (§4) form a complete Brouwerian lattice. It will be proved in Theorem VI.9 that the congruence relations on any lattice also form a complete Brouwerian lattice.

† G. Birkhoff, Ann. of Math. **38** (1937), 57–60.

‡ P. Crawley, Proc. AMS **13** (1962), 748–52.

THEOREM 25 (M. H. STONE). *The ideals of any distributive lattice L form a complete Brouwerian lattice.*

PROOF. As shown in §2, the ideals of *any* lattice form a complete lattice. Hence it suffices to prove that any two ideals A and B of L have a relative pseudocomplement $B : A$. But the elements $c \in L$ such that $a \wedge c \in B$ for every $a \in A$ form such an ideal $C = B : A$. For (i) $A \wedge X \subset B$ and $x \in X$ imply that for every $a \in A$, $a \wedge x \in B$, whence $X \subset C$; while conversely (ii) $a \wedge c \leqq b$ and $a \wedge c_1 \leqq b$ imply

$$a \wedge (c \vee c_1) \leqq (a \wedge c) \vee (a \wedge c_1) \leqq b,$$

whence C is an ideal.

11.* Theorem of Glivenko

For any fixed element c of any Brouwerian lattice L, the lemma at the end of §8 shows that the function

$$f_c\colon a \to c : a = a^c \tag{17}$$

defines a symmetric Galois connection on L. Hence:

$$\begin{gathered} a \leqq a^{cc},\ a^c = a^{ccc},\ a \leqq b \text{ implies } a^c \geqq b^c, \\ (a \vee b)^c = a^c \wedge b^c, \text{ and } (a \wedge b)^c \geqq a^c \vee b^c. \end{gathered} \tag{18}$$

Further, the set of "closed" elements satisfying $a = a^{cc}$ forms a complete lattice C, in which joins are given by the new binary operation $a \nabla b = (a \vee b)^{cc}$, while the meet operation is the same as in L. We now prove a less obvious result.

LEMMA. *In any Brouwerian lattice,*

$$(a \wedge b)^c = a^c \nabla b^c. \tag{19}$$

PROOF. By the definition of a^c as $c : a$, evidently $a \wedge a^c \leqq c$ for all $a \in L$. Replacing a by a^c, we obtain

$$a^c \wedge a^{cc} \leqq c \quad \text{for all } a \in L. \tag{20}$$

Now suppose $x \wedge a \wedge b \leqq c$ in L; define $y = x \wedge a^{cc} \wedge b^{cc}$. Clearly $y \leqq x$, whence $y \wedge a \wedge b \leqq c$, which implies $y \wedge a \leqq c : b = b^c$. But $y \wedge a \leqq y \leqq b^{cc}$ by definition of y; hence $y \wedge a \leqq b^c \wedge b^{cc} \leqq c$ by (20). This implies $y \leqq c : a = a^c$; but $y \leqq a^{cc}$ by its definition; hence $y \leqq a^c \wedge a^{cc} \leqq c$, by (20). In summary, $x \wedge a \wedge b \leqq c$ implies $x \wedge a^{cc} \wedge b^{cc} \leqq c$, or $(a \wedge b)^c \leqq (a^{cc} \wedge b^{cc})^c$. But the reverse inequality is obvious; hence $(a \wedge b)^c = (a^{cc} \wedge b^{cc})^c$.

On the other hand, $(a^c \wedge b^c)^c = a^{cc} \wedge b^{cc}$, and so

$$(a \wedge b)^c = (a^{cc} \wedge b^{cc})^c = ((a^c \vee b^c)^c)^c = a^c \nabla b^c,$$

proving the lemma.

Specializing to the case $c = O$, we obtain the preceding results with a^c replaced by a^*. In this case, since $(a \vee a^*)^* = a^* \wedge a^{**} = O$, $a \nabla a^* = O^* = I$:

$$a \wedge a^* = O \quad \text{and} \quad a \nabla a^* = I. \tag{21}$$

Therefore the closed elements form a Boolean lattice.

From this fact, there follows the first statement of the following remarkable theorem, due essentially to V. Glivenko.

THEOREM 26. *If L is a Brouwerian lattice, then the correspondence $a \to a^{**}$ is a closure operation on L, and a lattice-epimorphism of L onto the* (*complete*) *Boolean lattice C of "closed" elements of L. Moreover $a^{**} = b^{**}$ if and only if $a \wedge d = b \wedge d$ for some "dense" $d \in L$ such that $d^{**} = I$.*

To prove the second statement, suppose that $a \wedge d = b \wedge d$, where $d^{**} = I$. Then

$$a^{**} = a^{**} \wedge I = a^{**} \wedge d^{**} = (a \wedge d)^{**}.$$

Similarly, $b^{**} = (b \wedge d)^{**}$; but $a \wedge d = b \wedge d$; hence $a^{**} = b^{**}$. Conversely, suppose a^{**}; set $d = (a \vee b^*) \wedge (a^* \vee b)$. Then

$$d^{**} = (a^{**} \nabla b^*) \wedge (a^* \nabla b^{**}) = (b^{**} \nabla b^*) \wedge (a^* \vee a^{**}) = I,$$

by what was proved earlier. Moreover

$$\begin{aligned} a \wedge d &= a \wedge (a \vee b^*) \wedge (a^* \vee b) = a \wedge (a^* \vee b) \\ &= (a \wedge a^*) \vee (a \wedge b) = a \wedge b. \end{aligned}$$

Similarly, $b \wedge d = a \wedge b$, whence $a \wedge d = b \wedge d$, completing the proof.

In Theorem 26, C is complete if L is.

The preceding result can be applied to the complete Brouwerian lattice $L(A)$ of all ideals of a *given* Boolean algebra A. In this case, $a \wedge x' = O$ for all $x \in X$, an arbitrary subset of A, if and only if $a \leqq x$ for all $x \in X$—i.e., if and only if a belongs to the *closed* ideal consisting of the lower bounds of X (cf. §10). From this result there follows

THEOREM 27 (GLIVENKO–STONE).† *The completion by cuts $C(A)$ of any Boolean algebra A is a Boolean algebra; moreover the correspondence $J \to J^{**}$ is a lattice-epimorphism of the lattice $\hat{A}$ of all ideals of A onto $C(A)$.*

PROOF. We have just shown that the closed elements of $L(A)$ correspond to the closed ideals of A, where $J \subset K$ in $L(A)$ means that $J \leqq K$ in the completion of A by cuts. Hence the Boolean algebra C of Theorem 26 is isomorphic to $C(A)$, and the function $J \to J^{**}$ is lattice-epimorphism. Q.E.D.

Stone lattices. It is not hard to prove‡ that the following conditions are equivalent in any Brouwerian lattice: (i) $a^* \vee a^{**} = I$ for all a, (ii) $a^* \vee b^* = (a \wedge b)^*$ for all a, b, (iii) the Boolean algebra of all closed elements is a sublattice, (iv) every element a^* ($a \in L$) is complemented.

† V. Glivenko, Bull. Acad. Sci. Belgique **15** (1929), 183–8; M. H. Stone [**3**]; A. Monteiro, Revista Union Mat. Argentina **17** (1955), 149–60.

‡ O. Frink, Duke Math. J. **29** (1962), 505–14; J. Varlet, Mém. Soc. Roy. Sci. Liège 8 (1963), 1–71. Stone lattices were first studied seriously by G. Grätzer and E. T. Schmidt, Acta Math. Acad. Sci. Hungar. **8** (1957), 455–60.

Brouwerian lattices which satisfy any (hence all) of conditions (i)–(iv) are called "Stone lattices"; it is known that the lattice of all ideals of any complete Boolean algebra (or Stone lattice!) is itself a complete Stone lattice.

Exercises for §§10–11:

1. In each of the two nondistributive lattices of five elements, find a pair a, b such that $a:b$ does not exist.
2. Show that the nonmodular lattice N_5 and all its interval-sublattices are pseudo-complemented, but N_5 is not relatively pseudo-complemented.
3. In the lattice graphed to the right, find a, b such that

$$(a \wedge b)^* = a^* \vee b^*.$$

4. Show that a Brouwerian lattice is a Boolean algebra if either: (i) $(x^*)^* = x$ for all x, or (ii) $x \wedge x^* = O$ for all x.
5. (a) Prove that every chain is a Brouwerian lattice.
   (b) Construct a relatively complemented lattice which is not relatively pseudo-complemented.
6. Show that any distributive lattice is isomorphic with a sublattice of a relatively pseudo-complemented lattice.

*7. Show that a Brouwerian lattice is a Stone lattice if and only if any two distinct minimal prime ideals are coprime. (Grätzer-Schmidt)

PROBLEMS (See also Chapter IX)

31. Characterize to within isomorphism the (Noetherian) lattice of all algebraic varieties in affine and projective n-space over: (a) the complex field, (b) the real field, (c) the rational field, (d) a general field.
32. Characterize the class of (complete) lattices whose lattice-epimorphic images are all complete.
33. What are the complete distributive lattices which satisfy both laws of (2) (i.e., are Brouwerian and dually Brouwerian)? Which are countable?†
34. Is the center of any complete lattice a closed sublattice? (S. Holland)
35. Is the completion by cuts of every uniquely complemented lattice uniquely complemented? (A. G. Waterman)
36. Is the completion by cuts of every orthomodular lattice an orthomodular lattice? (A. Ramsay)
37. Does there exist a nontrivial complete Boolean algebra which has no proper automorphism? (B. Jonsson)

---

† For a related problem, see G. Raney, Proc. AMS **4** (1953), 518–22, and Trans. AMS **97** (1960), 418–26. For Problem **34**, see Theorem XI.14′.

CHAPTER VI

UNIVERSAL ALGEBRA

1. Algebra

"Universal algebra" provides *general* theorems about *algebras* with single-valued, universally defined, finitary operations. This concept may be defined as follows.†

DEFINITION. An *algebra* A is a pair $[S, F]$, where S is a nonempty set of elements, and F is a specified set of *operations* f_α, each mapping a power $S^{n(\alpha)}$ of S into S, for some appropriate nonnegative *finite* integer $n(\alpha)$.

Otherwise stated, each operation f_α assigns to every $n(\alpha)$-ple $(x_1, \cdots, x_{n(\alpha)})$ of elements of S, a value $f_\alpha(x_1, \cdots, x_{n(\alpha)})$ in S, the result of performing the operation f_α on the sequence $x_1, \cdots, x_{n(\alpha)}$. If $n(\alpha) = 1$, the operation f_α is called *unary*; if $n(\alpha) = 2$, it is called *binary*; if $n(\alpha) = 3$, it is called *ternary*, etc. When $n(\alpha) = 0$, the operation f_α is called *nullary*; it selects a fixed element of S (e.g., a group identity, or a lattice O or I).

Example 1. A group‡ is a set with one binary operation $f(x, y) = xy$ and one unary operation $g(x) = x^{-1}$, satisfying the identities

$$f(x, f(y, z)) = f(f(x, y), z), \qquad \text{(Associative Law)}, \tag{1a}$$

$$f(f(g(x), x), y) = f(y, f(g(x), x)) = y. \tag{1b}$$

Example 2. A lattice is a set with two binary operations $f(x, y) = x \wedge y$ and $g(x, y) = x \vee y$ satisfying identities L1–L4 of Chapter I: $f(x, x) = g(x, x) = x$, $f(x, y) = f(y, x)$, etc.

Groups, lattices, and rings are all *families* of algebras: all algebras of each family are assumed to have the same set F of operations and to satisfy a specified set of postulates. Such "families" of algebras will be studied in §§6–12 below; for the present, we consider them only as providing examples of individual algebras.

Example 3. A vector space over a division ring D is a set with one binary operation $f(x, y) = x + y$ and a unary operation $f_\lambda(x) = \lambda x$ for each $\lambda \in D$, called (left-) multiplication by λ. These operations are assumed to satisfy certain commutative, associative and distributive laws [**B-M**, Chapter VII, §2].

Note that, in Example 3, the number of distinct operations is usually infinite, a possibility allowed by our definition. Note also that the set of operations depends on D; hence the class of all vector spaces (with unspecified D), is *not* a "family" of algebras.

† G. Birkhoff [**1**], [**3**], and Proc. First Canadian Math. Congress (1945), 310–26.

‡ Other, equivalent, definitions of a group are given in §12 below.

Example 4. A field, defined as a set S with two binary operations $+$ and $\cdot$ and two unary operations $x \to -x$ and $x \to x^{-1}$, is *not* an algebra in the sense defined above, because 0^{-1} is not defined. We could make any field into an algebra, by defining 0^{-1} as 0, but we would then sacrifice the familiar identity $(xx^{-1})y = y$ for all x, y.

Example 5. A σ-lattice, defined as closed under two *countable* operations $\bigwedge_{k=1}^{\infty} x_k$ and $\bigvee_{k=1}^{\infty} x_k$, subject to certain identities, is also *not* an abstract algebra, because its operations are not finitary; they apply to infinite sets of elements.

Many results of universal algebra can be extended to sets with infinitary operations ("infinitary algebras"), and to operations which are not universally defined ("partial algebras"). For example, one can consider many topological spaces (Chapter IX) as specified by a single "convergence" operation $x_n \to x$. The definitions of subalgebra, morphism, and direct product given below extend to the topological concepts of closed subspaces, continuous mapping, and Cartesian product, respectively. But we shall not discuss these extensions here.

Universal algebra falls somewhere between mathematics and mathematical logic. Some of its results are "theorems about theorems", and in this sense they are *metamathematical.* We shall concentrate here on the mathematical aspects of universal algebra, referring the reader to the literature† for the metamathematical aspects.

2. Subalgebras

By a *subalgebra* of an abstract algebra $A = [S, F]$ is meant a (possibly void‡) subset T of S which is closed under the operations of F, or F-*closed.* That is, we require that if $f_\alpha \in F$ and $x_1, \cdots, x_{n(\alpha)} \in T$, then $f_\alpha(x_1, \cdots, x_{n(\alpha)}) \in T$. The couple $[T, F]$ is then also an abstract algebra. Evidently, if $[T, F]$ is a subalgebra of $[S, F]$, and $[U, F]$ is a subalgebra of $[T, F]$, then $[U, F]$ is a subalgebra of $[S, F]$.

When applied to Examples 1–3 above, the preceding definition specializes to the concepts of subgroup, sublattice, and (vector) subspace, respectively. Note that, with the subspace O, "different" operations of multiplication by scalars are the *same* as functions. Also, the "same" operation defines "different" functions (with different domains and codomains) when restricted to a subalgebra. For applications to algebraic problems, it turns out to be simpler to adopt the conventions used below, than to use those now standard for functions; see also §6.

We now show that the subalgebras of any abstract algebra A form a *Moore family* of subsets of A (§V.1).

THEOREM 1. *Any intersection $\bigcap T_\tau$ of subalgebras T_τ of A is a subalgebra, and A is a subalgebra of itself.*

For, if $x_1, \cdots, x_{n(\alpha)} \in \bigcap T_\tau$, then $x_1, \cdots, x_{n(\alpha)} \in T_\tau$ for all subalgebras T_τ of the given family, hence $f_\alpha(x_1, \cdots, x_{n(\alpha)}) \in T_\tau$ for all T_τ and so $f_\alpha(x, \cdots, x_{n(\alpha)}) \in \bigcap T_\tau$.

† A. Robinson, *On the metamathematics of algebra*, Amsterdam, 1951; A. Tarski, Proc. Internat. Congr. Math. Cambridge, 1950, Vol. I, pp. 705–20; P. M. Cohn [UA].

‡ If A contains a least nonvoid subalgebra (such as the identity if A is a group), the void set is not considered to be a subalgebra.

COROLLARY 1. *The subalgebras of any abstract algebra A form a complete lattice.*

Conversely, any lattice L is isomorphic with the lattice $\hat{L}$ of all its principal ideals. If L is finite, then these are the "subalgebras" of $A = [L, F]$ relative to the binary join operation $a \vee b$ and the unary projection operations $\psi_c: a \to a \wedge c$. Hence any *finite* (complete) lattice is isomorphic with the lattice of all subalgebras of a suitable algebra.

The analogous facts for infinite complete lattices are much less simple; see §VIII.5.

The intersection $\bar{T} = \bigcap S_\alpha$ of all subalgebras S_γ of A which contain a given subset T of $A = [S, F]$ is called the subalgebra *generated* by T. It follows from Theorem 1 and the results of §V.1, that the correspondence $T \to \bar{T}$ is a *closure* operation on the subsets of A.

One can also define $\bar{T}$ recursively in a (finitary) algebra as the set of all values of *compound operations* or *polynomials* with arguments in T, which is itself called a set of *generators* of $\bar{T}$. This approach will be developed in §8 below.

Exercises for §§1–2:

1. Let $A = [S, F]$ be a *unary algebra*, having only unary and 0-ary operations. Show that $X \cup Y = \bar{X} \cup \bar{Y}$ for any two subalgebras of A, but that the lattice of all subalgebras of A need not be distributive.
2. Let $A = [S, F]$ have only 0-ary operations. Show that the subalgebras of A form a Boolean lattice, which is a principal dual ideal in the lattice $\mathbf{2}^S$.
3. Consider a group as an algebra $G = [S, F]$ with a single binary operation xy^{-1}. Obtain a set of postulates equivalent to (1a)–(1b).
4. Construct a complete set of postulates for commutative rings, as algebras $R = [S, F]$ with binary operations $-$, $\cdot$. Show that subalgebras are subrings.
5. In Exercise 4, augment F by the unary operations $f_a(x) = ax$ ("translations"), for each $a \in S$. What are "subalgebras"?
6. Extend the results of Exercises 4–5 to linear associative algebras.
7. Show that any complete lattice is isomorphic with the lattice of all subalgebras of some "infinitary algebra". (G. Birkhoff [**1**, Theorem 5.1])

3. Morphisms

In §§3–11, we shall fix the set F of operations f_α and the corresponding integers $n(\alpha)$, and shall consider algebras $A = [S, F]$ with variable S. We shall call such algebras *similar*. Our considerations will therefore be applicable to groups, lattices, rings, left-modules over a fixed ring R, etc.; we begin with the concept of morphism.

DEFINITION. Let $A = [S, F]$ and $B = [T, F]$ be similar abstract algebras. A function $\phi: S \to T$ is a *morphism*† of A into B if and only if, for all $f_\alpha \in F$ and $x_i \in S$,

$$f_\alpha(x_1\phi, \cdots, x_n\phi) = (f_\alpha(x_1, \cdots, x_{n(\alpha)}))\phi. \tag{2}$$

A morphism of A *onto* B is called an *epimorphism*; a one-one morphism is called a *monomorphism*. An *isomorphism* of A with B is defined to be a one-one morphism

† Also often called homomorphism.

from A onto B; an *automorphism* is an isomorphism of an algebra with itself, and an *endomorphism* is a morphism of an algebra into itself.

If $\phi: A \to B$ and $\psi: B \to C$ are morphisms, then their composite $\phi \circ \psi: A \to C$ is also a morphism, as follows by direct substitution into (2). In particular, this is true of the endomorphisms of an algebra into itself. Since the composition of mappings (functions) is associative, we have

THEOREM 2. *The endomorphisms of any abstract algebra form a monoid with unit.*

If a morphism $\phi: A \to B$ is a *bijection*, then its inverse ϕ^{-1} is also an automorphism. This follows since

$$f_\alpha(x_1, \cdots, x_n)\phi^{-1} = [f(x_1\phi^{-1}\phi, \cdots, x_n\phi^{-1}\phi)]\phi^{-1}$$
$$= [f(x_1\phi^{-1}, \cdots, x_n\phi^{-1})]\phi\phi^{-1} = f(x_1\phi^{-1}, \cdots, x_n\phi^{-1}).$$

COROLLARY. *The automorphisms of any abstract algebra A form a group*, Aut A.

Conversely, one can show that any group G is isomorphic with the group of all automorphisms of a suitable abstract algebra—in fact,† $G \cong$ Aut L for a suitable distributive lattice L. Likewise, every semigroup with unit is the semigroup of all endomorphisms of a suitable unary algebra, as has been shown by Armbrust and Schmidt.

THEOREM 3. *Under any morphism ϕ of A into B,* (i) *if T is a subalgebra of A, then $\phi(T)$ is a subalgebra of B, and* (ii) *if U is a subalgebra of B, then $\phi^{-1}(U)$ is a subalgebra of A.*

PROOF OF (i). Given $f \in F$ and $y_1, \cdots, y_n(\alpha) \in \phi(T)$, choose $x_1, \cdots, x_{n(\alpha)} \in T$ such that $\phi(x_i) = y_i$. Since T is a subalgebra, $f_\alpha(x_1, \cdots, x_{n(\alpha)}) \in T$; hence, using (2), $f_\alpha(y_1, \cdots, y_{n(\alpha)}) = f_\alpha(\phi(x_1), \cdots, \phi(x_{n(\alpha)})) = \phi(f_\alpha(x_i, \cdots, x_{n(\alpha)})) \in \phi(T)$ which shows that $\phi(T)$ is a subalgebra.

PROOF OF (ii). Given $f_\alpha \in F$ and $x_1, \cdots, x_{n(\alpha)} \in \phi^{-1}(U)$, by definition of ϕ^{-1}, every $y_i = \phi(x_i) \in U$. Hence, by (2),

$$\phi(f_\alpha(x_1, \cdots, x_{n(\alpha)})) = f_\alpha(\phi(x_1), \cdots, \phi(x_{n(\alpha)})) = f_\alpha(y_1, \cdots, y_{n(\alpha)}) \in U$$

since U is a subalgebra. Hence $f_\alpha(x_1, \cdots, x_{n(\alpha)}) \in \phi^{-1}(U)$, completing the proof.

In particular, if T is generated by k elements $x_1, \cdots, x_k$, then $\phi(T)$ is generated by $\phi(x_1), \cdots, \phi(x_k)$.

One can augment the set F of original operations of an algebra $A = [S, F]$ by adding either (i) the set F_1 of all automorphisms of A, (ii) the set F_2 of all endomorphisms of A, or (iii) the set F_3 of all endomorphisms of A onto A. Note that $F_1 \leqq F_3 \leqq F_2$, with $F_3 = F_1$ if A is finite. The subalgebras of $A_1 = [S, F \cup F_1]$ are called *characteristic* subalgebras of A (in symbols, $S \trianglelefteq A$); those of $A_2 = [S, F \cup F_2]$, *fully characteristic* subalgebras of A; and those of $A_3 = [S, F \cup F_3]$, *strictly characteristic*‡ subalgebras of A.

† G. Birkhoff, Revista Union Mat. Argentina **11** (1946), No. 4, R. Frucht, Canad. J. Math. **2** (1950), 417–9, No. 1; M. Armbrust and J. Schmidt, Math. Ann. **154** (1964), 70–72.

‡ Generalizing the corresponding notions for subgroups; see R. Baer, Bull. AMS **50** (1944), 143–60.

The operations of F_1, F_2, and F_3 all being unary, it follows that the lattices of characteristic, fully characteristic, and strictly characteristic subalgebras of A are all *closed sublattices* of the complete lattice of all subalgebras of A.

Most of the results in §§1–3 are true also for partial and infinitary algebras.

Exercises:

1. Show that if A has a unique one-element subalgebra A, then this subalgebra is fully characteristic.
2. Define the ϕ-subalgebra of $A = [S, F]$ as the intersection of the maximal proper subalgebras of A. Show that M is a characteristic subalgebra of A (in symbols, $M \lhd A$).
3. Show that if $B \lhd A$ and $C \lhd B$, then $C \lhd A$.
4. Prove analogs of the results of Exercise 3 for fully and for strictly characteristic subalgebras.
5. Show that, in the group D_4 of the square, though the center Z and rotation-subgroup $\{R\} \supset Z$ are characteristic subgroups, $\{R\}/Z$ is not a characteristic subgroup of D_4/Z.
6. Show that the monomorphisms of any algebra A into itself form a semigroup.
7. Let A be any algebra, and let $G(A)$ be the group of its automorphisms. From the relation $a \rho \gamma$ meaning $a = \gamma(a)$ $(a \in A, \gamma \in G(A))$, use polarity to set up a "Galois theory" relating certain subalgebras of A to certain subgroups of $G(A)$.

4. Congruence Relations

We now show that the epimorphic images $\phi(A)$ of an abstract algebra A can all be determined, up to isomorphism, by studying the equivalence relations on A. We recall that an "equivalence relation" is a binary relation $x\theta y$, also written $x \equiv y \pmod{\theta}$, which is reflexive, symmetric and transitive; A/θ will denote the set of all equivalence classes on A.

DEFINITION. A *congruence relation* on an algebra $A = [S, F]$ is an equivalence relation θ of A such that, for all $f_\alpha \in F$, $x_i \equiv y_i \pmod{\theta}$, $i = 1, \cdots, n(\alpha)$, implies $f_\alpha(x_1, \cdots, x_{n(\alpha)}) \equiv f_\alpha(y_1, \cdots, y_{n(\alpha)}) \pmod{\theta}$. (Substitution Property)

THEOREM 4. *Let ϕ be a morphism of A into B. Then if $x\theta y$ means $\phi(x) = \phi(y)$, θ is a congruence relation on A.*

PROOF. Since equality is reflexive, symmetric, and transitive, θ is an equivalence relation. Moreover the identity (2) asserts that if $x_i\theta y_i$ for $i = 1, \cdots, n(\alpha)$, then $f_\alpha(x_1, \cdots, x_{n(\alpha)}) \equiv f_\alpha(y_1, \cdots, y_{n(\alpha)}) \pmod{\theta}$—i.e., that the equivalence relation is a congruence relation. Conversely we have

THEOREM 5. *Let θ be a congruence relation on an abstract algebra $A = [S, F]$, and let $x \to P_\theta(x)$ be the mapping carrying each $x \in S$ into its equivalence class in S/θ. Then the operations f_α on S/θ defined by the formula*

$$f_\alpha(P_\theta(x_1), \cdots, P_\theta(x_{n(\alpha)})) = P_\theta(f_\alpha(x_1, \cdots, x_{n(\alpha)})), \tag{3}$$

define an algebra $B = [S/\theta, F]$ similar to A. Moreover the mapping $x \to P_\theta(x)$ is an epimorphism from A onto B.

PROOF. By the Substitution Property, the functions defined by (3) are single-valued and defined for all $n(\alpha)$-ples of appropriate length in S/θ.

The algebra B of Theorem 5 will be denoted by A/θ; by (3), the mapping

$x \to P_\theta(x)$ is the epimorphism from A onto A/θ. Conversely, if $\phi: A \to B$ is any epimorphism, then by Theorem 4 the elements of B are the equivalence classes of A/θ, and by (2) the operations of B are those of $[S/\theta, F]$. This completes the proof of the following result.

THEOREM 6. *The epimorphic images of any abstract algebra A are the algebras A/θ defined by the congruence relations θ on A.*

In case A is a group, the congruence relations θ on A are the partitions of A into the cosets of its normal subgroups. If A is a ring, its congruence relations are the partitions of A into the residue classes of its ideals.

The congruence relations on an abstract algebra A are *partially* ordered as equivalence relations: $\theta \leqq \theta'$ means that $x \equiv y \pmod{\theta}$ implies $x \equiv y \pmod{\theta'}$. In the poset $\Theta(A)$ of all congruence relations on A, universal bounds O and I are defined by:

(4) $$x \equiv y \pmod{O} \quad \text{if and only if } x = y.$$

(4′) $$x \equiv y \pmod{I} \quad \text{for all } x, y;$$

these are trivial congruence relations.

THEOREM 7. *Let $B = A/\theta$ be any epimorphic image of the abstract algebra A. Then the congruence relations on B are the partitions of B defined by the congruence relations $\theta' \geqq \theta$ on A.*

SKETCH OF PROOF. If $\phi: A \to B$ is the epimorphism defined by θ, and ψ is any congruence relation on B, define $x\theta'y$ in A to mean that $\phi(x) \equiv \phi(y) \pmod{\psi}$ in B. Then θ' is an equivalence relation on A with $\theta \geqq \theta$, enjoying the Substitution Property. Conversely, if $\theta' \geqq \theta$ is a congruence relation on A, and $u\psi v$ in B is defined to mean that $x\theta'y$ for one and hence every $x \in \phi^{-1}(u)$ and $y \in \phi^{-1}(v)$ in A, one gets an equivalence relation ψ on B having the Substitution Property.

The reader should have no difficulty in verifying in detail the statements made above.† Theorem 7 contains as a special case the Second Isomorphism Theorem of group theory.

Next, we define a *translation* of an algebra $A = [S, F]$ as a *unary* operation of the form (for suitable constant elements c_i):

(5) $$g_{\alpha,c,k}(x) = f_\alpha(c_1, \cdots, c_{k-1}, x, c_{k+1}, \cdots, c_{n(\alpha)}), \cdots, f_\alpha \in F.$$

LEMMA. *An equivalence relation on A is a congruence relation if and only if it has the Substitution Property for every translation of A.*

PROOF. Trivially, the condition stated is necessary. It is sufficient, since if $x_k\theta y_k$ for $k = 1, \cdots, n(\alpha)$, then

(5′) $$f_\alpha(y_1, \cdots, y_{k-1}, x_k, \cdots, x_n)\theta f_\alpha(y_1, \cdots, y_k, x_{k+1}, \cdots, x_n)$$

† For a more elaborate discussion of isomorphism theorems, see P. M. Cohn [UA, Chapter II, §6].

for $k = 1, \cdots, n - 1$, by (5) whence by transitivity there follows

$$f_\alpha(x_1, \cdots, x_n)\theta f_\alpha(y_1, \cdots, y_n),$$

as claimed.

THEOREM 8. *The congruence relations on an algebra $A = [S, F]$ form a closed sublattice*† $\Theta(A)$ *of the complete lattice* $E(S)$ *of all equivalence relations on* S.

PROOF. By the lemma, it suffices to consider *unary* operations $g_\gamma(x)$. Now let B be any set of congruence relations θ_β on A. In $E(S)$, $\theta = \bigwedge_B \theta_\beta$ is defined by the condition

$$\text{(6)} \qquad x\theta y \text{ if and only if } x\theta_\beta y \text{ for all } \theta_\beta \in B.$$

But by (5), $x\theta_\beta y$ for all θ_β. Since every θ_β has the Substitution Property, this implies $g_\gamma(x) \equiv g_\gamma(y)$ for all $\theta_\beta \in B$. Therefore, by (6), $g_\gamma(x) \equiv g_\gamma(y) \pmod{\theta}$, whence θ has the substitution property for all g_γ and is a congruence relation.

Dually, $\phi = \bigvee_B \theta_\beta$ is defined by the condition

$$\text{(6')} \qquad \begin{array}{ll} x\phi y \text{ if and only if, for some finite chain} & x = z_0, z_1, \cdots, z_m = y, \\ \text{and associated } \beta(j) \in B, & z_{j-1}\theta_{\beta(j)}z_j \quad \text{for } j = 1, \cdots, m. \end{array}$$

Hence, if $x\phi y$, we will have $z_{j-1} \equiv z_j \pmod{\theta_{\beta(j)}}$ and so $g_\gamma(z_{j-1}) \equiv g_\gamma(z_j) \pmod{\theta_{\beta(j)}}$ for $j = 1, \cdots, m$, if each $\theta_{\beta(j)}$ has the Substitution Property. But this implies $g_\gamma(z_{j-1}) \equiv g_\gamma(z_j) \pmod{\phi}$, by (6′) applied to the $g_\gamma(z_j)$. Therefore, γ being transitive, $g_\gamma(x) \equiv g_\gamma(y) \pmod{\phi}$ and ϕ has the Substitution Property as claimed.

The complete lattice $\Theta(A)$ is called the *structure lattice* of A. When $\Theta(A) \cong \mathbf{2}$, that is, when A has only trivial congruence relations, A is called *simple*.

THEOREM 9 (FUNAYAMA–NAKAYAMA‡). *For any lattice* L, $\Theta(L)$ *is a complete Brouwerian lattice.*

PROOF. That $\Theta(L)$ is a complete lattice follows from Theorem 8. Hence (see §V.10), it suffices to prove that

$$\text{L6*.} \qquad a \equiv b \bmod \theta \wedge \bigvee_C \theta_\gamma \text{ implies } a \equiv b \bmod \bigvee_C (\theta \wedge \theta_\gamma).$$

But $a \equiv b\ (\theta \wedge \bigvee_C \theta_\gamma)$ implies that $y \equiv x(\theta)$ for all $y, z \in [a \wedge b, a \vee b]$ *and* that, for some finite sequence $a = x_0, x_1, \cdots, x_n = b$ of $x_i \in L$, $x_{i-1} \equiv x_i\ (\theta_i)$, where $\theta_i \in C$. We now define $y_i = [(a \wedge b) \vee x_i] \wedge (a \vee b)$; clearly $y_0 = a$, $y_n = b$, and $y_{i-1} \equiv y_i\ (\theta \wedge \theta_i)$. Hence L6* follows by transitivity. Q.E.D.

Exercises:

1. Prove that the congruence relations on a chain (considered as a distributive lattice) are just its partitions into nonoverlapping segments.
2. Prove that any congruence relation on an algebra A induces a congruence relation on every subalgebra of A.

---

† G. Birkhoff [**3**, Theorem 24] proved that $\Theta(A)$ was a sublattice; V. S. Krishnan, J. Madras Univ., Vol. **16**B, p. 16, proved that this sublattice was closed.

‡ N. Funayama and T. Nakayama, Proc. Imp. Acad. Tokyo **18** (1942), 553–4.

3. Prove that, if θ and θ_1 are any two congruence relations on a relatively complemented lattice, then $\theta\theta_1 = \theta_1\theta$.
4. Show that, if G is a group, then θ is characteristic if and only if its kernel is a characteristic subgroup of G.
5. Show that if C is a characteristic subalgebra of A/θ, then its inverse image is a characteristic subalgebra of A.
6. Let G be a group acting on a set S. Show that an equivalence relation on S is a congruence relation on $A = [S, G]$ if and only if it divides G into "sets of imprimitivity".
7. Let A be any algebra, T any subalgebra of A, and θ any congruence relation on A. Show that the set of $x \in A$ such that $x\theta t$ for at least one $t \in T$ forms a subalgebra.
8. For any group G, let $A = [G, F]$, where F is the set of all left-translations $f_a(x) = ax$, $a \in G$. Show that the structure lattice of A is isomorphic to the lattice of all subgroups of G.
9. Show by a counterexample that Theorem 8 fails in infinitary algebras.

5. Direct and subdirect products

From any two similar abstract algebras, $A = [X, F]$ and $B = [Y, F]$ one can construct the *direct product* $A \times B = [X \times Y, F]$ from the elements $[x, y]$ of the Cartesian product $X \times Y$, by defining for any $f_\alpha \in F$ and for $n = n(\alpha)$:

$$f_\alpha([x_1, y_1], \cdots, [x_n, y_n]) = [f_\alpha(x_1, \cdots, x_n), f_\alpha(y_1, \cdots, y_n)].$$

This construction contains as special cases the usual definitions of the direct product of two (multiplicative) groups, the direct sum of two rings, the cardinal product of two lattices, and the direct sum ("biproduct") of any two R-modules over the same ring R.

It is easy to establish the following isomorphisms:

$$A \times B \cong B \times A, \qquad A \times (B \times C) \cong (A \times B) \times C, \tag{7}$$

by using the one-one correspondences $[x, y] = [y, x]$, etc. More generally, let Γ be any set of similar algebras $A_\gamma = [S_\gamma, F]$; define the (unrestricted) direct product $\prod_\Gamma A_\gamma$ as the set of all functions a: $\gamma \to a(\gamma) \in A_\gamma$, with

$$f_\alpha(a_1, \cdots, a_{n(\alpha)}) = b\colon\ \gamma \to f_\alpha(a_1(\gamma), \cdots, a_{n(\alpha)}(\gamma)) \in \prod_\Gamma A_\gamma.$$

This direct product $\prod_\Gamma A_\gamma$ is, clearly, independent of the order and order of combination of the factors.

One can construct congruence relations on the direct product $A \times B$ from congruence relations of the factors as follows.

THEOREM 10. *Let θ_A and θ_B be congruence relations on similar abstract algebras A and B. Then the definition*

$$[a, b] \equiv [a_1, b_1] \quad \textit{if and only if } a\ \theta_A\ a_1 \textit{ and } b\ \theta_B\ b_1, \tag{8}$$

gives a congruence relation on $A \times B$.

PROOF. It is trivial to verify the reflexive, symmetric and transitive laws and the Substitution Property for the relation (8) by applying the law in question to each component separately.

We shall denote the congruence relations (8) by $\theta_A \times \theta_B$. It is easy to prove that $(A/\theta_A) \times (B/\theta_B) = (A \times B)/(\theta_A \times \theta_B)$; we omit the proof.

In general, not every congruence relation on $A \times B$ is of the form $\theta_A \times \theta_B$. For example, if $G = \{0, 1\}$ is the additive group of integers mod 2, then the congruence relation on the four-group $G \times G$ with the equivalence classes $\{[0, 0], [1, 1]\}$ and $\{[1, 0], [0, 1]\}$ is a congruence relation not a product congruence.

Analogous results can be proved for direct products $\prod A_\gamma$ generally; one defines $x \equiv y \pmod{\prod \theta_\gamma}$ if and only if $x_\gamma = y_\gamma \pmod{\theta_\gamma}$ for all γ, that is, if and only if each pair of corresponding components is congruent.

DEFINITION. A subalgebra $C = [S, F]$ of a direct product of similar algebras $A_\gamma = [X_\gamma, F]$ is called a *subdirect product* of the A_γ if, given $x_\gamma \in X_\gamma$, there exists an element $c \in S$ having x_γ for its component in A_γ. An epimorphism $\phi: A \to C$ of an algebra $A = [X, F]$ onto a subdirect product of similar algebras $A_\gamma = [X_\gamma, F]$ is called a *representation* of A as a subdirect product of the A_γ.

THEOREM 11. *If ϕ is a representation of A as a subdirect product C of similar algebras A_γ, then $C \cong A/\bigwedge \theta_\gamma$, where θ_γ is the congruence relation associated with the morphism ϕ_γ mapping A onto $A_\gamma = A/\theta_\gamma$. Conversely, every choice of congruence relations θ_γ on A yields a representation of A as a subdirect product $C = A/\bigwedge \theta_\gamma$ of the algebras $A_\gamma = A/\theta_\gamma$.*

PROOF. By (6), the morphism $\phi: A \to C$ defines a morphism of A onto each A_γ, and by Theorem 4, each such epimorphism is defined by a congruence relation θ_γ, such that $A_\gamma \cong A/\theta_\gamma$. Two elements of A are mapped onto the same element of C if and only if they are congruent modulo every θ_γ; hence $C \cong A/\bigwedge A_\gamma$. Conversely, given an indexed set $\{\theta_\gamma\}$ of congruence relations on A, the natural epimorphisms $\phi_\gamma: A \to A_\gamma$ of A onto the $A_\gamma = A/\theta_\gamma$ define a morphism of A onto a subalgebra C of $\prod A_\gamma$, which is a subdirect product of the A_γ.

COROLLARY 1. *The isomorphic representations of an algebra A as a subdirect product correspond one-one to the sets of congruence relations θ_γ on A such that $\bigwedge \theta_\gamma = O$, the equality relation.*

DEFINITION. An algebra A is called *subdirectly irreducible* when, in any isomorphic representation of A as a subdirect product of algebras A_γ, one of the associated natural epimorphisms $A \to A_\gamma$ is an isomorphism.

Loosely speaking, this means that an algebra is subdirectly irreducible when it cannot be represented as a subdirect product of "smaller" algebras (i.e., proper epimorphic images).

COROLLARY 2. *Let P be the poset of all congruence relations $\theta_\gamma > O$ on an algebra A. Then A is subdirectly irreducible if and only if P has a least element θ_m.*

PROOF. If θ_m exists, then $\theta_\gamma > O$ implies $\theta \geqq \theta_m$; hence $\bigwedge_P \theta_\gamma = O < \theta_m$ implies that some $\theta_\gamma = O$. It follows, by the preceding definition and Corollary 1, that A is subdirectly irreducible. Conversely, if θ_m does not exist, then $\bigwedge_P \theta_\gamma = O$ where all $\theta_\gamma > O$, and so A is not subdirectly irreducible.

Example 6. The only subdirectly irreducibly distributive lattice is the ordinal **2**. (One-element algebras, for which $O = I$ in $\Gamma(A)$, are excluded from consideration here.) For, if $O < a < I$, then the endomorphisms $\phi_a: x \to x \wedge a$ and $\phi'_a: x \to x \vee a$ define proper congruence relations θ_a and θ'_a with $\theta_a \wedge \theta'_a = O$. A similar result holds for Boolean algebras.

The fundamental theorem on subdirect decompositions, to be proved in Chapter VIII, asserts that *every* algebra having more than one element is a subdirect product of subdirectly irreducible algebras. Unique factorization theorems for direct decompositions will be proved in Chapter VII (see also Theorem III.11).

Exercises:

1. Prove that every Boolean algebra having more than two elements is subdirectly reducible.
2. Show that the structure lattice $\Theta(A)$ of any direct product $A = \Pi A_\gamma$ of algebras contains $\Pi\Theta(A_\gamma)$ as a sublattice.
3. Let A, B be the algebras with two elements 0, 1, under addition mod 2 and unary operation defined by $x' = x$ in A and $x' = 1 - x$ in B. Prove that $A \times B \cong B \times B$, though A and B are not isomorphic. (McKinsey)
4. (a) Show that, if A is a group or Boolean algebra, then $\theta \wedge \theta' = O$ and $\theta \vee \theta' = I$ in $\Theta(A)$ imply that $A \cong (A/\theta) \times (A/\theta')$.

   (b) Show that this is not true in every distributive lattice. (*Hint*. Let $A = \mathbf{3}$.)

*5. Show that if A and B are two finite algebras with one unary operation, then $A^2 \cong B^2$ implies $A \cong B$. (Marica-Bryant)

*6. Show that if $A^n \cong B^n$ in Exercise 4, then $A \cong B$.†

6. Free Word Algebras

We now consider the "species" of all algebras having a given set F of operations. More precisely, we suppose given a fixed set of indices a and associated nonnegative integers $n(\alpha)$. We then consider the class ("species") of all algebras $A = [S, F]$ which have for each given index α and $n(\alpha)$-ary operation $f_\alpha: S^{n(\alpha)} \to S$.

Any such "species" of algebras is closed under the formation of subalgebras and direct products, as defined earlier. Furthermore, morphisms are only defined between algebras of the same species (but see §§11–12 below).

For each choice of F and cardinal number r, finite or infinite, we can construct a (free) *word algebra* $W_r(F) = [W_r, F]$ (sometimes called "primitive algebra") having r generators or *letters* x_i, as follows. (The set X of x_i is often called the *alphabet* of $W_r(F)$.)

Call each letter x_i an F-polynomial of *rank* 0. For any positive integer ρ, define an F-polynomial of rank ρ recursively, as any expression ("word") of the form $f_\alpha(u_1, \cdots, u_{n(\alpha)})$, where one $u_j = p_j(x_1, \cdots, x_r)$ is an F-polynomial of rank $\rho - 1$, and every u_j is an F-polynomial of rank $\leqq \rho - 1$. Equality in $W_r(F)$ is defined to mean formal identity: $x_i = x_j$ means $i = j$, and

$$f_\alpha(u_1, \cdots, u_{n(\alpha)}) = f_\beta(v_1, \cdots, v_{n(\beta)}) \tag{9}$$

means that $\alpha = \beta$ and $u_k \equiv v_k$ for all $k = 1, \cdots, n(\alpha) = n(\beta)$.

† O. A. Ivanova, Vestnik Moskov. Ser. I Mat. Meh. No. 3 (1964), 31–8 (Math. Revs. **29**, #2207).

For example, let F consist of two binary operations $\wedge$ and $\vee$ and let $r = 2$. For simplicity, denote x_1 and x_2 by x and y, respectively. Then the elements of rank one in $W_2(F)$ are eight in number.

$$x \wedge x, x \wedge y, y \wedge x, y \wedge y; \quad x \vee x, x \vee y, y \vee x, y \vee y.$$

Those of rank two are $64 + 128 = 192$ in number; there are 64 combinations of x and y with elements of rank one:

$$x \wedge (x \wedge x), \cdots, y \vee (y \vee y); \quad (x \wedge x) \wedge x, \cdots, (y \vee y) \vee y,$$

and 128 combinations of pairs of elements of rank one:

$$(x \wedge x) \wedge (x \wedge x), \cdots, (x \wedge x) \vee (y \wedge y), \cdots, (y \vee y) \wedge (x \vee x), \cdots, (y \vee y) \vee (y \vee y).$$

A more interesting example is the following.

Example 7. Let F consist of a single unary operation f. Then $W_1(F)$ contains for each nonnegative integer ρ just one F-polynomial $f(p_\rho) = p_{\rho+1}$. Hence $W_1(F)$ is isomorphic to the set $\mathbf{N}$ of all nonnegative integers under Peano's successor function $\sigma(n) = n + 1$.

Because of this fact, word algebras are also called "Peano algebras", and a set of axioms for $W_r(F)$ resembling those of Peano for $\mathbf{N}$ can be formulated.† The basic property of (free) word algebras is contained in

THEOREM 12. *Any injection $\delta: X \to A$ of the alphabet X in to an F-algebra $A = [S, F]$ can be (uniquely) extended to a morphism from $W_r(F)$ to A.*

PROOF. By induction on rank, each $p \in W_r(F)$ of rank ρ is mapped onto a unique element $q = p\phi \in A$, so that

(9′) If $p = f_\alpha(u_1, \cdots, u_{n(\alpha)})$, then $q = f_\alpha(u_1\phi, \cdots, u_{n(\alpha)}\phi)$.

By (9)–(9′), the mapping is a morphism (or "F-morphism").

COROLLARY. *If $A = [S, F]$ has r generators, then $A \cong W_r(F)/\theta$ is an epimorphic image of $W_r(F)$.*

Exercises:

1. In the case of two binary operations, how many elements of rank three does W_2 contain?
2. Show that every algebra with countable many generators and operations is countable.
3. Show that $W_r(F)$ cannot be finite unless F is void.
4. (a) Show that in any word algebra $W_r(F)$, the maximal subalgebras are the subsets which omit one generator.

   (b) Prove that the set of generators of any word algebra is the complement of its ϕ-subalgebra.
5. Show that the group of automorphisms of any word algebra is the symmetric group on its generators.
6. (a) Prove that the word algebra $W_1(\sigma)$ with one generator and one unary operator is isomorphic with $\mathbf{Z}^+$ under Peano's successor function.

† See J. Schmidt, Z. Math. Logik Grundlagen Math. **11** (1965), 227–39, where earlier work of Lewig and Slominski is discussed.

(b) Show that the semigroup of endomorphisms of $W_1(\sigma)$ is isomorphic with $\mathbf{Z}^+$ under addition.

7. Prove that the word algebra $W_r(\sigma)$ is isomorphic with the set-union of r copies of $W_1(\sigma)$. Describe the monoid of endomorphisms of $W_r(\sigma)$.

8. (a) Prove that the word algebra with one generator and m unary operators $\sigma_1, \cdots, \sigma_m$ is isomorphic with the right-regular representation of the free monoid FS_m with m generators.

(b) Prove that the endomorphism-monoid of $W_1(\sigma_1, \cdots, \sigma_m)$ is isomorphic with FS_m.

9. Same questions for $W_r(\sigma_1, \cdots, \sigma_m)$.

7. Free Algebras

The morphism property of Theorem 12 is shared (for suitable families of algebras) by many algebras other than the free word algebras discussed above. In general, any algebra $C = [T, F]$ which has this morphism property with respect to a family Γ of similar algebras $A_\gamma = [S_\gamma, F]$ is called a "free algebra" of Γ. The present section will deal with such free algebras and their properties.

The free Boolean algebras constructed in §III.5 are free in this sense, within the family of Boolean algebras. A similar result holds for the free distributive lattices constructed in §III.4, and for the free modular lattice M_{28} constructed in §III.6. It is also shared by the free group with r generators, in the family of groups.† We now define the notion in general.

DEFINITION. Let Γ be any class of similar algebras $A_\gamma = [S_\gamma, F]$. An algebra $C \in \Gamma$ is *freely generated* by a subset $X \subset C$ when: (i) X generates C, and (ii) any function $f: X \to S_\gamma$ can be extended to a morphism $\phi: C \to A_\gamma$.

Since X generates C, it is obvious that this extension is unique. Moreover, if C_1 and C_2 are any two algebras of Γ, "freely generated" by subsets $X_i \subset C_i$ of the same cardinality, then any bijection $\beta: X_1 \leftrightarrow X_2$ can be (uniquely) extended to an (iso)morphism $\mu: C_1 \leftrightarrow C_2$. This proves

THEOREM 13. *In a class Γ of similar algebras, the "free algebra with r generators" is uniquely determined (up to isomorphism) by the cardinal number r, and Γ.*

We now show to construct such free algebras.

DEFINITION. Let Γ be any class of similar algebras $A_\gamma = [S_\gamma, F]$, and let r be any cardinal number. Let Δ be the set of all functions $\delta: x_i \to \delta(x_i) \in A_{\gamma(\delta)}$ from a fixed set X of r symbols x_i into the A_γ. For any polynomial $p(x_1, \cdots, x_r) \in [W_r, F]$, define

$$\delta(p) = p(\delta(x_1), \cdots, \delta(x_r)) \in A_{\gamma(\delta)}. \tag{10}$$

Define $p \equiv q \pmod{\Gamma}$ to mean that $\delta(p) = \delta(q)$ for all such δ; this is a congruence relation $\theta(\Gamma)$ on $[W_r, F]$. Then $[W_r, F]/\theta(\Gamma)$ is called the free algebra with r generators associated with Γ; we denote it $F_r(\Gamma)$.

We now ask: when will $F_r(\Gamma)$ belong to Γ? Though by construction, every injection $\theta: X \to A$ can be extended to a (unique) morphism, this does not imply that $F_r(\Gamma)$ is "freely generated" by X unless $F_r(\Gamma) \in \Gamma$.

Clearly, the preceding construction expresses $F_r(\Gamma)$ as a subalgebra of products of copies of the $A_\gamma \in \Gamma$. Hence

† M. Hall, *Theory of groups*, MacMillan, 1959, Theorem 7.1.2.

LEMMA 1. *Given a class* Γ *of similar algebras* A_γ, $F_r(\Gamma)$ *is a subdirect product of copies of subalgebras of the* A_γ.

COROLLARY. *If* Γ *is closed with respect to formation of subalgebras and direct products, then* $F_r(\Gamma) \in \Gamma$.

In view of our earlier remark, this result gives a sufficient condition for $F_r(\Gamma)$ to be "freely generated" by X.

THEOREM 13'. *Let* Γ *be any family of algebras which is closed under formation of subalgebras and direct products. Then* $F_r(\Gamma)$ *is freely generated by the set* X *of* x_i.

COROLLARY 1. *Every injection* $f\colon X \to F_r(\Gamma)$ *of the generators of* $F_r(\Gamma)$ *into* $F_r(\Gamma)$ *can be extended to an endomorphism of* $F_r(\Gamma)$.

COROLLARY 2. *Let* θ *be any mapping of the set* X *of generators of* $F_r(\Gamma)$ *into the set* Y *of generators of* $F_s(\Gamma)$, $r \leqq s$. *Then* θ *can be extended to a monomorphism* $\bar{\theta}\colon F_r(\Gamma) \to F_s(\Gamma)$.

For, θ has a right-inverse projection $\psi\colon Y \to X$, which can be extended to an epimorphism $\bar{\psi}\colon F_s(\Gamma) \to F_s(\Gamma)$. The extension of θ to a morphism $\bar{\theta}\colon F_r(\Gamma) \to F_s(\Gamma)$ has in consequence a right-inverse $\bar{\psi}$, and is the desired monomorphism.

Since the composite of any two morphisms is a morphism, $F_r(\Gamma)$ remains "free" if Γ is extended to include all epimorphic images of its members. Hence we have

COROLLARY 3. *In the class* $\mathfrak{A}$ *of all algebras constructible from a given algebra* A *by formation of subalgebra, direct product, and epimorphic image, the free algebra* F_r *with* r *generators is a subalgebra of* $A^{o(A)^r}$.†

More precisely, let X be the set of all elements $\bar{x}_i \in A^{o(A)^r}$ whose δ-coordinate is $\delta(x_i) \in A$ for each δ. Then F_r is the subalgebra of $A^{o(A)^r}$ generated by $X = \{\bar{x}_i\}$.

Algebraic independence. Closely related to the above concept of a free algebra is the following notion of algebraic independence, due to Marczewski.‡

DEFINITION. A set of elements $a_1, \cdots, a_r$ of an algebra $A = [S, F]$ is *algebraically independent* if, for any polynomials $p, q \in P_r(F)$, the equation $p(a_1, \cdots, a_r) = q(a_1, \cdots, a_r)$ implies the identity $p(x_1, \cdots, x_r) = q(x_1, \cdots, x_r)$ for all $x_i \in A$.

Thus, an algebra $A = [S, F]$ is freely generated if and only if it has a set of algebraically independent generators (the images of the generators of $F_r(A)$).

Exercises:

1. (a) Why don't "regular" rings form a family of algebras?
   (b) Show that groups are not a "family" of monoids.
2. Show that the free commutative ring with r generators is $\mathbf{Z}(x_1, \cdots, x_r)$, which is an integral domain.
3. Prove that an algebra A is freely generated if and only if it has a set of algebraically independent generators.

---

† Here $o(A)$ is the cardinality of A. For a generalization of Corollary 5, see G. Birkhoff [**3**, p. 441, Corollary 2]; also Proc. First Canadian Math. Congress (1945), p. 321.

‡ E. Marczewski, Fund. Math. **48** (1959), 135–45 (Math. Revs. **22** #2569); see also Math. Revs. 23 #A3106, 23 #A3107, 21 #3364, and 28 #5020.

4. Prove that, for any commutative ring R, the free R-module with r generators is R^r.

5. (a) Show that if A is finite of order r, then the order of the free algebra $F_n(A)$ is at most r^{r^n}.

(b) For arbitrary finite r, construct an algebra A for which the preceding upper bound is attained.

6. A set G of generators of an algebra A is called "independent" when no proper subset of G generates A. The intersection of the maximal subalgebras of A may be called the "ϕ-subalgebra" of A. Show that, if A is finite, its ϕ-subalgebra consists precisely of those elements occurring in *no* set of independent generators of A.

*7. Extend the theory of free algebras to algebras with infinitary operations, illustrating by the case that F consists of just one operation on countable sequences.

8. Why is the result of Exercise 7 compatible with the nonexistence of a "free complete lattice" (§XI.4) with three generators?

9. Show that the free semilattice F_r with r generators satisfies $F_{r+s} \cong F_r \times F_s$.

8. Free Lattices

The preceding notions apply in particular to the class of all lattices. In this case, an explicit test for the relations $p \leqq q$ and $p = q$ between polynomials has been derived by Whitman.† It applies to $W_r(\wedge, \vee)$ by induction on rank ρ, through recursive application of the following four basic rules:

$$p \vee q \leqq a \quad \text{if} \quad p \leqq a \text{ and } q \leqq a, \tag{11}$$

$$b \leqq p \wedge q \quad \text{if} \quad b \leqq p \text{ and } b \leqq q, \tag{11'}$$

$$p \wedge q \leqq a \quad \text{if} \quad p \leqq a \text{ or } q \leqq a, \tag{12}$$

$$b \leqq p \vee q \quad \text{if} \quad b \leqq p \text{ or } b \leqq q. \tag{12'}$$

Clearly (11) and (11′) are dual, and so are (12) and (12′). For instance, if we wish to find out whether $p \vee q \leqq r \vee s$, we first apply test (11) to $p \vee q$, treating $r \vee s$ as a, and then (12′) to $r \vee s$, treating $p \vee s$ as b. Hence $p \vee q \leqq r \vee s$ if and only if (i) $p \leqq r \vee s$ and $q \leqq r \vee s$, or (ii) $p \vee q \leqq r$, or (iii) $p \vee q \leqq s$. In this way the verification of $p \vee q \leqq r \vee s$ is reduced to combinations of four or fewer questions, in each of which the total rank of the expressions involved is reduced by one. Hence, repeating this process of reduction if the sum of the ranks of a and b is $w + w'$, we can test for $a \leqq b$ by $4^{w+w'}$ or fewer elementary tests, of the form $x_i \leqq x_j$ (which is true if and only if $i = j$).

THEOREM 14. *If we define $a\theta b$ in $W_r(\wedge, \vee)$ to mean that $a \leqq b$ and $b \leqq a$, then, under $\leqq$, $W_r(\wedge, \vee)/\theta$ is the free lattice $FL(r)$ with r generators.*

The proof depends on two lemmas.

LEMMA 1. *$FL(r)$ is a quasi-ordered set.*

PROOF. Inductively, $p \leqq p$ and $q \leqq q$ imply $p \leqq p \vee q$ and $q \leqq p \vee q$ by (12′), whence $p \vee q \leqq p \vee q$ by (11). Hence, by duality and induction, the relation is reflexive. We now prove transitivity: that $a \leqq b$ and $b \leqq c$ imply

† P. Whitman [1], [2]. For more recent results, see R. A. Dean, [Symp., pp. 31–42].

$a \leqq c$. We first consider the case that one of the extreme terms a, c is involved in reducing one or both of the two inequalities.

If $p \vee q \leqq b$ by (11) and $b \leqq c$, then $p \leqq b \leqq c$ and $q \leqq b \leqq c$, whence by induction $p \leqq c$ and $q \leqq c$, and so by (11) $p \vee q \leqq c$. Again, if $p \wedge q \leqq b$ by (12) and $b \leqq c$, then $p \leqq b \leqq c$ or $q \leqq b \leqq c$, whence by induction $p \leqq c$ or $q \leqq c$, and so by (12) $p \wedge q \leqq c$. Dually, if $a \leqq b \leqq p \wedge q$ or $a \leqq b \leqq p \vee q$ follows by a decomposition of $p \wedge q$ resp. $p \vee q$, we can prove $a \leqq p \wedge q$ resp. $a \leqq p \vee q$.

There remains the case that both $a \leqq b$ and $b \leqq c$ are first reduced by writing $b = p \vee q$ (or $b = p \wedge q$). If $a \leqq p \vee q$ by (12′) and $p \vee q \leqq c$ by (11), then $a \leqq p$ or $a \leqq q$, and $p \leqq c$ and $q \leqq c$. Hence $a \leqq p \leqq c$ or $a \leqq q \leqq c$; in either case, by induction, $a \leqq c$. The case $a \leqq p \wedge q \leqq c$ is dual, completing the proof. Now using Theorem I.3, we infer the following

Corollary. *If we define $a = b$ in $FL(r)$ to mean $a \leqq b$ and $b \leqq a$, then $FL(r)$ is a partly ordered set.*

Lemma 2. *$FL(r)$ is a lattice, in which the expression $a \wedge b$ is g.l.b. (a, b) and $a \vee b$ is l.u.b. (a, b).*

Proof. By (12) and Lemma 1, $a \wedge b \leqq a$ and $a \wedge b \leqq b$. By (11′), $x \leqq a$ and $x \leqq b$ imply $x \leqq a \wedge b$. Hence $a \wedge b$ is g.l.b. (a, b). Dually, $a \vee b$ is l.u.b. (a, b).

Proof of theorem. Choose elements g_c in any lattice L, one corresponding to each x_α. By direct substitution, each $a \in FL(r)$ determines a unique element $a^* \in L$. By Lemma 2, this correspondence is epimorphic, from $FL(r)$ onto the sublattice of L generated by the g_α.

Using similar techniques, Dilworth [1] has proved a remarkable extension of Whitman's theorem, which shows moreover that Theorem IV.14 cannot be extended to nonatomic lattices.

Theorem 15. *In the free complemented lattice with r generators, $F_r(\wedge, \vee, ')$, every element is uniquely complemented; moreover $F_2(\wedge, \vee, ')$ contains $F_d(\wedge, \vee, ')$ (d = countable infinity) as a sublattice.*

(In the preceding theorem, L1–L4 and universal bounds O, I are assumed, together with $x \wedge x' = I$, and $(x')' = x$. It is *not* assumed that $x \leqq y$ implies $x' \geqq y'$.) In the same paper, Dilworth has shown further that any lattice can be extended to a lattice with unique complements.

Word problem. The problem of deciding, in a finite number of steps, whether or not two polynomials $p, q \in W_r(F)$ have the same value $p(\mathbf{x})$ for all choices of $x_1, \cdots, x_r$ in every member of a family Γ of algebras A_γ, is called the "word problem" for Γ. It is clearly equivalent to the problem of deciding whether or not $p(\mathbf{x}) = q(\mathbf{x})$ in $F_r(\Gamma)$. Theorem 14 solves the word problem for lattices; Dilworth (op. cit. supra) has solved the word problem for complemented lattices. The word problems for distributive lattices, Boolean algebras, and polynomials ("words") in $r \leqq 3$ letters for modular lattices were solved in Chapter III, loc. cit. supra.†

It is easy to see that, after one has constructed $F_r(\Gamma)$ for all finite r, $F_{\aleph}(\Gamma)$ for any cardinal number $\aleph$ can be constructed by transfinite extension as an "inductive limit" of the simplest kind (cf. Chapter VIII). The reason for this is as follows.

THEOREM 16. *Let θ be any insertion of the set X of generators of $F_r(\Gamma)$ into the set Y of generators of $F_s(\Gamma)$, $r \leqq s$. Then θ can be extended to a monomorphism μ: $F_r(\Gamma) \subset F_s(\Gamma)$.*

For, θ has a right-inverse projection $\psi: Y \to X$, which can be extended to an epimorphism $\bar{\psi}: F_s(\Gamma) \to F_r(\Gamma)$, by definition of "free algebra". The extension of θ to a morphism $\bar{\theta}$: $F_r(\Gamma)$ therefore has a right-inverse $\bar{\psi}$, and is the desired monomorphism.

Example 8. Using the preceding lemma, one can easily construct the *free Boolean algebra* $F_\omega(\Gamma)$ with countably many generators. Consider the set 2^ω of all infinite *dyadic decimals* .010010111..., etc. Let S_k be the set of all such decimals with k-th digit 1, and S'_k its complement. Using the disjunctive normal form, one can represent $F_r(\Gamma)$—where Γ is the class of Boolean algebras,—as the ring of subsets of 2^ω whose first r digits belong to some specified set of sequences $\mathbf{n} = (n_1, \cdots, n_r)$. Clearly $F_1(\Gamma) \subset F_2(\Gamma) \subset F_3(\Gamma) \subset \cdots$, and so by the lemma $F_\omega(\Gamma) = \bigcup F_r(\Gamma)$. We shall see in Chapters IX–X that this is isomorphic to both (i) the field of *all* open-and-closed subsets of the Cantor ternary set (i.e., the set of triadic decimals 0, $n_1 n_2 n_3 \cdots$ whose d digits are 0 or 2), and (ii) the field of subsets of [0, 1] generated by intervals with dyadic rational endpoints, modulo finite sets.

Free modular lattices. The free modular lattice M_{28} with three generators was discussed in detail in §III.6, and we have also seen (§III.7, especially Exercise 8) that the free modular lattice generated by any two finite chains is distributive. However, the word problem for the free modular lattice with $n = 4$ generators is unsolved, and very difficult. It is known that every such lattice is infinite, and of infinite length.

Rather than try to summarize what else is known about this problem we shall refer the reader to the excellent review article by Whitman in [**Symp.**, pp. 17–22].

Exercises:

1. (a) Show that $FL(\mathbf{1} + \mathbf{2})$ has 9 elements and a plane diagram.
   (b) Show that $FL(\mathbf{1} + \mathbf{3})$ has 20 elements and a plane diagram.

*2. Show that the free *modular* lattice generated by $\mathbf{2} + \mathbf{1} + \mathbf{1}$ contains exactly 238 elements.‡

*3. Show that $FL(\mathbf{2} + \mathbf{2})$ and $FL(\mathbf{1} + \mathbf{4})$ are infinite. (H. L. Rolf, Yu I. Sorkin)

4. (a) Show that $FL(\mathbf{3})$ is infinite.
   (b) Show that $FL(\mathbf{3})$ contains an infinite chain.

† For the word problem in other algebraic systems, see P. Hall, J. London Math. Soc. **33** (1958), 482–96.

‡ K. Takeuchi, Tôhoku Math. J. **11** (1959), 1–12. Exercise 3 is due to Yu I. Sorkin, Mat. Sb. **30** (1952), 677–94.

5. (a) In $FL(4)$, let $p_i = \bigvee_{j \neq i} x_j$. Show that $p_1 \vee p_2 < (p_1 \vee p_2 \vee p_3) \wedge (p_1 \vee p_2 \vee p_4)$.

(b) Show that the atoms of $FL(n)$ generate a distributive sublattice if and only if $n \leqq 3$.

(c) Show that the atoms of $FL(4)$ generate a sublattice of 22 elements.

(Exercises 5a–5b, 6, 9 state results of P. Whitman [1].)

6. Show that the automorphism group of $FL(n)$ is the symmetric group of degree n, and that the ϕ-sublattice of $FL(n)$ simply omits its generators.

7. (a) Show that $FL(3)$ contains $FL(n)$ as a sublattice, for any finite or countable n.

(*b) Show that any infinite sublattice of a free lattice has infinite length.†

*8. (a) Show that the free lattice $FL(P)$ generated by any finite or countable poset P is monomorphic to $FL(3)$.

(b) Show that if copies of *all* lattices with three generators can be embedded in a lattice L, then L is uncountable.‡

*9. Show that of all the elements in $W_{\aleph}(\wedge, \vee)$ equal to a given element in $FL(\aleph)$, any two of shortest length are equivalent under L2–L3 alone.

10. Show that the free modular ortholattice generated by a chain of length r is $\mathbf{2}^r$. (See also §IV.3, Exercise 10).

*11. Show that, in any finite lattice which is monomorphic to a free lattice, $x \wedge \bigvee_{i=1}^{5} y_i = \bigvee_{j=1}^{5} (x \wedge \bigvee_{i=1, j\neq i}^{5} y_i)$, but that this identity does not hold in all lattices. (B. Jonsson)

9. Postulates

As was remarked in §1, the more common families of abstract algebras are defined as sets of elements, closed under certain operations, and satisfying certain *postulates*. These postulates vary considerably in their nature, at least four different types of statements being distinguishable.§

1. *Identity*. Let $[A, F]$ be an algebra, and let p, q be any two polynomials of $[P_r, F]$. Then the statement

$$p(a_1, \cdots, a_r) = q(a_1, \cdots, a_r) \text{ for all } a_i \in A \tag{13}$$

is called an *identity*. For any binary operation, the commutative law ($r = 2$) and associative law ($r = 3$) are identities. So are the idempotent law ($r = 1$) and, for a system with two binary operations $\wedge$ and $\vee$, the laws of contraction ($r = 2$):

$$x \wedge (x \wedge y) = x \vee (x \wedge y) = x.$$

2. *Identical implications*. Let $p_0, p_1, \cdots, p_s$ and $q_0, q_1, \cdots, q_s$ be two finite sets of polynomials of $[P_r, F]$. The statement

$$\begin{aligned} &\text{If } p_i(a_1, \cdots, a_r) = q_i(a_1, \cdots, a_r) \text{ for } i = 1, \cdots, s, \\ &\text{then } p_0(a_1, \cdots, a_r) = q_0(a_1, \cdots, a_r), \end{aligned} \tag{14}$$

is called an *identical implication*. When $s = 0$, statement (14) reduces to (13); hence every identity is an identical implication.

---

† See B. Jonsson, Canad. J. Math. **13** (1961), 256–64, B. Jonsson and J. E. Kiefer, ibid. **14** (1962), 487–97, F. Galvin and B. Jonsson, ibid., 265–72; for this and related results. Also, R. A. Dean in [**Symp.**, pp. 31–42], Theorems 6, 8.

‡ For Exercise 8, see P. Crawley and R. A. Dean, Trans. AMS **92** (1959), 35–47, Theorems 6, 8.

§ See Roger Lyndon, Bull. AMS **65** (1959), 287–99, for a discussion of this point from the standpoint of the propositional calculus.

The cancellation law in groups ("if $ax = ay$, then $x = y$") is such an identical implication. Another is: $a \wedge x = a \wedge y$ and $a \vee x = a \vee y$ imply $x = y$, which holds in any distributive lattice. Still another is: $a \wedge x = a \wedge y = O$ and $a \vee x = a \vee y = I$ imply $x = y$; this holds in any distributive lattice and in any lattice with unique complements.

3. *Disjunctive implication.* Under the hypotheses of (14), writing $(a_1, \cdots, a_r) = \mathbf{a}$ for short, the statement

$$\text{If } p_i(\mathbf{a}) = q_i(\mathbf{a}) \text{ for } i = 2, \cdots, s, \text{ then} \\ p_0(\mathbf{a}) = q_0(\mathbf{a}) \text{ or } p_1(\mathbf{a}) = q_1(\mathbf{a}), \tag{15}$$

may be *called a disjunctive implication.* Thus, the cancellation law for integral domains ("if $ax = ay$, then $x = y$ or $a = 0$") is a disjunctive implication.

4. *Existential identity.* Under the hypotheses of (15), a statement of the form "$(Qa_1), \cdots, (Qa_r), p(a_1, \cdots, a_r) = q(a_1, \cdots, a_r)$," where each Qa_i is either "for all $a_i \in A$" or "there exists an $a_i \in A$", and at least one Qa_i is of the latter sort, may be called an *existential identity.* As an example, the statement that "for all a, b, the equation $xa = b$ has a solution" is an existential identity.

The different kinds of postulates described above differ considerably as regards being preserved under the formation of subalgebras, epimorphic images, and direct products.

THEOREM 17. *Identities are preserved under the formation of subalgebras, epimorphic images, and direct products.*

PROOF. If (13) holds in A, and S is a subalgebra of A, then (13) holds in S *a fortiori.* The same is true if A is replaced by A/θ—so that $=$ is replaced by θ (i.e., by $\equiv \pmod{\theta}$). Finally, if (13) holds in each factor of the direct product $\prod A_\gamma$, then it holds in $\prod A_\gamma$, since it holds for each component $a_{\iota\gamma}$ of every element $a_\iota \in \prod A_\gamma$.

THEOREM 18. *Identical implications are preserved under the formation of subalgebras and direct products.*

PROOF. If (14) holds in A, and S is a subalgebra of A, then (14) holds in S *a fortiori.* Likewise, if (14) holds in each factor A_γ of $\prod A_\gamma$, and $p_i(\mathbf{a}) = q_i(\mathbf{a})$ in $\prod A_\gamma$ for $i = 1, \cdots, s$, then $p_i(\mathbf{a}_\gamma)$ in each A_γ for $i = 1, \cdots, s$, by definition of $\prod A_\gamma$. Hence $p_0(\mathbf{a}_\gamma) = q_0(\mathbf{a}_\gamma)$ in each A_γ by (14), applied to A_γ. Hence $p_0(\mathbf{a}) = q_0(\mathbf{a})$ in $\prod A_\gamma$, by definition of $\prod A_\gamma$.

COROLLARY. *Any identity or identical implication which holds in a family of similar algebras A_γ, holds in the associated free algebra with r generators, for any cardinal number r.*

Identical implications are not necessarily preserved under epimorphisms. For example, let A be the additive semigroup of nonnegative integers, and let $\theta = \theta(N, m)$ identify n with $n + km$ if $n \geqq N$. Then the cancellation law ($a + x = a + y$ implies $x = y$) holds in A, but not in A_θ if $N > 0$.

Disjunctive implications are preserved under the formation of subalgebras, but not of epimorphic images *or* direct products. Thus the cancellation law for integral domains holds in the ring $\mathbf{Z}$ of positive integers, but not in the ring $\mathbf{Z}_n$ of integers mod n unless n is a prime, nor does it hold in the direct product $\mathbf{Z} \times \mathbf{Z}$.

Existential identities are not preserved under the formation of subalgebras, unless additional operations are introduced. Indeed, Kirby Baker has constructed an (infinite) system with binary (nonassociative) multiplication, in which $xa = b$ has a solution for all a, b, but in which the set of subalgebras satisfying this condition is *not* closed under intersection.

Exercises:

1. Show that any subdirect product of copies of the five-element nonmodular lattice N_5 satisfies:
   (a) $(x \wedge y) \vee (x \wedge z) = x \wedge [(y \wedge z) \vee (z \wedge x) \vee (x \wedge y)]$, (H. Lowig)
   (b) $x \vee y = y \vee z = z \vee x$ implies $(x \wedge y) \vee (y \wedge z)$. (A. Waterman)

*2. Show that the free lattice with three generators satisfying all identities true in N_5 contains 99 elements. (A. Waterman)

3. Consider the condition that, in a commutative ring R, for each $a \in R$, $a^n = 0$ for some $n = n(a) \in \mathbf{Z}^+$. Is this condition preserved under subalgebra? Finite direct product? Unrestricted direct product? Epimorphic image?

*4. Show that neither the symmetric partition lattices nor general subgroup lattices satisfy any identical implication on lattice polynomials which is not implied by L1–L4.

5. Describe the free Newman algebras with one and with two generators, respectively.

6. In a lattice L with more than one element, let $\alpha: x \geqq z$ imply $p(x, y, z) = q(x, y, z)$. Show that either α holds in every lattice or α implies L5. (*Hint.* Study Exercise 1 for §8.)

7. In a modular lattice L with more than one element, let β: $f(x, y, z) = g(x, y, z)$ be any identity. Show that either β holds in every modular lattice, or β implies L6.

*8. Show that there exists an identity not implied by L5 which holds in every normal-subgroup lattice.†

*9. Exhibit a seven-element algebra which possesses no finite set of identities implying all identities.

*10. Show that, if the identity $p = q$ holds in a lattice L, then $p = q$ is also an identity in the complete lattice $\hat{L}$ of all ideals of L.

10. Families of Algebras

In each "species" (§6) $\mathscr{F}$ of algebras with a specified set F of operations, one can distinguish various "families" of algebras, each such family being characterized by the set of polynomial identities $p = q$ which its members satisfy. To make this connection precise, we introduce a basic *polarity*.‡

DEFINITION. For any F-equation $p = q$ and F-algebra $A = [S, F]$, define $(p = q) \rho A$ to mean that $p = q$ holds identically in A. Under the *polarity* defined by this relation, for a given F, call the *closed* subsets $\Gamma = (\Gamma^\dagger)^*$ of F-algebras and $\Gamma^\dagger$ of F-equations *families* of algebras and equations, respectively.

† See B. Jonsson, Math. Scand. **1** (1953), 193–206. Exercise 9 is a result of R. C. Lyndon, Proc. AMS **5** (1954), 8–9; see also Bull. AMS **65** (1959), 287–99.

‡ G. Birkhoff [**3**, §10]. Theorem 22 below was proved there as Theorem 10; the proof is valid also for infinitary operations. Note that, although $\mathscr{F}$ is a "class" (the cardinality of its members is unrestricted), F is assumed to be a set.

From the results of §§7, 9, it follows that any "family" Γ of algebras: (i) is closed under the formation of subalgebras, direct products, and epimorphic images, and (ii) contains for any cardinal number r a "free algebra" $F_r(\Gamma)$ with r generators. Further, the *identities* in r letters of $\Gamma^\dagger$ (the polar of Γ) are just the *equations* between polynomials in the generators of $F_r(\Gamma)$. In the present section, we shall prove some sharp converses to these results.

Now let $A = [S, F]$ be *any* algebra with r generators which satisfies all identities of Γ. By the result just stated, $F_r(\Gamma)$ is epimorphic to A; moreover $W_r(F)$ is epimorphic to $F_r(\Gamma)$. In summary, we have proved

THEOREM 19. *Let A be any F-algebra with r generators which satisfies all identities valid in a class of F-algebras closed under formation of subalgebras and direct products. Then A is an epimorphic image of $F_r(\Gamma)$. Moreover*

$$A \cong W_r(F)/\theta_1 \quad \text{and} \quad F_r(\Gamma) \cong W_r(F)/\theta, \qquad \theta_1 \supseteq \theta. \tag{16}$$

(The proof also uses Corollary 1 of Theorem 13.)

Having postulated a set Δ of equations on a given set F of operations, let $\Delta^* = \Gamma$ be the corresponding family of algebras. The *free* algebra with r generators $x_1, \cdots, x_r$ can then often be constructed by using the following result.

THEOREM 20. *Let it be possible to reduce every element $p = p(x_1, \cdots, x_r)$ of $W_r(F)$ to one of the polynomials p_j of the subset $C \subset W_r(F)$ by repeated use of identities of Δ, and let there exist an algebra $A = [S, F]$ in Δ^* with generators $x_1, \cdots, x_r$ such that $p_j \neq p_k$ and $\{p_j, p_k\} \subset C$ imply that $p_j(x_1, \cdots, x_r) \neq p_k(x_1, \cdots, x_r)$. Then A is the free algebra with r generators of Δ^*.*

PROOF. Let $B = [T, F]$ be given in Δ^*, and let a function $\phi: X \to T$ be given, where $X = \{x_1, \cdots, x_r\}$. We *define* the extension $\bar{\phi}$ of ϕ by:

$$\bar{\phi}(a_j) = p_j(x_1\phi, \cdots, x_r\phi), \tag{17}$$

where p_j is the *unique* F-polynomial such that $p_j(x_1, \cdots, x_r) = a_j$ in $[S, F]$. We must show that $\bar{\phi}$ is a morphism: that for any $f_i \in F$,

$$f_\alpha(\bar{\phi}(a_1), \cdots, \bar{\phi}(a_{n(\alpha)})) = \bar{\phi}(f_\alpha(a_1, \cdots, a_{n(\alpha)})). \tag{18}$$

By (17), the left side of (18) is $f_\alpha(p_1(\mathbf{x}\phi), \cdots, p_{n(\alpha)}(\mathbf{x}\phi))$, where $\mathbf{x} = (x_1, \cdots, x_r)$. The right side of (18) is however $\bar{\phi}(q(\mathbf{x}))$, where $q(\mathbf{x}) = f_\alpha(p_1(\mathbf{x}), \cdots, p_{n(\alpha)}(\mathbf{x}))$ is the unique polynomial in C to which the expression can be reduced by Δ.

Moreover every equation $f_\alpha(\mathbf{a}) = c$ in any "multiplication table" of A can be deduced from Δ, as:

$$f_\alpha(p_1(\mathbf{x}), \cdots, p_{n(\alpha)}(\mathbf{x})) = q(\mathbf{x}), \qquad \mathbf{x} = (x_1, \cdots, x_r), \tag{19}$$

where the p_l and q are the unique F-polynomials in C such that $p_l(\mathbf{x}) = a_l$ and $q(x) = \mathbf{c}$. The desired identity (18) is therefore equivalent to

$$f_\alpha(p_1(\mathbf{x}\phi), \cdots, p_{n(\alpha)}(\mathbf{x}\phi)) = q(x_1\phi, \cdots, x_r\phi), \tag{20}$$

which, since $[T, F] \in \Delta^*$, follows by applying the identities of Δ to polynomials in the $y_i = x_i\phi$. Q.E.D.

We now prove a beautiful theorem of B. H. Neumann.†

THEOREM 21. *Let A be a free algebra with generators x_i. Then A/θ is freely generated by the equivalence classes $\tilde{x}_i$ of the x_i if and only if θ is a fully characteristic congruence relation.*

The proof of this result depends on the following obvious lemma. (Note that the $\tilde{x}_i$ generate A/θ in any case.)

LEMMA 1. *An endomorphism ϕ of an algebra A induces an endomorphism on A/θ if and only if $x\theta y$ implies $(x\phi)\theta(y\phi)$.*

PROOF OF THEOREM. Suppose that θ is a fully characteristic congruence relation on A. If $\tilde{x}_i \to \tilde{y}_i$ is any mapping of the $\tilde{x}_i$ into A/θ, let $x_i \to y_i$ be *some* corresponding mapping $A \to A$, and let ϕ be the unique extension of the latter mapping to an endomorphism of A. For any elements $p(x_1, \cdots, x_n)$ and $q(x_1, \cdots, x_r)$ of A, $p\theta q$ in A implies $[p(x_1\phi, \cdots, x_r\phi)]\theta[q(x_1\phi, \cdots, x_r\phi)]$ since θ is fully characteristic. By Lemma 1, this is just the condition that ϕ induce an endomorphism $\tilde{\phi}$ on A/θ, hence A/θ is free.

Conversely, if A/θ is freely generated by the $\tilde{x}_i$, there is an endomorphism ϕ of A which maps the $\tilde{x}_i$ into the $\tilde{y}_i$, and induces an endomorphism $\tilde{\phi}$ on A/θ. Therefore, for any polynomials p and q, $p\theta q$ in A (i.e., $[p(x_1, \cdots, x_r)]\theta[q(x_1, \cdots, x_r)]$) must imply

$$\begin{aligned}\phi(p(x_1, \cdots, x_r)) &= p(x_1\phi, \cdots, x_r\phi) = p(y_1, \cdots, y_r)\ \theta\ q(y_1, \cdots, y_r)\\ &= q(y_1\phi, \cdots, y_r\phi) = \phi(q(x_1, \cdots, x_r)).\end{aligned}$$

Hence, by Lemma 1, $x\theta y$ implies $(x\phi)\theta(y\phi)$. But since the y_i are arbitrary, every endomorphism ϕ of A must be induced in this way, and so (by definition), θ is fully characteristic, completing the proof.

COROLLARY. *For given F and r, the free algebras with r generators are the quotient-algebras $W_r(F)/\theta$ of the word algebra $W_r(F)$ by fully characteristic congruence relations.*

LEMMA 2. *The set $\Gamma^\dagger$ of all identities true in any algebra A or set Γ of similar algebras $A_\nu = [S_\nu, F]$ is closed under the following rules of inference:*

(i) *If $p_h = q_h$ is an identity for $h = 1, \cdots, n$, then $f_i(p_1, \cdots, p_n) = f_i(q_1, \cdots, q_n)$ for any n-ary operation $f_i \in F$.*

(ii) *If $p = q$ is an identity for Γ, then any substitution of a polynomial $\eta(x_j)$ for all occurrences of the primitive symbol x_j in $p = q$ yields an identity for Γ.*

The preceding rules of inference are obvious; it is also obvious that they are statements about the relation of Γ to the word algebras $W_r(F)$. Specifically, rule (i) of inference asserts that any equivalence relation θ is a congruence relation on

† B. H. Neumann, *Special topics in algebra*, mimeographed notes, New York University, 1962, pp. 58–62. See also J. Schmidt, Math. Ann. **158** (1965), 131–57.

any $W_r(F)$. Moreover, the substitutions of the form (ii) are just the endomorphisms of $W_r(F)$. Hence, by Theorem 21, closure under (ii) asserts that $W_r(F)/\theta$ is a *free algebra*. That is, we have proved

THEOREM 22. *For any given set F of operations, consider the polarity between algebras $A = [S, F]$ and identities $p \equiv q$ defined by the following relation ρ:*

$$(p = q)\ \rho\ A \quad \text{means that} \quad p \equiv q \text{ in } A. \tag{21}$$

Under this polarity, the "closed" sets of algebras are those closed under the formation of subalgebra, direct product, and epimorphic image; the "closed" sets of identities are the sets of equivalent relations closed under the rules of inference (i)–(ii).

Tarski calls families of algebras defined by the preceding polarity "equationally definable".

11. Polymorphisms; Crypto-isomorphisms

Ideally, universal algebra should lead to a systematic classification and enumeration of all conceivable algebras. To carry out this program, a logical beginning consists in classifying algebras into *species* according to the number of their distinct nullary, unary, binary, ternary, ... operations and the minimum number of their generators. (This number always exists, because the cardinal numbers are well ordered; see Chapter VIII.) One thus associates with each algebra $A = [S, F]$ a unique word algebra $W_r(F)$ and congruence relation θ.

In this spirit, one can easily enumerate all algebras† with *one* generator and *one unary* operation σ. We first recall Example 7 of §6, which asserts

LEMMA 1. *The (free) word algebra $W_1(\sigma)$ with one unary operation σ and one generator x is isomorphic to the set* N *of nonnegative integers, with $\sigma(k) = k + 1$. Moreover $W_r(\sigma)$ is the disjoint sum of r copies of* N.

LEMMA 2. *The most general proper congruence relation θ on $W_1(\sigma)$ is specified by choosing two positive integers m and n, and asserting that*

$$k\ \theta\ l \quad \text{if and only if} \quad k \equiv l\ (n) \quad \text{and} \quad k, l \geqq m. \tag{22}$$

The classification of unary algebras with r generators x_i proceeds similarly. To each i are associated either the void set $\varnothing$ or two positive integers $m(i)$ and $n(i)$, the smallest integers such that $\sigma^m(x_i) = \sigma^{m+n}(x_i)$. In addition, to each pair i, j are associated the void set or the smallest pair of positive integers h, k such that $\sigma^h(x_i) = \sigma^h(x_k)$.

The corresponding problem for *binary* algebras, even those having just one operation, is enormously difficult. It will not be considered here. Instead, we will conclude this section by remarking pessimistically that even the solution of the analogous problem for *every* $W_r(F)$ would not completely solve the classification problem. This is because of two complications, associated with the concepts

† A detailed discussion of algebras with one unary operation would essentially repeat much of Dedekind's classic monograph *Was sind und was sollen die Zahlen*, Braunschweig, 1888.

of polymorphism and cryptomorphism, respectively. We now discuss these concepts; the first is due to Bourbaki.

DEFINITION. A *polymorphism* from an algebra $A = [S, F]$ to an algebra $B = [T, G]$ is a function-pair consisting of a bijection $\theta: F \leftrightarrow G$ such that

$$(23) \qquad g_\beta = \theta(f_\alpha) \quad \text{implies} \quad n(\beta) = n(\alpha),$$

and a function $\phi: S \to T$ such that

$$(24) \qquad g_\beta(\phi(x_1), \cdots, \phi(x_n)) \equiv \phi(f_\alpha(x_1), \cdots, (x_n)), n = n(\alpha) = n(\beta).$$

A polymorphism with ϕ a bijection is a *polyisomorphism*.

For example, any dual isomorphism between lattices is a polyisomorphism with $\theta(\wedge) = \bigvee$ and $\theta(\vee) = \bigwedge$.

The notion of a polyisomorphism can be specialized in an obvious way to the notion of a polyautomorphism. An important class of polyautomorphisms is provided by nonsingular *semilinear* transformations of the n-dimensional (right) vector space over a division ring D into itself. By this is meant a mapping $\mathbf{x} \to \mathbf{y}$ of the form

$$(25) \qquad y_i = \sigma\left(\sum a_{ij}x_j\right),$$

where σ is an automorphism of D and $\|a_{ij}\|$ is a nonsingular square matrix, with entries in D.

A much more serious complication is associated with the fact that the *same* abstract algebra can often be defined in several nonpolyisomorphic ways. For example, a group was defined in §1 as an algebra with one binary and one unary operation. It can also be defined as a set with one binary operation xy, one unary operation $x \to x^{-1}$, *and* one nullary operation which picks a constant "identity element" e, and satisfies the identities

$$(26) \qquad x(yz) = (xy)z, \qquad xx^{-1} = x^{-1}x = e, \qquad ey = ye = y.$$

Third, a group can be defined as a set with *two* binary operations $x/y = xy^{-1}$ and $x\backslash y = x^{-1}y$ (like rings and lattices), subject to such identities as

$$(27) \qquad x/x = y\backslash y, \qquad y/(y\backslash x) = x, \qquad x\backslash(y/z) = (x\backslash y)/z, \text{ etc.},$$

for all x, y, z. Finally, a group can be defined as a set with a single associative binary operation xy, in which the equations $xa = b$ and $ay = b$ can always be solved for x and y (Existential Identity).

Therefore, given any group $[X, F]$ with $f = \{\cdot, {}^{-1}\}$, one can construct from it three others $[X, F_i]$ $(i = 1, 2, 3)$ with different sets of operations, which are not polyisomorphic but cryptoisomorphic ("crypto" = hidden) in the following sense.

DEFINITION. Let $A = [X, F]$ and $B = [X, G]$ be two algebras with the same elements. A *cryptoisomorphism* between A and B consists of two sets of equation-sets: (i) for each $f_\alpha \in F$, a set of G-polynomial equations $q_i^\alpha(\mathbf{x}, y) = \tilde{q}_i^\alpha(\mathbf{x}, y)$ equivalent to $y = f_\alpha(\mathbf{x})$ in A, and (ii) for each $g_\beta \in G$, a set of F-polynomial equations $p_i^\beta(\mathbf{x}, y) = \tilde{p}_i^\beta(\mathbf{x}, y)$ equivalent to $y = g_\beta(\mathbf{x})$ in B.

For example, every Boolean lattice is cryptoisomorphic to an appropriate Boolean algebra and (trivially) conversely. Under a slightly more general definition, any lattice L is cryptoisomorphic to the *semilattice* defined by the same elements under $\vee$ alone. This is because $y = x_1 \wedge x_2$ is equivalent to the following set of two equations and one conditional implication: (i) $x_1 \vee y = x_1$, (ii) $x_2 \vee y = x_2$, and (iii) If $x_1 \vee z = x_1$ and $x_2 \vee z = x_2$, then $y \vee z = y$.

Exercises for §§10–11:

1. Show that, in the complete lattice of "closed" sets of (finitary) identities, I is the closure of $x \equiv y$. Describe the dual "family" of $I^\dagger$ of algebras, for any given F.

2. Given a congruence relation θ on an algebra A, let $\bar{\theta}$: $A \to A/\theta$ be the associated "natural" epimorphism. Given a morphism ϕ: $A \to B$, show that: (i) there is a morphism ψ: $(A/\theta) \to B$ such that $\bar{\theta}\psi = \phi$ if and only if $a \; \theta \; a_1$ in A implies $a\phi = a_1\phi$ (ϕ "respects" θ), and (ii) in this event, there is only one such ψ.

3. Verify Lemma 1 of §11.

4. Show that the lattice of all families of lattices is distributive. (*Hint*. Use Theorems 9 and 21.)

5. (a) Prove that the lattice of all families of groups is modular.

(b) Show that every family of Abelian groups consists of those in which $nx \equiv 0$ for some natural integer n.

(c) Show that the lattice of all families of Abelian groups is distributive.

6. Show that the lattice of all families of algebras with one unary operator and one generator is distributive.

7. Show that if A and B are cryptoisomorphic, then Aut $A \cong$ Aut B, but that A and B need not have isomorphic subalgebra or structure lattices.

8. (a) For any group G, let $A = [G, T]$, where T is the set of all left-translations $f_a(x) = ax$, $a \in G$. Show that the structure lattice of A is isomorphic to the subgroup-lattice of G.

(b) Show that A is not crytoisomorphic to G.

9. Show that the correspondence between a distributive lattice $(L, \wedge, \vee)$ and the ternary algebra A defined from it by the median operation is sometimes a cryptoisomorphism, but that this is unusual.

10. If A is a finitely generated algebra, show that every set of generators of A contains a finite subset which generates A.

11. Prove that polyisomorphism and cryptoisomorphism are equivalence relations between algebras.

12. Define cryptoisomorphisms from any Boolean algebra to (i) a Boolean ring, and (ii) a Sheffer algebra.

13. Solve the word problem for the free Sheffer algebra with r generators.

14. Show that any two cryptoisomorphic algebras $[S, F]$ and $[T, G]$ have cryptoisomorphic free algebras.

*15. Show that, in the lattice of families of lattice identities, the family of those holding in the nonmodular five-element lattice N_5 is covered by the family of those satisfied in every distributive lattice. (*Hint*. Every proper sublattice or epimorphic image of N_5 is distributive.)

*16. Establish a "cryptomonomorphism" from any Lie algebra into an "enveloping" linear associative algebra.†

*12. Functors and Categories

Many of the preceding constructions define *functors* from a given species or family of algebra to another species of algebra (or the same species). By this is

† This is the Birkhoff-Witt Theorem of [**UA**, p. 294].

meant† a function ψ on sets to sets *and* functions to functions which is a morphism for composition and the identity function:

$$(24) \qquad \psi(1_S) = 1_{\psi(S)} \quad \text{and} \quad \psi(f \circ g) = \psi(f) \circ \psi(g),$$

for any set S and function $f: S \to T$.

More can be said: most of the preceding constructions are functions which carry morphisms in a given species Γ of algebra into morphisms of the image species $\psi(\Gamma)$ (some only carry isomorphisms into isomorphisms).

Likewise any species (or family) of algebras defines a *concrete category*, whose objects are the algebras of the given family, with hom (A, B) for any two algebras of the given species or family the set of all morphisms $\phi: A \to B$. It is trivial that 1_A is an (iso)morphism, and that the set of morphisms is closed under composition of functions.

Just as the concept of a poset generalizes (by abstraction) the concept of a family of *subsets* of a given set, so the category concept defines a natural generalization of the concept of a family of *functions* between sets. Unlike posets, categories need not be "sets" (their cardinality may be undefinable); moreover even when they are sets, they are only "partial" algebras. Hence the techniques of this chapter are not applicable to categories.

From the rapidly developing theory of categories, we shall borrow only two related concepts: the concepts of injective and projective category. The concrete category of all sets and functions (mappings) has the following two properties:

(i) Given a *mono*morphism (injection) $\mu: A \to B$, there exists a right-inverse *epi*morphism (surjection) $\nu: B \to A$ such that $\mu\nu = 1_A$,

(ii) Given an *epi*morphism $\nu: B \to A$, there exists a left-inverse *mono*morphism $\mu: A \to B$ such that $\mu\nu = 1_A$.

These properties are not possessed, however, by the category of all Abelian groups and their morphisms. Thus the monomorphism $\mu: \mathbf{Z}_2 \to \mathbf{Z}_4$ with $\mu(n) = 2n$ has no right-inverse, and the epimorphism $\nu: \mathbf{Z}_4 \to \mathbf{Z}_2$ with $\nu(n) \equiv n \pmod 2$ has no left-inverse.

DEFINITION. A (concrete) category Γ is *injective* when statement (i) holds in Γ; it is *projective* when statement (ii) holds in Γ.

Two basic questions about any family Φ of algebras are: is Φ injective? is Φ projective? It is classic that the category of all vector spaces (or "D-modules") over any fixed field or division ring D is injective and projective. So is the category of all finite Boolean algebras‡—and likewise, so is that of all chains. (What about that of all finite distributive lattices? of all Boolean lattices?)

† For the general theory of functors and categories, see S. Mac Lane, Bull. AMS **71** (1965), 40–106; P. Freyd, *Abelian Categories*, Harper and Row, 1964. Some applications of these concepts to universal algebra are given in [**UA**], many of the deeper duality theories about categories apply only to special algebraic structures over Abelian groups.

‡ Any *invertible* functor, such as that from $\mathbf{Z}_2$-modules to generalized Boolean algebras, trivially takes injective (resp. projective) concrete categories into categories with the same property.

THEOREM 23. *Any free algebra F is projective.*

This means that, given any morphism $\mu: F \to B$ and an epimorphism $\nu: A \to B$, there exists a morphism $\theta: F \to A$ such that $\mu = \nu\theta$. Let $\{x_\alpha\}$ be the canonical generators of F, and choose $y_\alpha \in A$ so that $\nu(y_\alpha) = \mu(x_\alpha)$; this is possible since ν is an epimorphism. Then the mapping which carries each x_α into the chosen y_α can be extended to a morphism $\theta: F \to A$, since F is free. Since $\mu(x_\alpha) = \nu(y_\alpha) = \nu(\theta(x_\alpha))$, and the x_α generate F, the result follows.

PROBLEMS

38. Does Funayama's Theorem hold in commutative monoids in which $a + b = 0$ implies $a = b = 0$, i.e., which are "centerless" in the sense of Jonsson and Tarski? [The answer is no. (B. Jonsson)]

39. Find necessary and sufficient conditions on a lattice L for its congruence relations to form a Boolean algebra.†

40. Which finite lattices are monomorphic to $FL(3)$? Which countable lattices have this property? Which finite lattices are interval sublattices of $FL(n)$?

41. Describe the sublattice of $FL(n)$ consisting of the elements invariant under all its automorphisms. What are the modular pairs of $FL(n)$? (A. G. Waterman)

*42. Solve the decision problem for the free modular lattice with four generators—with n generators.

43. Determine the free modular lattice generated by four elements a, a', b, b' with $a \wedge a' = b \wedge b' = 0$ and $a \vee a' = b \vee b' = I$.

44. Determine the free modular lattice generated by $\mathbf{1} + \mathbf{1} + \mathbf{n}$. (Cf. Exercise 2 after §8.)

45. Show that a modular lattice is a subdirect product of copies of $\mathbf{2}$ and M_5 if and only if it satisfies the identical inequality $a \vee (x \wedge b) \vee (y \wedge b) \geqq b \wedge (a \vee x) \wedge (a \vee y) \wedge (x \vee y)$, and its dual.

46. Find the free lattices with four generators in the families of lattices generated by M_5 and N_5, respectively.

47. For each n, p, determine the free (modular) lattice with four generators in the family of algebras generated by $PG_{n-1}(\mathbf{Z}_p)$.

48. Solve the word problem for the free ortholattice with 2 generators—with n generators—with n orthogonal generators.

49. Let $F(m, k)$ be the free primitive algebra with m unary operations and k generators. Is $F(m, k)$ crypto-isomorphic to the free semigroup with k generators when $m = 1$? What is the structure lattice of $F(m, k)$? Which congruence relations are strictly characteristic?

50. Characterize, as a lattice, the complete lattice of all strictly characteristic subalgebras of the word algebra with two generators and one binary operation.‡

---

† See G. Grätzer and E. T. Schmidt, Acta Math. Acad. Sci. Hungar. **9** (1958), 137–75; and Iqbalunissa, J. Madras Univ. **33** (1963), 113–28 (Math. Revs. **30** #4698). See also P. Crawley, Pacific J. Math. **10** (1960), 787–95.

‡ See Kalicki–Scott, Indag. Math. **17** (1955), 650–2.

*51. Describe the lattice of all families of group identities.

*52. Describe the lattice of all families of lattice identities—of all those which include L5.

*53. Let I be the class of all algebras with a single binary operation; let I^* be the set of all identities on a single binary operation. Let $A \rho L$ ($A \in I$, $L \in I^*$) mean that L is identically true in A. Discuss the complete lattices defined by the resulting polarity.

54. Using "clones" [**UA**, Chapter III], decide which (free) word algebras can be embedded isomorphically in which others.

55. For what definitions of an operation $x \circ x^*$ in terms of a binary relation ρ, it is true that, in the cardinal product XY of two sets with ρ defined by $(x,y) \rho (x^*,y^*)$ means $x \rho x^*$ and $y \rho y^*$, we will have $(x,y) \circ (x^*,y^*) = (x \circ x^*, y \circ y^*)$?

*56. Solve the word problem for the operations of addition, multiplication and composition in $\mathbf{R}^{\mathbf{R}}$ and $\mathbf{Z}^{\mathbf{Z}}$.

CHAPTER VII

APPLICATIONS TO ALGEBRA

1. Modules; Groups with Operators

Abelian groups, vector spaces, hypercomplex algebras, rings, and representations of groups and rings by linear operators on vector spaces can all be considered as special kinds of "modules", in a sense to be defined below. The first half of this chapter will be devoted to analyzing the structure of such modules,† and the structure of more general groups and loops with operators. The basic tools used will be the theory of modular lattices and "universal algebra", as developed in Chapter VI.

By a *module* (or "Ω-module") is meant an additive Abelian group A with a (possibly void) set Ω of endomorphisms of A. Such a module is an "algebra" in the sense of Chapter VI: one can consider the set of operations to be $F = \{-, \Omega\}$ (i.e., substraction and "multiplication by any $\omega \in \Omega$") or $F^* = \{+, -, \Omega\}$. The structure of modules is distinguished by a very special property.

THEOREM 1. *The structure lattice and the subalgebra lattice of any module M are isomorphic modular lattices.*

SKETCH OF PROOF. The congruence relations on M are the partitions of M into cosets of the different Ω-submodules of M—i.e., of the additive subgroups $S \subset M$ such that $x \in S$ and $\omega x \in S$ for all $\omega \in \Omega$. But these Ω-submodules are just the "subalgebras" of M, considered as an "algebra"

Finally, these "subalgebras" form a sublattice of the modular lattice (Theorem I.11) of all normal subgroups of M. This is true since, in any algebra A, the **subalgebras closed under any set of endomorphisms (by Theorem 7 of Chapter VI).**

Example 1. Let R be any ring with unity 1. If we choose Ω as the set of left-multiplications $\alpha: x \to ax$, then the lattice of Theorem 1 is the lattice of left-ideals of R. The lattices of right-ideals and of two-sided ideals of R can be treated as submodules similarly, by varying the choice of Ω in obvious ways.

Example 2. Let $f: G \to L_n(D)$ be any endomorphism from a group G to the full linear group of nonsingular $n \times n$ matrices over a division ring D. Take for Ω the set of linear operators T_g on D^n which correspond to elements of G under f, *and* multiplication by scalars $\lambda \in D$. This defines a module, whose Ω-submodules

† Obtaining such structure and decomposition theorems was one of the early aims of lattice theory. See for example R. Dedekind [**1**], [**2**]; G. Birkhoff [**1**], [**3**], and O. Ore [**1**]. Note that the definition of "module" given below is more general than that commonly used today, which requires Ω to be a ring of endomorphisms.

are the subspaces of D^n which are mapped into themselves by all the given linear operators.

Decompositions of representations of rings can be handled in the same way.

Groups with operators. Much of Theorem 1 remains true even for *non*-Abelian groups with operators (i.e., with a given set of endomorphisms). Thus we have†

THEOREM 1′. *Let G be any group, with a set Ω of operators (endomorphisms of G). Then the Ω-invariant normal subgroups of G form a modular lattice.*

The proof is like that of the first part of Theorem 1. Clearly, if Ω includes all inner automorphisms, then the lattices of the Ω-subalgebras of $\{G, -, \Omega\}$ is isomorphic to the lattice of its congruence relations. However, this is not true in general for non-Abelian groups without operators.

2. Quasigroups and Loops

One can generalize some of the results of §1, and their deeper consequences, to loops with operators. We will now lay the foundations for this generalization.

DEFINITION. A *quasigroup* is an algebra with a binary multiplication, in which any two of the three terms of the equation $ab = c$ uniquely determine the third. This means that division is single-valued, so that the following two-sided *cancellation law* holds:

$$ax = ay \quad \text{or} \quad xb = yb \quad \text{implies} \quad x = y. \tag{1}$$

Thus the class of quasigroups is definable by identical implications.

If one defines left-division $\backslash$ and right-division $/$ as additional binary operations, by letting $x = a\backslash c$ mean that $ax = c$ and $y = c/a$ mean that $ya = c$, one easily verifies the following identities:

EQ1. $$a(a\backslash c) = c \quad \text{and} \quad (c/b)b = c.$$

EQ2. $$a\backslash(ab) = b \quad \text{and} \quad (ab)/b = a.$$

EQ3. $$c/(a\backslash c) = a \quad \text{and} \quad (c/b)\backslash c = b.$$

Conversely, any *equasigroup* $[S, \cdot, \backslash, /]$ which satisfies identities EQ1–EQ3 is a quasigroup as defined above. That is, quasigroups and equasigroups are *crypto-isomorphic* classes of algebras.

From a general algebraic standpoint, the concept of equasigroup seems more natural than that of quasigroup for two reasons. First, a nonvoid subset of a quasigroup Q is itself a quasigroup (a "subquasigroup") if and only if it is a subalgebra (in the sense of Chapter VI) of the *equasigroup* $(Q, \cdot, \backslash, /)$. And second, if θ is a congruence relation on Q, Q/θ is a quasigroup if and only if θ is a congruence relation on Q regarded as an equasigroup—i.e., if and only if θ has the substitution property for $\backslash$ and $/$ as well as $\cdot$.

† The important concept of a "group with operators" was developed by Emmy Noether and her co-workers, to unify the structure theory of the kinds of systems enumerated above; see W. Krull, Math. Z. **23** (1925), 161–96, and Vander Waerden [**MA**, §38 ff].

A *loop* is a quasigroup with a two-sided "identity" 1, satisfying $1x = x1 = x$ for all x, so that $x/x = x\backslash x = 1$. Loops can also be defined in two ways, corresponding to the two ways of defining a quasigroup given above.

Definition. A congruence relation θ with respect to multiplication ab on a loop G is *normal* if and only if, for all x, $ux \equiv x\ (\theta)$ or $xu \equiv x\ (\theta)$ implies $u \equiv 1\ (\theta)$. A loop is *normal* when all its congruence relations (for multiplication) are normal.

Any equivalence relation on a loop which has the Substitution Property with respect to all three equasigroup operations is clearly normal when considered with respect to multiplication alone, since $ux \equiv x$ implies $u \equiv x/x = 1$.

Theorem 2. *A loop G is normal if either* (i) *G is a group, or* (ii) $(xy)x^{-1} = y = x^{-1}(xy)$ *for all x, y in G and some x^{-1}, or* (iii) *G is finite.*

Proof. Case (ii) includes case (i). For case (ii), $ux = x$ implies $u = (ux)x^{-1} \equiv xx^{-1} = 1\ (\theta)$, and symmetrically. Finally, suppose (iii) holds. In any loop, $x \equiv y\ (\theta)$ implies $xb \equiv yb\ (\theta)$. Since $x \to xb$ carries distinct elements into distinct elements, the number of elements in the residue class containing xb is at least as great as the number in the residue class containing x. But for any x, $xb = y$ has a solution; so all residue classes have the same cardinal number. If G is finite, this implies that θ is normal: the sets of xu resp. ux for $u \equiv 1\ (\theta)$ must include *all* $y \equiv x\ (\theta)$.

Loops with operators. Just as with groups, an endomorphism of a loop may be called an "operator". Moreover the structure lattice of a loop with operators is a closed sublattice of the structure lattice of the loop *per se*.

3. Permutable Congruence Relations

The notion of a congruence relation was defined for algebras generally in §VI.4. It was shown there that the congruence relations on any algebra $A = [S, F]$ form a closed sublattice of the lattice $E(S)$ of all equivalence relations on its elements. We now consider the *permutability*† of such congruence relations; this notion will play a key role in many of the developments of this chapter.

In general (cf. §XIV.13), the *product* $\theta\theta'$ of two binary relations θ and θ' on a set S is defined by the following rule:

(2) $a \equiv b\ (\theta\theta')$ if and only if $x \in S$ exists such that $a \equiv x\ (\theta)$ and $x \equiv b\ (\theta')$.

Multiplication of relations is associative. Moreover, since equivalence relations are reflexive and transitive, it is also idempotent on equivalence relations, in the sense that $\theta^n = \theta$ for any $n = 2, 3, 4, \cdots$.

Definition. Two relations θ and θ' on S are *permutable* if and only if $\theta\theta' = \theta'\theta$. This means that if $a \equiv x\ (\theta)$ and $x \equiv b\ (\theta')$ for some $x \in S$, then $a \equiv y\ (\theta')$ and $y \equiv b\ (\theta)$ for some $y \in S$, and conversely.

† Permutable equivalence relations were first studied by P. Dubreil and M.-L. Dubreil-Jacotin, who called them "associable"; see J. de Math. **18** (1939), 63–95.

LEMMA. *Any two congruence relations on an equasigroup are permutable.*

PROOF. Let θ and ϕ be any two congruence relations on an equasigroup, and let $a\theta x$ and $x\phi b$. Define y by the formula

$$y = (b/u)\cdot((x/u)a),$$

then

$$a = (b/u)\cdot((b/u)\backslash a)\phi y, \qquad y\theta(b/u)\cdot((a/u)\backslash a) = (b/u)\cdot u = b.$$

Furthermore, any congruence relation on any equasigroup is determined by any one equivalence class.

To prove analogous results for loops (and quasigroups), we must confine our attention to normal congruence relations as defined in §2.

THEOREM 3. *A normal congruence relation on a loop is permutable with any congruence relation.*†

PROOF. Suppose that $a \equiv x\ (\theta)$ and $x \equiv b\ (\theta')$; set $a = ux$, $b = xv$, and $y = u(xv) = ub$. Since $x \equiv b\ (\theta')$, $y = ub \equiv ux = a\ (\theta')$. Again, since $ux = a \equiv x = 1x\ (\theta)$, and θ is normal, $u \equiv 1\ (\theta)$ and so $y = ub \equiv 1b = b\ (\theta)$. This proves that $\theta\theta' \leqq \theta'\theta$; to prove the converse, we simply reverse the order of multiplication.

COROLLARY. *In any normal loop, all congruence relations are permutable.*

For, if we adjoin new operations to an algebra, then the lattice of congruence relations becomes smaller. Hence if the congruence relations on a group (or other loop) G are normal, they are normal on the same group as a group with operators.

THEOREM 4. *The congruence relations on any algebra with permutable congruence relations form a modular lattice in which $\theta \vee \theta' = \theta\theta' = \theta'\theta$.*

INDIRECT PROOF. By Theorems VI.8 and IV.13, any two permutable congruence relations on an algebra A form a modular pair in the dual of the structure lattice of A. Theorem 4 above is an immediate corollary of this result.

DIRECT PROOF. By definition, $a \equiv b\ (\theta \vee \theta')$ if and only if

$$a = x_0\theta x_1\theta' x_2\theta x_3\cdots x_n = b$$

for some finite chain. That is, $\theta \vee \theta'$ is the union of all finite products $\theta\theta'\theta\theta'\cdots$. But since the congruence relations are permutable, the multiplication is idempotent, commutative, and associative; it defines a semilattice. Hence all these products are equal to $\theta\theta' = \theta'\theta$, proving the last equations.

By the one-sided modular law, to prove that the lattice of congruence relations is modular, it suffices to prove

$$\theta'' \leqq \theta' \quad \text{implies} \quad \theta' \wedge (\theta\theta'') = (\theta' \wedge \theta)\theta''. \tag{3}$$

But suppose $a \equiv b\ (\theta')$ and $a \equiv b\ (\theta\theta'')$, i.e., that $a \equiv b\ (\theta' \wedge (\theta\theta''))$. Then, for some x, $a \equiv x\ (\theta)$ and $x \equiv b\ (\theta'')$. But $\theta'' \leqq \theta'$; hence $x \equiv b\ (\theta')$ also. Since $a \equiv b\ (\theta')$ and θ' is transitive, this implies $a \equiv x\ (\theta \wedge \theta')$. Since $x \equiv b\ (\theta'')$, we infer (3). Q.E.D.

† G. Trevisan, Rend. Sem. Mat. Univ. Padova **19** (1950), 367–70.

Actually, as has been observed by B. Jonsson, it is enough to assume that $\theta\theta'\theta = \theta \vee \theta'$ for all θ, θ'.

Generalizations. Families of algebras having permutable congruence relations are exceptional, as is shown by the following remarkable generalization of Theorem 3 due to Malcev.†

THEOREM OF MALCEV. *For all pairs of congruence relations on a family ("primitive class") of algebras to be permutable, it is necessary and sufficient that there exist a ternary polynomial $p(x, y, z)$ such that $p(x, x, y) \equiv p(y, x, x) \equiv y$.*

Thus, as shown by Trevisan‡, there exist nonpermutable congruence relations on some quasigroups. Likewise, although by Funayama's Theorem VI.9 the congruence relations on any lattice form a distributive lattice, the congruence relations on a lattice are not generally permutable. J. Jakubik (cf. Math. Revs. **21**, 6337) has found necessary and sufficient conditions on a lattice for all its congruence relations to be permutable; see Exercise 10 below.

Exercises for §§1–3:

1. Show that any associative quasigroup is a group.
2. Show that a group can be defined (up to cryptoisomorphism) as an equasigroup which satisfies the identity $(x/y)\backslash z = y/(z\backslash x)$. (*Hint.* See [**UA**, p. 165, Exs. 4–6].)
3. Show that the class of loops is closed under direct product, but not under subalgebra or epimorphic image. (Kiokemeister and Bates)
4. Show that, if all congruence relations on an algebra A are permutable, the same is true of any epimorph of A.
5. Show that, in an algebra with only unary operations, all congruence relations are permutable if and only if the number of congruence relations is at most three.
6. Show that the structure lattice of the chain **4** is a Boolean lattice, though the congruence relations on the lattice **4** are not permutable.

Exercises 7–8 state results of S.-C. Wang, Acta Math. Sinica **3** (1953), 133–41. (See Math. Revs. **17**, p. 121.)

7. Show that the congruence relations on any relatively complemented lattice are permutable.

*8. Show that the congruence relations on a complemented modular lattice L form a Boolean algebra if and only if all neutral ideals of L are principal.

9. Show that if the congruence relations on a distributive lattice are permutable, it is relatively complemented. (J. Hashimoto)
10. Show that all congruence relations on a lattice L are permutable if and only if, given $\theta, \phi \in \Theta(L)$ and $a < x < b$ with $a\theta x$ and $x\phi b$, there exists in $L/(\theta \wedge \phi)$ a relative complement of x in $[a, b]$. (J. Jakubik)
11. Show that the Corollary of Theorem 4 need not hold in infinitary partial algebras (e.g., in topological Abelian groups).

4. Direct Decompositions

We will now show how to determine the direct decompositions of any algebra with permutable congruence relations from a knowledge of its structure lattice.

---

† A. I. Malcev, Mat. Sb. **35** (1954), 3–20; see also H. A. Thurston, Proc. London Math. Soc. **8** (1958), 127–34 (Math. Revs. **19**, p. 1033); and A. F. Pixley, Proc. AMS **14** (1963), 105–9.

‡ G. Trevisan, Rend. Sem. Mat. Univ. Padova **22** (1953), 11–22.

LEMMA. *Let θ_1, θ_2 be permutable congruence relations on an algebra A, with $\theta_1 \wedge \theta_2 = O$ and $\theta_1 \vee \theta_2 = I$. Then A is isomorphic with the direct product $A_1 \times A_2$, where A_i is the epimorphic image A/θ_i $(i = 1, 2)$.*

PROOF. Let $x\theta_i$ be the residue class of A mod θ_i containing x. Then $x \to (x\theta_1, x\theta_2)$ is an epimorphism of A onto a subalgebra of $A_1 \times A_2$. If $x\theta_1 = y\theta_1$ and $x\theta_2 = y\theta_2$, then $x = y$ since $\theta_1 \wedge \theta_2 = O$, so the correspondence is one-one. Given x and y in A, since $\theta_1\theta_2 = \theta_1 \vee \theta_2 = I$, there is a z such that $x \equiv z\ (\theta_1)$ and $z \equiv y\ (\theta_2)$, so $z \to (x\theta_1, y\theta_2)$, hence the correspondence is onto.

THEOREM 5. *The direct decompositions $A \cong A_1 \times \cdots \times A_n$ of an algebra A are given by the sets of congruence relations θ_i such that $\bigwedge_{i=1}^n \theta_i = O$ and, for $i = 2, \cdots, n$:*

$$\theta_1 \wedge \cdots \wedge \theta_{i-1} \text{ is permutable with } \theta_i, \tag{4a}$$

$$(\theta_1 \wedge \cdots \wedge \theta_{i-1}) \vee \theta_i = I. \tag{4b}$$

PROOF. Given such a direct decomposition, if $x\theta_i y$ means that x and y have the same i-components, the stated conditions are obvious. Conversely, if the conditions are fulfilled, then $B = A/(\theta_1 \wedge \cdots \wedge \theta_{n-1}) \cong A_1 \times \cdots \times A_{n-1}$ by induction on n, and $A \cong B \times A_n$ by the preceding lemma.

COROLLARY 1. *If all congruence relations on A are permutable, then the direct decompositions of A are given by the sets of dually independent elements in the (modular) structure lattice of A, whose meet is O.*

Caution. In Theorem 1, it is not sufficient that the θ_i be permutable, that $\bigwedge \theta_i = O$, and that (4b) hold. For, let A be the (trivial) algebra with five elements 0, 1, 2, 3, 4 and no operations. Let

$$\theta_1 = (0, 1)(2, 3, 4), \qquad \theta_2 = (0, 2)(1, 3, 4), \qquad \theta_3 = (0, 3)(1, 2, 4).$$

Then $\theta_i\theta_j = \theta_j\theta_i = I$ for $i \neq j$, $\theta_1 \wedge \theta_2 \wedge \theta_3 = O$, and yet $A \cong (A/\theta_1) \times (A/\theta_2) \times (A/\theta_3)$ is false.

We now recall that all congruence relations on a normal loop, with or without operators, are permutable. There follows

COROLLARY 2. *The direct decompositions of any group or loop G, with or without operators, are given by the sets of dually independent elements $\theta_i \in \Theta(G)$ such that $\bigwedge \theta_i = O$.*

5. Jordan–Hölder Theorems

The Jordan–Dedekind chain condition (§II.8) and the notions of perspectivity and projectivity (Chapter III, §§11–12) evolved from ideas originally used to prove the Jordan–Hölder Theorem for groups. We now prove a generalized Jordan–Hölder Theorem for groups and loops with operators.

The proof applies to any algebra A whose congruence relations are permutable, *and* which contains a one-element subalgebra 1. Given this 1, one can associate with any congruence relation θ on A the subalgebra $S(\theta)$ of all $u \equiv 1\ (\theta)$, as well as the quotient-algebra A/θ. We will use this notation below.

THEOREM 6. *Let A be any algebra with a one-element subalgebra* 1, *all of whose congruence relations are permutable. Then, for any congruence relations θ_1, θ_2 on A, $S(\theta_1 \vee \theta_2)/\theta_2$ and $S(\theta_1)/(\theta_1 \wedge \theta_2)$ are isomorphic.*

PROOF. Consider $S(\theta_1 \vee \theta_2)/(\theta_1 \wedge \theta_2) = T$ as an algebra. On T, θ_1 and θ_2 are congruence relations which satisfy the conditions of the lemma preceding Theorem 5. Hence Theorem 6 asserts that, if $B = B_1 \times B_2$, then B/B_2 is isomorphic with B_1.

The preceding result generalizes what is sometimes called the First Isomorphism Theorem. If A is a group or normal loop (with or without operators), it can be given a sharper formulation. If a^{-1} is the right-inverse of a, then the "translations" $T_a(x) = a(xa^{-1})$ and $U_a(x) = (ax)a^{-1}$ (which are the inner automorphisms in groups) clearly leave 1 invariant. Hence, if θ is a congruence relation on A, and so one for the unary operations T_a and U_a, the latter must leave the normal subloop of $x\theta 1$ invariant.

It follows that the isomorphism of Theorem 6 must preserve the operations T_a and U_a for all $a \in A$, and not just for all $a \in S(\theta_1 \vee \theta_2)$. Hence, if we define two quotient-loops to be *centrally isomorphic* when they are isomorphic for all T_a and U_a (and all operators) as well as for multiplication, we obtain

COROLLARY 1. *If, in Theorem* 6, *A is a group or normal loop, then $S(\theta_1 \vee \theta_2)/\theta_2$ and $S(\theta_1)/(\theta_1 \wedge \theta_2)$ are centrally isomorphic.*

If $\theta' \leqq \theta$ in Theorem 6, the quotient-algebra $S(\theta)/\theta'$ has as lattice-theoretic analog the interval $[\theta', \theta]$. Thus Theorem 6 states that quotient-algebras which correspond to transposed intervals $[\theta_2, \theta_1 \vee \theta_2]$ and $[\theta_1 \wedge \theta_2, \theta_1]$ are isomorphic. By definition of perspectivity, since isomorphism is transitive, there follows

COROLLARY 2. *If A is an algebra with one-element subalgebra and permutable congruence relations, then projective intervals in the structure lattice of A correspond to isomorphic quotient-algebras.*

Applying the preceding correspondence to the projectivity established in the Corollary of Theorem III.9, we obtain

THEOREM 7. *Let A be an algebra with a one-element subalgebra and permutable congruence relations. If*

$$O = \theta_0 < \cdots < \theta_m = I \quad \text{and} \quad O = \theta_0' < \cdots < \theta_n' = I$$

are finite chains of congruence relations on A, then these chains can be refined by interpolating terms $\theta_{i,j} = \theta_{i-1} \vee (\theta_j' \wedge \theta_i)$ and $\theta_{i,j}' = \theta_{j-1}' \vee (\theta_i \wedge \theta_j')$ so that corresponding quotients $S(\theta_{i,j})/\theta_{i,j-1}$ and $S(\theta_{i,j}')/\theta_{i-1,j}'$ are isomorphic.

COROLLARY 1. *Let $O = \theta_0 < \cdots < \theta_m = I$ and $O = \theta_0' < \cdots < \theta_n' = I$ be finite maximal chains of congruence relations on an algebra A with a one-element subalgebra whose congruence relations are permutable. Then $m = n$, and the $S(\theta_i)/\theta_{i-1}$ are pairwise isomorphic with the $S(\theta_j')/\theta_{j-1}'$.*

Corollary 2. *In Theorem 7 and Corollary 1, if A is a group or normal loop with operators, then the isomorphisms are central isomorphisms.*

Further results. The preceding result can be generalized in many ways. Thus, the oldest form of the Jordan–Hölder Theorem refers to so-called *composition series*, in which each θ_{i-1} is a maximal congruence relation on the subalgebra $S(\theta_i)$, assuming that all congruence relations on $S(\theta_i)$ as well as those on A are permutable. Since the families of algebras discussed in Theorem 2 are closed under subalgebra formation, this is true of them. The results of §II.8 yield

Corollary 3. *If A is a group (with or without operators), or a loop satisfying one of the conditions of Theorem 2, then the quotient-groups occurring in any two finite composition series are pairwise isomorphic.*

An important theorem of Wielandt† states that the "composition subgroups" of any finite group which occur in some composition series, form a sublattice of the lattice of all subgroups.

For still other generalizations, see the refs. listed in footnotes 5–8 of [**LT2**, pp. 88–9]. Generalizations to transfinite chains of normal subgroups will be given in Chapter VIII below.

Exercises for §§4–5:

1. Show that every loop has a unique one-element subalgebra.
2. Give direct proofs of Theorems 5 and 7 for rings, from first principles.
3. Apply Theorem 5 and Corollary 2 of Theorem 7 specifically to: (a) rings, (b) vector spaces over division rings, (c) invariant subspaces of ring representations.
4. Deduce the "invariance of the dimension number" of any vector space from Corollary 1 of Theorem 7.
5. Show that the correspondence $\theta \to S(\theta)$ defined at the beginning of §5 is a monomorphism from $\Theta(A)$ into the subalgebra-lattice of A.
6. Deduce the lemma of §3 from Corollary 2 of Theorem IV.13.
7. Exhibit an algebra A with a one-element subalgebra and permutable congruence relations, but having congruence relations $\theta \neq \theta'$ such that $S(\theta) = S(\theta')$.‡
8. Let A be any algebra having only unary operations in pairs α_i, β_i, such that

$$\alpha_i(\beta_i(x)) = \beta_i(\overline{\alpha_i(x)}) = x \quad \text{for all } x.$$

Show that A is simple if and only if the group of permutations generated by the α_i is primitive.

9. (Malcev) Call a translation of an algebra A "invertible" if it has an inverse translation. Show that the invertible translations of A form a group Γ, and that if Γ acts transitively on A, then all congruence relations on A commute.

6. Kurosh–Ore Theorem; Remak's Principle

We now recall from §VI.5 the notion of subdirect product. Repeating Corollary 1 of Theorem VI.11, we see that the subdirect decompositions of any abstract algebra A are given by the sets of congruence relations on A which satisfy $\bigwedge \theta_i = O$.

† For a proof, see M. Hall, *The theory of groups*, Macmillan, 1959, pp. 133–4.

‡ J. Jakubic, Czechoslovak Math. J. 4 (1954), 314–17.

For the special case of algebras with permutable congruence relations, Theorem 5 shows that, when r is finite, one gets a direct product if and only if

$$(\theta_1 \wedge \cdots \wedge \theta_{k-1}) \vee \theta_k = I$$

for $k = 2, \cdots, r$.

THEOREM 8. *Let A be any algebra whose congruence relations are permutable. The number of factors in any irredundant representation of A as a finite subdirect product of subdirectly irreducible algebras is the same for all such representations.*

PROOF. This is immediate from the Kurosh–Ore Theorem of §III.12, and the fact (Theorem 4) that the congruence relations on any algebra with permutable congruence relations form a modular lattice.

By Theorems 2–3, the above result holds for normal loops, including all finite loops, loops satisfying $x^{-1}(xu) = (ux)x^{-1} = u$, and groups with operators. Since it holds for groups with operators, it holds in particular for the normal subgroups of any group, and the ideals of any ring.

Remak's Principle. Again, let x_1, x_2, x_3 be any three congruence relations on an algebra A with a one-element subalgebra and permutable congruence relations. Referring to the diagram in §III.7 of the free modular lattice M_{28} with three generators, we see that

$$e \wedge f = f \wedge g = g \wedge e = o, \qquad e \vee f = f \vee g = g \vee e = i.$$

Hence e/o and g/o are both perspective to i/f, and so by Corollary 1 of Theorem 6 they are isomorphic. Similarly, e/o and f/o are isomorphic, and by the lemma of §4, the quotient-algebra $S(i)/o$ is the direct product of any two of these. We conclude

THEOREM 9 (REMAK'S PRINCIPLE). *Let θ_1, θ_2, θ_3 be congruence relations on an algebra A which has a one-element subalgebra and permutable congruence relations. Further, let $\alpha_1 = (\theta_2 \vee \theta_3) \wedge [\theta_1 \vee (\theta_2 \wedge \theta_3)]$ and cyclically for α_2, α_3; let $\delta = (\theta_1 \wedge \theta_2) \vee (\theta_2 \wedge \theta_3) \vee (\theta_3 \wedge \theta_1)$, and let γ be dual to δ. Then the six quotient-algebras $S(\alpha_j)/\delta$ and $S(\gamma)/\alpha_j$ are isomorphic; moreover $S(\gamma)/\delta$ is isomorphic to the square of any one of these.*

Now observe that in the case of normal loops, the above isomorphisms are *central* by Corollary 2 of Theorem 7. Also, for any elements $[a, 1]$ of $S(\alpha_1)/\delta$ and $[1, y]$ of $S(\alpha_3)/\delta$ in $S(\gamma)/\delta \times S(\alpha_3)/\delta$, clearly

$$([a, 1]^{-1}[1, y])[a, 1] = [(a^{-1}1)a, 1^{-1}y1] = [1, y].$$

By central isomorphism, $[(a^{-1}x)a, 1^{-1}1] = [x, 1]$ for the element $[x, 1]$ of $S(\alpha_i)/\delta$ corresponding to $[1, y]$ in α_3/δ. But every $[x, 1]$ corresponds to some $[1, y]$; hence $(a^{-1}x)a = x$ for all x, a. We conclude

COROLLARY 1. *In any normal loop, the $S(\alpha_i)/\delta$ of Theorem 9 are loops whose translations $x \to (a^{-1}x)a$ and $x \to a^{-1}(xa)$ all reduce to the identity.*

COROLLARY 2. *In any group, the $S(\alpha_i)/\delta$ of Theorem 9 are all commutative.*

Theorem 9 has many other similar corollaries, of which we prove only one.

COROLLARY 3. *Let S, T, and U be three invariant subalgebras of a Lie algebra. If the sublattice of invariant Lie subalgebras which they generate is not distributive, then*

$$(S \vee T) \wedge (T \vee U) \wedge (U \vee S)/(S \wedge T) \vee (T \wedge U) \vee (U \wedge S)$$

is a zero algebra.

PROOF. The invariant subalgebras of a Lie algebra satisfy the hypotheses of Theorem 9, with $S = S(\alpha_1)$, $T = S(\alpha_2)$, $U = S(\alpha_3)$. Any element of $S(\alpha_2)/\delta$ can be written in the form $[x, x]$ as an element of $S(\gamma)/\delta = S(\alpha_1)/\delta \times S(\alpha_3)/\delta$. But now, if $[a, 0]$ is an element of $S(\gamma)/\delta$ we see that $[a, 0][x, x] = [ax, 0]$; but this must have the form $[y, y]$ since it is also in $S(\alpha_2)/\delta$; hence $ax = 0$. This holds for all x and a in $S(\gamma)/\delta$, and so $S(\gamma)/\delta$ is a zero algebra.

7. Theorem of Ore

We now prove a theorem on modular lattices, due to Ore [**2**, p. 272], which implies that direct factorization of any group or normal loop with operators is unique, provided its structure lattice has finite length. To state the lattice-theoretic result, we call an element e of a modular lattice a *direct join* of elements $a_1, \cdots, a_n$ (in symbols $e = a_1 \times \cdots \times a_n$) when the a_i are independent and have the join e.

THEOREM 10 (ORE). *Let L be any modular lattice of finite length. If I has two representations $a_1 \times \cdots \times a_m$ and $b_1 \times \cdots \times b_n$ as a direct join of indecomposable elements, then $m = n$ and the a_i and b_j are projective in pairs.*

PROOF. Let $a'_i = a_1 \times \cdots \times a_{i-1} \times a_{i+1} \times \cdots \times a_m$ and $b'_j = b_1 \times \cdots \times b_{j-1} \times b_{j+1} \times \cdots \times b_n$; then $I = a_i \times a'_i = b_j \times b'_j$ for all i, j. If $I = b_j \times a'_i$ for some i, j we shall say that a_i is *replaceable* by b_j. We shall show that every a_i is replaceable by some b_j, using induction on the length of L. Without loss of generality, we assume $i = 1$.

Case I. Suppose $a_1 \vee b'_j < I$ for some j, say $j = 1$. Let q_h denote $(a_1 \vee b'_h) \wedge b_h$. Since $a_1 \vee b'_1 \geqq b_1$ would imply $a_1 \vee b'_1 \geqq b_1 \vee b'_1 = I$, contrary to hypothesis, clearly $q_1 = (a_1 \vee b'_1) \wedge b_1 < b_1$. Moreover, since the b_h are independent and $q_h \leqq b_h$, $c = \bigvee_{h=1}^{n} q_h$ is the *direct* join of the q_h, and so $d[c] = \sum d[q_h] < \sum d[b_h] = d[I]$, whence $c < I$.

Again, by induction on n, writing $a_1 \vee b'_h = d_h$,

$$\begin{aligned} c = \bigvee_{h=1}^{n} q_h &= (b_1 \wedge d_1) \vee \left[\bigvee_{k=2}^{n} b_k \wedge \bigwedge_{k=2}^{n} d_k\right] \\ &= d_1 \wedge \left(b_1 \vee \left[\bigvee_{k=2}^{n} b_k \wedge \bigwedge_{k=2}^{n} d_k\right]\right) \qquad \text{by L5,} \\ &= d_1 \wedge \left(\bigvee_{k=1}^{n} b_k\right) \wedge \bigwedge_{k=2}^{n} d_k, \qquad \text{by L5, since } \bigwedge_{k\neq 1} d_k \geqq b_1 \\ &= d_1 \wedge I \wedge \bigwedge_{k=2}^{n} d_k = \bigwedge_{k=1}^{n} (a_1 \vee b'_k) \geqq a_1. \end{aligned}$$

Therefore, $(c \wedge a'_1) \vee a_1 = c \wedge (a'_1 \vee a_1) = c$ by the modular law, and $c \wedge a'_1 \wedge a_1 = O$, whence $c = a_1 \times (c \wedge a'_1)$.

Hence by induction on the length of L, a_1 is replaceable by some factor $e_{h1} > O$ of some q_h, in any representation of $c = q_1 \times \cdots \times q_n = a_1 \times (c \wedge a'_1)$ as a direct join of indecomposable factors e_{hk} of the q_h; write $e = e_{h1}$ for short. Then $c = e \times (c \wedge a'_1)$ by definition, whence

$$\begin{aligned} e \vee a'_1 &= e \vee (c \wedge a'_1) \vee a'_1 \qquad \text{by L4} \\ &= a_1 \vee (c \wedge a'_1) \vee a'_1 = c \vee a'_1 = I \qquad \text{by the above.} \end{aligned}$$

But $d[e] = d[a_1]$, by replaceability in $c = a_1 \times (c \wedge a'_1)$; hence $e \wedge a'_1 = O$ and $I = e \times a_1$. Moreover $e = b_h$. For $e \wedge (b_h \vee a'_1) \leqq e \wedge a'_1 = O$ and by the modular law

$$e \vee (b_h \wedge a'_1) = (e \vee a'_1) \wedge b_h = I \wedge b_h = b_h,$$

whence e is a direct factor of $b_h = e \times (b_h \wedge a'_h)$. But by hypothesis, b_h is indecomposable; hence $e = b_h$ and $I = b_h \times a'_1$. Hence a_1 is replaceable by b_h, where $h \neq 1$.

Case II. Suppose $a_1 \vee b'_j = I$ for all j, yet $a'_1 \vee b_j < I$ for all j, the only possibility left to consider. Then as in Case I, but with the roles of a_1 and b_1 interchanged, b_1 can be replaced by some $a_k \neq a_1$, say a_m (renumbering subscripts). Therefore $x \to (x \vee a_m) \wedge a'_m$ is a perspectivity of $[O, b'_1]$ to $[O, a'_m]$. Hence if b^*_j denotes $(b_j \vee a_m) \wedge a'_m$, then $a'_m = a_1 \times \cdots \times a_{m-1} = b^*_2 \times \cdots \times b^*_n$. By induction on length, a_1 is replaceable by some b^*_j in $[O, a'_m]$. Moreover, $b_j \vee a'_1$ contains

$$\begin{aligned} b_j \vee a_m &= (b_j \vee a_m) \wedge (a'_m \vee a_m) \\ &= [(b_j \vee a_m) \wedge a'_m] \vee a_m = b^*_j \vee a_m. \end{aligned}$$

Hence $b_j \vee a'_1$ contains $b^*_j \vee a_2 \vee \cdots \vee a_{m-1} \vee a_m = a'_m \vee a_m$ (since a_1 is replaceable by b^*_j in a'_m); that is, $b_j \vee a'_1 = I$. But $d[a_1] = d[b^*_j] = d[b_j]$, the three being projective. Hence $I = b_j \times a'_1$, and a_1 is replaceable by b_j. This completes the proof.

Using Theorem 4 and the Corollary of Theorem 5, we infer

COROLLARY 1. *Let $A = A_1 \times \cdots \times A_m = B_1 \times \cdots \times B_n$ be any two representations of an algebra A as a direct product of indecomposable factors where* (i) *A has a one-element subalgebra,* (ii) *all congruence relations on A are permutable, and* (iii) *the lattice of congruence relations on A has finite length. Then $m = n$ and the A_i and B_j are pairwise isomorphic.*

COROLLARY 2. *Let G be any normal loop, with or without operators; suppose also that the lattice of congruence relations on G has finite length. Then in any two representations of G as a direct product of directly indecomposable factors, the factors are pairwise centrally isomorphic.*

Generalizations. Various important generalizations of Ore's Theorem 10 are known. Thus B. Jonsson and A. Tarski have proved a unique decomposition theorem for all algebras with a binary operation $+$, an element 0 such that $0 + x = x + 0 = x$ for all x, and a structure lattice of finite length. F. B. Thompson has shown that it is not sufficient that 0 be merely idempotent, however.†

Again, A. W. Goldie‡ has extended the usual proof for groups, based on Fitting's Lemma, to algebras whose congruence relations commute and whose structure lattice has finite length.

Exercises for §§6–7:

1. Generalize Corollary 3 of Theorem 9 to arbitrary hypercomplex algebras.

2. (a) Show that S, T, U are normal subloops of a normal loop G, with $S \cap T = T \cap U = U \cap S = 1$, $S \vee T = T \vee U = U \vee S = G$, if and only if there is an isomorphism $\phi: S \to T$ such that U consists of the couples $(s, \phi(s))$ in $S \times T$.

(*b) Find necessary conditions on a loop S, in order that the elements (s, s) form a normal subloop of $S \times S$.

3. (a) Show that, if L is distributive in Theorem 10, then the a_i and b_j are *equal* in pairs (not just projective).

(b) Show that the preceding conclusion holds more generally if the a_i and b_j are neutral elements of L.

4. Let A and B be two-element algebras with a unary operation $'$, such that $0' = 0$ and $1' = 1$ in A, whereas $0' = 1$ and $1' = 0$ in B. Show that $A \times B \cong B \times B$, yet A, B are directly indecomposable and nonisomorphic. (B. Jonsson)

5. Show that the unique factorization theorem is not valid for all finite algebras having a one-element subalgebra. (F. B. Thompson)

6. In the additive group of the ring $\mathbf{Z}[i]$ of Gaussian integers, let A, A', B, B' be the cyclic subgroups generated by $1, i, 3 + 4i, 4 + 5i$. Show that although $G = A \times A' = B \times B'$, where A, A', B, B' are directly indecomposable, A is not "replaceable" by B or B'.

7. Show that, if $\Theta(G)$ is isomorphic with a projective geometry of dimension $n > 1$, then G is Abelian. (M. Hall)

*8. Construct a 12-element commutative semigroup which does not have the unique factorization property. (R. McKinsey)

8. Subgroup Lattices

The "structure lattice" $\Theta(G)$ of all normal subgroups of a given group G, which determines its direct and subdirect decompositions, has been analyzed in the preceding sections. .We now turn our attention to the lattice $L(G)$ of all subgroups of G.

Whereas all simple groups have the same structure lattice, it seems likely that no two nonisomorphic finite simple groups have isomorphic subgroup lattices. Indeed, the "subgroup lattice" $L(G)$ characterizes G far better than $\Theta(G)$ does.§

For example, it is easy to show that $L(G)$ is finite if and only if G is finite. Suppose G is infinite. If any $a \in G$ has infinite order, then the cyclic subgroup

† See the Appendix by Jonsson and Tarski to A. Tarski, *Cardinal algebras*, Oxford, 1949; F. B. Thompson, Bull. AMS **55** (1949), 1137–41 (for Exercise 5 below). See also C. C. Chang, B. Jonsson and A. Tarski, Fund. Math. **55** (1964), 249–81; B. Jonsson, Colloq. Math. **14** (1966), 1–32.

‡ Proc. Cambridge Philos. Soc. **48** (1952), 1–34. See also R. Baer, Trans. AMS **62** (1947), 62–98, and Bull. AMS **54** (1948), 167–74.

§ The question of how far $L(G)$ characterizes G was first studied by R. Baer; see Bull. AMS **44** (1938), 817–20. For the current status of this question see [**Su**].

which it generates has infinitely many subgroups. If every element of G has finite order, chooose any a in G. The cyclic subgroup generated by a being finite, there must be infinitely many elements of G not in this cyclic subgroup. Pick one of them, and repeat the process. In this way infinitely many subgroups of G may be obtained. Thus whenever $L(G)$ is finite, G must be finite also. The converse is immediate.

A deeper result is the fact that subgroup lattices satisfy no lattice identities or identical implications not consequences of L1–L4. To this end, we first prove

THEOREM 11. *Any sublattice of the lattice of all subgroups of any group is isomorphic to a lattice of partitions, and conversely.*

PROOF. With each subgroup S of a group G, associate the partition $\pi(S)$ of G into right-cosets Sx of S. Then obviously, $\pi(S \wedge T) = \pi(s) \wedge \pi(T)$. Moreover $x \equiv y\ (\pi(S) \vee \pi(T))$ if and only if for some finite chain $x_0 = x, \cdots, x_{2n} = y$, $x_{2k+1} \in Sx_{2k}$ and $x_{2k} \in Tx_{2k-1}$, whence $y \in (TS \cdots TS)x$, which is to say $y \in (T \vee S)x$. Hence $\pi(S) \vee \pi(T) = \pi(S \vee T)$.

Conversely, with each partition π of any set I, associate the subgroup $K(\pi)$ of all those permutations f of I which: (i) displace only a finite number of symbols, and (ii) "respect" π in the sense that $x\pi f(x)$ for all $x \in I$. Then, obviously, $K(\pi \wedge \pi') = K(\pi) \wedge K(\pi')$ and $K(\pi) \vee K(\pi') \subset K(\pi \vee \pi')$. Moreover, except in the trivial case $\pi = \pi'$, $K(\pi) \vee K(\pi')$ contains one transposition and all three-cycles (pqr) respecting $\pi \vee \pi'$. For, if (pqr) respects $\pi \vee \pi'$, then either $p\pi q\pi r$ or $p\pi q\pi' r$, or similar relations hold with π and π' interchanged. In the first case, $(pqr) \in K(\pi)$, while in the second, $(pqr) = (pr)(rq) \in K(\pi) \vee K(\pi')$. Since $K(\pi \vee \pi')$ is generated by any transposition and its three-cycles,† it follows that $K(\pi \vee \pi') = K(\pi) \vee K(\pi')$, completing the proof.

We now state without proof a much deeper theorem due to Whitman.‡

WHITMAN'S THEOREM. *Every lattice is isomorphic with some partition lattice.*

From the above results, we infer immediately

COROLLARY 1. *Every lattice is isomorphic with a sublattice of the lattice of all subgroups of some group.*

A corollary to this result and Theorem VI.17 is

COROLLARY 2. *If* $p(x_1, \cdots, x_r) \equiv q(x_1, \cdots, x_r)$ *holds identically in all subgroup lattices, then it holds in every lattice.*

We now apply the construction of Theorem 11 to characterize the permutability of subgroups in a group.

Two subgroups S and T of a group G are called *permutable* when $ST = TS$; this is equivalent to the condition that $S \vee T = ST$, and hence (for finite groups) to the equation $o(S)o(T) = o(S \wedge T)o(S \vee T)$.

---

† M. Hall, *Theory of groups*, p. 61, Lemma 5.4.1; note that any permutation satisfying (i) is a finite product of cycles.

‡ P. M. Whitman, Bull. AMS **52** (1946), 507–22.

LEMMA 1. *Let S and T be two subgroups of a group G, and let $\pi(S)$ and $\pi(T)$ be the partitions of G into right-cosets of S and T, respectively. Then $\pi(S)$ and $\pi(T)$ are permutable if and only if $ST = TS$.*

PROOF. By definition, $x \equiv y \bmod \pi(S)\pi(T)$ if and only if $y \in Tz \subset TSx$ for some $z \in Sx$—hence, if and only if $y \in TSx$. Hence $\pi(S)\pi(T) = \pi(T)\pi(S)$ if and only if $TSx = STx$ for all $x \in G$—that is, if and only if $ST = TS$.

COROLLARY. *If S and T are permutable subgroups of a group G, then the corresponding elements of $L(G)$ are a dually modular pair.*

For, by the lemma and Theorem IV.12, $\pi(S)$ and $\pi(T)$ are a dually modular pair in $E(G)$, the lattice of *all* partitions of G. Since, as was shown in proving Theorem 11 above, $L(G)$ is a sublattice of $E(G)$, the corollary follows.

9. Subgroups of Abelian Groups

When G is Abelian, then by Theorem 1 $L(G) \cong \Theta(G)$: the structure lattice and subalgebra lattice of G coincide. We next study subgroup lattices of finite Abelian groups. Since any finite Abelian group is a direct product of cyclic groups (of prime-power orders), it is natural to begin this study with the following example.

Example 1. Let Z_r be a finite cyclic group of order r, with generator a. Then [**B-M**, p. 140] every subgroup of Z_r is cyclic, with generator a^s for some $s|r$. Hence the lattice $L(Z_r)$ is isomorphic with the ideal of divisors of r, in the distributive lattice of positive integers, under the relation $m|n$. This shows that the lattice of all subgroups of any finite cyclic group is distributive.

More precisely, if $r = p_1^{e_1} \cdots p_k^{e_k}$, then $L(Z_r)$ is the product $C_1 \cdots C_k$ of chains of lengths $e_1, \cdots, e_k$ (or, equivalently, $L(Z_r) \cong \mathbf{2}^{e_1 + \cdots + e_k}$). We now generalize this result.

THEOREM 12. *Let $G = M \times N$, where M and N are finite groups of relatively prime orders. Then $L(G) \cong L(M) \times L(N)$.*

PROOF. Since $M \wedge N = 1$ and $M \vee N = G$, it suffices to prove that $S = (S \wedge M) \vee (S \wedge N)$ for any subgroup S of G. But $(S \wedge M) \vee (S \wedge N) \subset S$ trivially; while since any element $g \in S$ is the product of powers $g^a \in M$ and $g^b \in N$ of itself, where $a|n$ and $b|m$, $S \subset (S \wedge M) \vee (S \vee N)$. Equality follows. Q.E.D.

The converse of Theorem 12 is also true;† see Theorem 16 below. Moreover, using Theorem 12, it is easy to prove

THEOREM 13. *Let G be a finite Abelian group. Then $L(G)$ admits an involution (i.e., a dual automorphism of period two).*

PROOF. Since G is a direct product of Sylow subgroups of relatively prime orders, Theorem 12 reduces the proof to the case of Sylow subgroups, each of which is a direct product of cyclic groups of orders $p^{h_1}, \cdots, p^{h_t}$, where $h_1 \geqq h_2 \geqq \cdots h_t$. In any such subgroup, let $g_1, \cdots, g_t$ be the generators and write

† G. Zappa, Rend. Accad. Sci. Fis. Mat. Napoli **18** (1951), 1–7.

$$(5) \qquad \begin{gathered} g_1^{x_1} \cdots g_i^{x_i} \perp g_1^{y_1} \cdots g_i^{y_i} \text{ if and only if} \\ x_1y_1 + x_2y_2p^{h_1-h_2} + \cdots + x_iy_ip^{h_1-h_i} = 0. \end{gathered}$$

Then the *polarity* induced by the relation $\perp$ is an involution of $L(G)$; we omit the details.†

Distributive subgroup lattices. It is of interest to classify groups according to their subgroup lattices. To this end, we first determine which finite groups have distributive subgroup lattices.

LEMMA 1. *In group G, let A, B, C be cyclic subgroups generated by a, b, and $c = ab$, respectively. If $(A \vee B) \wedge C = (A \wedge C) \vee (B \wedge C)$, then $ab = ba$.*

PROOF. Since $C = \{ab\}$, $C \leqq A \vee B$ and $(A \vee B) \wedge C = C$; hence the hypothesis is $C = (A \wedge C) \vee (B \wedge C)$. Since every subgroup of a cyclic group is cyclic, moreover, $A \wedge C$ is generated by some power $a^r = c^h$ of a which is also a power of c, and $B \wedge C$ likewise by some $b^s = c^k$. To make $C = (A \wedge C) \vee (B \wedge C)$, we must have $c \in \{c^h\} \vee \{c^k\}$ and so (h, k) can be made relatively prime: $mh + nk = 1$ for suitable integers m, n. Therefore

$$ab = c = c^{mh}c^{nk} = c^{nk}c^{mh} = b^{ns}a^{mr} = a^{mr}b^{ns},$$

and so $a^{mr-1} = b^{1-ns}$. Hence the commutator

$$\begin{aligned} aba^{-1}b^{-1} &= b^{ns}a^{mr}a^{-1}b^{-1} = b^{ns}a^{mr-1}b^{-1} \\ &= b^{ns}b^{1-ns}b^{-1} = 1, \end{aligned}$$

and so $ab = ba$ as claimed.

THEOREM 14. *The lattice $L(G)$ of all subgroups of a finite group is distributive if and only if G is cyclic.*

PROOF. The discussion of Example 1 makes it sufficient to show that if $L(G)$ is distributive, then G is cyclic. Again, Lemma 1 shows that if $L(G)$ is distributive, then G must be Abelian. Hence, by the basis theorem for finite Abelian groups, G must be the direct product of cyclic groups of prime-power orders $q_1 = p_1^{k_1}, \cdots,$ $q_r = p_r^{k_r}$ with generators $a_1, \cdots, a_r$. If two p_i were equal, then G would contain two elements $b_i = a_i^{q_i/p}$ and $b_j = a_j^{q_j/p}$, $p = p_i = p_j$. These would generate an elementary Abelian group whose subgroup-lattice (of length two) was not distributive. Hence the p_i are all distinct; but in this case, $a = a_1a_2 \cdots a_r$ is of order $q_1q_2 \cdots q_r$, and generates G, which is therefore cyclic. Q.E.D.

Exercises for §§8–9:

1. Show that, for any prime p, the dihedral group of order $2p$ and the noncyclic group of order p^2 have isomorphic subgroup lattices.
2. For G and H cyclic, find necessary and sufficient conditions that $L(G) \cong L(H)$.
3. Show that if G is the dihedral group of order 10, then $L(G)$ is modular, whereas this is not true of the dihedral group of order 30.
4. (a) Let G be defined by $x^7 = y^6 = 1$, $yx = x^3y$. Show that $L(G)$ satisfies the Jordan–Dedekind chain condition, but is neither semimodular nor dually so.

† See G. Birkhoff, Proc. London Math. Soc. **38** (1935), p. 389. The polarity in question also maps characteristic subgroups onto characteristic subgroups.

(b) Let G be defined by $x^5 = y^4 = 1, yx = x^2y$. Show that $L(G)$ is semimodular but not modular.

5. (a) Show that if $L(G)$ is modular, and H is any epimorphic image of any subgroup of G, then $L(H)$ is modular.

(b) Find G and H such that $L(G)$ and $L(H)$ are modular, whereas $L(G \times H)$ is not modular.

6. (a) Show that, if A is an Abelian group of odd order, the characteristic subgroups of A correspond to the neutral elements of $L(A)$.

(b) Show that the lattice of characteristic subgroups of A is generated by two chains: the subgroups of x such that $p^\alpha x = 0$, and of y such that $y = p^\beta t$, where the exponents α and β characterize the subgroups.†

7. Let an element q of a lattice L with O be called *primary* when $[O, q]$ is a chain. Show that, if G is a finite group, every element of $L(G)$ is the join of suitable primary elements.

8. Show that, if an element $a \in L(G)$ is invariant under all lattice-automorphisms of $L(G)$, then it corresponds to a characteristic subgroup of the group G.

9. (a) Show that, if the order of a group G is the product of $k \leqq 4$ prime factors, then the length of $L(G)$ is k.

(b) Exhibit a group of order 1092, the length of whose subgroup lattice is only four. (E. T. Parker)

10. Show that, if G and H are finite Abelian groups without cyclic Sylow subgroup, then $L(G) \cong L(H)$ implies $G \cong H$.

10. Neutral Elements; Center

To illustrate subgroup lattices of finite non-Abelian groups, we now exhibit two examples.

Example 2. The quaternion group is the multiplicative group of the quaterion units ± 1, $\pm i$, $\pm j$, $\pm k$; it contains eight elements. All its subgroups are normal; its (modular) subgroup lattice is sketched in Figure 17a.

Example 3. The lattice of all subgroups of the octic group is the lower semi-modular lattice whose diagram is sketched in Figure 17b [**B-M**, p. 141].

(a) (b)

FIGURE 17

The following result is easily proved.

LEMMA. *If $L(G) \cong \mathbf{2}^2$ is the Boolean algebra of order four, then G is cyclic of order pq, where p and q are distinct primes.*

PROOF. Since $L(G)$ is distributive, G is cyclic. Since G is Abelian, the length of $L(G)$ equals the number of (equal or distinct) prime factors of the order $o(G)$ of G. Hence $o(G) = pq$ or p^2. But for neither of the two Abelian groups of order p^2 is $L(G) \cong \mathbf{2}^2$, whence the conclusion follows.

† G. Birkhoff, Proc. London Math. Soc. **38** (1935), 389.

THEOREM 15. *If a subgroup A of G corresponds to a neutral element $a \in L(G)$, then A is a characteristic subgroup of G.*

PROOF. By induction on $o(G)$, we can assume that the assertion holds for any proper subgroup or epimorphic image of G, and also for any neutral element $c < a$ of $L(G)$. We will prove Theorem 15 under these assumptions by contradiction.

Indeed suppose that $\alpha(A) = B \neq A$ for some automorphism α of G. Since automorphisms of G induce automorphisms on $L(G)$, the element $b \in L(G)$ which corresponds to B must also be neutral. Moreover, since the neutral elements of any lattice form a (distributive) sublattice, $a \wedge b = c < a$ must also be neutral. Hence, by induction, the subgroup $C = A \wedge B$ of G corresponding to $c \in L(G)$ must be a *characteristic* (hence normal) subgroup of G. Finally, α maps the interval sublattice $[c, a]$ of $L(G)$ onto $[c, b]$, isomorphically: each is isomorphic to

$$L(A/C) \cong L(B/C).$$

Now let $P \leqq A$ cover C in $L(G)$; and let $\alpha(P) = Q \leqq B$. Since $A \wedge Q \leqq A \wedge B = C < A$, A cannot contain Q. Also, since a and b are neutral elements of $L(G)$ independent over c, $(a \vee b)/c \cong (a/c) \times (b/c)$ and so $(p \vee q)/c \cong (p/c) \times (q/c) \cong \mathbf{2}^2$. Hence, by the preceding lemma, $(P \vee Q)/C$ must be cyclic. This contradicts the existence of distinct subgroups P/C and $Q/C = \alpha(P/C)$ having the same order, and proves the theorem.

THEOREM 16. *The center of $L(G)$ consists of those characteristic subgroups of G whose order and index are relatively prime.*

PROOF. If c is in the center of $L(G)$, then it is neutral and has a neutral complement c'. The corresponding subgroups C and C' of G are complementary characteristic subgroups of G, whence $G \cong C \times C'$. If $o(C)$ and $o(C')$ had a common prime factor p, then C and C' would contain by Cauchy's Theorem subgroups P and P_1 of order p; hence $o(P \vee P_1) = p^2$. But since $L(G) \cong L(C) \times L(C')$, we must have $L(P \vee P_1) = \mathbf{2}^2$, whence by Lemma 1 $o(P \vee P_1) \neq p^2$. This contradiction shows that $o(C)$ and $o(C')$ are relatively prime.

The converse was proved in Theorem 12.

11. Modular Subgroup Lattices

Obviously, the subgroup lattice $L(A)$ of any Abelian group A is modular. A few non-Abelian groups also share this property, as we shall now see.

By the Corollary of the Lemma of §8, $L(G)$ will be modular if any two subgroups of G are permutable. Hence, defining a group G to be *quasi-Hamiltonian* when all its subgroups are permutable, we have

THEOREM 17. *The subgroup lattice of any quasi-Hamiltonian group is modular.*

A group G is called *Hamiltonian* when all its subgroups are normal. Since a normal subgroup is permutable with any subgroup, evidently any Hamiltonian group is quasi-Hamiltonian. The quaternion group of Example 2, §10, is Hamiltonian; and conversely,† any Hamiltonian group is either Abelian or the direct product of the quaternion group with an Abelian group.

THEOREM 18. *Let G be any finite group; and let $L(G)$ be (upper) semimodular. Then G is solvable.*

PROOF. We will prove a sharper result: that for the largest prime divisor p of $o(G)$, the elements of order p constitute (with 1), a characteristic elementary Abelian subgroup P of G. Indeed, let $X = \{x\}$ and $Y = \{y\}$ be any two distinct cyclic subgroups of order p in G (the case that only one exists is trivial). Then $o(X \vee Y)$ has length two. We now use the

LEMMA. *If $L(G)$ has length two, then $o(G)$ is the product of two primes.*

PROOF. If $o(G)$ is a product of distinct primes ("square-free"), then G is solvable (Hall, op. cit., Corollary 9.4.1), and the length of $L(G)$ is the number of prime factors of $o(G)$ (Hall, Theorem 9.2.3), whence the result follows for square-free $o(G)$. If $o(G) = mp^2$ is not square-free, then G contains a subgroup S of order p^2, $L(S)$ has length two, whence $G = S$ and $o(G) = p^2$. Q.E.D.

By this lemma, $o(X \vee Y) \geqq p^2$ is the product of two primes, neither of which can exceed p since $o(X \vee Y) | o(G)$ and p is the largest prime divisor of G. It follows that $o(X \vee Y) = p^2$, whence $X \vee Y$ is *Abelian*. This shows that any two elements of G of order p are permutable, whence $P > 1$ exists as described. But $L(G/P)$ is an interval sublattice of $L(G)$, and so also semimodular. By induction on $o(G/P) < o(G)$, it follows that G/P and hence G is solvable.

THEOREM 19. *If G is finite and $L(G)$ is modular, then G is solvable.*

The preceding theorem opens the door to the determination of all finite groups having modular subgroup lattices, a problem solved by Iwasawa and Jones.‡ The solution is complicated (see Exercises 6–8 after §12). We will only state here the solution for the smaller class of finite groups G which have a *complemented modular subgroup lattice* $L(G)$.

Since any such $L(G)$ is a direct product of irreducible complemented modular lattice, it suffices by Theorem 15 to determine those G for which $L(G)$ is an irreducible complemented modular lattice. Such groups (called P-groups) are:

(i) Direct products of cyclic groups of prime order p; and

(ii) Semidirect products C_qS_p of a cyclic group of order q with a normal subgroup of type (i) above, in which $c^{-1}ac = a^r$ ($r \neq 1$, $r^q \equiv 1 \pmod{p}$) for some generator c of C_q.

More general conditions for $L(G)$ to be complemented, but not modular, have been found by Zacher.§

† M. Hall, *The theory of groups*, Theorem 12.5.4.

‡ K. Iwasawa, J. Fac. Sci. Tokyo **4** (1941), 171–94, and Japan J. Math. **18** (1943), 709–28; A. W. Jones, Duke J. Math. **12** (1945), 541–60. See Suzuki [**1**, pp. 11–18].

§ G. Zacher, Rend. Sem. Mat. Univ. Padova **22** (1953), 111–22.

12. Jordan–Dedekind Condition and Supersolvability

The following result is also easily established.

THEOREM 20. *If G is a finite p-group of order p^n, then $L(G)$ is dually semimodular of length n.*

PROOF. Any maximal subgroup S of G has prime index p. If S and T are any two distinct maximal subgroups of G, then since $S \vee T \geqq ST$ and $o(ST) = o(S)o(T)/o(S \wedge T)$,

$$p^{2n-2} = o(S)o(T) \leqq o(S \wedge T)o(S \vee T) = p^n o(S \wedge T),$$

clearly $o(S \wedge T) \geqq p^{n-2}$. Hence S and T cover $S \wedge T$. The proof is completed by the remark that any subgroup of a p-group is itself a p-group, whence the covering condition (ξ) which defines dual semimodularity is satisfied.

COROLLARY 1. *If G is a finite nilpotent group, then $L(G)$ is dually semimodular, and the dimension in $L(G)$ of any subgroup S of G is the number of prime divisors of $o(S)$.*

Indeed, $L(G) \cong L(P_1) \times \cdots \times L(P_r)$, by Theorem 17 and (Hall, Theorem 10.3.4), where the P_i are the Sylow subgroups of G.

COROLLARY 2. *If G is a finite nilpotent group, then $L(G)$ satisfies the Jordan–Dedekind condition.*

The preceding result can be generalized.

THEOREM 21. *If G is a finite group, then $L(G)$ satisfies the Jordan–Dedekind chain condition if and only if G is supersolvable.*

We recall (M. Hall, *op. cit.*, p. 149) that G is *supersolvable* when it contains a chain of normal subgroups N_i with cyclic N_{i-1}/N_{ij}; and that if G is finite, these quotient-groups may be taken of prime order. From this, it follows easily that for any subgroup S of G, the $S \vee N_i$ form a chain whose length is the number of (equal or distinct) prime divisors of $[G : S]$, and that S is maximal if and only if $[G : S]$ is a prime. The Jordan–Dedekind chain condition follows immediately from this.

For the proof of the much more difficult converse, due to Iwasawa, see Suzuki, op. cit. The converse is also related to a Theorem of Huppert (Hall, Theorem 10.5.8).

Example 4. Let A_4 be the solvable (but not supersolvable) alternating group on four letters. Then its subgroup lattice does not satisfy the Jordan–Dedekind chain condition; see Fig. 18a.

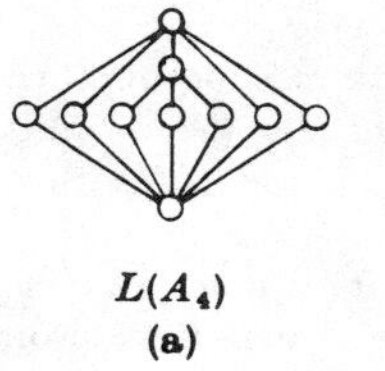

$L(A_4)$
(a)

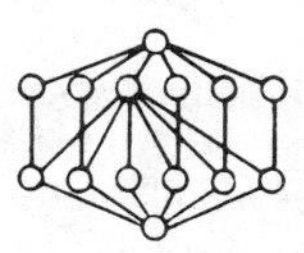

$L(H_{20})$
(b)

FIGURE 18

Example 5. Let H_{20} be the (metacyclic, supersolvable) holomorph of the cyclic group of order 5, with defining relations $a^5 = b^4 = 1$, $b^{-1}ab = a^2$. Then $L(H_{20})$ is not lower **semimodular; see Fig. 18b.**

Exercises for §§10–12:

1. (a) Determine all groups G such that $L(G)$ has length 3.
   (b) Determine all those such that $L(G)$ has length 4.
   (c) Show that the lattice whose diagram is sketched to the right is monomorphic to the subgroup lattice of $S_4{}^3$. (A. G. Waterman)

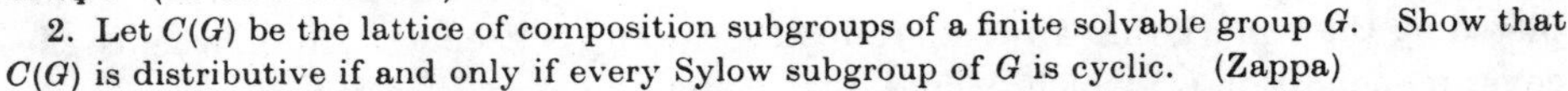

2. Let $C(G)$ be the lattice of composition subgroups of a finite solvable group G. Show that $C(G)$ is distributive if and only if every Sylow subgroup of G is cyclic. (Zappa)
3. Show that, for any finite distributive lattice $\mathbf{2}^P$, there exists a finite group with structure lattice $\Theta(G) = \mathbf{2}^P$. (Kuntzmann)
4. (a) Show that the dual of the lattice of all subgroups of any p-group of finite order is semimodular.
   (b) Show that it is not a matroid lattice unless the p-group is elementary Abelian.
5. Show that, if G is a p-group with more than one subgroup of order p, then all prime intervals of $L(G)$ are projective. (D. W. Barnes)
6. Define a "modular p-group" as a finite group $G = \{M, b\}$, generated by an Abelian normal subgroup M whose elements satisfy $y^{p^m} = 1$, p a prime, and an element b of order p^r, where $xb = bx^{1+p^s}$ for all $x \in M$ and $s > 1$ if $p = 2$. Show that $L(G)$ is modular for any modular p-group G.
7. Define a "modular pq-group" as a finite group $H = \{N, a\}$ generated by an Abelian normal subgroup N whose elements satisfy $x^p = 1$ for some prime p, and an element a of prime-power order q^β which satisfies $a^{-1}xa = x^n$ for all $x \in N$, where $n^q \equiv 1 \pmod p$. Show that $L(H)$ is modular for any modular pq-group H.

*8. Show that, if $L(G)$ is modular, then G is a direct product of Hamiltonian groups, modular pq-groups, and modular p-groups. (Iwasawa)

*9. Let G be a finite non-Abelian p-group. Show that if $L(G) \cong L(H)$, then H is also a p-group.

*10. (E. T. Parker) Let G be any finite group whose subgroup lattice $L(G)$ is relatively complemented. Prove:
   (a) Every element of G has square-free order,
   (b) If S is a subgroup of G, M a normal subgroup of S, and N a normal subgroup of M, then N is a normal subgroup of S,
   (c) Each Sylow subgroup of G is elementary,
   (d) G is metabelian, and its commutator-subgroup is the direct product of all Sylow subgroups not in the centrals of their normalizers.

11. Let G be a finite simple group of composite order. Show that the dual of $L(G)$ is not the subgroup lattice of any group. (Zacher)

*12. Let G be a solvable finite group. Show that if $L(G) \cong L(H)$, then H is solvable. (M. Suzuki)

13. Show that if G coincides with its commutator-subgroup, then any isomorphism $L(G) \cong L(H)$ carries the center of G into the center of H. (M. Suzuki)
14. Show that if G is simple and noncyclic, and $L(G) \cong L(H)$, then the order of H is at most the number of automorphisms of $L(G)$.
15. (a) Show that the right-cosets of any group G form a lattice (the "coset-lattice" of G).
   (b) Exhibit two nonisomorphic finite groups which have isomorphic coset-lattices. (M. Curzio)

PROBLEMS

57. (See Problem 38.) Does the unique factorization theorem hold in every finite semigroup? in every finite monoid?†

58. Is Ore's Theorem true for modular elements‡ in any semimodular lattice of finite length? in any geometric lattice?

59. Find necessary and sufficient conditions on a group G for $L(G)$ to be (a) self-dual, (b) complemented.

60. For which finite groups do the cosets of normal subgroups form a semimodular lattice?

61. In a finite group G, to which subgroups do the "standard" elements of $L(G)$ correspond?

62. For which equasigroups is the lattice of all sub-equasigroups distributive? modular? semimodular?

63. Is the structure lattice of every group monomorphic to the structure lattice of some Abelian group? (B. Jonsson)

† See C.-C. Chang, B. Jonsson, A. Tarski, Fund. Math. **55** (1964), 249–81.

‡ I.e., for elements a such that (a, b) is a modular pair for all b.

CHAPTER VIII

TRANSFINITE INDUCTION

1. Ascending and Descending Chain Conditions

The limiting processes of topology and analysis, which have a direct intuitive appeal, provide necessary tools for treating infinite sets and functions on them. They relate to infinite posets in various ways, which will be studied in Chapters IX–X below.

Even more basic, but also more uncertain (see §16), are various general *induction* arguments which refer to concrete posets (i.e., subsets ordered by inclusion) having infinite length. These arguments include Zorn's Lemma and the Axiom of Choice, which are now standard, and the subtler and less well-known concepts of join-inaccessibility and of being "compactly generated". Such induction arguments, and their applications to posets and lattices, are the theme of the present chapter. We begin with the simplest: the ascending and descending chain conditions.

DEFINITION. A poset P satisfies the *ascending* chain condition (ACC), and is called *Noetherian*,† when every nonvoid subset of P has a maximal element. It satisfies the *descending* chain condition (DCC) when its dual satisfies the ACC.

Evidently, any subset of a Noetherian poset is Noetherian. Dually, if a poset satisfies the DCC, then so do all its subsets (under the same partial ordering). Similarly, if Q is obtained from a Noetherian poset P by *weakening* its order, and P is Noetherian, then so is Q.

Evidently also, the positive integers satisfy the DCC under $\leqq$. Since $m|n$ implies $m \leqq n$, they also satisfy the DCC under the divisibility relation $m|n$. Since the lattice of ideals of the integral domain $\mathbf{Z}$ is dual to the preceding, it follows that the lattice of ideals of $\mathbf{Z}$ is Noetherian.

Of especial interest are *chains* which satisfy the ACC or DCC. These are, by definition, posets in which every nonvoid subset has a *greatest* resp. *least* element. Nonvoid chains which satisfy the DCC are called *well-ordered*‡ sets, or *ordinals*. The most familiar ordinal is $\boldsymbol{\omega}$, the order-type of the positive integers in their natural order; others will be constructed in §13. It is easy to prove

THEOREM 1. *The following statements about a poset P are equivalent:* (i) *P is Noetherian,* (ii) *every chain in the dual $\breve{P}$ of P is well-ordered,* (iii) *every well-ordered subset of P is finite.*

† Though Hilbert applied equivalent concepts as tools in ideal theory, it was Emmy Noether who first recognized the power of the ACC and DCC. See [**MA**, §80]. The ACC in quasi-ordered sets has been considered by G. Higman, Proc. London Math. Soc. **2** (1952), 326–36.

‡ The concept of a well-ordered set is due to G. Cantor. Strictly speaking, an ordinal is an equivalence class of well-ordered sets, or "order-type".

PROOF. If (i) fails, then P contains a nonvoid subset X which does not have a maximal element. Choose $a_1 \in X$; since a_1 is not maximal we can choose $a_2 \in X$, $a_2 > a_1$; since a_2 is not maximal, we can choose $a_3 \in X$, $a_3 > a_1$. The process can be continued to get an infinite ascending chain $a_1 < a_2 < a_3 < \cdots$ isomorphic to the ordinal $\boldsymbol{\omega}$; and thus (iii) fails. If (iii) fails, P contains a chain isomorphic to $\boldsymbol{\omega}$; the dual of $\boldsymbol{\omega}$ is not well ordered and (ii) fails. Finally if (ii) fails, some chain of $\breve{P}$ is not well ordered; therefore some subchain does not have a least element or equivalently, since we are dealing with chains, some subchain does not have a minimal element and $\breve{P}$ fails to satisfy the DCC. Thus P fails to be Noetherian, i.e., (i) fails, and the circle of proof is complete.

Closely related to the ascending chain condition (ACC) is the following

Extended Induction Principle. Let P be a Noetherian poset and Φ any property. Then, in proving that every $a \in P$ has property Φ, we can assume that every $x > a$ has property Φ.

PROOF. Suppose that some $a \in P$ fails to have property Φ. Then, by the ACC, there must be a *maximal* $a \in P$ not having property Φ. Since this is maximal, every $x > a$ has property Φ. But the last two statements contradict the hypothesis. Q.E.D.

COROLLARY. *In any Noetherian lattice L, every element can be expressed as a meet of a finite number of meet-irreducibles, and dually.*

PROOF. Given $a \in L$, either a is meet-irreducible, in which case the conclusion is trivial, or $a = b \wedge c$ for some $b > a$, $c > a$. Applying the Extended Induction Principle to b and c, the conclusion again follows.

The Kurosh–Ore Theorem, which states that the number $r = r(a)$ is the same in all *irredundant* expressions $a = x_1 \wedge \cdots \wedge x_r$ as a meet of meet-irreducible x_i, holds in any modular Noetherian lattice.†

Exercises:

1. (a) Show that any Noetherian lattice with O is complete.
   (b) Let L be any Noetherian lattice. Show that every nonvoid subset of L has a least upper bound.
2. Let P be a poset in which (i) every chain is finite, and (ii) every totally unordered subset is finite. Show that P is finite.
3. Show that any join-epimorphic image of a Noetherian lattice is also Noetherian.
4. Show that if P and Q are posets, then $P + Q$ and PQ are each Noetherian if and only if P and Q both are.
5. Show that every Noetherian Boolean lattice is finite.
6. (a) Show that any finite poset satisfies both chain conditions.
   (b) Show that a poset P satisfies both chain conditions if a $\acute{P}$ is Noetherian.
7. Show that the two conditions (i) P satisfies the DCC and (ii) P contains no infinite totally unordered set, are together equivalent to (iii) any infinite subset of P contains an infinite ascending chain. (I. Kaplansky)
8. Show that if the lattice of all congruence relations of an algebra A satisfies the ACC or the DCC, then any representation of A as a direct product can be refined to one of A as a direct product of indecomposable factors.

† For extensions to semimodular Noetherian lattices, see R. P. Dilworth in [**Symp**, p. 40].

9. In a modular lattice L, define $x \equiv y\ (\theta)$ when there is a finite connected chain joining $x \wedge y$ and $x \vee y$. Show that this defines a congruence relation on L, and that two elements are congruent under it if and only if they are in the same connected component of the diagram of L.†

*10. Let M be a modular lattice with maximal elements m_α, and let $\bigwedge m_\alpha = O$. If M satisfies the DCC, prove M of finite length.

2. Noetherian Distributive Lattices

Any Noetherian *distributive* lattice L is determined by the Noetherian poset $P(L)$ of its meet-irreducibles. The construction generalizes the bijection $L \leftrightarrow \mathbf{2}^P$ between distributive lattices L of length n and posets of order n, already established in §III.3.

DEFINITION. Given a poset P, the *restricted power* $\mathbf{2}^{(P)}$ is the family of all semi-ideals (M-closed subsets) of P which have *finite* sets of maximal elements (or "crowns"), ordered by set-inclusion.

Clearly, $\mathbf{2}^{(P)}$ is isomorphic to the poset of finite sets C of *incomparable* elements of P (also called "crowns"), where $C \leqq \tilde{C}$ means that for any $x \in C$ there exists $\tilde{x} \in \tilde{C}$ with $x \leqq \tilde{x}$. We now prove a nontrivial result.

THEOREM 2. *If P satisfies the* DCC, *then so does* $\mathbf{2}^{(P)}$.

PROOF. Let $C_1 \geqq C_2 \geqq C_3 \geqq \cdots$ be any descending chain of crowns in P. Each $x_j^i \in C_i$ is either ultimately replaced by some $x_{j'}^{i'} < x_j^i$, or it is not. (In the latter case, either $x_j^i \in C_l$ for all $l \geqq i$, or x_j^i simply disappears ultimately; we do not care which.) In the former case, set $\nu(i,j)$ equal to the *least* i' such that some $x_{j'}^{i'} < x_j^i$; in the latter, set $\nu(i,j) = 0$.

Now set $i(0) = 1$, set $i(n+1) = \sup_j \nu(i(n), j)$, and $\Gamma_n = C_{i(n)}$. Then $\Gamma_0 > \Gamma_1 > \Gamma_2 > \cdots$. Further, define $\lambda(m,j)$ as the largest λ such that

$$y_j^m > y_{f(1,j)}^{m+1} > \cdots > y_{f(\lambda,j)}^{m+\lambda}, \quad \text{where} \quad y_{f(k,j)}^{m+k} \in \Gamma_{m+k}.$$

For "terminal" elements (the "latter" case of the preceding paragraph), $\lambda(m,j) = 0$. Otherwise, clearly

$$\lambda(m,j) = \sup_k \lambda(m+1,k) \quad \text{such that } y_j^m > y_k^{m+1}; \tag{1}$$

and this number is *finite* if $\lambda(m+1,k)$ is finite for all $y_k^{m+1} < y_j^m$. We now prove

LEMMA 1. *If P satisfies the* DCC, *then every* $\lambda(m,j) < +\infty$.

PROOF. If $\lambda(m,j) = +\infty$, then since Γ_{m+1} is finite, $\lambda(m+1,k) = \infty$ for some $y_k^{m+1} < y_j^m$. Hence, recursively, we could construct an infinite descending chain $y_j^m > y_k^{m+1} > y_l^{m+2} > \cdots$, contrary to hypothesis.

We next set $\lambda = \max \lambda(0,j)$; Γ_λ will consist only of "terminal" elements, say s of them. Hence if $\Gamma_\lambda \equiv C_{i(\lambda)} > C_{r(1)} > \cdots > C_{r(q)}$, we must have $q < s$. This proves Theorem 2.

† Cf. Ore [1, pp. 421–4], for algebraic applications of Exercise 9.

We now prove the remarkable result that every distributive lattice which satisfies the DCC can be obtained by the above construction, first recalling a result from §III.3.

LEMMA 2. *In a distributive lattice, if p is join-irreducible, then $p \leqq \bigvee_{i=1}^{r} x_i$ implies $p \leqq x_i$ for some $i = 1, \cdots, k$.*

We now sharpen the Corollary of this result which was noted in §III.3.

LEMMA 3. *In a distributive lattice L, no element can have more than one finite representation as an irredundant join of join-irreducible elements (or dually).*

PROOF. Consider two such representations

$$a = x_1 \vee \cdots \vee x_r = y_1 \vee \cdots \vee y_s \tag{2}$$

of the same $a \in L$ as such a join. By Lemma 2, given x_i, some $y_j \geqq x_i$, and similarly some $x_k \geqq y_j$. Hence, unless x_i is redundant, $k = i$ and $x_i = y_j = x_k$. Therefore, if the representations are irredundant, the x_i and y_j are equal in pairs. Q.E.D.

THEOREM 3. *In a distributive lattice L which satisfies the* DCC, *let P be the poset of all join-irreducible elements. Then $L = \mathbf{2}^{(P)}$.*

PROOF. By the last corollary of §1, every $a \in L$ has at least one representation of the form (2); hence by Lemma 3 it has exactly one. Hence there is a one-one correspondence between the "crowns" of P and the elements of L, given by $(x_1, \cdots, x_r) \leftrightarrow x_1 \vee \cdots \vee x_r$. It remains to prove that $x_1 \vee \cdots \vee x_r \leqq y_1 \vee \cdots \vee y_s$ for two such joins if and only if every x_i is contained in some y_j, which follows from Lemma 2. This completes the proof.

Dualizing Theorem 2, we obtain the following

COROLLARY. *Let L be any Noetherian distributive lattice. Then each $a \in L$ can be uniquely represented as an irredundant finite meet of meet-irreducible elements.*

3. Finitely Generated Subalgebras

In the 1920's, it was proved that Hilbert's Finite Basis Theorem for ideals in polynomial rings was equivalent to various important extensions of the Principle of Finite Induction.† More generally, it is equivalent in any algebra to the condition that *every subalgebra is finitely generated* (i.e., has a finite set of generators).

THEOREM 4. *The subalgebra lattice of an algebra A is Noetherian if and only if every subalgebra S of A is finitely generated.*

PROOF. If A is Noetherian, define recursively subalgebras S_k ($k = 1, 2, 3, \cdots$) by setting $S_0 = \varnothing$, and adjoining to S_{k-1} some x_k in S but not in S_{k-1}; this is

† See [**MA**, §80]. The generalization to Theorem 4 is due to Bruce Crabtree and the author; cf. Proc. First Canad. Math. Congress (1945), p. 313.

possible unless $S_{k-1} = S$. Now consider the chain $\sigma: S_0 < S_1 < S_2 < \cdots$; by the ACC σ is finite; hence $S = S_n = \{x_1, \cdots, x_n\}$ is finitely generated.

Conversely, if A is not Noetherian, then it possesses an infinite sequence of subalgebras $S_1 < S_2 < S_3 < \cdots$. Since the operations of A are finitary, $S = \bigcup S_k$ is a subalgebra. If $S = \{x_1, \cdots, x_n\}$ were finitely generated, with $x_1 \in S_{k(i)}$, then S_K would contain every x_i for $K = \max k(i)$; hence $S_K = S$, a contradiction. In summary, if A is not Noetherian, it contains a subalgebra S which is not finitely generated, completing the proof.

COROLLARY. *A lattice L is Noetherian if and only if every ideal of L is principal (i.e., if and only if $\hat{L} = L$). Hence every Noetherian lattice is complete.*

Applications. Combining the Corollary of Theorem 3 with Corollary 1 of Theorem VI.11, we obtain

THEOREM 5. *Let A be any algebra whose congruence relations form a Noetherian distributive lattice. Then A and its epimorphic images have unique representations as irredundant finite subdirect products of subdirectly irreducible factors.*

We have a more novel application to algebraic geometry. An *algebraic variety* V in an affine n-space over a field F is defined (cf. Example 8 of §V.7) as the set of all points $\mathbf{x} = (x_1, \cdots, x_n)$ which satisfy some given set $J(V)$ of polynomial equations $p(x_1, \cdots, x_n) = 0$ with coefficients in F. If V and W are algebraic varieties defined by r equations $p_i = 0$ and $q_j = 0$ respectively, then it is easily shown that the rs equations $p_i q_j = 0$ are satisfied by points in $V \cup W$ and no others, while the $r + s$ equations $p_i = 0$, $q_j = 0$ are satisfied by points in $V \cap W$ and no others. Hence the algebraic varieties form a ring of sets (distributive lattice).

Furthermore, it is classic [**MA**, vol. 2, p. 25] that the ideals of the polynomial ring $F[x_1, \cdots, x_n]$ satisfy the ACC; hence, by the results of §V.7, the "polar" lattice of algebraic varieties satisfies the DCC. Consequently, the hypotheses of Theorem 3 apply, and so *every algebraic variety has a unique expression as an irredundant sum of a finite number of irreducible components.*

Finally, let L be any modular lattice which satisfies the DCC. Applying repeatedly the construction of Exercise 9, §1, we obtain a sequence of epimorphic images $L \succ L_1 \succ L_2 \succ \cdots$ and dual congruence relations θ_k which were first discussed by Ore [**1**, pp. 421–4]. In the case of algebraic varieties, one may show that the *dimension* of any V is the least k such that $V \equiv 0 \bmod \theta_k$. Dually, if $J(V)$ is the ideal of polynomial equations satisfied by all $\mathbf{x} \in V$, and $F[x_1, \cdots, x_n]$ is the ring of all polynomials in the x_i with coefficients in F, then k is the *transcendence degree* of the quotient-ring $F[x_1, \cdots, x_n]/J(V)$.

Exercises for §§2–3:

1. Show that, in $\mathbf{2}^{(P)}$, the crowns which cover a given crown $C = (x_1, \cdots, x_r)$ can all be obtained by adjoining to C some element y which either covers some x_i or is minimal, and then discarding redundant elements.
2. Show that the subgroups of the infinite cyclic group Z satisfy the ACC, whereas those of the "generalized" cyclic group of rational numbers under addition mod 1 do not.

*3. Show that P is isomorphic with the lattice of prime ideals of $\mathbf{2}^{(P)}$.

*4. (a) Let L be a modular, nondistributive lattice satisfying the DCC. Show that Lemma 3 of §2 does not hold.

(b) Show that Lemma 3 holds in the nonmodular lattice shown below.

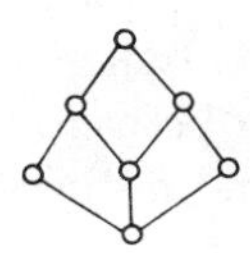

(c) Show that, if L is a lattice of finite length, then Lemma 3 holds in L if and only if L is semimodular and every modular sublattice of L is distributive.†

5. (a) Show that if the congruence relations on an algebra A are permutable and satisfy the ACC, but the lattice which they form is not distributive, then the conclusion of Theorem 5 fails in A. (*Hint.* See Exercise 4a.)

(b) Let A be a group with operators, whose "structure" lattice of congruence relations has finite length. Show that the conclusion of Theorem 5 holds unless A has an epimorphic image with two "independent" Ω-subgroups S and T which are Ω-isomorphic. (See [**LT1**, §103, end].)

6. Let P be the set of all finite sequences $\xi = (x_1, \cdots, x_m)$, $\eta = (y_1, \cdots, y_n)$, $\cdots$ of positive integers. Define $\xi \leqq \eta$ to mean that ξ can be obtained from η be a finite number of "reductions", each of which either deletes a component y_j or replaces it by a smaller integer. Prove that P satisfies the DCC.

7. How much of the discussion of algebraic varieties given in §3 is valid over integral domains as well as fields?

*8. Show that the ideals of any principal ideal domain form a Noetherian distributive lattice.

9. (a) Show that the subgroups of any finitely generated Abelian group form a Noetherian modular lattice.

(b) Construct a group with two generators, whose subgroups do not satisfy the ACC. (*Hint.* Every finite symmetric group is generated by two suitably chosen generators.)

*10. Show that Ore's Theorem (Theorem VII.10) does not hold in all Noetherian modular lattices.

4. Finitary Closure Properties

A closure property Φ, associated with a closure operation $S \to \bar{S}$ on the subsets of a set I, is called *finitary* when the condition $S \in \Phi$ is equivalent to the condition that $K \subset S$ and K *finite* imply $\bar{K} \subset S$. Therefore, it is equivalent to the condition that $\bar{S} = \bigcup \bar{K}_\gamma$ is the set-union of the closures $\bar{K}_\gamma$ of the *finite* subsets K_γ of S.

As examples of finitary closure properties may be mentioned: (i) being a *subalgebra* of *a* (finitary) algebra A, and (ii) being a *congruence relation* on A, considering each relation θ on A as a subset of $A \times A$ which consists of all (a, b) with $a\theta b$. On the other hand, topological closure is *never* finitary on a T_1-space, except in the trivial case of a discrete space; see Chapter IX. The converse of (i) is also true, as we now show.

† R. P. Dilworth, Ann. of Math. **41** (1940), 771–7; also Trans. AMS **49** (1941), 325–53. For Exercise 4a, see [**LT2**, p. 143].

THEOREM 6. *Let Φ be any finitary closure property on a set S. Then there exists an algebra $A = [S, F]$, whose subalgebras are precisely the Φ-closed subsets of S.*

PROOF. To construct $A = [S, F]$, let each pair $\alpha = (F, y)$ with $F = \{x_1, \cdots, x_n\}$ and $y \in \bar{F}$ define $f_\alpha: f_\alpha(x_1, \cdots, x_n) = y$, and $f_\alpha(z_1, \cdots, z_n) = z_1$, if $\mathbf{z} \neq \mathbf{x}$. Then a subset X of S is a *subalgebra* if and only if $F \subset X$ and $y \in \bar{F}$ implies $y \in X$.

We now define a *directed set* as a poset in which any two elements, and hence any finite subset, has an upper bound in the set.

LEMMA 1. *For any finitary closure operation, if a set D of closed subsets S_δ is directed by set-inclusion, then their set-union $\bigcup_D S_\delta$ is closed.*

PROOF. For any finite subset K of elements of the set-union, there is some one S_K which contains them all; but this S_K *is* in the set-union, which is therefore closed.

DEFINITION. An element a of a lattice L is *join-inaccessible*† when $\bigvee_D x_\delta = a$, with D directed, implies that some individual $x_\delta = a$. It is called *compact* when $\bigvee_J x_j \geqq a$ implies $\bigvee_F x_\phi \geqq a$ for some finite subset $F \subset J$.

LEMMA 2. *In any complete lattice, every compact element c is join-inaccessible.*

PROOF. By definition of compactness, c compact D directed and $\bigvee_D x_\delta = c$ imply that $\bigvee_F x_\phi \geqq c$ for some finite subset F of D. But, if α is any upper bound of F in D (such an upper bound exists, since D is directed), this implies

$$c \geqq \bigvee_D x_\delta \geqq x_\alpha \geqq \bigvee_F x_\phi \geqq c.$$

Hence $x_\alpha = c$, and c is join-inaccessible.

The converse is true in a very important case.

THEOREM 7. *Let L be the lattice of all closed subsets of I for a finitary closure property Φ. Then the following conditions on an element $c \in L$ are equivalent:* (i) *c is compact,* (ii) *c is join-inaccessible,* (iii) *c represents the closure $C = \bar{F}$ of a finite set F.*

PROOF. We have already shown that (i) implies (ii) in any complete lattice. We next prove by contradiction that (ii) implies (iii). Indeed, if C is not finitely generated, consider the directed set of $\bar{F}_\delta$ for finite subsets F_δ of C. The $\bar{F}_\delta$ form an upward directed set and $\bigvee \bar{F}_\delta = C$, yet by hypothesis no $\bar{F}_\delta = C$; hence c is not join-inaccessible. Finally, to prove that (iii) implies (i), we suppose that $C = \bar{F}$ is finitely generated, and that $\bigvee_K \bar{S}_k \supset C$. Form the $U_\Delta = \bigvee_\Delta \bar{S}_k$ for finite subsets $\Delta \subset K$; then $C \subset \bigvee_K \bar{S}_k = \bigcup U_\Delta$. Hence every $x \in F$ is in some U_Δ. The union of these finitely many U_Δ is again a U_Δ, which contains F and hence $\bar{F} = C$. This proves that C is compact, whence (iii) implies (i). Q.E.D.

† The concept of join-inaccessibility is due to G. Birkhoff and O. Frink [1]; that of compactness to L. Nachbin, Fund. Math. **36** (1949), 137–42; note the analogy with the Heine–Borel covering property for compact sets.

5. Complete Algebraic Lattices

We now define a lattice L to be "compactly generated" or *algebraic* when every element of L is a join of compact elements of L. The present section will be devoted to the study of complete algebraic lattices. We first prove

LEMMA 1. *The subalgebra-lattice $L(A)$ of any (finitary) algebra A is a complete algebraic lattice.*†

PROOF. Evidently, every subalgebra of A is the join of those "principal" subalgebras (with one generator) which it contains. By Corollary 2 of Theorem 6, however, every principal subalgebra of A corresponds to a compact element of $L(A)$. Since $L(A)$ is complete, the result follows.

THEOREM 8. *Let L be any complete algebraic lattice, and K the set of its compact elements. Then K is a join-semilattice, and L is isomorphic with the lattice $\hat{K}$ of all nonvoid ideals of K.*

PROOF. Let c and k be compact elements of L. Then $c \vee k \leqq \bigvee_S x_i$ implies $c \leqq \bigvee_S x_i$ and $k \leqq \bigvee_S x_i$, hence $c \leqq \bigvee_F x_i$ and $k \leqq \bigvee_G x_i$ for some finite subsets F and G of S. Therefore $c \vee k \leqq \bigvee_H x_i$ for the finite subset $H = F \cup G$ of S, whence $c \vee k$ is compact. This shows that K is a join-semilattice.

For each $a \in K$, the set K_a of all compact $c \leqq a$ is trivially an ideal of K, nonvoid since $O \leqq a$; moreover the mapping $a \to K_a$ is isotone. Conversely, if J is an ideal of the join-semilattice K, let $a(J) = \bigvee_J k_j$ in L. Then: $a(K_a) = a$ since L is algebraic, $J \subset K_a$ trivially, and the mapping $J \to a(J)$ is isotone. Further, if $c \leqq a_J$ is compact, then $c \leqq \bigvee_F k_j$ for some finite subset F of J by the definition of compactness. But since J is an ideal, this implies $c \in J$, whence $K_{a(J)} = J$. In conclusion, the isotone functions $a \to K_a$ and $J \to a(J)$ are inverses of each other, and so both isomorphisms. Q.E.D.

With Lemma 1, Theorem 8 gives as a corollary the following important result.

THEOREM 8′. *A lattice L is isomorphic to the lattice of all subalgebras of a suitable (finitary) algebra if and only if it is complete and algebraic.*

DEFINITION. A complete lattice L is said to be *meet-continuous* when, for any directed set $D \subset L$

$$a \wedge \bigvee_D x_\delta = \bigvee_D (a \wedge x_\delta). \tag{3}$$

By definition, any complete Brouwerian lattice is meet-continuous. We now prove a stronger result.

LEMMA 2. *Any complete algebraic lattice L is meet-continuous.*

PROOF. Since $a \wedge \bigvee_D x_\delta \geqq \bigvee_D (a \wedge x_\delta)$ trivially in (3), it suffices to prove that $a \wedge \bigvee_D x_\delta \leqq \bigvee_D (a \wedge x_\delta)$. Again, since L is compactly generated, $a \wedge \bigvee_D x_\delta = \bigvee_S c_\sigma$, where each c_σ is compact. Hence, to prove the inequality

† The same is true of the structure lattice of A; this was first proved by Birkhoff–Frink [1, Theorem 6].

in question, it suffices to prove that for any *compact* $c = c_\sigma$ contained in $a \wedge \bigvee_D x_\delta$, $\bigvee_D (a \wedge x_\delta) \geqq c$. On the other hand, by definition of compactness, $c \leqq a \wedge \bigvee_D x_\delta \leqq \bigvee_D x_\delta$ implies $c \leqq \bigvee_F x_\phi$ for some *finite* subset F of D. And, since D is directed, this implies that $c \leqq x_\phi$ for some one $x_\phi \in D$—and hence $c \leqq a \wedge x_\phi$ since $c \leqq a \wedge \bigvee_D x_\delta \leqq a$. But this implies $c \leqq \bigvee_D (a \wedge x_\delta)$, since $x_\phi \in D$, which completes the proof of Lemma 2.

LEMMA 3 (A. G. WATERMAN). *In a complete meet-continuous lattice L, an element c is compact if and only if it is join-inaccessible.*

PROOF. For D directed, $c \leqq \bigvee_D x_\delta$ implies, by (3),

$$c = c \wedge \bigvee_D x_\delta = \bigvee_D (c \wedge x_\delta).$$

Hence, if c is join-inaccessible, there exists $x_0 \in D$ such that $c = c \wedge x_0$, that is, $c \leqq x_0$, and so c is compact. Conversely, $c = \bigvee_D x_\delta$ with D directed implies that $c \leqq \bigvee_D x_\delta$; hence, if c is compact, there must exist $x_0 \in D$ with $c \leqq x_0$. Since $x_0 \in D$, $x_0 \leqq \bigvee_D x_\delta = c$ and so $c = x_0$; hence c must be join-inaccessible.

Lemmas 2–3 show the equivalence of Theorem 8 with the following result.

THEOREM 9. *A lattice L is isomorphic with a subalgebra-lattice if and only if L is complete, meet-continuous, and every element of L is a join of join-inaccessible elements.*

PROOF. By Lemmas 2–3, any complete algebraic lattice has the properties listed. And by Lemma 3 alone, any lattice with the properties listed in Theorem 9 is (complete and) algebraic.

HISTORICAL NOTE. Theorem 9 was the main result of Birkhoff–Frink [**1**, Theorem 2]. Slightly later, Nachbin† introduced the notion of a "compact" element, and proved most of Theorem 8. The phrase "compactly generated" lattice was introduced by Dilworth and Crawley [**1**], with special reference to the atomic, semimodular case. The equivalence of this concept with the conditions of Theorem 9 was established by A. G. Waterman and the author. The present formulation of Theorem 8 is due to K. Baker and the author.

Much more difficult than Theorems 8–8′–9 is the following very important analogous result.

THEOREM OF GRÄTZER-SCHMIDT. *A lattice L is isomorphic to the structure lattice $\Theta(A)$ of all congruence relations on a finitary algebra A if and only if it is complete and algebraic.*‡

---

† L. Nachbin, Fund. Math. **36** (1949), 137–42, Theorem 1. For related results, see P. Crawley, Proc. AMS **13** (1962), 748–52. For supplements to the results of Birkhoff–Frink, see K. Balachandran, Proc. AMS **9** (1955), 548–53, and K.-H. Diener, Arch Math. **7** (1956), 339–45.

‡ G. Grätzer and E. T. Schmidt, Acta Sci. Math. Szeged **24** (1963), 34–59; also E. T. Schmidt, ibid., 251–4. For an important extension, see E. T. Schmidt, Acta Math. Acad. Sci. Hungar **15** (1964), 37–45.

Exercises for §§4–5:

1. (a) Show that the ideal completion $\hat{L}$ of any lattice L is algebraic.

(b) Show that every lattice is isomorphic with a sublattice of some complete algebraic lattice.

2. (a) Prove from first principles that, in the subalgebra lattice $L(A)$ of any (finitary) algebra A, $S_\alpha \uparrow S$ (as a directed set) implies $S_\alpha \cap T \uparrow S \cap T$.

(b) Show that a complete lattice is meet-continuous if (3) holds for D any chain. (O. Frink)

3. Prove that a closure property on subsets of a set I is finitary if and only if the set-union of any (upward) directed family of closed sets is itself closed.

4. (a) Show that, in the subalgebra lattice of an algebra A, S is join-inaccessible if and only if it is finitely generated.

(b) Show that a complete algebraic lattice is isomorphic with the subalgebra lattice of a unary algebra if and only if it is completely distributive.

5. Prove that an ideal J of a lattice L is principal if and only if it is join-inaccessible in the lattice of all ideals of L.

6. Show that every complete lattice is isomorphic with the lattice of all subalgebras of some *infinitary* algebra.

7. Let $\Theta(A)$ be the lattice of all congruence relations on a (finitary) algebra A.

(a) Show that $\Theta(A)$ is a complete algebraic lattice.

(b) Represent $\Theta(A)$ as a subalgebra lattice.

*8. Let L be a complete algebraic lattice, and let $[a \wedge b, b] \cong [a, a \vee b]$ for all $a, b \in L$. Show that L is modular. (P. Crawley)

*9. Show that any complete algebraic distributive lattice is Brouwerian.

10. Let $\Theta(A)$ be the lattice of all congruence relations on a (finitary) algebra A. Show that if B is obtained from A by introducing further finitary operations, then $\Theta(B)$ is a closed sublattice of $\Theta(A)$.

*11. Prove that, if the subalgebra lattice of a (finitary) algebra is Boolean, then it is isomorphic with the Boolean algebra of all subsets of some set (see §V.5).

*12. Given a poset P, show that there exists a section complemented lattice L whose structure lattice $\Theta(L) \cong \mathbf{2}^P$.†

6. Regular Rings

The class of "regular rings" was invented by von Neumann to "coordinatize" complemented modular lattices of unrestricted length. Their theory illustrates some of the preceding ideas, and will be sketched here.

DEFINITION. A *regular ring* is an associative ring R with unity 1 in which every element a has a pseudo-inverse x such that $axa = a$.

It follows that ax and xa are idempotents; neither is 0 unless $a = 0$. Hence the radical of any regular ring is 0; more, every principal left or right ideal is idempotent. We will now consider the lattices of left ideals, right ideals, and two-sided ideals of an arbitrary regular ring R, assuming without proof the following result [**CG**, Theorem 2.3].

PRINCIPAL IDEAL THEOREM. *The join of any two principal right (left) ideals of a regular ring R is itself a principal right (left) ideal.*

In other words, in a regular ring R, any two elements a, b have a "greatest common left-divisor" $d = ar + bs$ $(r, s \in R)$ such that: (i) $a = dx$, $b = dy$, and

† G. Grätzer and E. T. Schmidt, Acta Math. Acad. Sci. Hungar **13** (1962), 179–85.

(ii) if $a = cx'$, $b = cy'$, for some $x', y' \in R$, then necessarily $c = dz$ for some $z \in R$. Likewise, any $a, b \in R$ have a greatest common right-divisor in R.

By straightforward induction, this yields the

COROLLARY. *A right (left) ideal of a regular ring is finitely generated if and only if it is principal.*

By applying Theorem 7, we obtain the following basic result as a further corollary.

THEOREM 10. *In the lattices of all right (left) ideals of any regular ring, an element is compact (join-inaccessible) if and only if it corresponds to a principal ideal.*

On the other hand, for any element $a \in R$, the principal right ideal $aR = axaR \subset axR \subset aR$; hence $aR = eR$ where $e = ax$ is an *idempotent* satisfying $ee = axax = ax = e$. Further, for any idempotent e, $b \in eR$ if and only if $eb = eex = ex = b$, and likewise $b \in (1 - e)R$ if and only if $eb = e(1 - e)x = 0x = 0$; hence $eR \in (1 - e)R = 0$. Moreover $b = eb + (1 - e)b$ for all $b \in R$; hence $eR + (1 - e)R = R$. We conclude that every principal right ideal eR has a *complement* $(1 - e)R$ in the lattice of all right ideals of R. Clearly, the same holds for left ideals.

We now consider the *polarity* (§V.7) associated with the relation $xy = 0$. For any set X, the set X^* of all $y \in R$ such that $xy = 0$ for all $x \in X$ is a right ideal since $xy = 0$ implies $xyR = 0$. Likewise, for any set Y, the set $Y^\dagger$ of all $x \in R$ such that $xy = 0$ for all $y \in Y$ is a left ideal. Hence the polarity establishes a dual isomorphism between certain left ideals and certain right ideals.

Further, for any idempotent e, we have as above:

$$(1 - e)R = e^* \quad \text{and} \quad e = R(1 - e). \tag{4}$$

Therefore, since $e^* = (Re)^*$, the polarity in question maps *principal* left ideals RE into principal right ideals $(1 - e)$. Since, by Chapter V, (11):

$$(J + K)^* = J^* \cap K^* \quad \text{and} \quad (J + K)^\dagger = J^\dagger \cap K^\dagger, \tag{5}$$

we conclude that the principal right (and left) ideals are closed not only under join (by the Principal Ideal Theorem) but also under meet: they form *sublattices* of the lattice of all such ideals. That is, in a regular ring, any two elements have a least common left-multiple and a least common right-multiple. Keeping Theorem 10 in mind, we therefore conclude

THEOREM 11. *In the lattices of Theorem 10, the compact elements form sublattices which are dually isomorphic to each other under the polarity defined by $xy = 0$.*

We now recall that the right and left ideals of any ring form modular lattices. Also, for idempotents e, f in a ring R, $eR \subset fR$ if and only if $e = fx$ for some $x \in R$ (i.e., f left-divides e). We therefore have

THEOREM 12. *The idempotents of any regular ring form dually isomorphic complemented modular lattices under the relations of left- and right-divisibility; moreover $(1 - e)$ is a complement of e in both.*

(*Caution.* We can have $eR = fR$ without $e = f$; Theorem 12 refers to the poset defined by the quasi-orderings of left- and right-divisibility.)

In the special case that the above lattices have *finite length*, much more can be said. Then (trivially) every element in Theorem 10 is "compact", hence every left or right ideal is principal. In this case, the *center* $Z(R)$ of R is a direct sum of subfields F_k; the unities 1_S of these subfields (which are idempotent) constitute the center of *both* of the complemented modular lattices of Theorem 11; they generate the *two-sided* ideals of R, which is therefore the direct sum $R = \bigoplus R_k$ of these ideals. Each R_k is *simple*, and (by a well-known result of Wedderburn) isomorphic to the full matrix algebra $M_n(D_k)$ of all $n \times n$ matrices with entries in a division ring D_k with center F_k.

In this special case, we have already seen (Chapter IV, §§7, 13, 14) that every Desarguesian complemented modular lattice of finite length is isomorphic with one of the lattices just described. It is a deep problem of lattice theory to determine which (Desarguesian) complemented modular lattices of unrestricted length stand in the same relation to some (otherwise unspecified) regular ring. A monumental achievement of von Neumann [**CG**, Theorem 14.4] is the result that it is sufficient that the lattice have a "basis" of four or more pairwise perspective elements, which are independent with join I.

Exercises for §6: (See also Exercises after §IV.15.)

1. Show that the finite-dimensional subspaces of Hilbert space $\mathfrak{H}$ form a relatively complemented modular lattice which satisfies the DCC. Show that these are the compact elements in the (ortho)lattice $L(\mathfrak{H})$ of *all* subspaces of $\mathfrak{H}$.
2. Consider $L(\mathfrak{H})$ under the polarity induced by the relation $x \perp y$.
3. Formulate and prove analogs of Exercises 1–2 for the lattice $L_c(\mathfrak{H})$ of all closed subspaces of $\mathfrak{H}$.

Define a *Baer semigroup* as a multiplicative semigroup with 0 in which the right (resp. left) annihilator of each element is a principal right (resp. left) ideal generated by an idempotent.†

*4. Show that the posets of right and left annihilators of elements of any Baer semigroup form lattices.

*5. Show that, if every right annihilator eS contains an idempotent e_0 such that $eS = e_0S$ and Se_0 is a left annihilator, then the lattices of Exercise 4 are dually isomorphic.

7. Zorn's Property; Hausdorff's Axiom

Intermediate between the notion of a finitary closure property, and the unblushing assumption of the Axiom of Choice (whose use we have avoided hitherto) stands the following Property of Zorn.

DEFINITION. A family Φ of subsets of a set I has *Zorn's Property* when the set-union $\bigcup_C S_\alpha$ of any chain C of sets $S_\alpha \in \Phi$ again belongs to Φ.

Zorn's Property is trivially verified if Φ satisfies the ascending chain condition, since C must then have a largest member S_μ, and $S_\mu = \bigcup_C S_\alpha$ is then automatically in C.

† This definition is due to M. F. Janowitz, Duke Math. J. **32** (1963), 85–96, from which Exercises 4–5 are taken.

Remark 1. It is evident that any intersection $\bigcap \Phi_\tau$ of families of subsets of I having Zorn's Property itself has Zorn's Property. Thus, Zorn's Property is a *closure* property in the sense of §V.1. Most applications of Zorn's Property depend on the following related result.

THEOREM 13. *The family of all subsets of a set I which have any given finitary closure property has Zorn's Property.*†

PROOF. Let C be any chain of sets S_α having a given finitary closure property; let $U = \bigcup_C S_\alpha$, and let K be any finite subset of elements $x_1, \cdots, x_n$ of U. Then each $x_i \in S_{\alpha(i)}$ for some $S_{\alpha(i)} \in C$; let S_μ be the largest of these $S_{\alpha(i)}$, $i = 1, \cdots, n$. Clearly $K \leqq S_\mu$; hence $\bar{K} \leqq \bar{S}_\mu = S_\mu \leqq U$. Hence $U \in \Phi$, the closure property being finitary. Q.E.D.

COROLLARY 1. *The subalgebras of any algebra A have Zorn's Property. So do the congruence relations on A, considered as subsets of* $A \times A$.

COROLLARY 2. *For any subset X and subalgebra T of an algebra A, the family of subalgebras S of A such that* $S \cap X \subset T$ *has Zorn's Property.*

Of particular interest are the cases $X = 1$, $H = \varnothing$ (the void set) for a ring R with unity. The ideals S such that $S \wedge 1 = \varnothing$ are then the ideals $S < R$.

COROLLARY 3. *Given elements* $a \neq b$ *of an algebra A, the family of congruence relations* θ *such that* $a \equiv b\ (\theta)$ *has Zorn's Property.*

Zorn's Property has many other algebraic applications. For instance, it can be used to extend the Jordan–Hölder Theorem, by transfinite induction, to well-ordered *ascending* normal subgroup series.‡ Much more sophisticated generalizations of the Theorem of Ore have been proved by Kurosh, Baer, and others.§

The preceding ideas can be generalized to posets in various ways.¶

For example, let B be any *complete lattice*. A subset P of B will be said to have Property Z when $C \subset P$ and C a chain imply $\bigvee_C c_\alpha \in P$. Evidently, Property Z specializes to Zorn's Property when B is the complete Boolean algebra of all subsets of a set I, ordered by set-inclusion.

LEMMA 1. *If P has Property Z, then every chain in P has an upper bound in P.*

Posets having the preceding property are called *inductive*. This property has many consequences, if one is willing to assume the following variant of the Axiom of Choice.

HAUSDORFF'S MAXIMAL PRINCIPLE. Every chain C in a poset P can be extended to a *maximal* chain M in P.

---

† The converse is also true [**UA** Theorem 1.2, p. 45,]. Cohn speaks of "algebraic" where we use "finitary".

‡ See G. Birkhoff, Bull. AMS **40** (1934), 847–50; A. Kurosh, Math. Ann. **111** (1935), 13–18. For Zorn's original formulation, see Bull. AMS **41** (1935), 667–70.

§ See for example R. Baer, Trans. AMS **62** (1947), 62–98, and **64** (1948), 519–51.

¶ See [**LT2**, §III.6]; also [**Kel**, pp. 33–6].

Using this assumption, it is easy to prove the following result. (The result would be trivial if P were Noetherian.)

THEOREM 14. *If P is inductive, then P contains a maximal element.*

PROOF. Let M be *any* maximal chain in P, and form $m = \bigvee_M x_\alpha$. Then $m \vee M$ is a maximal chain with greatest element m, which is therefore maximal in P.

COROLLARY 1. *If P has Property Z, then P contains a maximal element.*

Combining this result with Corollaries 1 and 2 of Theorem 13, respectively, we get two important conclusions.

COROLLARY 2. *Given a subset X and Y of an algebra A, there is a maximal subalgebra M of A such that $M \cap X \leqq Y$ (i.e., $M \leqq Y \cup X'$).*

For example, let R be a ring with unity 1, and let J be a proper ideal of R. Then there is a *maximal* ideal $M \geqq J$ of R such that $M \cap 1 = \varnothing$ (i.e., $1 \notin M$): any proper ideal of R can be extended to a maximal proper ideal.

COROLLARY 3. *Given elements $a \neq b$ of an algebra A, there exists a maximal congruence relation θ such that $a \equiv b\ (\theta)$.*

8. **Subdirect Decomposition Theorem**

Using the last corollary, one can prove a fundamental decomposition theorem for abstract algebras. We first prove

LEMMA 1. *The congruence relation θ in Corollary 3 of Theorem 14 is strictly irreducible: the set Θ of congruence relations $\phi > 0$ on A has a least member λ.*

PROOF. If $\phi > \theta$, then $a \equiv b\ (\phi)$. Hence, if λ is the intersection of all congruence relations μ such that $a\mu b$ and $\mu \geqq \theta$ (cf. Theorem VI.8), then $\phi \geqq \lambda > \theta$. This construction exhibits the desired λ.

COROLLARY. *For any $a \neq b$ in an algebra A, there exists a congruence relation $\theta = \theta(a, b)$ such that A/θ is subdirectly irreducible (strictly) and such that $a \not\equiv b\ (\theta)$.*

Applying the preceding result to the set of *all* pairs $a \neq b$ of distinct elements, we get a set of congruence relations $\theta(a, b)$ whose meet is clearly O. By Corollary 1 of Theorem VI.11, this implies the following basic subdirect decomposition theorem.†

THEOREM 15. *Any algebra A is a subdirect product of subdirectly irreducible algebras.*

COROLLARY 1. *Any distributive lattice is isomorphic with a subdirect product of copies of* **2**.

PROOF. It suffices to show that **2** is the only nontrivial subdirectly irreducible

† G. Birkhoff, Bull. AMS **50** (1944), 764–8.

distributive lattice (**1** is trivial). But unless a distributive lattice L is **1** or **2**, it contains an element c not O nor I. The endomorphisms $x \to x \wedge c$ and $x \to x \vee c$ define proper congruence relations θ and θ' with $\theta \wedge \theta' = O$ (by Theorem I.10); hence L is subdirectly reducible.

COROLLARY 2. *Any distributive lattice L is isomorphic with a ring of sets.*

For, considering L as a sublattice of $\mathbf{2}^I$ for some set I of congruence relations on L, map each $c \in L$ onto the subset $S(c)$ of I which consists of those mappings ϕ such that $\phi(c)$ is the larger element of **2**. One easily verifies that $c \to S(c)$ is an isomorphism.

In particular, if L is complemented (a Boolean lattice), then so is the ring of sets. This proves

COROLLARY 3. *Any Boolean lattice is isomorphic with a field of sets.*

It follows that each of the postulate systems for Boolean algebra described in Chapter II is a complete set of axioms for the algebra of classes with respect to finite union, finite intersection, and complementation.

COROLLARY 4. *Any Boolean ring is isomorphic with a subdirect product of copies of* $\mathbf{Z}_2$.

PROOF. Any Boolean ring is cryptoisomorphic with a generalized Boolean algebra (§II.12).

Likewise, any vector space V over any division ring R can be proved isomorphic with a subdirect product of copies of R, using Theorem 15 alone. Actually, V is isomorphic with a restricted *direct* product of copies of R, for reasons to be indicated in the next section.

The argument used to prove Theorem 15 can be applied to complete algebraic lattices generally, as will now be shown.

DEFINITION. In a lattice, an element a is called *strictly meet-irreducible* when the set of all $x > a$ has a least element.

THEOREM 16. *In any complete algebraic lattice L, every element is a meet of strictly meet-irreducible elements.*

PROOF. Let a and $b \nleq a$ be given in L. Then, since L is algebraic, there exists a *compact* element $c \in L$ contained in b but not in a. Let K be the set of all $k \in L$ which contains a but not c. Then $a \in K$; moreover the join of any directed subset of K is in K, since c is compact. Hence K contains a *maximal* element m. By definition, *any* $x > m$ must contain c; hence $m \vee c$ is the *least* element of L which properly contains c, and so c is *strictly irreducible*.

For given $a \in L$, now consider the meet $\bigwedge m_\alpha$ of all strictly irreducible elements which contain a. Trivially, $\bigwedge m_\alpha$ contains a. But by the preceding paragraph, for any $b > a$, there exists a strictly irreducible m which contains a but not some $c \leq b$, hence which does not contain b. Hence $\bigwedge m_\alpha = a$ (i.e., $\bigwedge m_\alpha > a$ is impossible). Q.E.D.

Finally, the construction used to prove Theorem 15 has been refined by G. Bruns† to prove the following representation theorem for Stone lattices (§V.11): Any Stone lattice S is isomorphic to a pseudo-complemented sublattice (or "*-sublattice") of the lattice $\hat{A}$ of all ideals of a complete atomic Boolean algebra A. Namely, for any lattice-monomorphism $\alpha: S \to P(E)$ representing S as a field of subsets of a set E such that $a \nleq b$ $(a, b \in L)$ implies that $(a\alpha) \cap (b\alpha)'$ is infinite, one can take A to be the power set $P(E)$.

Exercises for §§7–8:

1. Prove in detail Corollaries 2–3 of Theorem 14.
2. Prove that a lattice L is complete if every maximal chain of L is complete. (B. Rennie)
3. Show that, if (3) holds for any chain D in a complete lattice, then it holds for any directed set.‡
4. (a) Show that every partial ordering can be strengthened to a simple ordering. (E. Szpilrajn, Fund. Math. **16** (1930), 386–9.)

   (b) Show that a partial ordering can be strengthened to a well-ordering if and only if it satisfies the DCC.
5. Show that a distributive lattice L is relatively complemented if and only if no two distinct prime ideals of L are comparable. (Grätzer–Schmidt)
6. Show that any ideal of a Boolean lattice A is the intersection of the prime ideals which contain it.
7. Show that a distributive lattice with O and I is Boolean if and only if every prime ideal is maximal. (L. Nachbin)
8. Show that, in the distributive lattice $\mathbf{2}^{(\omega)}$ of all finite subsets of an infinite set, every prime dual ideal is maximal.
9. Show that an ideal J of a lattice L is completely meet-irreducible if and only if, for some subset $S \subset L$, J is maximal subject to the condition $J \cap S = \varnothing$.
10. Show that a complete lattice is isomorphic with a complete ring of sets (§V.5) if and only if it is completely distributive, and every element is a join of completely join-irreducible elements. (Raney–Balachandran)
11. (a) In the product $\mathbf{2R}$ ($\mathbf{R}$ the reals) show that the ideal of all pairs $(0, x)$ $(x \in \mathbf{R})$ cannot be extended to a maximal proper sublattice. (K. Takeuchi)

    (b) Prove that every proper sublattice of a relatively complemented distributive lattice can be extended to a maximal proper sublattice. (J. Hashimoto [**1**])

*12. (a) Show that, if (e) is a principal ideal of a Boolean algebra A, then there exists a subalgebra S of A which contains exactly one representative from each residue class of (e).

(b) Find a Boolean algebra A and a nonprincipal ideal $J \subset A$, for which no such subalgebra exists.§

13. (a) Let $\phi: S \to B$ be a morphism of Boolean algebras from a subalgebra $S \subset A$ of a Boolean algebra A into B. Show that, given $x \in A$, ϕ can be extended to a morphism $\phi^*: \{S, x\} \to B$ if and only if, for some $y \in B$: $\phi([0, x] \cap S) \subset [0, y] \cap \phi(S)$, $\phi([x, I] \cap S) \subset [y, I] \cap \phi(S)$.

    (b) Show that, if A is complete, such a y always exists. (*Hint.* We have
    $$\sup \phi([0, x] \cap S) \leqq \inf \phi([x, I] \cap S).)$$

    (c) Show that, if A is complete, then any morphism $\phi: S \to B$ from a subalgebra of A into B can be extended to a morphism $\bar{\phi}: A \to B$.

---

† G. Bruns, Duke Math. J. **32** (1965), 555–6; cf. also G. Grätzer, ibid. **30** (1963), 469–74.

‡ U. Sasaki, J. Sci. Hiroshima Univ. Ser. A **14** (1950), 100–1.

§ Exercise 12 describes results of J. von Neumann and M. H. Stone, Fund. Math **25** (1935), 353–78. Exercise 13 describes results of R. Sikorski, Ann. Polon. Math. **21** (1948), 332–5; it states that complete Boolean algebras are *injective*.

14. Let A be any algebra with a binary operation which satisfies the identities $x(yx) = x$, $x(xy) = y(yx)$, $(xy)z = (xz)(yz)$. Show that A is isomorphic with an algebra of sets under subtraction.†

15. Show that, if a family Φ of subsets of a set I has Zorn's property, then it contains the set-union of any directed family of its members‡.

9. Atomic Algebraic Lattices

By definition, an *atomic* lattice is a lattice L in which every element is a join of atoms (points), and hence of the atoms which it contains. Theorem 16 has the following immediate corollary.

COROLLARY. *The dual of any relatively complemented, complete algebraic lattice is atomic.*

We also know (Theorem V.18) that any complete atomic Boolean lattice is isomorphic with $\mathbf{2}^{\aleph}$, where $\aleph$ is the cardinality of the set of its points. As corollaries, we infer: (i) in any (complete) atomic Boolean lattice, every point is compact, and (ii) any (complete) atomic Boolean lattice is algebraic.

In general, complete atomic lattices need not be algebraic: thus meet-continuity (§5, (3)) fails in the (atomic) orthomodular lattice of all closed subspaces of Hilbert space; and it fails to hold in the (atomic) complemented modular lattice of all closed subspaces of Hilbert space which have finite dimension *or* finite co-dimension; and it fails in T_1-lattices (§IX.5).

We now define a generalization of the class of geometric lattices, which also includes all atomic Boolean lattices.

DEFINITION. An *atomic matroid lattice* is a complete, algebraic, atomic, semi-modular lattice.

The theory of atomic matroid lattices was originated (in other terminologies) by F. Maeda [**1**], Dilworth-Crawley [**1**], and U. Sasaki.§ The modular case had been treated previously by O. Frink [**2**]. We first observe a simple corollary of Theorem 8.

LEMMA 1. *In a complete atomic algebraic lattice L, the compact elements are the joins of finite sets of points.*

PROOF. Since L is atomic, every compact $c \in L$ is a join of points; such a (possibly infinite) set of points can be replaced by a finite set since c is compact. Conversely, if $a = \bigvee_F p_\phi$ is a finite join of points, then since each p_ϕ is compact and the compact elements of L form a join-semilattice, a is compact.

---

† J. A. Kalman, Indag. Math. **22** (1960), 402–5.

‡ J. Mayer Kalkschmidt and E. Steiner, Duke Math. J., **31** (1964), 287–9.

§ J. Sci. Hiroshima Univ. **A16** (1953), 223–8 and 409–16; see also F. Maeda, ibid. **A27** (1963), 73–84 and 85–96. For completions by cuts of atomic modular lattices, see J. E. McLaughlin in [**Symp**, pp. 78–80].

We now use the hypothesis of semimodularity.

LEMMA 2. *In any semimodular lattice L:*
(π) *For any $a \in L$ and any point $p \in L$, either $p \leqq a$ or $p \vee a$ covers a.*

PROOF. In any lattice L, (b, p) is a modular pair for all $b \in L$ and any point $p \in L$. For, if $p \wedge b = O$, for $x = O$ and p (the only elements of $[p \wedge b, p]$) it is trivial that $x = (x \vee b) \wedge p$; for $p \leqq b$, the verification is even more trivial. Now suppose that $p \vee a$ fails to cover a, and let $a < b < p \vee a$; then $a \in [O, b]$, yet $(a \vee p) \wedge b > a$; hence (p, b) is not a modular pair, contrary to hypothesis.

LEMMA 3. *In any semimodular atomic lattice L, the set J of joins $x = \bigvee_F p_\phi$ of finite sets of points forms an ideal: the ideal of all ("compact") elements of finite height in L.*

PROOF. By Lemma 2, any $x \in J$ has finite height, and J is trivially join-closed. Conversely, since L is atomic, any $a \in L$ is a join of points. If $h[a] = r$ is finite, then by induction there exist r points $p_i \leqq a$ such that $(\bigvee_{i=1}^{k-1} p_i) \wedge p_k = O$ for $k = 1, \cdots, r$; since $h = [\bigvee_{i=1}^{r} p_i] = r$ by Lemma 1, and $\bigvee_{i=1}^{r} p_i \leqq a$ trivially, it follows that $a = \bigvee_{i=1}^{r} p_i \in J$. The fact that J is an ideal is a corollary.

As an immediate corollary of Lemma 3 and Theorem 8, we obtain

THEOREM 18. *Any atomic matroid lattice L is isomorphic to the lattice $\hat{J}$ of all ideals of the ideal J of all elements of L having finite height* (*i.e., of all finite joins of points*).

That is, L is isomorphic to the lattice of all subsets of the points $p_\alpha \in L$ which are "closed" with respect to some finitary abstract dependence relation having the Steinitz-Mac Lane Exchange Property (Chapter IV, (5)).

We now turn our attention to the modular case. Just as in §IV.6, we can divide the points of any *modular* atomic lattice M into equivalence classes E_k of mutually perspective points. The "flats" (i.e., closed subsets) of any class E_k of mutually perspective points satisfy PG3, and form what may be called an *atomic projective geometry*. One can show quite easily that M is an ideal in the direct product of the P_k, thereby proving

THEOREM 19. *Any modular atomic matroid lattice is isomorphic to an ideal of a direct product of atomic projective geometries.*

Further, apart from certain finite projective lines and non-Desarguesian projective planes discussed in Chapter IV, one can introduce coordinates in a division ring D into the atomic projective geometries defined above, by using von Staudt's "algebra of throws" (§IV.14). Frink **[2]** has shown that in the case of infinite length, the atomic projective geometry is just the lattice $PG_{\aleph}(D)$ of all subspaces of the vector space $D^{\aleph}$. This gives

THEOREM 20. *Every atomic projective geometry of height $h > 3$ is isomorphic with the lattice of all subspaces of $D^{\aleph}$, for some division ring D and cardinal number $\aleph$.*

Finally, let L be *any* complemented modular lattice, not necessarily atomic or complete. We first recall that, by Theorem V.12, the lattice of all ideals of L is modular—hence so is the lattice of all dual ideals of L. Next, we note that to be a proper dual ideal of L is equivalent to not containing O—and that this is a condition of finite character. Hence every proper dual ideal can be extended to a maximal proper dual ideal, which is "covered" by the improper dual ideal L. In particular, for any $a \neq I$ of L, the principal dual ideal of $x \geqq a'$ can be extended to a maximal proper ideal Q, which cannot contain a (or it would contain $a \wedge a' = O$). Hence the intersection of all maximal dual ideals is I, the least dual ideal. We conclude by such reasoning that the *dual* ideals of *any* complemented modular lattice M, partially ordered by the *dual* of inclusion, form an *atomic* modular lattice $\check{M}$.

We cannot however conclude that (§5, (3)) holds (though its dual does). But if we define a "point" as a maximal dual ideal, and a "line" as the intersection of two distinct maximal dual ideals, then PG1–PG2 are easy to prove. The "flats" form a complete atomic modular lattice.

We now associate with each element $a \in L$ the set $S(a)$ of all maximal dual ideals P, such that $a \in P$. It is easy to show that $S(a)$ is always a "flat", that $a > b$ implies $S(a) > S(b)$, and that $S(a \wedge b)$ is the intersection of $S(a)$ and $S(b)$. Frink has shown further that $S(a \vee b)$ is the join of $S(a)$ and $S(b)$, the difficult point is to show that if $a \vee b \in P$, then there exist Q, R with $a \in Q$, $b \in R$, and such that P contains the set-intersection of Q and R.

Since the "flats" form a direct union of atomic projective geometries by what we have already shown, we get the following final result.

THEOREM 21. *Any atomic complemented modular lattice is isomorphic with a sublattice of a direct union of atomic projective geometries.*

The theory of complete *non*atomic algebraic modular lattices is far less well developed.†

10. Ordinal Sums and Products

We next define two binary operations of *ordinal* addition and multiplication, analogous to the corresponding *cardinal* operations defined in §III.1, but not to be confused with them. Under these operations, the set of all isomorphism-types of finite posets forms an algebra, in which the finite chains form a subalgebra isomorphic with $[\mathbf{Z}^+; +, \cdot]$. More relevant to us is the fact that countable ordinals also form an algebra, which is a subalgebra of the algebra of all isomorphism-types of countable posets. The operations are defined as follows.‡

DEFINITION. Let X and Y be any two (disjoint) posets. The *ordinal sum* $X \oplus Y$ of X and Y is the set of all $x \in X$ and $y \in Y$; $x < y$ for all $x \in X$ and $y \in Y$;

† See R. E. Johnson, Trans. AMS **84** (1957), 508–49, where applications to ideal theory are considered.

‡ For a deeper study of generalizations of these operations, see A. Tarski, "Ordinal Algebras", with Appendices by C.-C. Chang and B. Jonsson, North-Holland, Amsterdam, 1956.

the relations $x \leqq x_1$ and $y \leqq y_1$ ($x, x_1 \in X$; $y, y_1 \in Y$) have unchanged meanings. The *ordinal product* $X \cdot Y$ of X and Y is the set of all ordered pairs (x, y) ($x \in X$, $y \in Y$), ordered *lexicographically* by the rule that $(x, y) < (x_1, y_1)$ if and only if $x < x_1$, or $x = x_1$ and $y < y_1$.

One can easily prove that the ordinal sum and ordinal product of any two chains are again chains, and that the ordinal sum and ordinal product of any two ordinals are again ordinals. (Consider, for example, $\boldsymbol{\omega} \cdot \boldsymbol{\omega}$ or $\mathbf{2} \cdot \boldsymbol{\omega} = \boldsymbol{\omega} \oplus \boldsymbol{\omega}$.) Ordinal addition and multiplication are not commutative in general. Thus:

$$\mathbf{1} \oplus \boldsymbol{\omega} = \boldsymbol{\omega} \neq \boldsymbol{\omega} \oplus \mathbf{1},$$

$$\boldsymbol{\omega} \cdot \mathbf{2} = \boldsymbol{\omega} \neq \mathbf{2} \cdot \boldsymbol{\omega} = \boldsymbol{\omega} \oplus \boldsymbol{\omega}.$$

For finite posets, the diagram of $X \oplus Y$ can be constructed by laying the diagram of Y above the diagram of X and drawing lines from all minimal elements of Y to all maximal elements of X, as in Figure 19a.

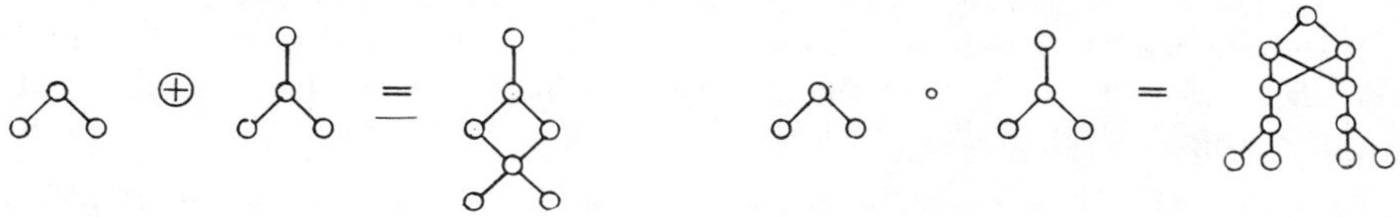

(a) (b)

FIGURE 19

Similarly, for finite sets X and Y, the diagram for $X \circ Y$ can be constructed by using the following easily proved result. In $X \circ Y$, (x_1, y_1) covers (x, y) if and only if (i) $x = x_1$ and y_1 covers y, or (ii) x_1 covers x and y_1 is minimal and y maximal in Y. For an example, see Figure 19b. It is clear from these two examples that the operations $\oplus$ and $\circ$ are not commutative in general. However, one can easily verify the following identities:

$$X \oplus (Y \oplus Z) = (X \oplus Y) \oplus Z \quad \text{and} \quad X \circ (Y \circ Z) = (X \circ Y) \circ Z, \tag{7}$$

$$(X \oplus Y) \circ Z = (X \circ Z) \oplus (Y \circ Z), \tag{8}$$

$$\breve{(X \oplus Y)} = \breve{Y} \oplus \breve{X} \quad \text{and} \quad \breve{(X \circ Y)} = \breve{X} \circ \breve{Y}, \tag{9}$$

where in (9) $\breve{}$ refers to the unary *dualization* operation which carries each poset P into its dual $\breve{P}$.

11. Subchains of Q and R

The classification of finite chains is trivial, by Theorem I.5. We now consider countable chains, and more generally arbitrary subchains of the chain **R** of all real numbers. As before, we let $\boldsymbol{\omega}$ denote the chain of all positive integers, **Z** the chain of all integers, and **Q** the chain of all rational numbers. Many chains can be represented as subchains of **Q** or **R**, and hence can be pictured as sets of points on the line. We develop these possibilities in the remainder of this section.

THEOREM 22. *Any countable chain is order-monomorphic to (isomorphic with a subchain of) the chain* $\mathbf{Q}$.

PROOF. Let $r_1, r_2, r_3, \cdots$ be an enumeration of the rational numbers. Let $a_1, a_2, a_3, \cdots$ be an enumeration of the given chain C. Set $f(a_1) = 0$. By induction, define $f(a_n)$ as follows. Since C is a chain, a_n must satisfy one of the following: (i) a_n exceeds $a_1, \cdots, a_{n-1}$, (ii) a_n is less than $a_1, \cdots, a_{n-1}$, or (iii) a_n is between a_i and a_j for some greatest a_i and least a_j with $i, j < n$. In case (i) set $f(a_n) = n$, in case (ii), set $f(a_n) = -n$, in case (iii) define $f(a_n)$ as the first r_k in the same order relation to $f(a_1), \cdots, f(a_{n-1})$ as a_n is to $a_1, \cdots, a_{n-1}$. Then f is clearly an isotone one-one map of C into $\mathbf{Q}$.

Since $\mathbf{Q}$ is itself countable, Theorem 22 asserts that $\mathbf{Q}$ is a "universal" countable chain.

We now turn our attention to the topological notion of "denseness". The topological aspects of this notion will be discussed in §X.8.

DEFINITION. A chain C is called *dense-in-itself* if and only if, given $a < b$ in C, there exists $c \in C$ which satisfies $a < c < b$. A subset S of a chain C is called *order-dense* in C if and only if, for every $a < b$ in C, not in S, an element s of S can be found satisfying $a < s < b$.

THEOREM 23. *Any countable chain C which is dense-in-itself is isomorphic to either* $\mathbf{Q}$, $1 \oplus \mathbf{Q}$, $\mathbf{Q} \oplus 1$, *or* $1 \oplus \mathbf{Q} \oplus 1$.

PROOF. Let D consist of all elements of C except $\sup C$ and $\inf C$, if these exist. Since C is dense-in-itself, D has no sup or inf, and D is also dense-in-itself. Map D into $\mathbf{Q}$ by the construction used in the proof of Theorem 22. Let m, n be integers in the range of f, where f is the mapping of D into $\mathbf{Q}$. Then every rational number between m and n is in the range of f. If this were not true, there would be a first k with r_k not in the range. We could then find a_i and a_j satisfying $m \leqq f(a_i) < r_k < f(a_j) \leqq n$, and having the additional property that $f(a_i) < r_l < f(a_j)$ implies $l \geqq k$. Since D is dense-in-itself, there would exist a first a_s with $a_i < a_s < a_j$; but this would imply $f(a_s) = r_k$, a contradiction.

Since D is unbounded, there exist arbitrarily large integers m and n such that $-m$ and n are in the range of f. Thus f is an isomorphism of D onto $\mathbf{Q}$; C is isomorphic to D, $1 \oplus D$, $D \oplus 1$, or $1 \oplus D \oplus 1$; hence C is isomorphic to $\mathbf{Q}$, $1 \oplus \mathbf{Q}$, $\mathbf{Q} \oplus 1$, or $1 \oplus \mathbf{Q} \oplus 1$.

Note that $\mathbf{Q}$ is order-dense in $\mathbf{R}$. This result can be generalized as follows.

THEOREM 24. *A chain C is isomorphic with a subchain of* $\mathbf{R}$ *if and only if C contains a countable order-dense subset.*

PROOF. Suppose C contains a countable order-dense subset $A = \{a_1, a_2, \cdots\}$; without losing generality, we may suppose that A contains the greatest and least elements of C if they exist. By Theorem 22, we can map A onto a subchain of the chain $\mathbf{Q}$. By definition of "order-dense", every c in C but not in A is uniquely determined by the cut which it defines in A, i.e., by the partition of A into the set of $a_i < c$ and the set of $a_i > c$. Let $r_1 =$ l.u.b. $f(a_i)$ with $a_i < c$, and $r_2 =$ g.l.b.

$f(a_i)$ with $a_i > c$; let $2f(c) = r_1 + r_2$. This defines an isomorphism between C and a subset of $\mathbf{R}$. We omit details, because they are easy to supply.

Conversely, if C is isomorphic with a subset of $\mathbf{R}$, we can find a countable dense subset of C by enumerating the intervals I_i: $[m_i/n_i, m_i'/n_i']$ of $\mathbf{R}$ having rational endpoints, and choosing one c_i from each I_i except when no element of C corresponds to an element of I_i.

Exercises for §§10–11:

1. Prove identities (7)–(9).
2. (a) Show that any chain not an ordinal contains the dual $\breve{\omega}$ of ω as a subchain.

   (b) Show that, if a chain C and its dual $\breve{C}$ are both well-ordered, then C is finite.

   (c) Let C be a conditionally complete chain in which every element has an immediate predecessor and an immediate successor. Show that $C \cong \mathbf{Z}$.
3. Show that every well-ordered subchain of $\mathbf{R}$ is countable.
4. Show that $\mathbf{R} \circ \mathbf{R}$ is not order-monomorphic to $\mathbf{R}$.
5. What is the least ordinal α such that $\alpha \circ \omega = \alpha$.
6. Prove in detail that, if both V and W are chains or ordinals, respectively, then so are $V \oplus W$ and $V \circ W$.
7. Show that the poset $P \oplus Q$ is Noetherian if and only if both P and Q are Noetherian—and similarly for $P \circ Q$.
8. Show that $\mathbf{Q} \oplus \mathbf{Q} = \mathbf{Q}$, in the sense of isomorphism.
9. Show that any dense subset T of a dense subset S of a chain C which is dense-in-itself, is dense in C.
10. Show that, if L and M are complete lattices, then so are $L \oplus M$ and $L \circ M$.
11. Obtain necessary and sufficient conditions on posets P and Q, for $P \oplus Q$ to be a complete lattice.
12. Show that the ordinal product $L \circ M$ is a lattice if and only if L, M are lattices and, in addition, L is a chain or M has universal bounds. (M. M. Day)

*13. Show that, if A, B, C are complete chains, then $A \circ B = A \circ C$ implies $B = C$. (A. Gleason)

12*. Homogeneous Continua; Souslin Problem

The chain $\mathbf{Z}$ of all integers and the chain $\mathbf{R}$ of all real numbers are remarkable for their *homogeneity*: each possesses a transitive group of automorphisms. Though one can construct other homogeneous chains (as ordered groups, see Chapter XIII), very few are conditionally complete. In fact, we have

THEOREM 25. *The chains* $\mathbf{Z}$ *and* $\mathbf{R}$ *are the only two conditionally complete chains which are* (i) *homogeneous* (*i.e., possess transitive automorphism groups*) *and* (ii) *have a countable dense subset.*

PROOF. By Theorem 24, C is isomorphic with a subset in $\mathbf{R}$; we can therefore assume $C \subset \mathbf{R}$. The interior† S of the complement $C' \cap \mathbf{R}$ of C in $\mathbf{R}$ is then open, and consists of a countable family of nonoverlapping open intervals. Condition (i) implies that at least one endpoint of each finite such open interval must belong to C. If in one case, both end points a, b belong to C, then some element $a \in C$ is covered by another element $b \in C$; it follows by (i) that *every* $X \in C$ must be covered by some $y \in C$, and similarly that every x *covers* some $z \in C$. We can hence

† We are here anticipating topological ideas to be treated in Chapters IX–X.

find a subinterval J of C isomorphic with $\mathbf{Z}$; we omit the details. Finally, if some $a \in C$ were not in J, we could by (i) form $\sup J$ or $\inf J$; but such elements could not both cover and be covered by other elements; hence $C = J \cong \mathbf{Z}$.

Otherwise, just one end point of each open interval belongs to C. In this case, if we move each point p of C towards the origin by an amount exactly equal to the sum of the lengths of the intervals between p and the origin, we get an isomorphic map of C on $\mathbf{R}$ or an interval of $\mathbf{R}$. Since if one $x \in C$ were greatest or least, every element would be so (by (i)), clearly either C consists of 0 alone (exceptional case), or C is isomorphic to an open interval of $\mathbf{R}$—and hence to $\mathbf{R}$ itself.

Various modifications of the preceding conditions are possible. Thus we can replace (i) by a hypothesis (i′) that *no* covering relation holds (i.e., that the chain is dense-in-itself), and get $\mathbf{R}$; or we can replace it by the hypothesis (i″) that every $a \in C$ covers and is covered by other elements, and get $\mathbf{Z}$.

A very interesting modification of (ii) is Souslin's Condition: (ii′) Every set of disjoint open intervals of C is countable. It is easily shown, using Theorem 23 or 24, that (ii) implies (ii′). For every positive integer n, there are at most $n + 1$ disjoint intervals of length $1/n$ or greater whose midpoint x satisfies $m \leqq x \leqq m + 1$ —hence only countably many disjoint intervals of positive length in $\mathbf{R}$ are possible. Souslin's Hypothesis is that, conversely, (ii′) implies (ii) in any chain.†

Another interesting set of conditions on chains has been proposed by G. D. Birkhoff. These include Souslin's Condition (ii′), the condition of *strong* homogeneity: (i*) all nonvoid intervals of $\mathbf{R}$ are isomorphic, and the condition that the union of any countable sequence $I_1 < I_2 < I_3 < \cdots$ of open intervals is an open interval. Examples of homogeneous linear continua not isomorphic to $\mathbf{R}$ have been found by Vazquez and Zubieta, and by Arens.‡

13. Well-ordering; Ordinals

We recall from §1 that a *well-ordered set*, or *ordinal*, is a poset W in which every nonvoid subset has a least member. Equivalently, it is a chain satisfying the DCC. It is immediate that any well-ordered set is a chain, that any finite chain is well-ordered, and that any subset of a well-ordered set is well ordered.

The set $\boldsymbol{\omega}$ of all positive integers in their natural order is the least infinite ordinal; it is contained (monomorphically) in all others. Among other countable ordinals, if we define ${}^{\mathbf{n}}W$ by

$$(10) \qquad {}^{\mathbf{n}}W = W \cdot {}^{\mathbf{n}-\mathbf{1}}W = W \cdot \cdots \cdot W \quad \text{(to } n \text{ factors)},$$

we may write down all *ordinal polynomials* in $\boldsymbol{\omega}$, of the form:

$$(11) \qquad (\mathbf{m}_n \circ {}^{\mathbf{n}}\boldsymbol{\omega}) \oplus (\mathbf{m}_{n-1} \circ {}^{\mathbf{n}-\mathbf{1}}\boldsymbol{\omega}) \oplus \cdots \oplus (\mathbf{m}_1 \circ \boldsymbol{\omega}) \oplus \mathbf{m}_0,$$

where $\mathbf{m}_n, \cdots, \mathbf{m}_0$ are any n finite ordinals, $\mathbf{m}_n \neq 0$. This set of ordinals is closed under ordinal addition and multiplication, as one easily shows using (7)–(8).

† Proposed by M. Souslin, Fund. Math **1** (1920), 223.

‡ See Bol. Soc. Mat. Mex. **1** (1944), 1–18 and **2** (1945), 91–4; R. Arens, ibid. **2** (1946), 33.

The Extended Induction Principle of §1, dualized (since the ACC is replaced by the DCC) specializes to the following extension of the Principle of Finite Induction.

First Principle of Transfinite Induction. Let $\{P_\alpha\}$ be any well-ordered set of propositions. To prove each P_α is true, it is enough to prove that each P_α is true on the assumption that P_β is true for every $\beta < \alpha$.

More interesting and instructive is the specialization of the preceding notion obtained by using limit numbers.

In a well-ordered set, we define a *limit number* to be an element α such that, given any $\beta < \alpha$, there exists γ satisfying $\beta < \gamma < \alpha$. By definition, we get as a corollary of the First Principle above, the

Second Principle of Transfinite Induction. Let $\{P_\alpha\}$ be any well-ordered set of propositions. To prove that each P_α is true it is enough to prove (i) P_1 is true, (ii) if P_α is true then $P_{\alpha+1}$ is true, (iii) if α is a limit number and all P_β with $\beta < \alpha$ are true, then P_α is true.

But the most fascinating results about ordinals cannot be obtained so easily; they make essential use of the Axiom of Choice. Among these is the fact that *the ordinals themselves are well ordered*. More precisely, we can prove:†

Theorem 26. *Any set of ordinals is well ordered, if $V \leqq W$ is defined to mean that V is isomorphic with an ideal of W.*

Proof. Note first that an "ideal" of a chain is just an initial interval. We now develop the proof in a sequence of lemmas.

Lemma 1. *The ideals of an ordinal W consists of W itself and, for each $a \in W$, the set of all $x < a$.*

Proof. For each a in any chain, the set of $x < a$ is an ideal. Conversely, if J is a proper ideal of an ordinal W, then there must be a least $a \in W$ not in J; clearly J contains all $x < a$, but not a or any $c > a$.

Lemma 2. *There is at most one isomorphism between given ideals J and K of ordinals V, W.*

Proof. Suppose θ and θ^* are any two such. Since J is well ordered, there must be a first $a \in J$ such that $\theta(a) \neq \theta^*(a)$, say $\theta(a) > \theta^*(a)$. The image $\theta(J)$ cannot contain $\theta^*(a)$, since $\theta(b) > \theta^*(a)$ if $b \geqq a$, while $\theta(b) = \theta^*(b) < \theta^*(a)$ if $b < a$. Yet $\theta(a) > \theta^*(a)$; hence $\theta(V)$ is not an ideal, contrary to hypothesis.

Now observe that if θ_α is, for each α, a function from a subset $S_\alpha \subset X$ into Y, and if $\theta_\alpha(x) = \theta_\beta(x)$ for all α, β whenever both are defined, then the θ_α have a greatest common extension θ, defined by making $\theta(x) = \theta_\alpha(x)$ if the latter is defined for one or more α. It follows from Lemma 2 that there is a most extensive isomorphism between ideals of any two given ordinals V and W. By Lemma 1, this involves either all of V, or all of W, or is between the set of $x < a$ and that of

† [Hau, §13]; Hausdorff attributes the result to G. Hessenberg. Note that the class of all ordinals, though well-ordered, is not itself an ordinal because it is not a set.

$y < b$ for some $a \in V, b \in W$. But in the last case it could be properly extended by making $\theta(a) = b$. Hence we conclude

LEMMA 3. *Of any two well-ordered sets V, W, one is isomorphic with an ideal (initial segment) of the other.*

It follows that the relation $\leqq$ of Theorem 26 satisfies P4. But P1, P3 are obvious, and P2 (with $=$ meaning isomorphism) is a corollary of Lemma 2. Finally, if Ω is any nonvoid set of ordinals, select $W \in \Omega$. The set of $V \not> W$ consists by Lemma 3 of ordinals isomorphic with ideals of W, and these form a well-ordered set by Lemma 1. This completes the proof of Theorem 26.

Exercises for §§12–13:

1. (a) Show that in any nonvoid chain C satisfying (i″), given $a < b$, there exists x such that $a < x < b$.

(b) Infer that for any a, $x < a$ and $x > a$ have solutions. (*Hint.* Use the isomorphism between $(-\infty, +\infty)$ and $(a, +\infty)$.)

2. (a) Show that any linear homogeneous continuum C is "sequentially compact", in the usual topological sense that every sequence of elements of C contains a convergent subsequence.

(b) Show that any nonvoid C has at least the power of continuum. (Zubieta–Vazquez)

3. Prove that if a poset P is isomorphic with a subset of an ordinal W, then it is isomorphic with an ideal of W.

4. (a) Show that, for ordinals, $V < W$ is equivalent to $W = V \oplus X$ for some X, and that this X is unique.

(b) Infer that we have one-sided cancellation and subtraction.

(c) Show that, however, $X \oplus V = Y \oplus V$ does not imply $X = Y$.

5. Show that $X \circ V = Y \circ V$ implies $X = Y$ unless $V = 0$, but that $\boldsymbol{\omega} \circ \mathbf{2} = \boldsymbol{\omega} \circ \mathbf{1}$.

6. (a) Which ordinal polynomials (11) are isomorphic with all their dual ideals? (Such ordinals are called indecomposable.)

(b) Show that every ordinal has a unique representation as an ordinal sum of non-increasing indecomposables. (Sierpinski)

7. (a) Which ordinal polynomials (11) have the property that, for all $X \leqq W$, $W = T \circ X$ for some T?

(b) Which have the property that $W = W \circ X$ for all $X < W$?

(c) Show that no ordinal polynomial is isomorphic with all its cofinal subsets, but that the least uncountable ordinal is.

8. (a) Show that $\mathbf{2}$ and $\boldsymbol{\omega}$ have no common right-multiple.

(b) Show that the set of all left-divisors of any ordinal is finite.

9. (a) Let θ be any order-endomorphism of an ordinal W. Show that $\theta(x) \geqq x$ for all x. (Hausdorff)

(b) Show that, for ordinals, $V < W$ implies ${}^{V}\mathbf{2} < {}^{W}\mathbf{2}$.

10. (a) Let V, W be ordinals with $V < W$. Show that unique ordinals X, R exist such that $W = (X \circ V) \oplus R$, $O \leqq R < V$. (This is a left-division algorithm.)

(b) Show that any ordinal can be represented as an ordinal product of a finite number of factors not further factorable.

(c) Show that $\boldsymbol{\omega} \circ (\boldsymbol{\omega} \oplus \mathbf{1}) = \boldsymbol{\omega} \circ \boldsymbol{\omega}$: no unique factorization theorem is true.

*11. (a) Using Transfinite Induction, define an ordinal arithmetic by Cantor's rules:

$$V \oplus (W \oplus 1) = (V \oplus W) \oplus 1, \qquad (V \oplus 1) \circ W = V \circ W \oplus W,$$

and, for limit-ordinals,

$$V \oplus (\mathrm{Lim}\ W_\alpha) = \mathrm{Lim}\ (V \oplus W_\alpha), \qquad (\mathrm{Lim}\ V_\alpha) \circ W = \mathrm{Lim}\ (V_\alpha \circ W).$$

Show that this is identical with that defined in §9.

(b) Show that $(\mathrm{Lim}\, V_\alpha) \oplus W \geqq \mathrm{Lim}\,(V_\alpha \oplus W)$ and $V \circ (\mathrm{Lim}\, W_\alpha) \geqq \mathrm{Lim}\,(V \circ W_\alpha)$, but that equality need not hold.

14. Axiom of Choice

We now show that Hausdorff's Maximal Principle (§7) is equivalent to the Axiom of Choice in various forms, and present some results which require this axiom for their proof. We first note (with M. Zorn) that HMP implies the following lattice-theoretic generalization of Zorn's Property:

(Z) if every chain C of a poset P has an upper bound $u(C) \in P$, then P contains a maximal element.

Indeed, any upper bound $u(M)$ of any maximal chain M of P will be such an element, and HMP asserts the existence of such M in P.

Now consider the original Axiom of Choice of Zermelo:†

(AC) Let I be any set. Then there exists a function f which selects from each nonvoid subset $S \subset I$ a member $f(S) \in S$.

Lemma 1. *Hausdorff's Maximal Principle implies Zermelo's Axiom of Choice.*

Proof. Consider the class Γ of all functions which assign to *some* nonvoid subsets $S \subset I$ a member $g(s) \in S$. Define $g \leqq h$ if h is an extension of g; this makes Γ a poset. Choose a *maximal chain* $M \subset \Gamma$, assuming HMP, and form the common extension g_m of the $g \in \Gamma$ (§11, before Lemma 3). Then g_m must be defined for all nonvoid S; otherwise there would exist a nonvoid T_0 with some element t_0, yet $g_m(T_0)$ undefined. Set

$$g^*(S) = g_m(S) \quad \text{if } g_m(S) \text{ is defined}$$
$$= t_0 \in T_0 \quad \text{if } S = T_0.$$

This would be a proper extension of g_m, contrary to the latter's definition. Hence $g_m = f$ will work. Q.E.D.

Lemma 2. *Zermelo's Axiom of Choice implies that every set I can be well ordered.*

Proof. Consider the relations ρ which well-order *subsets* $S(\rho)$ of I, *consistently* with the f of (AC), in the sense that under ρ, the first element of the *complement* A' of any ρ-ideal $A \subset S(\rho)$ is $f(A')$. As in Lemma 3 of §11, these form a chain. Their maximum common extension $\bar{\rho}$ well-orders I; we omit the details.

Lemma 3. *The possibility of well ordering implies Hausdorff's Maximal Principle.*

Proof. Given a chain C in a poset P, well-order $P - C = P \cap C' = W$. Let ψ assign to each $a \in W$ the value $\psi(a) = 1$ if, for all $b < a$ such that $\psi(b) = 1$ and for all $b \in C$, either $a \geqq b$ or $a \leqq b$, and set $\psi(a) = 0$ otherwise. This function is uniquely and consistently defined (by transfinite induction); by augmenting C with all $a \in W$ such that $\psi(a) = 1$, we get a maximal chain.

† E. Zermelo, Math. Ann. **59** (1904), 514 and **65** (1908), 261. A more thorough discussion of variants of the Axiom of Choice may be found in [**LT2**, pp. 42–4]; See also [**Kel**, pp. 32–6].

Combining Lemmas 1–3 above, we obtain

THEOREM 27. *Hausdorff's Maximal Principle, Zermelo's Axiom of Choice, and the statement that every set can be well ordered, are all equivalent.*

THEOREM 28. *If we define $S \leqq T$ to mean that there is an injection of S into T, cardinal numbers are well ordered.*

PROOF. Consider the mapping $W \to n(W)$ which assigns to each ordinal its cardinal number. It is isotone, and by Lemma 2 it is onto. But the ordinals are well ordered (Theorem 26); hence the same is true for cardinal numbers.

Using the Axiom of Choice, we can also prove that

$$\alpha + \beta = \alpha\beta \operatorname{Max}(\alpha, \beta) \tag{11}$$

for cardinal numbers, if α or β is infinite. This means that transfinite cardinal arithmetic (unlike the cardinal arithmetic of infinite posets!) is trivial. As corollaries, we see that $\alpha + \alpha = \alpha^2 = \alpha$ for all infinite cardinal numbers α. Various other results are stated as Exercises.

Exercises:

1. Without using the Axiom of Choice, prove that the cardinal numbers satisfy P2. (Schroeder–Bernstein Theorem)

2. Show that the "first" ordinal having a given infinite cardinal number is always a limit-ordinal.

3. Show that every chain has a well-ordered cofinal subset.

*4. Let P be any poset in which every well-ordered subset has a l.u.b. in P. Let f be any function on P to P such that $f(x) \geqq x$ for all $x \in P$. Without assuming any form of the Axiom of Choice, show that $f(m) = m$ for some $m \in P$. (N. Bourbaki)

*5. Show that, in any complete algebraic lattice, any $a < b$ can be connected by a chain in which the covering relations are dense.

*6. Show that the Axiom of Choice is implied by the hypothesis that the cardinal numbers are simply ordered.†

*7. Show that the laws $\alpha + \alpha = \alpha$ and $\alpha^2 = \alpha$ for infinite cardinals cannot be proved without the Axiom of Choice.

8. Let C be any conditionally complete chain with O in which: (i) every nonvoid set bounded below has an infimum; (ii) every element a not an upper bound of C is covered by some $S(a) \in C$. Show that C is well ordered. (A. V. Lemmon–A. G. Waterman)

9. (a) In any chain C, compare the following three numbers: (i) the smallest cardinal number of a dense set (Cantor); (ii) the least upper bound of the cardinal numbers of sets of disjoint open intervals (Souslin); (iii) the least upper bound of the cardinal numbers nested (transfinite) sequences

$$(a_1, b_1) > (a_2, b_2) > (a_3, b_3) > \cdots > (a_w, b_w) > (a_{w+1}, b_{w+1}) > \cdots \text{ of open intervals.}$$

(b) Observe that (ii) gives a greater number than (iii), in the case of $\mathbf{R} \circ \mathbf{R}$.

*10. Show that a chain C is order-monomorphic to $\mathbf{R}$ if and only if every uncountable set of intervals contains an uncountable subset, of which any two have a common element.‡

† Exercise 6 is due to F. Hartogs, Math. Ann. **76** (1915), 443; Exercise 7 is due to Tarski, Fund. Math. **5** (1924), 147–54.

‡ B. Knaster, Mat. Sb. (58) **16** (1945), 281–90.

15*. Ordinal Powers

It is not clear how to define ordinal powers of posets of infinite length. Cantor did this by induction on the exponent (see Exercise 3 below). However, this definition makes ${}^{\omega}\mathbf{2}$ countable; whereas it seems much more natural to identify ${}^{\omega}\mathbf{2}$ with the Cantor discontinuum of binary decimals, in their usual lexicographic ordering. This definition can be generalized as follows.†

Definition. The *ordinal power* ${}^{X}Y$ of two posets X and Y is the set of all functions $f\colon X \to Y$, with $f \leqq g$ defined to mean that for any $a \in X$ such that $f(a) \not\leqq g(a)$, there is a $b < a$ such that $f(b) < g(b)$ *and* $f(x) \leqq g(x)$ for all $x \leqq b$.

If $X = W$ is an ordinal, this is equivalent to making $f < g$ in ${}^{W}Y$ mean that $f(m) < g(m)$ for the *least* $m \in X$ such that $f(x) \neq g(x)$. If X satisfies the DCC, it is equivalent to making $f < g$ in ${}^{X}Y$ mean that $f(m) < g(m)$ for every minimal $m \in X$ such that $f(x) \neq g(x)$. In ${}^{\mathbf{R}}\mathbf{R}$ (for real functions), it is equivalent to making $f < g$ mean that $f(m) < g(m)$ for some m *and* $f(x) \leqq g(x)$ for all $x \leqq m$. More generally, this is true in ${}^{C}Y$ for any chain C.

Ordinal exponentiation satisfies a number of simple identities ("laws of exponents"), such as

$$(12) \qquad {}^{X\oplus Y}Z = ({}^{X}Z)\cdot({}^{Y}Z), \qquad {}^{X+Y}Z = ({}^{X}Z)({}^{Y}Z), \qquad \breve{{}^{X}Y} = {}^{X}\breve{Y}.$$

Again ${}^{X}({}^{Y}Z)$ is order-epimorphic to ${}^{(X\cdot Y)}Z$, though not in general order-isomorphic to it. For a further discussion of the identities of ordinal (and cardinal) arithmetic, the reader is referred to the references cited above.‡

For any ordinal W and chain C, ${}^{W}C$ is a chain: if $f \neq g$, there is a first $x = x_0$ where $f(x_0) \neq g(x_0)$, and $f \gtrless g$ according as $f(x_0) \gtrless g(x_0)$. However, one cannot simply order ${}^{\mathbf{Z}}\mathbf{Z}$, ${}^{\mathbf{Q}}\mathbf{Q}$, or ${}^{\mathbf{R}}\mathbf{R}$ in a natural way, as the following result shows.

Theorem 29 (K. Baker). *There is no simple ordering of the set of all functions $f\colon \mathbf{Z} \to \mathbf{Z}$ which is "natural"—i.e., invariant under the group $A =$ Aut $\mathbf{Z}$ of all order-automorphisms of $\mathbf{Z}$. The same is true if $\mathbf{Z}$ is replaced by $\mathbf{Q}$ or $\mathbf{R}$.*

Proof. Let μ be the automorphism $n \to n + 1$, and consider the function f defined by

$$f(2k) = 2k + 1, \qquad f(2k + 1) = 2k, \quad \text{all } k \in \mathbf{Z}.$$

Let $g = \mu f \mu^{-1}$; clearly $g \neq f$. Hence, under any simple ordering, either $f < g$ or $f > g$. But if the order is "natural", this implies $g = \mu f \mu^{-1} < \mu g \mu^{-1} = \mu^2 f \mu^{-2} = f$, a clear contradiction.

To cover the cases of $\mathbf{Q}$ and $\mathbf{R}$, it suffices to extend f to a function f_1 with $f_1(x) \equiv x$ for x not an integer. It would be interesting to prove the corresponding result for continuous functions from $\mathbf{R}$ to $\mathbf{R}$, if true.

---

† See [**Hau**, §16]; [**W–R**]; G. Birkhoff [**5**, §§8–9]; and M. M. Day [**1**, §7]. The present definition differs from these for general posets, but all are equivalent (and very satisfactory!) for ordinals.

‡ See also H. Bachmann, *Transfinite Zahlentheorie*, Springer, 1955; A. Tarski, *Ordinal algebras*, North-Holland Publ. Co., Amsterdam, 1956; C. Kuratowski, *Cardinal and ordinal numbers*, Warsaw, 1958.

HAHN'S ORDINAL POWER. Though there is no natural way to simply order $\mathbf{R}^{\mathbf{R}}$, one can simply order a fairly large subset of ${}^{C}\mathbf{R}$ for any chain C, by the following construction of H. Hahn [**1**]. This construction is important for the study of ordered and lattice-ordered Abelian groups (including vector lattices).

DEFINITION. Hahn's ordinal power ${}^{(C)}\mathbf{R}$ is the set of all functions $f: C \to \mathbf{R}$, C a chain, such that $f(x) = 0$ except on a well-ordered subset $W(f) \subset C$. The meaning of $f \leqq g$ is the same as in ${}^{C}\mathbf{R}$.

Exercises:

1. Prove the validity of (12).
2. Show that the congruence relation of Exercise 9 after §1 yields the real field $\mathbf{R}$ from ${}^{\omega}\mathbf{2}$.
3. Define ordinal exponentiation inductively, following Cantor, by

$$ {}^{1}V = W, \qquad {}^{W\oplus 1}V = {}^{W}V \circ V, \qquad {}^{(\mathrm{Lim}\, W_\alpha)}V = \mathrm{Lim}\, ({}^{W_\alpha}V). $$

Show that the inductive definitions of ${}^{\omega}\mathbf{2}$ and ${}^{\omega}\boldsymbol{\omega}$ are not equivalent to those of the text.

4. For functions $f: X \to Y$ (X, Y posets), define $f \prec g$ to mean that if $f(a) \nleqq g(a)$, then there exists $x < a$ such that $f(x) < g(x)$. Is $\prec$ a partial ordering?

5. Show that ${}^{X}L$ is a lattice if and only if one of the following three conditions holds: (i) L is a lattice and X is unordered, (ii) L is a lattice with O and I, (iii) L is a chain and X is a dual tree (i.e., for any $a \in X$, $[a, \infty)$ is a chain).

16*. Continuum Hypothesis; Some Uncertainties

The study of the foundations of set theory has brought to light three questionable hypotheses: the Axiom of Choice, the Continuum Hypothesis, and Souslin's Hypothesis. These have implications also for lattice theory, and so it seems desirable to review their status here.

Today, the Axiom of Choice is considered to be harmless in principle and necessary in practice. Halmos† has reduced it to a hypothesis which seems unavoidable: "The Cartesian product of a nonempty family of nonempty sets is nonempty".

Nevertheless its acceptance leads one to the very specific conclusion that $\mathbf{R}$ can be well ordered, and this seems impossible to do in any constructive sense. Though one can construct numerous *countable* ordinal "polynomials" in $\boldsymbol{\omega}$, by ordinal addition and multiplication (using finite ordinal exponents),‡ nobody has ever "constructed" an explicit function which well-orders an uncountable subset; we have no way of imagining what an uncountable well-ordered set "looks" like. *The problem of "constructively" well-ordering an uncountable well-ordered set is the most fundamental problem of set theory.*

Our ignorance goes much further. Every infinite set contains a countable subset; hence $\aleph_0$ is the *least* infinite cardinal number. But we have no idea what the second smallest infinite cardinal number is; it exists (assuming the Axiom of Choice) by Theorem 27. A celebrated conjecture is the following

† P. Halmos, *Naive set theory*, Van Nostrand, 1960, p. 59.

‡ For a thoughtful review of this problem, see D. L. Kreider and Hartley Rogers, Jr., Trans. AMS **100** (1961), 325–69.

CONTINUUM HYPOTHESIS. The second infinite cardinal number $\aleph_1$ is the power of the continuum.

Sierpinski has shown that this conjecture is equivalent to numerous other interesting (unproved and undisproved) propositions,† Gödel has shown that it is consistent with the usual formal axiom systems of set theory, and even that this is true of the Generalized Continuum Hypothesis $\aleph_{n+1} = 2^{\aleph_n}$. Recently Paul Cohen has shown that it is independent of them.‡

In other words, existing formalizations of logic and the axioms of set theory are incapable of deciding whether the Continuum Hypothesis is true or false (if, indeed, it is "meaningful"). The same seems to be true of Souslin's Hypothesis; it seems likely that this also is independent of the other axioms of set theory, and very likely of the Axiom of Choice and the Continuum Hypothesis as well.§

These questions will be kept largely in the background for the remainder of this book. Thus, much of the discussion of Chapter IX–XI is restricted to countable operations, and hence (in some sense) is more "constructive". (Curiously, Theorem XI.13 depends on the Continuum Hypothesis.) However, it should be remembered that the Axiom of Choice is needed to prove the subdirect decomposition theorem (§8), and that this will be invoked essentially in Chapter XIII ff.

Additional Exercises:

*1. Let L be the Boolean lattice of all subsets of a countable set, modulo finite subsets. Show that any maximal chain in L contains a well-ordered uncountable subchain. (A. Gleason, E. E. Moise)

*2. Show that a countably infinite Boolean algebra cannot be Noetherian. (*Hint.* If Noetherian, it would have to contain an infinite set of disjoint atoms.)

*3. Show that every complete, countably infinite Boolean algebra is monomorphic to $\mathbf{2}^{\omega}$.

*4. Let $B_0 \subset B$ be Boolean algebras, and A a *complete* Boolean algebra. Show that every morphism $\mu: B_0 \to A$ can be extended to a morphism $\mu: B \to A$. (Sikorski¶)

*5. Find a distributive lattice L having a chain which cannot be extended to a maximal sublattice.||

*6. Describe the free distributive lattice with $\aleph$ generators, $\aleph$ any cardinal number.

*7. Show that a distributive lattice can be embedded in a Boolean algebra with preservation of infinite joins and meets if and only if it is completely distributive.††

*8. Show that there is no identity which is satisfied in every finite partition lattice, not a consequence of L1–L4.‡‡

---

† W. Sierpinski, *L'hypothese du continu*, Paris, 1927.

‡ K. Gödel, "The consistency of the continuum hypothesis", Princeton, 1940; Paul Cohen, Proc. Nat. Acad. Sci. **50** (1963), 1143–8 and **51** (1964), 105–10. Note that, with the current formalizations of logic and set theory, this can be demonstrated by *countable* models!

§ In a forthcoming paper, S. Tennenbaum will show that there exist "models" of set theory (including the axiom of choice) in which every Souslin continuum *is* isomorphic to the reals, and in which this is not so.

¶ See W. A. J. Luxemburg, Fund. Math. **55** (1964), 239–47.

|| K. Takeuchi, J. Math. Soc. Japan **2** (1951), 228–30. For Exercise 6, see G. Ya Areskin, Mat. Sb. **33** (1953), 133–56, and A. Nerode, Trans. AMS **91** (1959), 139–51.

†† N. Funayama, Nagoya Math. J. **15** (1959), 71–81.

‡‡ See D. Sachs, Proc. AMS **12** (1961), 944–5.

*9. (Iwamura's Lemma) Show that every directed set D is the union $D = \bigcup D_{\aleph}$ of a chain of directed sets $D_{\aleph}$, each of smaller cardinality than D.

PROBLEMS

64. Does there exist an infinite group for which $L(G)$ has finite length? (Kaplansky)

65. Is there any identity satisfied in every finite modular lattice which is not satisfied in every modular lattice?

66. Is every equational class of lattices generated by its finite members? (B. Jonsson)

67. Construct two nonisomorphic, finite, subdirectly irreducible algebras which generate the same family of algebras. (B. Jonsson)

68. Develop a unique factorization theorem for "canonical" ordinal products. (C.-C. Chang [**Symp**., p. 125].)

69. Find a direct construction which represents any epimorphic image of a lattice of partitions (i.e., sublattices of Π_n) as lattice of partitions.

*70. Construct an injection—a bijection—from $\mathbf{R}$ to the set of all countable ordinals (i.e., of ordinals α_0 such that the set of all $\alpha < \alpha_0$ is countable). (Hilbert's First Problem)

*71. A cardinal number $\aleph$ is called *inaccessible* from below if there are $\aleph$ distinct cardinal numbers less than $\aleph$. Does there exist an uncountable inaccessible cardinal number? (Sierpinski)

Chapter IX

APPLICATIONS TO GENERAL TOPOLOGY

1. Metric Spaces

So far we have considered primarily *discrete* lattices, and their generalizations to *algebraic* lattices. Many of the deepest results of lattice theory are concerned with interconnections between the notion of *limit* and that of *order*, in *nonalgebraic* lattices. Broadly speaking, the next three chapters will deal with such interconnections and their applications. Building on Chapters V and VIII, they will give perspectives into lattice theory entirely different from those provided by earlier chapters.

The present chapter will deal with various applications of lattice concepts to general topology—i.e., to the general theory of topological spaces. The ideas of general topology can be most simply introduced through the concept of a metric space.

Definition. A *metric space* is a collection of elements ("points"), related by a real-valued distance function $\delta(x, y)$ which satisfies

M1. $\delta(x, x) = 0$, while $\delta(x, y) > 0$ if $x \neq y$,

M2. $\delta(x, y) = \delta(y, x)$ (symmetry),

M3. $\delta(x, y) + \delta(y, z) \geqq \delta(y, z)$ (triangle inequality).

In a metric space M, a sequence $\{x_n\}$ of points is said to *converge* to the limit point a (in symbols, $x_n \to a$), if and only if $\text{Lim}_{n\to\infty} \delta(x_n, a) = 0$. One calls a set $S \subset M$ *closed* (in M) when $\{x_n\} \subset S$ and $x_n \to a$ imply $a \in S$. One calls a set $U \subset M$ *open* when its complement is closed. An equivalent condition is that, for any $a \in U$, a positive constant $\epsilon = \epsilon(a, U)$ can be found, so small that $\delta(x, a) < \epsilon$ implies $x \in U$.

Conversely, one can show that $x_n \to a$ in a metric space M if and only if every open set U of M which contains a contains all x_n with $n > N(U)$, some positive integer depending on U (and on $\{x_n\}$). Equivalently, $x_n \to a$ if and only if every closed set $S \subset M$ which contains an infinite subsequence of $\{x_n\}$ contains a.

The preceding observations show that, in a metric space M, the concepts of closed set, open set, and convergent sequence are interchangeable: one can define each of them in terms of any of the others, without reference to the concept of distance. Properties of M which are definable in terms of any (hence all) of these three concepts are called *topological* properties. Distance itself is *not* a topological property.

Taking the limit of a convergent sequence can be considered as performing a sometimes defined (i.e., "partial") infinitary operation, in the sense of §VI.1. That is, any metric space M can be considered as an *infinitary partial algebra*, whose "subalgebras" are just the closed sets of M. The analog of congruence relations is provided by the "fiberings" of M into disjoint closed subsets; from these one can construct all continuous surjections ("epimorphisms") $\alpha: M \to X$ into other metric and topological spaces.†

By Theorem VI.1, it follows that being metrically closed is a *closure* property in the sense of §V.1: conditions C1–C3 stated there are satisfied, and the closed subsets of any metric space M form a *Moore family*. We will now study in more detail the complete lattice $L(M)$ of all closed subsets of M, and the dual lattice $\check{L}(M)$ of all *open* sets—i.e., of all complements of closed sets.

LEMMA. *In any metric space*:

(α) *the sum of any two closed sets is closed*;
(β) *any intersection of closed sets is closed*;
(γ) *any point is a closed set.*

PROOF. Conclusion (β) follows since the closed subsets form a Moore family. Condition (α) follows from the fact that any sequence of points of $S \cup T$ must contain an infinite subsequence of points of S or one of points of T, and any such sequence must converge to the same limit point. Condition (γ) follows since (by M1) the sequence $\{f, f, f, \cdots\}$ converges to f, and to no other point.

2. Topological Spaces

Instead of considering a metric space as a partial algebra for the infinitary "convergence" operation applied to sequences, one can consider closure as an undefined unary operation on the Boolean algebra (power set) $\mathbf{2}^M$ of all subsets of M, yielding a Moore family of closed subsets. This viewpoint leads to the definition of a topological space already given in §V.4.

DEFINITION. A topological space is a set X together with a family of closed subsets of X such that

R1. The sum of any two closed sets is closed.
R2. Any intersection of closed sets is closed.
R3. The set and the empty set $\varnothing$ are closed.

We have seen (Lemma 1, §1) that any metric space is a topological space in which, by (γ):

S1. Any point is a closed set (Riesz condition).

A T_0-space which satisfies S1 is called a T_1-space.

† However, because M is only a "partial algebra", the image $\alpha(M)$ is not determined up to homeomorphism by the fibering.

Any metric space M satisfies the following three additional separation conditions; we omit the proofs.

S2. If p and q are any two distinct points, then M contains disjoint open sets U and V such that $p \in U$ and $q \in V$ (Hausdorff condition).

S3. If U is an open set containing p, then there is an open set V containing p such that $\bar{V} \subset U$ (Regularity condition).

S4. If A and B are disjoint closed sets of M, then M contains open sets $U \supset A$ and $V \supset B$ such that $U \cap V = \varnothing$ (Normality condition).

A topological space which satisfies S2 is called a *Hausdorff* space; if it also satisfies S3, it is called *regular*; if it satisfies S4, it is called *normal*. It is proved in [**Kel**] that S2 implies S1; that S3 and S1 imply S2; and that S4 and S1 imply S3. We have thus stated that any metric space is a *normal Hausdorff* space (hence a regular one).

As will be made more precise in §§4–5, the preceding separation conditions (especially S4) very nearly characterize rings of "closed" sets in metric spaces. The discussion of §§4–5 below will thus supplement and deepen that of §V.4, where some general properties of the lattices $L(X)$ of closed and $\check{L}(X)$ of open sets of T_1-spaces have already been established.

3. Directed Sets and Nets

The problem of treating T_1-spaces generally as infinitary partial algebras for suitable "convergence" operators can be solved by using *directed sets* of indices.†

Accordingly, we define a *net* $\{x_\alpha\}$ of points of a (perhaps topological) space X as a function from a directed set A of indices α into X. (The case $A = \boldsymbol{\omega}$ gives ordinary infinite sequences.) The concept of convergence for sequences in a metric space is then generalized as follows:

Definition. In a T_1-space X, the net $\{x_\alpha\}$ *converges* to the point a (in symbols, $x_\alpha \to a$) when, for each open set U containing a, $\gamma(U)$ exists such that $x_\alpha \in U$ for all $\alpha \geqq \gamma(U)$.

The appropriateness of the preceding definition of convergence is due to the following basic result.

Theorem 1. *In a T_1-space X, the closure $\bar{S}$ of any set S consists of the limits of convergent nets of points of S.*

Proof. Suppose $\{x_\alpha\} \subset S$ and $x_\alpha \to a$. Then $\bar{S}' = S^{-\prime} \subset S$ is an open set which contains no x_α; hence, by definition of $x_\alpha \to a$, it cannot contain a. Therefore $a \in \bar{S}$. Conversely, given $a \in \bar{S}$, we use the different open sets U_α which contain a as an index set, ordered by *dual* inclusion. Then $U_\alpha \cap S$ is nonempty

† Limits of directed sets were first considered by E. H. Moore, Proc. Nat. Acad. Sci. (1915), 628–32; see also E. H. Moore and L. H. Smith, Amer. J. Math. **44** (1922), 102–21. They were first applied to general topology in G. Birkhoff [**4**]; the present treatment includes improvements of Kelley [**1**].

(otherwise $U'_\alpha \supset S$, whence $U'_\alpha \supset \bar{S}$ contrary to the assumption $a \in U_\alpha$); choose $x_\alpha \in U_\alpha \cap S$ (this requires the Axiom of Choice). Clearly $x_\alpha \to a$, completing the proof.

This shows that the closed subsets of any T_1-space are its "subalgebras", when considered as an infinitary partial algebra with respect to convergence "operations", over suitable directed sets, as in Theorem VI.1.

We now ask the converse question: for which "convergence" operators (on nets over suitable directed sets) do the "subalgebras" define T_1-spaces? The author [**4**] originally tried to answer this question by generalizing the familiar concept of "subsequence" to that of a *cofinal subset* of a net $\{x_\alpha\}$, defined as the net $\{x_\beta\}$ for a subset $B \subset A$ of the directed set A of indices of $\{x_\alpha\}$, which contained for any α_0 some $\beta \geqq \alpha_0$. Kelley (see [**1**] and [**Kel**, p. 70]) has proposed generalizing the concept of subsequence to a much wider class of "subnets", defined as follows.

DEFINITION. Given nets $\{x_\alpha\}$ over A and $\{y_\beta\}$ over B, $\{y_\beta\}$ is a *subnet* of $\{x_\alpha\}$ when there exists a mapping π: $B \to A$ such that: (i) $y_\beta = x_{\pi(\beta)}$ for all $\beta \in B$, and (ii) to each $\alpha_0 \in A$, there exists a $\beta_0 \in B$ such that $\beta \geqq \beta_0$ implies $\pi(\beta) \geqq \alpha_0$.

Evidently, every cofinal subset of a net is a "subnet", with π the insertion function. Subnets have the advantage that the usual definition of compactness (§7 below) is equivalent to the following: a T_1-space is compact if and only if every net has a convergent subnet.†

LEMMA 1. *The nets of any T_1-space satisfy:*

T1. *If $x_\alpha = a$ for all $\alpha \in A$, then $x_\alpha \to a$.*

T2. *If $x_\alpha \to a$ and $\{x_\beta\}$ is a subnet of $\{x_\alpha\}$, then $x_\beta \to a$.*

The proof is trivial (see [**Kel**, p. 70]).

We next show that, from any "convergence" which satisfies conditions T1–T2, one gets a topological space by taking the "subalgebras" as closed sets. Condition R1 ((β) of the lemma of §1) follows as in Theorem VI.1; condition R3 is trivial (condition (γ) follows from T1). To complete the proof, we now establish

LEMMA 2. *Condition* T2 *implies condition* (α) *of the lemma of* §1.

PROOF. Let $\{x_\alpha\} \subset S \cup T$, where S and T are closed. Then either $\{x_\alpha\} \cap S$ or $\{x_\alpha\} \cap T$ is cofinal, thus a subnet of x_α. Hence, if $x_\alpha \to a$, there is a subnet of $\{x_\alpha\}$ in S or in T, which by T2 also converges to a. Since X and Y are closed, $a \in (S \cup T)$ in either case. Hence $S \cup T$ is closed if S and T are closed, as claimed.

Therefore, for any concept of convergence which satisfies T1–T2, the closed sets define a possible topology. However, in general topological spaces, this convergence has very few properties. For example, consider the degenerate "topological space" X having a continuum of points, but $\varnothing$ and X as its only closed (or open) sets. In this space, $x_\alpha \to a$ for *all* $\{x_\alpha\}$ and a!

To avoid such pathological cases, one must assume one or more of the "separation" conditions S1–S4 mentioned at the end of §1. For example, we have

† See [**Kel.**, p. 136]. For another advantage of subnets, see [**Kel.**, p. 77, Example E].

LEMMA 3. *A topological space X is a Hausdorff space if and only if its convergent nets satisfy also*

T3. *If $x_\alpha \to a$ and $x_\alpha \to b$, then $a = b$.*

For the proof, see [**Kel**, p. 67]. A related result is the following.

LEMMA 4. *A Hausdorff space is regular if and only if its convergent nets satisfy*

T4. *If $x_\beta^\alpha \to x_\alpha$ for every α, and $x_{\beta(\alpha)}^\alpha \to x$ for every choice of $\beta(\alpha)$, then $x_\alpha \to x$.*

For the proof, see G. Birkhoff [**4**, Theorem 7a].†

Even T1–T4 are not sufficient to imply that $\bar{\bar{S}} = \bar{S}$—that is, that the "closure" $\bar{S}$ of ("subalgebra" generated by) a set S should consist of the limits of convergent nets $\{x_\alpha\} \subset S$. For this to be the case one must postulate (Birkhoff [**4**, Theorem 5], [**Kel**, p. 69]) the

Law of iterated limits. Let A be a directed set, and $B_\alpha = \{x_{\alpha,\beta}\}$ a directed set for each $\alpha \in A$, and let C be the Cartesian product $C = A \prod_A B_\alpha$, where $(\alpha, \boldsymbol{\beta}) \geqq (\alpha', \boldsymbol{\beta}')$ means that $\alpha \geqq \alpha'$ and $\beta'(\alpha) \geqq \beta(\alpha)$ for all $\alpha \in A$. Let $x_\alpha \to p$ and, for each fixed α, let $x_{\alpha,\beta} \to x_\alpha$. Then $x_{\alpha,\beta(\alpha)} \to p$.

(Note that, even if A and all B_α are copies of $\boldsymbol{\omega}$, C is not: the preceding law cannot be formulated in terms of sequences alone.)

Exercises for §§1–3:

1. Show that if $X = X_1 + X_2$ is the sum of complementary closed sets, then $L(X) \cong L(X_1)L(X_2)$ (cardinal product).

2. Specialize Lemma 1 to Fréchet "L-spaces", that is, sets with a sequential "convergence operator" $x_n \to x$ such that:

L1. If every $x_n = a$, then $x_n \to a$.

L2. If $\{x_{n(i)}\}$ is a subsequence of $\{x_n\}$, and $x_n \to a$, then $x_{n(i)} \to a$.

3. Show that if $x_n \to a$ in a metric space and $\phi(n)$ is any permutation of the subscripts, then $x_{\phi(n)} \to a$.

4. Show that, if X is a finite T_1-space, then $L(X) \cong \mathbf{2}^{n(X)}$.

5. Show that, if D is a directed set, then the terminal segments $[\alpha, \infty)$ of D form a *filter* base of subsets.

6. Show that, if a subset of a directed set is not cofinal, its complement must be.

7. (a) Show that, in a topological space X, $a \in \bar{S}$ if and only if every open set $U = \overline{T'}$ containing a contains a point of S.

(b) Show that this result requires only $R2$. (Ore)

8. Let X be a T_0-space in which a set S is closed if it contains all limits of *countable* nets of points of S. Show that a set $T \subset X$ is closed when it contains all limits of convergent *sequences* of points of T, with index-set $\mathbf{Z}^+$.

*9. Define a *tower* as a poset isomorphic with the directed set $\mathbf{2}^{(C)}$ of all finite subsets of some set C. Show that every directed set contains a cofinal subset $\{x_\alpha\}$, where the index is a "tower" and $\alpha > \beta$ implies $x_\alpha \geqq x_\beta$. (M. Krasner)

10. In Example 1 of §6, from every finite set $F \subset X$, select any $y_F \in F'$. Show that, if the F are directed by set-inclusion, then $y_F \to a$ for every $a \in P$.

---

† See also G. Grimeisen, Math. Ann. **141** (1960), 318–42; **144** (1961), 386–417; **147** (1962), 95–109.

4. Regular Open Sets

By Theorem V.12, the closed subsets of any T_1-space X form a complete atomic distributive lattice $L(X)$; as in §V.10, the dual lattice $\breve{L}(X)$ of all open sets of X is a complete (nonatomic) *Brouwerian* lattice. We shall now study this lattice more closely, noting first that since the dual of the closure $\bar{S}$ of a set S (under complementation) is its interior $S'^{-\prime}$, a set S is open if and only if $S = S'^{-\prime}$.

DEFINITION. An open set S is *regular* when it is the interior of its closure—i.e., when $S = \bar{S}'^{-\prime} = S^{-\prime-\prime}$, a set T is *nowhere dense* when $T^{-\prime-} = X$.

A nowhere dense open set S in a T_1-space is necessarily void, since it must satisfy $S \subset \bar{S}'^{-\prime} = X' = \varnothing$.

LEMMA 1. *In the lattice $\breve{L}(X)$ of open sets of any T-space X, $\bar{S}'$ is the pseudo-complement S^* of S (in the sense of* §V.10).

PROOF. Since $S \subset \bar{S}$, $S \cap \bar{S}' \subset \bar{S} \cap \bar{S}' = \varnothing$; while conversely, $S \cap T = \varnothing$ (T open) implies $\bar{S} \cap T = \varnothing$, whence $T \subset \bar{S}'$. Hence $\bar{S}' = S^*$ is the largest open set which is disjoint from S, as claimed.

COROLLARY. *An open set S is regular if and only if $S = (S^*)^*$—that is, if and only if S is closed in $\breve{L}(X)$ under the symmetric polarity defined by the relation $S \cap T = \varnothing$.*

Note that, in §2, a T_1-space was defined to be *regular* when, given $a \in A$, A an open set, an open set B could be found such that $a \in B \subset \bar{B} \subset A$. But since the operation $S \to S'^{-\prime}$ is isotone, this implies $\bar{B}'^{-\prime} \subset A'^{-\prime} = A$: that A should contain a *regular* open set $B = B^{-\prime-\prime} = B^{**}$ such that $a \in B$.

Glivenko's Theorem V.26 can be restated for (Brouwerian) lattices of open sets as follows.

THEOREM 2. *Let X be any topological space. Then the correspondence $S \to \bar{S}'^{-\prime}$ is a lattice epimorphism of the lattice $\breve{L}(X)$ of all open sets of X onto the complete Boolean algebra $B(X)$ of all regular open sets of X.*

More generally, the preceding result is valid in any closure algebras, since the latter's "open" elements form a Brouwerian lattice.

Glivenko's Theorem has other corollaries. Thus, the dense open sets which satisfy $S^{**} = X$ form a dual ideal in $\breve{L}(X)$. Also, a closed set T is nowhere dense if and only if $T'^{-} = X$—i.e., if and only if its open complement T' is dense. Hence nowhere dense closed sets form an ideal in $\breve{L}(X)$. Finally, a set T is nowhere dense if and only if its closure is; hence nowhere dense sets form an ideal.

LEMMA 2. *If S and T are open sets, then $S^{**} = T^{**}$ if and only if the difference between S and T is nowhere dense.*

PROOF. By Glivenko's Theorem, $S^{**} = T^{**}$ if and only if $S \cap D = T \cap D$, where D is some dense open set. By Boolean ring algebra, this means that $S \cap D + T \cap D = (S + T) \cap D$, i.e., that the (symmetric) difference $S + T$

between S and T lies in the nowhere dense complement of D—or, equivalently, that $S + T$ is nowhere dense.

COROLLARY. *No nonvoid open set X is nowhere dense.*

THEOREM 3. *If X is a subset of Euclidean space without isolated points, then the Boolean algebra A of "regular" open sets of X is the completion by cuts of the free Boolean algebra B_∞ with countable generators.*

PROOF. The proof will apply generally to any T_1-space with a countable basis of regular open sets, a_i. Indeed, the a_i and their pseudo-complements a_i^* generate a copy of the free Boolean algebra B_∞ with countable generators, which may also be defined (by Corollary 4 of Theorem VI.13) as the limit of $\mathbf{2} \subset \mathbf{2}^2 \subset \mathbf{2}^4 \subset \cdots \subset \mathbf{2}^{2^n} \subset \cdots$. Clearly every $a \in A$ defines a cut in B_∞; x is in the lower half of the cut if and only if $x \leqq a$, and in the upper half if and only if $x \geqq a$. Again, unless $b \leqq a$ in A, $b - \bar{a}$ is a nonvoid (since a, b are regular) open set which contains some $a_i > 0$ in b but not in a; hence different elements of A correspond to different cuts in B_∞. Finally, every cut L, U in B_∞ corresponds to some $a \in A$. For take the set of all $x_i \leqq b$ for all $b \in U$, the upper half of the cut. Then the regular hull $(\bigvee x_i)^{**} \leqq b$ for all $b \in U$. But $x_i \in L$ in B_∞ if and only if $x \leqq b$ for $b \in U$—i.e., if and only if $x \leqq (\bigvee_L x_i)^{**} \in A$, completing the proof.

Compact elements. In $\breve{L}(X)$, an element u is "compact" (§VIII.4) if and only if the corresponding open set U is compact (Heine–Borel condition). Hence any U which corresponds to a "compact" element of $\breve{L}(X)$ of a T_2-space must have an open complement $U' \subset X$: every "compact" element $u \in \breve{L}(X)$ is complemented. Therefore, if $\breve{L}(X)$ is a complete *algebraic* lattice, then every open set must be a union of compact open sets. The converse is also true, so that $\breve{L}(X)$ is a complete algebraic lattice if and only if X is a Stone space (§9). In this case, Theorem VIII.8 applies nicely: K is the Boolean algebra of open-and-closed sets, and $\hat{K} \cong \breve{L}(X)$ is the lattice of all ideals of K.

5. T_1-lattices

It is almost evident that any T_1-space X is determined up to homeomorphism by the lattice $L(X)$ of its closed sets (and hence also by $\breve{L}(X)$).

THEOREM 4. *Any T_1-space† X is determined up to homeomorphism by the atomic dually Brouwerian lattice $L(X)$ of all its closed sets, ordered by inclusion.*

PROOF. The points of X are the elements which cover 0 (the "atoms") of $L(X)$. And the closure $\bar{S}$ of any subset S of X consists of the $p \in X$ represented by atoms of $L(X)$ contained in the join (in $L(X)$) of the atoms in S.

LEMMA 1. *For each point p in a T_1-space X, the principal dual ideal $J(p)$ of all closed sets which contain p is a maximal ideal of $L(X)$, and conversely.*

† For extension of this result to T_0-spaces, see W. J. Thron, Duke Math. J. **29** (1962), 671–80; D. Drake and W. J. Thron, Trans. AMS **120** (1965), 57–71. For T_2-spaces etc., see J. Kerstan, Math. Nachr. **17** (1958), 16–18 and 27–46.

For, if K is any larger dual ideal, then K must contain $a \wedge p = 0$ for some a not containing p. Conversely, let J be any principal dual ideal in $L(X)$, with least element m corresponding to the closed set $S(J)$. Unless $S(J)$ is a single point for any point $p \in S(J)$, the dual ideal of all closed $T \geqq p$ will be larger than J; hence J cannot be maximal.

In view of the preceding remarks, it is easy to characterize those lattices which are isomorphic with the lattice of all closed subsets of some T_1-space. They are simply complete atomic lattices whose *duals* are Brouwerian. Except in trivial cases, they are *not* algebraic (although atomic Brouwerian lattices are!).

We therefore define a *T_1-lattice* as a complete atomic lattice whose dual is Brouwerian. We have just seen that a lattice is a T_1-lattice if and only if it is isomorphic with the lattice $L(X)$ of all closed subsets of a suitable T_1-space X. It is also evident that any T_1-lattice L has Wallman's Disjunction Property:

(W) Given $s > t$, there exists a $p \in L$ such that $s \wedge p > 0$, yet $t \wedge p = 0$.

Closure algebras. If one deletes the assumption of atomicity, one gets an interesting generalization of the concept of a T_1-lattice, named "closure algebra" by Mc Kinsey and Tarski and studied by them.†

DEFINITION. A *closure algebra* is a Boolean algebra with a closure operation which satisfies conditions C1, C2, C3* of §V.4, and also $\bar{0} = 0$.

Closure algebras have many interesting properties, a few of which are summarized in [**LT2**, Chapter XI, §7]. Especially relevant is

THEOREM 5. *The "open" elements of any closure algebra C form a Brouwerian lattice.*

PROOF. Given $a, b \in C$, clearly $a \wedge x \leqq b$ implies $x \leqq b \vee a'$. For $x = x'^{-\prime}$ *open*, since $y \to y'^{-\prime}$ is an isotone function, this is equivalent to the inequality $x \leqq (b \vee a')'^{-\prime} = (b' \wedge a)^{-\prime}$. Hence the open elements of C form a Brouwerian lattice, in which $b : a = (a \wedge b')^{-\prime}$.

The notion of a "closure algebra" is defined by identities on finitary operations. Hence (cf. Chapter VI) one can construct the *free* closure algebra with any number of generators. Tarski and Mc Kinsey (loc. cit) have shown that the free closure algebras with countable generators can be realized in Euclidean space. Kuratowski had shown many years earlier‡ that the free closure algebra with one generator is infinite, whereas that for the unary complementation and closure operations alone has exactly 14 elements. As a poset, its diagram has the form sketched in Figure 20.

† Ann. of Math. **45** (1944), 141–91, and **47** (1946), 122–62. Independently, H. Terasaka had earlier made a similar study (Math. Revs. **11** (1950), p. 310). For "closure algebras" on Boolean σ-algebras, see R. Sikorski, Fund. Math. **36** (1949), 165–206.

‡ Fund. Math. **3** (1922), 182–99.

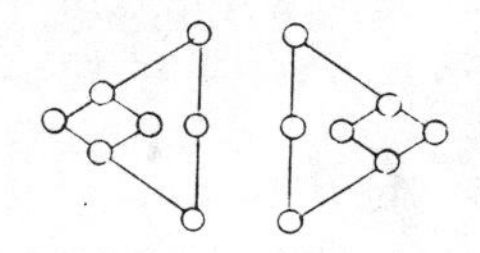

FIGURE 20

Exercises for §§4–6:

1. Show that, in the complete lattice $L(\mathbf{R})$ of all closed subsets of the real line, $c_\alpha \downarrow c$ implies $a \vee c_\alpha \downarrow a \vee c$ for any $a \in L(\mathbf{R})$, but not dually.

2. Show that a topological space X is a (Hausdorff) T_2-space if and only if, in $L(X)$, given atoms p and q, there exist elements $a, b \in L(X)$ such that $a \wedge q = O$, $b \wedge q = O$, $a \vee b = I$.

3. Find analogous conditions for X to be normal and regular (a T_3-space and a T_4-space), respectively.

4. Show that a set N is nowhere dense if and only if, whenever S has a nonvoid interior, $S \cap N'$ has a nonvoid interior.

5. Show that the intersection of two "regular" open sets must be regular, but that their set-union need not be.

6. Show that a Boolean algebra is isomorphic with the lattice of all regular open sets of a suitable T_1-space if and only if it is complete.

7. Show that a set S in a topological space X satisfies $S^{-\prime-\prime} = S^{\prime-\prime-}$ if and only if S differs from a "clopen" (closed and open) set by a nowhere dense set.†

*8. Show that the free Boolean algebra with countably many generators is isomorphic with the field of all "clopen" subsets of the Cantor ternary set. (M. H. Stone [**2**, p. 303])

9. (a) Show that, in the T_1-lattice $L(X)$, a point p is a "compact" element if and only if it is "isolated" ($p \not\subset p'^{-}$).

(b) Infer that $L(X)$ is not an algebraic lattice unless X is discrete, so that $L(X) \cong \mathbf{2}^{n(X)}$.

*10. Show that there exist nonhomeomorphic T_0-spaces X and Y with, nevertheless, $L(X) \cong L(Y)$.‡

6. Lattice of Topologies; Arnold's Theorem

Given any two topologies on a set ("space") X, defined by $\bigcap$-rings Γ and Δ of "closed sets", one says that the first topology is *weaker* than the second when $\Gamma \subset \Delta$. The analogous (see §7) relation for convergence is to say that σ is *stronger* than τ when $x_\alpha \to^\sigma a$ implies $x_\alpha \to^\tau a$ (the stronger the topology, the more sets are closed, and the fewer sequences converge).

Since $\bigcap$-rings of subsets of X are *subalgebras* of $\mathbf{2}^X$, regarded as a $\bigcap$-lattice, we see the different *topologies* on a set X form a *closed dual ideal*.§ Moreover, the Hausdorff topologies form a *join-semilattice*, a subset which contains with any topology all stronger topologies.

Example 1. In the weakest T_1-topology on X, the closed sets are just the finite subsets and X itself.

† N. Levine, Amer. Math. Monthly **68** (1961), 474–7.

‡ R. L. Blair, Duke, Math. J. **22** (1955), 271–80.

§ These notions were introduced by the author in Fund. Math. **26** (1936), 156–66, which unfortunately contains several slips corrected here. For further extensions, see E. Hewitt, Duke Math. J. **10** (1943), 309–33.

Example 2. For any pseudo-metric [**Kel**, p. 119] $\delta(x, y)$, define $x_\alpha \to^\delta a$ to mean that $\delta(x_\alpha, a) \to 0$. Then the join of a set Φ of the resulting topologies is defined by the rule $x_\alpha \to^\delta a$ for all $\delta \in \Phi$.

A third example is furnished by the notion of a *Fréchet L-space*, which represents a pioneer attempt to characterize topologies in terms of convergence. This notion is defined by the following sequential analogs of conditions T1–T3 of §3:

F1. If every $x_n = a$, then $x_n \to a$.

F2. If $x_n \to a$ and $\{x_{n(k)}\}$ is an infinite subsequence of $\{x_n\}$, then $x_{n(k)} \to a$.

F3. If $x_n \to a$ and $x_n \to b$, then $a = b$.

Urysohn† has shown that one can extend any definition of convergence satisfying F1–F3 to a *star-convergence* also satisfying F1–F3, as follows:

URYSOHN'S THEOREM. *Let $x_n \to *a$ mean that every subsequence $\{x_{n(k)}\}$ of (x_n) contains a subsequence $\{x_{n(k(i))}\}$ converging to a. Then $x_n \to *a$ satisfies* F1–F3, *and also:*

F4. *If every subsequence of $\{x_n\}$ contains a sequence star-converging to a, then $x_n \to *a$.*

We omit the proof of Urysohn's Theorem, which is only stated for purposes of orientation.‡ It can be generalized to cofinal subsets of nets. Moreover, if $f: X^r \to Y$ is any function of r variables $x_1, \cdots, x_r$ in a Fréchet L-space X to a Fréchet L-space Y, which is *continuous* in the sense that $x_\alpha^j \to x_j$ for $j = 1, \cdots, r$ implies $f(x^1, \cdots, x^r) \to f(x_1, \cdots, x_r)$, then it remains continuous for L^*-convergence (i.e., if $\to$ is replaced by $\to^*$ throughout).§

If, in any Fréchet L-space F, we define a set X to be closed when $\{x_n\} \subset X$ and $x_n \to a$ imply $a \in X$, we evidently make F into a topological space. Since the preceding definition, applied to the relation $x_n \to *a$ of star-convergence, yields exactly the same class of closed sets, it is often said that F and F^* have the same topology. An alternative view is to consider families of "closed sets" and of "convergent nets" on a given space X relative to the following binary relation:

(A) A subset $S \subset X$ is *closed* relative to a convergence statement $x_\alpha \to a$ ($\{x_\alpha\}$ a net)—in symbols, $S\ \rho\ (x_\alpha \to a)$—means that if S contains a cofinal subset of $\{x_\alpha\}$, then S contains a.

One can show that the class of all convergent nets of any topological space enjoys, besides T1–T2, also the following general properties:¶

† P. Urysohn, Enseignement Math. **25** (1926), 77–83.

‡ For recent extensions of Urysohn's Theorem, see J. Kisynski, Colloq. Math. **7** (1959–1960), 205–11; also R. M. Dudley, Trans. AMS **112** (1964), 483–507.

§ This result is contained in Gerald Edgar's Harvard Doctoral Thesis, *Convergence from an algebraic point of view*, 1973.

¶ See G. Birkhoff [**4**], conditions (4ε), (4δ), and B. H. Arnold, Ann. of Math. **54** (1951), 320.

T2′. If from each cofinal subset $\{x_\beta\}$ of $\{x_\alpha\}$ one can extract a set of points which after suitable rearrangement converges to a, then $x_\alpha \to a$.

T3′. If $x^\beta \to a$ and, for each index β, $x^\beta_\alpha \to x^\beta$ for a directed set of indices A_β, then there exists a "diagonal" net $\{x^{\beta(\gamma)}_{\alpha(\gamma)}\}$ with a directed set C of indices γ, which converges to a. (Law of Iterated Limits)

Evidently, condition T2′ is a kind of converse to T2, which is closely related to the idea of star-convergence.

Condition T3′ is related to C2, which states that the closure of X consists of X and its limit points. But the real significance of the preceding conditions is expressed in the following result of Arnold.

ARNOLD'S THEOREM. *For the binary relation ρ of (A), the polars $\Gamma^\dagger$ of families Γ of "convergent" nets are the $\bigcap$-rings of "closed" sets, and the polars Δ^* of families of sets are the families of convergent nets satisfying* T1–T2–T2′–T3′. *Moreover the one-one correspondence $\Delta^* \rightleftharpoons (\Delta^*)^\dagger$ with inverse $\Gamma^\dagger \rightleftharpoons (\Gamma^\dagger)^*$ is that between convergent nets and closed sets in topological spaces.*

Kelly [**1**, §13] has proved a closely related theorem, which refers to subnets instead of cofinal subsets. In both theorems, the precise class of directed sets involved is unclear.

Exercises:

1. Prove in detail that the different T_1-topologies on a set X form a complete lattice, in which O consists of X and all its finite subsets.
2. Prove that the *metrizible* topologies on X form a complete meet-semilattice, in which $(\delta \wedge \delta')(x, y) = \max(\delta(x, y), \delta'(x, y))$.
3. Let M be the set of all metric "distance functions" on X, and let $\delta \leqq \delta'$ mean that $\delta(x, y) \leqq \delta'(x, y)$ for all $x, y \in X$. Show that M is a conditionally complete join-semilattice, dually epimorphic to that of Exercise 2.
4. Show that the g.l.b. of any set of *regular* topologies on a space X is itself regular. (M. J. Norris)†
5. Prove the same result for *completely* regular topologies. (N. Levine).

*6. Show that, among all the locally connected (locally arc-connected) topologies containing a given topology, one is least. (A. Kennison)

7. Bases and Subbases; Compactness

We recall from §V.10 that a complete Brouwerian lattice is a complete lattice in which

L6*. $a \wedge \bigvee x_\sigma = \bigvee (a \wedge x_\sigma)$ for any set of x_σ.

We now develop the theory of such lattices further.

DEFINITION. In a complete Brouwerian lattice L, a *basis* is a subset B of L such that every element of L is a join of elements of B; a *subbasis* is a subset S of

† Proc. AMS **1** (1950), 754–5. For Exercise 5 see Amer. Math. Monthly (1963), 284. For Exercise 6, see Kennison's Harvard Ph. D. Thesis (1963).

L such that every $a \in L$ is a join of *finite* meets of elements of L. The concepts of basis and subbasis are defined dually in the dual of a complete Brouwerian lattice.†

When L is the Brouwerian lattice $\breve{L}(X)$ of all open subsets of a T_1-space X, for example, the preceding definition of a "basis" specializes to the usual notion of a *neighborhood-basis* of open sets. For example, in ordinary Euclidean space, B might consists of all spheres with rational radii whose centers have rational coordinates (a *countable* neighborhood-basis).

Dually, one can define a *basis* of closed sets of a topological space X as any family B of closed sets such that

(B*) Any closed set is an intersection of closed sets of B.

Likewise, a *subbasis* of closed sets is defined as a family G of closed sets, such that every closed set is an intersection of finite unions of members S_α of G. For the real numbers, the closed intervals $[a, b]$ together with the semi-infinite intervals $[a, +\infty)$ and $(-\infty, b]$ form a basis of closed sets in the usual topology.

LEMMA. *A subset S of a complete Brouwerian lattice L is a subbasis of L if and only if S generates L under binary meet and arbitrary join.*

PROOF. Paraphrasing Chapter VI, but admitting the *infinitary* join operation, clearly the subalgebra $\bar{S}$ generated by S contains all joins of finite meets:

$$\bigvee_{\Sigma}\left\{\bigwedge_{F(\sigma)} x_{\sigma i}\right\} \in \bar{S} \text{ if all } x_{\sigma i} \in S, \text{ and } \Sigma \text{ indexes a collection of finite sets } F(\sigma) \text{ of } x_{\sigma i} \quad (i = 1, \cdots, n(\sigma)). \tag{1}$$

Again, by the generalized idempotent law L^*, the set of elements of the form (1) is closed under infinitary join. Finally, the meet of two expressions like (1), with $f_\sigma = \bigwedge_{F(\sigma)} x_{\sigma i}$ and $g_\tau = \bigvee_{G(\tau)} y_{\tau j}$ $(y_{\tau j} \in S)$, also has the form (1) since

$$\cdot\left[\bigvee_{\Sigma} f_\sigma \wedge \bigvee_{T} g_\tau\right] = \bigvee_{\Sigma \times T} (f_\sigma \wedge g_\tau), \tag{2}$$

much as in Chapter V, (4′).

It follows that the intersections of finite unions of any family of sets form a $\bigcap$-ring. Hence any family of sets of a space Q, which contains Q and whose intersection is void, is a subbasis of closed sets for *some* topology on Q.

Compactness. The most important applications of the notion of subbasis are to the problem of "compactification": embedding a given topological space in a compact closure.

A topological space X is called *compact* when, from any family Ψ of open sets S_ψ such that $\bigcup_\Psi S_\psi = X$, one can extract a *finite* subfamily Φ from Ψ such that

† This leads to some ambiguity in complete Brouwerian lattices whose duals are also Brouwerian (e.g., in complete Boolean algebras).

$\bigcup_\Phi S_\phi = X$ (Heine–Borel–Lebesgue (H–B–L) covering property). Clearly, this amounts to asserting that X is a compact element (or, in this case equivalently, by Lemma 3 of §VIII.5, $\uparrow$-inaccessible) in the (algebraic) Brouwerian lattice $\check{L}(X)$ of all open sets of X.

Dually, X is compact when $L(X)$ has the property that $\bigwedge_\Psi y_\psi = O$ in $L(X)$ implies that $\bigwedge_\Phi y_\phi = O$ for some finite subfamily $\Phi \subset \Psi$. This is equivalent to saying that, if the dual ideal $D(\Psi)$ generated by a subset Ψ of $L(X)$ is proper, then $\bigwedge_\Psi y_\psi > O$. For $D(\Psi)$ consists of the elements x such that $x \geqq \bigwedge_\Phi y_\phi$ for some finite subfamily Φ of Ψ.

THEOREM 6. *A T_1-space X is compact if and only if, in $L(X)$, every maximal (proper) dual ideal is principal.*

PROOF. Suppose that, in $L(X)$, every maximal proper dual ideal is principal, and let Ψ be any family of closed sets having the finite intersection property; extend the dual ideal generated by Ψ to a maximal proper dual ideal (this is possible since $L(X)$ has a O; see §VIII.7). This will be principal, and represent a point contained in every set of Ψ, which proves that X is compact. Conversely, if X is a compact T_1-space, then any proper dual ideal H in $L(X)$ represents a family of closed sets with the finite intersection property. If p is any point common to the sets, then the principal dual ideal $J(p)$ is a proper dual ideal containing H; if H is maximal, then $H = J(p)$, which completes the proof.

Local compactness. A T_1-space is called *locally compact* when every point p has a neighborhood whose closure is compact. In a *regular* T_1-space, this amounts to requiring that p be contained in the interior of a regular closed set c (one which is the closure of its interior), which is compact. This gives us the

COROLLARY. *A regular T_1-space is locally compact if and only if, in $L(X)$, every atom is interior to some regular element $c = c^{**}$ such that, in the ideal of $x \leqq c$, every maximal dual ideal is principal.*

8. Alexander and Tychonoff Theorems; Compactification

We now say, dually, that a subset K of a complete Brouwerian lattice L has the *Finite Join Property* when $\bigvee_F x_\alpha < I$ for every *finite* subset F of K. This is obviously a property of finite character. We say that a subset S of L has the *H–B–L Property* if $\bigvee_T x_\alpha = I$ and $T \subset S$ imply that $\bigvee_F x_\alpha = I$ for some finite subset $F \subset T$.

THEOREM 7. *If a subbasis S of a complete Brouwerian lattice L has the H–B–L Property, then so does L.*

PROOF. It suffices to show that if $K \subset L$ has the Finite Join Property, then $\bigvee_K x_\alpha < I$ (i.e., that $\bigvee_K x_\alpha = I$ implies that $\bigvee_F x_\alpha = I$ for some finite $F \subset K$). But any $K \subset L$ with the Finite Join Property can be extended to a *maximal* M with this property, $K \subset M \subset L$, since the property is of finite character. Since S is a subbasis, each $x_\mu \in M$ satisfies $x_\mu = \bigwedge_{F(\mu)} s_{\mu,i}$ of $s_{\mu,i} \in S$.

Moreover $s_{\mu,i} \vee \bigvee_{G(i)} x_{i,j} = I$ for *all* $s_{\mu,i} \in \Gamma(\mu)$ and a suitable finite subset $G(i) \subset M$ would imply, for $u_i = \bigvee_{G(i)} x_{i,j}$,

$$x_\mu \vee u_i = \left\{\bigwedge_{F(\mu)} s_{\mu,i}\right\} \vee u_i = \bigwedge_{F(\mu)} \{s_{\mu,i} \vee u_i\},$$

by L6, where $s_{\mu,i} \vee u_i = I$ by hypothesis. Substituting above, we see that it would imply $x_\mu \vee u_i = x_\mu \vee \bigvee_{G(i)} x_{i,j} = I$. This is impossible, since $x_\mu \in M$, all $x_{i,j} \in G(i) \subset M$, the set of $x_{i,j}$ is finite, and M has the Finite Join Property.

Hence it must be possible, given $x_\mu \in M$, to add *some* element $s_{\mu,i} \in F(\mu)$ to M without destroying the Finite Join Property. Since M is maximal, it follows that this $s_{\mu,i} \in M$, where $s_{\mu,i} \geqq \bigwedge_{F(\mu)} s_{\mu,j} = x_\mu$. That is, for every $x_{\mu,i} \in M$, we can find $s_\mu = s_{\mu,i} \in M \wedge S$ such that $s_\mu \geqq x_\mu$. Since the $s_\mu \in M$, they have the Finite Join Property; since the $s_\mu \in S$ and S has the Heine–Borel property, it follows that $\bigvee_M s_\mu < I$. A fortiori, since $K \subset M$, $\bigvee_K x_\alpha \leqq \bigvee_M x_\mu \leqq \bigvee_M s_\mu < I$, completing the proof.

Theorem 7 has as its most important consequence† the following

COROLLARY (ALEXANDER'S THEOREM). *A T_1-space X is compact if it contains a subbasis Σ of closed sets K_β with the following compactness property:*

(C) *If $\bigcap_\Psi K_\psi = \varnothing$ for some set $\Psi \subset \Sigma$, then*
$\bigcap_F K_\phi = \varnothing$ for some finite subset $F \subset \Psi$.

PROOF. By hypothesis, $\breve{L}(X)$ has a subbasis S (consisting of the K'_β) with the H–B–L property. Hence $\breve{L}(X)$ has the H–B–L property, by Theorem 7. Dualizing again, we see that X is compact.

DEFINITION. The *topological product* $P = \prod X_\alpha$ of a family of topological spaces X_α is the Cartesian product of the X_α, with the Cartesian products $\prod C_\alpha$ of closed sets of the X_α for a *subbasis* of closed sets.

Example 3. For any cardinal number $\aleph$, the *Cantor $\aleph$-space* is the topological product of $\aleph$ copies of the discrete space of two points. The Cantor $\aleph_0$-space is homeomorphic to the *Cantor ternary set* in $[0, 1] \subset \mathbf{R}$, consisting of all numbers whose ternary decimal expansion consists of 0's and 1's alone; it is thus metrizible.

Example 4. The topological product $[0, 1]^\omega$ of countable copies of the interval $[0, 1]$ is called the *Hilbert cube*. It is a universal separable metric space.

In general, a net $\{\mathbf{x}_\beta\} \subset \prod X_\alpha$ converges to the limit $\mathbf{c}$ when, for each α, the α-components $x_{\beta,\alpha} \to c_\alpha$. This conforms to the idea, mentioned in §1, that a topological space is an infinitary partial algebra under suitable operations of convergence.

From Alexander's Theorem, one gets as an almost immediate consequence the following important result.

THEOREM 8 (TYCHONOFF). *Any product of compact topological spaces is itself compact.*

† This generalizes a classic theorem of J. W. Alexander, Proc. Nat. Acad. Sci. **25** (1939), 52–4 and 296–8, see also Ann. of Math. **39** (1938), 883–912.

Caution. Though the product of any two locally compact topological spaces is locally compact, the product of countably many copies of the real line L is not locally compact, even though L is.

When X is *compact*, Theorem 2 can be considerably extended as follows.

THEOREM 9. *Any compact T_1-space X is determined to within homeomorphism by any ring B of closed subsets of X which constitutes a basis of closed sets.*

PROOF. Each point of X corresponds to a maximal dual ideal in B, as in §5. Conversely, let M be any maximal (proper) dual ideal in B. Being proper, M must have the finite intersection property; indeed, one easily shows that to be a maximal dual ideal of a lattice is *equivalent* to being maximal subject to the finite intersection property. Since X is compact, this implies that there is a point $p \in X$ common to all (closed) subsets of M. But there cannot be more than one such point or M would not be maximal: given $p \neq q$, B (being a basis) must contain a closed set which contains p but not q, and one could adjoin this set to M without destroying the finite intersection property.

THEOREM 10 (WALLMAN†). *Let L be any distributive lattice with the Disjunction Property (W) of §5. Then L is isomorphic with a sublattice of the lattice $L(X)$ of all closed subsets of a compact T_1-space X.*

PROOF. We take as the points of X the maximal dual ideals of L. We take for a basis B of closed subsets of X, the sets $S(a)$ consisting of those points $p = p(M)$ such that $a \in M$. The following facts are easily verified: (i) $S(a \wedge b) = S(a) \cap S(b)$, (ii) $S(a \vee b) = S(a) \cup S(b)$, (iii) the $S(a)$ have the compactness property. Indeed, (i) and (ii) are general properties of prime ideals in distributive lattices, and every maximal ideal is prime. The Disjunction Property shows that X, so defined, is a T_1-space.

To prove (iii), let J be any family of sets of B having the finite intersection property; and let M be the extension of the corresponding $a > 0$ in L to a maximal proper dual ideal M of L. Such an extension exists, since being a proper dual ideal in a lattice with 0 is a finitary property (Chapter VIII). Clearly $p(M) \in S(a)$ for any $a \in M$—hence the $S \in J$ have a nonvoid intersection. This completes the proof of Theorem 10.

Now let L be the distributive lattice defined by the subsets of any *basis* of closed subsets of a (normal) T_1-space X which is also a *ring* of subsets of X. We may then use Theorem 10 to construct a compact space X^* from L, containing X as a dense subset. This will be the Wallman-Cech compactification of X; X^* has the same (Cech) dimension and homology groups as X.‡

Exercises for §§7–8:

1. Show that Wallman's Disjunction Condition (W) is equivalent to the condition that $s > t$ implies $s^* > t^*$ in $\check{L}(X)$.

† H. Wallman, Ann. of Math. **39** (1938), 112–26; see also P. Samuel, Trans. AMS **94** (1948), 100–32.

‡ O. Frink, Amer. J. Math. **86** (1964), 602–7.

2. Show that if a topological space is regular, then its regular open sets constitute a neighborhood-basis.

3. Let $\mathcal{N}$ and $\mathcal{U}$ be open neighborhood-bases for topologies on a space X. Show that the topology defined by $\mathcal{N}$ is weaker than that defined by $\mathcal{U}$ if and only if, given $N \in \mathcal{N}$ and $x \in N$, there exists $U \in \mathcal{U}$ with $x \in U \subset N$.

4. (a) Show that every compact subset of a T_2-space is closed.

(b) Show that every closed subset of a compact topological space is compact.

5. Show that every compact Hausdorff space is normal.

*6. Show that a normal or completely regular topology on a set X is compact if and only if it is minimal in the poset of all such topologies on X. (M. P. Berri)†

7. Show that the continuous images ("epimorphs") of any compact Hausdorff space X are defined by its partitions into (disjoint) closed subsets ("subalgebras").

8. Show that a Hausdorff space X is compact if and only if every net in X has a convergent subnet. (J. L. Kelley)

9. Let $X = \prod X_\iota$ (Cartesian product). Prove that $x_\alpha \to a$ in X if and only if $x_{\alpha,\iota} \to a_\iota$ in every X_ι.

*10. Define a net $\{x_\alpha\}$ in a set X to be *universal* when, for any $S \subset X$, the net is ultimately (for all $\alpha \geqq \sigma(S)$) either in S or in S'. Prove that every net has a universal subnet. (J. L. Kelley)

*11. From a topological space X, form a set $X^* = X \cup \{\infty\}$ whose open sets are the open subsets of X and the subsets Y whose complement in X^* is a closed compact subset of X. Show that X^* is compact, and a Hausdorff space if and only if X is a locally compact Hausdorff space. (P. Alexandroff)

12. Show that, in the poset of all completely regular topologies on a set X, an element is minimal if and only if the topology is compact.‡

13. Show that the product of any two locally compact topological spaces is itself locally compact.

9. Stone Representation Theorem

By Theorem 9, a compact T_1-space X is determined up to homeomorphism by any *ring* of closed sets which forms a basis, considered as a distributive lattice. On the other hand, a closed subset of X has a closed complement if and only if it disconnects X. Hence X has a *field* of closed sets which forms a basis if and only if it is *totally disconnected*, in the sense that any two distinct points lie in complementary closed sets.

Conversely, the set of *all* open-and-closed subsets of any totally disconnected compact T_1-space forms a field of closed sets which is a basis. This shows that the discussion of §8 leads to Boolean algebras in the case of totally disconnected spaces. We now characterize these spaces in another way.

LEMMA. *A compact space X is totally disconnected if and only if it is zero-dimensional.*

PROOF. Let X be totally disconnected, let $p \in X$ be given, and let $U(p)$ be any open set which contains p. For each $q \in U'$, there exists an open-and-closed set $S(q)$ containing q but not p. But U' is compact (since X is); hence some finite union $S(q_1) \cup \cdots \cup S(q_n)$ must contain U'. Its complement $V = \bigcap S'(q_i)$ will be

† Trans. AMS **108** (1963), 97–105.

‡ J. Wada, Osaka Math. J. **5** (1953), 1–12.

a neighborhood of p contained in U and having a void boundary. The existence of such a V for every $U(p)$ defines X as zero-dimensional. Conversely, any zero-dimensional space is trivially totally disconnected (and Hausdorff).

Any Boolean algebra trivially has Wallman's Disjunction Property (W); hence (Theorem 10) it can be regarded as a basis of closed sets of a (totally disconnected) compact T_1-space; the converse has been noted above. This proves the major part of the following classic result of Stone [**2**, Theorem 4].

THEOREM 11. *There is a one-one correspondence between Boolean algebras A and totally disconnected (i.e., zero-dimensional) compact T_2-spaces X, under which the elements of A correspond to the open-and-closed subsets of X, and the points of X to the prime ($=$ maximal) ideals of A.*

Because of this result, Stone called zero-dimensional compact T_2-spaces "Boolean spaces." However, to honor Stone's own fundamental work (Stone [**1**], [**2**]) on the subject, the "Boolean" space associated with a given Boolean algebra A is commonly called the *Stone space* of A, and Theorem 11 is called the Stone representation theorem.

The preceding considerations can be extended to zero-dimensional locally compact T_2-spaces. In any such space X, the *compact open* sets form a generalized Boolean algebra (§II.12). It has also been extended to compact T_0-spaces.†

10. Lattices of Continuous Functions

Given a *completely regular* T_1-space X, by definition there exists for any neighborhood U of any point $x \in X$ a function $f: X \to [0, 1]$ such that $f(x) = 0$ and $f \equiv 1$ on U', the complement of U. The set of all real continuous functions on a given completely regular T_1-space X forms a *lattice* $C(X)$ under the usual definition, that $f \leqq g$ in $C(X)$ means $f(x) \leqq g(x)$ for all $x \in X$. The set $C(X)$ is also a *ring* under the usual definitions of addition and multiplication, and this ring has been very intensively studied;‡ we shall study $C(X)$ as a lattice-ordered ring in Chapter XVII (and as a vector lattice in Chapter XV).

We note here only one, purely lattice-theoretic property of the $C(X)$.

THEOREM 12 (KAPLANSKY§). *Any compact Hausdorff space K is determined up to homeomorphism by the lattice $C(K)$ of its continuous functions.*

SKETCH OF PROOF. We shall say that a prime ideal $P \subset C(K)$ is *associated* with a point $x \in K$, when $f \in P$ and $g(x) < f(x)$ imply $g \in P$. The proof depends on the following three lemmas.

LEMMA 1. *Each prime ideal P is associated with one and only one point of K.*

LEMMA 2. *Two prime ideals P, Q are associated with the same point $x \in K$, if and only if $P \wedge Q$ contains a prime ideal.*

† Stone [**3**] and L. Rieger, Cas. Mat. Fys. **74** (1949), 56–61.

‡ See L. Gillman, and M. Jerison, *Rings of continuous functions*, van Nostrand, 1960.

§ I. Kaplansky, Bull. AMS **53** (1947), 617–22.

LEMMA 3. *Let f_0 be a fixed function in $C(K)$ and S a subset of K. Then a point x is in the closure $\bar{S}$ of S if and only if some prime ideal $P(x)$ associated with x contains the intersection $A(S)$ of the prime ideals containing f_0 which are associated with points of S.*

Granted these three lemmas (proved in [**LT2**, p. 175]), call two prime ideals in $C(K)$ "equivalent" when their intersection contains a third prime ideal. Lemmas 1–2 show that the classes of "equivalent" prime ideals may be interpreted as points of K. Lemma 3 shows how to describe the topology of K in terms of inclusion relations among these prime ideals.

Kaplansky's Theorem has been extended to noncompact spaces by Shirota and Henriksen.†

Exercises for §§9–10:

1. Show that any Boolean space is normal.
2. Characterize (up to Boolean isomorphism) the Boolean algebra of all regular open subsets of Hilbert space in (a) the "strong" and (b) the "weak" topology.
3. Call a topological space "extremally disconnected" when the closure of *any* open set is open. Prove that a "Boolean" space is the Stone space of a *complete* Boolean algebra if and only if it is extremally disconnected.‡
4. Show that, in an extremally disconnected compact space, an open set is regular if and only if it is also closed ("clopen"). (O. Frink)
5. Show that, if A is a Boolean algebra and $S(A)$ its Stone space, then Aut A is isomorphic to the group of homeomorphisms of $S(A)$. (M. H. Stone)
6. Let B be the complete Brouwerian lattice of all closed subsets of $\mathbf{Q}$, in its natural topology. Show that the center of B consists of all "clopen" subsets of $\mathbf{Q}$, and is *not* complete.
7. Show that, if the lattices of nonnegative upper semicontinuous functions on two completely regular topological spaces X and Y are isomorphic, then X and Y are homeomorphic.§
8. Show that the completion by cuts of the (vector) lattice $C_b(X)$ of all bounded continuous functions on a regular T_1-space X is isomorphic to $C(K)$, where K is the Stone space of the Boolean algebra of all regular open sets of X.

PROBLEMS

72. Find necessary and sufficient conditions on a Boolean lattice for it to be isomorphic with the lattice of all "regular" open sets of some metric space.¶

73. (a) Obtain conditions on the convergence of nets, necessary and sufficient for polarity to neighborhoods in a Hausdorff space.

(b) Same question for convergence of sequences in a metric space.

74. What is the relation between the continuity of a function $f(x, y)$ on the product $X \times Y$ of two Hausdorff spaces, and the condition that $x_\alpha \to x$ and $y_\alpha \to y$ imply $f(x_\alpha, y_\alpha) \to f(x, y)$ for nets?

---

† T. Shirota, Osaka Math. J. **4** (1952), 121–32; M. Henriksen, Proc. AMS **7** (1956), 959–60.

‡ A. Gleason, Illinois J. Math. **2** (1958), 482–9.

§ J. Nagata, Osaka Math. J. **1** (1949), 166–81. For Exercise 8, see R. P. Dilworth, Trans. AMS **68** (1950), 427–38; A. Horn, Pacific J. Math. **3** (1953), 143–52; R. S. Pierce, Canadian J. Math. **5** (1953), 95–100.

¶ This may be related to the Moore problem: is every normal Moore space metrizible. (For this problem, see R. H. Bing, Proc. AMS **16** (1965), 612–19.

75. How can one best represent an arbitrary complete Brouwerian lattice by sets? In this representation, to what do "compact" elements of the lattice correspond?

76. Find necessary and sufficient conditions on a Brouwerian lattice for it to be isomorphic with the lattice of all closed elements in a suitable closure algebra.† Does the class of all such lattices constitute a family?

77. Extend topological dimension theory to closure algebras. (Cf. [**LT2**, Problem 83].)

78. If one quasi-orders directed sets ("cofinal order types") by monomorphism, is the resulting poset (§II.1) directed? a complete lattice?‡

79. (a) Solve the word problem for the free Brouwerian lattice with n generators (especially for $n = 1, 2$).

(b) Same problem for the free closure algebra with n generators.

80. Is the lattice of all T_0-topologies on an uncountable set complemented?§

---

† See Mc Kinsey–Tarski, Ann. of Math. **45** (1944), esp. p. 146.

‡ J. W. Tukey, "Convergence and uniformity in topology", Ann. of Math. Study **2** (1940), p. 15. See J. R. Isbell, Trans. AMS **116** (1965), 394–416.

§ For countable sets, see H. Gaifman, Canad. J. Math. **18** (1966), 83–8. Added in proof: Solved affirmatively by A. K. Steiner, Trans. AMS **122** (1966), 379–98.

CHAPTER X

METRIC AND TOPOLOGICAL LATTICES

1. Valuations; Quasi-metric Lattices

Many of the most important applications of lattices to mathematics involve limiting processes like those of real analysis. Such processes can be defined in many ways, as will be shown in this chapter. The simplest way is in terms of "valuations", as defined below.

DEFINITION. By a *valuation* on a lattice L is meant a real-valued function (functional) $v[x]$ on L which satisfies

V1. $$v[x] + v[y] = v[x \vee y] + v[x \wedge y].$$

A valuation is *isotone* if and only if

V2. $$x \geqq y \quad \text{implies} \quad v[x] \geqq v[y]$$

and *positive* if and only if $x > y$ implies $v[x] > v[y]$.

Example 1. In any modular lattice of finite length, the *height* function $h[x]$ is a positive valuation, by §II.8, (21).

Example 2. In any real finite-dimensional vector space $\mathbf{R}^n$, lattice-ordered by letting $(x_1, \cdots, x_n) \leqq (y_1, \cdots, y_n)$ mean that $x_k \leqq y_k$ for all k, any *linear functional* $c[x] = c_1x_1 + \cdots + c_nx_n$ is a valuation. This valuation is positive if and only if all c_k are positive.

For applied mathematics, the most important valuations on lattices are *measure* and *probability* functions on fields of sets. These will be studied in greater depth in Chapter XI; the present chapter will treat primarily the general properties of limiting processes in lattices. From this standpoint, the first thing to observe is the relation between valuations and metric spaces.

THEOREM 1. *In any lattice L with an isotone valuation, the distance function*

(1) $$d(x, y) = v[x \vee y] - v[x \wedge y]$$

satisfies for all $x, y, z, a \in L$:

(2) $$d(x, x) = 0, \qquad d(x, y) \geqq 0, \qquad d(x, y) = d(y, x),$$

(3) $$d(x, y) + d(y, z) \geqq d(x, z) \qquad (\textit{triangle inequality}),$$

(4) $$d(a \vee x, a \vee y) + d(a \wedge x, a \wedge y) \leqq d(x, y).$$

PROOF. Using L1, the fact that v is isotone, and L2, respectively, we get the three relations of (2). We next prove (4). By definition, the left-hand of (4) is

$$v[a \vee x \vee y] - v[(a \vee x) \wedge (a \vee y)] + v[(a \wedge x) \vee (a \wedge y)] - v[a \wedge x \wedge y].$$

By the one-sided distributive law, this is at most

$$v[a \vee x \vee y] - v[a \vee (x \wedge y)] + v[a \wedge (x \vee y)] - v[a \wedge x \wedge y].$$

Transposing middle terms and using V1 twice, we see that this equals

$$v[a] + v[x \vee y] - v[a] - v[x \wedge y] = d(x, y),$$

which proves (4).

Finally, using (4), we can prove (3). In fact,

$$d(x, y) + d(y, z) = d(x \vee y, y) + d(y, x \wedge y) + d(y \vee z, y) + d(y, y \wedge z)$$

$$\geqq d(x \vee y \vee z, y \vee z) + d(y \vee z, y) + d(y, x \wedge y) + d(x \wedge y, x \wedge y \wedge z),$$

since by (4) $d(x \vee y \vee z, y \vee z) \leqq d(x \vee y, y)$ and $d(x \wedge y, x \wedge y \wedge z) \leqq d(y, y \wedge z)$. But the last sum is

$$d(x \vee y \vee z, x \wedge y \wedge z) \geqq d(x \vee y, x \wedge y) = d(x, y),$$

proving (3).

Any set M on which is defined a distance or metric $d(x, y)$ satisfying (2) and (3) is called a *pseudo-metric* (or quasi-metric) space; hence a lattice with isotone valuation is called a *pseudo-metric* (or quasi-metric) lattice.

The following test for a functional on a relatively complemented lattice to be a valuation is often convenient.

LEMMA. *A real-valued functional* $v[x]$ *on a relatively complemented lattice with* O *is a valuation provided*

V1*. $\quad v[x \vee y] = v[x] + v[y]$ *whenever* $x \wedge y = O$.

PROOF. For any x, y, let t be a relative complement of $x \wedge y$ in $[O, y]$. By definition, $(x \wedge y) \wedge t = O$ and $(x \wedge y) \vee t = y$; hence $v[y] = v[x \wedge y] + v[t]$. Moreover, since $t \leqq y$, $x \wedge t = x \wedge (y \wedge t) = (x \wedge y) \wedge t = O$, while

$$x \vee t = [x \vee (x \wedge y)] \vee t = x \vee [(x \wedge y) \vee t] = x \vee y.$$

Hence $v[x \vee y] = v[x] + v[t]$. Subtracting the two equations,

$$v[x \vee y] - v[y] = v[x] + v[t] - v[x \wedge y] - v[t],$$

proving V1.

2. Metric Lattices; Metric Completion

If $d(x, y) = 0$ implies $x = y$ in a pseudo-metric space M, then M is by definition a metric space. In a pseudo-metric lattice, this condition is equivalent by V1–V2 to the condition that $x \vee y > x \wedge y$ should imply $v[x \vee y] > v[x \wedge y]$. Hence a pseudo-metric lattice yields a *metric* space under (1) if and only if the valuation which defines it is *positive*. Lattices with a positive valuation are therefore called *metric lattices*. A basic property of metric lattices is the following converse of Example 1 of §1.

THEOREM 2. *Any metric lattice is modular.*

PROOF. If $x \leqq z$, using V1 repeatedly, we have

$$\begin{aligned} v[x \vee (y \wedge z)] &- v[(x \vee y) \wedge z] \\ &= v[x] - v[x \wedge y] + v[y \wedge z] + v[y \vee z] - v[x \vee y] - v[z] \\ &= v[x] - v[x \wedge y] - v[x \vee y] + v[y \wedge z] + v[y \vee z] - v[z] \\ &= -v[y] + v[y] = 0. \end{aligned}$$

We always have $x \vee (y \wedge z) \leqq (x \vee y) \wedge z$, and since v is positive, we have proved $x \vee (y \wedge z) = (x \vee y) \wedge z$. Q.E.D.

A simpler, less direct proof can be given, using the fact that any nonmodular L, must contain the five-element sublattice N_5 sketched below. Since $d = a \wedge b = a \wedge c$ and $e = a \vee b = a \vee c$ we have $v[a] + v[b] = v[e] + v[d] = v[a] + v[c]$, and hence $v[b] = v[c]$.

(The equations $v[a] = v[b] = v[c] = \frac{1}{2}$, $v[e] = 1$, $v[d] = 0$ give however a nontrivial isotone valuation on N_5.)

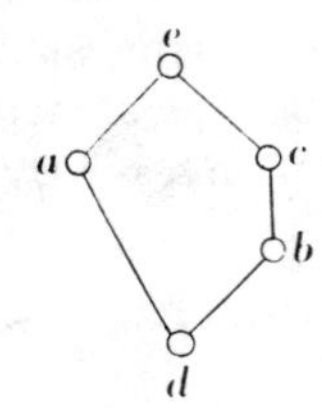

FIGURE 21

The preceding argument can be generalized as follows.

THEOREM 3. *In a metric lattice, no interval can be projective with a proper part of itself.*

PROOF. In a lattice with valuation, we first define the *value* of any interval $[a, b]$ to be $v[b] - v[a]$.

LEMMA. *In any lattice with valuation, all projective intervals have the same value.*

For, by V1, this is true of transposed intervals, and equality is transitive. It is a corollary that, if $[a, b]$ and $[c, d]$ are projective, then

(*) $$v[b] - v[a] = v[d] - v[c] \quad \text{under } any \text{ valuation.}$$

If also $[a, b] \subset [c, d]$, so that $c \leqq a \leqq b \leqq d$, then for any positive valuation $v[c] \leqq v[a]$ and $v[b] \leqq v[d]$, which is only compatible with (*) if all the preceding inequalities and inclusions are equalities, proving the theorem.

Given a pseudo-metric space M, there is a standard method of constructing from it a metric space in such a way that the metric properties are preserved; it goes as follows. Define a relation $x \sim y$ on M by the condition $d(x, y) = 0$. Then (2) shows that this relation is symmetric and reflexive, while (3) and $d(x, z) \geqq 0$ show it is transitive; hence it is an equivalence relation. Let M^* denote the set of equivalence classes of M and, if $x \in M$, let x^* denote its equivalence class. Define $d(x^*, y^*) = d(x, y)$. By (3), equivalent pairs of elements of M are equal distances apart; hence the mapping $M \to M^*$ given by $x \to x^*$ is isometric, and (2)–(3) hold in M^*. But $x^* \neq y^*$ implies $0 < d[x, y] = d[x^*, y^*]$, so that M^* is in fact a metric space.

We now apply the preceding construction to an arbitrary pseudo-metric lattice L.

THEOREM 4. *Any pseudo-metric lattice L is a pseudo-metric space, in which joins and meets are uniformly continuous. The relation $d[x, y] = 0$ is a congruence relation, mapping L isometrically and lattice-epimorphically onto a metric lattice.*

PROOF. It only remains to prove uniform continuity of meets and joins and that $d[x, y] = 0$ is a congruence relation on L. We shall prove uniform continuity of meets and joins simultaneously, in the sharp form

$$
\begin{aligned}
&d(a \vee b, c \vee d) + d(a \wedge b, c \wedge d) \\
(5)\qquad &\leqq d(a \vee b, c \vee b) + d(c \vee b, c \vee d) \\
&\quad + d(a \wedge b, c \wedge b) + d(c \wedge b, c \wedge d) \quad \text{by (3)} \\
&\leqq d(a, c) + d(b, d),
\end{aligned}
$$

applying (4) to terms one and three, and to two and four. Now (5) shows that the relation $d(x, y) = 0$ has the substitution property for meets and joins; hence it is a congruence relation modulo under which L has a lattice-epimorphic image. This epimorphic image is just the metric space L^* constructed from L as a pseudo-metric space.

Again, if M is any metric space, we may define

$$
(6)\qquad d(\{x_n\}, \{y_n\}) = \lim_{n \to \infty} d(x_n, y_n)
$$

for Cauchy sequences $\{x_n\}$, $\{y_n\}$, thus converting the set of all Cauchy sequences in M into a pseudo-metric space. Applying to this the construction preceding Theorem 4, we get a metric space $\bar{M}$ which is *complete*. Mapping x onto the equivalence class of $x, x, \cdots$ carries M isometrically onto a dense subset $\bar{M}$, and $\bar{M}$ is determined up to isometry by this fact and its completeness. If M is a metric lattice, and if $\{x_n\}$ and $\{y_n\}$ are Cauchy sequences, then $\{x_n \vee y_n\}$ and $\{x_n \wedge y_n\}$ will also be Cauchy sequences by (5); define them as $\{x_n\} \vee \{y_n\}$ and $\{x_n\} \wedge \{y_n\}$, respectively. Then L1–L4 can be proved by passage to the limit. Moreover $\{v[x_n]\}$ will converge to a limit which we can define as $v[\{x_n\}]$, and then prove that it is an isotone valuation. This makes $d(\{x_n\}, \{y_n\})$ as defined above the distance between two Cauchy sequences. Hence we obtain almost immediately from Theorem 4,

THEOREM 5. *Any metric lattice M has a unique metrically complete hull, in which it is (metrically) dense.*

The hull is also a conditionally complete lattice, as will be shown below in §10.

Exercises for §§1–2 (see also §III.12, Exercise 5):

1. Show that every real functional on a chain satisfies V1.
2. Show that, in a metric distributive lattice, $d(x, y) + d(y, z) = d(x, z)$ if and only if $y \in [x \wedge z, x \vee z]$. (Pitcher–Smiley)
3. (a) Let $v[x]$ and $w[y]$ be positive valuations on lattices X resp. Y. Show that the sum $v[x] + w[y]$ defines a positive valuation on XY.
   (b) Given a bounded positive valuation on a lattice X, show how to construct one on X^ω (ω countable infinity).
4. Let $v: L \to G$ be a "valuation" satisfying V1–V2, where L is a lattice and G is a (simply) ordered commutative group. Prove analogs of Theorems 1–2. Show just where the hypothesis of commutativity is used.
5. In two-dimensional space-time, with $r = t + x$, $s = t - x$, the usual partial ordering is to set $(x, t) \leqq (x_1, t)$ if and only if $r \leqq r_1$ and $s \leqq s_1$. Show that $v(x, t)$ satisfies the wave equation $v_{tt} = v_{xx}$ if and only if it is a valuation with respect to the lattice defined by the above partial ordering.

3. Distributive Valuation

We shall now characterize these valuations which determine *distributive* metric lattices.

This is easy: since $x \vee (y \wedge z) \leqq (x \vee y) \wedge (x \vee z)$ in any lattice, a metric lattice is distributive if and only if $v[x \vee (y \wedge z)] = v[(x \vee y) \wedge (x \vee z)]$ identically. But in any metric lattice, we have by V1

$$v[x \vee (y \wedge z)] = v[x] + v[y] + v[z] - v[y \vee z] - v[x \wedge y \wedge z],$$
$$v[(x \vee y) \wedge (x \vee z)] = v[x \vee y] + v[x \vee z] - v[x \vee y \vee z].$$

Substituting and transposing, we get the equivalent symmetric condition

$$v[x \vee y \vee z] - v[x \wedge y \wedge z] = v[x \vee y] + v[y \vee z] + v[x \vee z] - v[x] - v[y] - v[z].$$

This condition is not self-dual, but by V1 again,

$$\begin{aligned} v[x \vee y] + v[y \vee z] + v[x \vee z] - v[x] - v[y] - v[z] \\ = v[x] + v[y] + v[z] - v[x \wedge y] - v[y \wedge z] - v[x \wedge z]. \end{aligned}$$

Hence an equivalent self-dual symmetric condition is

$$(7) \quad \begin{aligned} 2\{v[x \vee y \vee z] - v[x \wedge y \wedge z]\} = v[x \vee y] + v[y \vee z] + v[x \vee z] \\ - v[x \wedge y] - v[y \wedge z] - v[x \wedge z]. \end{aligned}$$

Valuations satisfying (7) will be called *distributive* valuations. We conclude

THEOREM 6. *A metric lattice is distributive if and only if its valuation is distributive.*

Conversely, we note that *every distributive lattice L has a nontrivial isotone valuation.* For, let $\theta: x \to S(x)$ be the isomorphic representation of L by a ring of sets, constructed in Chapter VIII. For each point p, the function

(8)
$$v_p[x] = 1 \quad \text{if } p \in S(x)$$
$$= 0 \quad \text{otherwise}$$

is a nontrivial isotone valuation on L; cf. Exercise 4.

It is *not* true, however, that every modular lattice has a nontrivial isotone valuation. A counterexample will be given below (Exercise 4); cf. Example 3, p. 75.

4. Valuations on Modular Lattices

Using the methods of §II.8, we shall now determine all possible valuations on any modular lattice of finite length. Call an interval $[a, b]$ in a lattice *prime* when b covers a. By the lemma of §2, projective intervals have the same value. But the relation of projectivity between intervals is an equivalence relation. There follows

LEMMA 1. *Each valuation assigns a unique value λ_p to each class of projective prime intervals.*

Moreover, if $\gamma: O = x_0 < \cdots < x_n = x$ is any chain connecting 0, to x then $v[x] = v[O] + \sum v[x_{i-1}, x_i]$. Hence, if $p[x, \gamma]$ denotes the number of occurrences of a prime interval projective to p in γ, we have

(9)
$$v[x] = v[O] + \sum \gamma_p p[x, \gamma].$$

We shall now prove that $p[x, \gamma]$ is the same for all γ connecting O to x, the proof being by induction on the length of connecting chains, just like that of the Jordan-Hölder Theorem (§II.8).

Let $P(m)$ be the proposition that our assertion holds for all x connected to O by a chain of length m. Then $P(1)$ is obvious, so we suppose $P(m)$ and prove $P(m + 1)$. Suppose $\gamma: O = x_0 < \cdots < x_{m+1} = x$ and $\gamma': O = y_0 < \cdots < y_{m+1} = x$ are connected chains. Since the lattice is modular, and x covers y_m and x_m, x_m and y_m

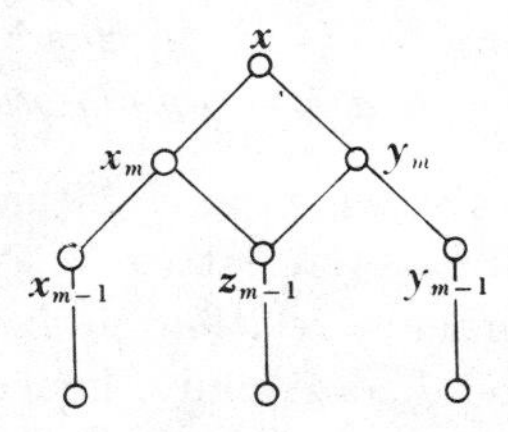

Figure 22

both cover $x_m \wedge y_m = z_{m-1}$. Now we can connect O to z_{m-1} by a chain $\gamma'': O = z_0 < \cdots < z_{m-1}$. Then the chains $\gamma_1: O = x_0 < \cdots < x_m$ and $\gamma_2: O = z_0 < \cdots < z_{m-1} < x_m$ connect O to x_m while $\gamma_1': O = y_0 < \cdots < y_m$ and $\gamma_2': O = z_0 < \cdots < z_{m-1} < y_m$ connect O to y_m. By $P(m)$, $p[x_m, \gamma_1] = p[x_m, \gamma_2]$ and $p[y_m, \gamma_1'] = p[y_m, \gamma_2']$. By noting that $[x_m, x]$ and $[z_{m-1}, y_{m-1}]$ are projective as are $[y_m, x]$ and $[z_{m-1}, x_m]$, and by using the obvious relations of $p[z_{m-1}, \gamma'']$ to $p[x_m, \gamma_2]$ and

$p[y_m, \gamma_2']$ (depending on the projectivity classes of $[z_{m-1}, x_m]$ and $[z_{m-1}, y_m]$ respectively), a simple calculation shows $p[x, \gamma] = p[x, \gamma']$.

We thus define $p[x]$ to be the common value of the $p[x, \gamma]$. Since L is modular, $[x, x \vee y]$ and $[x \wedge y, y]$ are isomorphic, for any x, y. Hence

$$p[x \vee y] - p[x] = p[y] - p[x \wedge y], \tag{10}$$

and we have

LEMMA 2. *In a modular lattice of finite length, if $p[x]$ denotes the number of prime intervals projective to p in some (and hence every) chain connecting 0 to x, then $p[x]$ is a valuation.*

Further, trivially,

LEMMA 3. *Any linear combination $\lambda_0 + \sum \lambda_k v_k[x]$ of valuations is itself a valuation.*

It is a corollary that (9) represents a valuation for any choice of the coefficients λ_p. We summarize in

THEOREM 7. *The different valuations on a modular lattice L of finite length correspond one-one to the choices of $v[0]$ and of the value λ_p assigned to the classes of projective prime intervals of L:*

$$v[x] = v[0] + \sum \lambda_p p[x],$$

where $p[x]$ is the number of prime intervals projective to p in any maximal chain joining 0 with x.

By combining Theorem 7 with the results of §III.10, one can describe completely the structure of finite-dimensional modular lattices.

THEOREM 8. *Let L be any modular lattice of finite length. The congruence relations on L correspond one-one to the sets of classes of projective prime intervals, which they annul. Hence they form a Boolean algebra.*

(Cf. Funayama's Theorem VI.9, which asserts that the congruence relations on any lattice form a complete Brouwerian lattice.)

PROOF. Let θ be any congruence relation on L. If $x \equiv x \vee y \pmod{\theta}$, then $x \wedge y \equiv (x \vee y) \wedge y = y \pmod{\theta}$ and dually; hence if θ annuls a (prime) interval, it annuls all projective (prime) intervals (we use the fact that θ is transitive). Conversely, let $S = S(\theta)$ be any set of equivalence classes of projective prime intervals; define $v_S[x]$ as the sum of the $p[x]$ for p not in S. By Theorem 6, this is a valuation; it is clearly isotone. By Theorem 4, this valuation defines a congruence relation $\theta(S)$. Moreover, $x \equiv y \pmod{\theta}$ if and only if every prime interval in any maximal chain joining $x \wedge y$ with $x \vee y$ is in S. Hence θ is determined by $S(\theta)$, and the correspondence $\theta \rightarrow S(\theta)$ is one-one. It is, moreover, clearly isotone; hence it is an isomorphism.

COROLLARY 1. *A modular lattice of finite length is "simple" (i.e., without proper congruence relation) if and only if all its prime intervals are projective.*

We now exhibit a (simple, complemented) modular lattice of infinite length which has no nontrivial valuation: the lattice $L(Z_2^\omega)$ of all subgroups of the direct product $G = Z_2^\omega$ of countably many copies of the cyclic group Z_2 of order two. In G, *any* two subgroups S and T with infinite order *and* index are projective; we omit the proof. Moreover in G, we can form $1 < S < U < G$ with $S/1$, U/S, and G/U all infinite. By the lemma of §2, these facts imply

$$v[1, S] = v[1, U] = v[1, S] + v[S, U] = 2v[1, S],$$

whence $v[1, S] = 0$ for any subgroup S with infinite order and index. The conclusion that $v[S] = v[1] = \text{const.}$ for *all* subgroups S of G now follows easily.

Exercises for §§3–4:

1. Generalize Theorem 8 to modular lattices in which all bounded chains are finite.
2. Generalize Theorem 8 to "valuations" with values in a commutative group.
3. Let $L(F^\omega)$ be the atomic complemented modular lattice of all subspaces of F^ω, a countable-dimensional vector space. Let $[O, a]$, $[a, I]$, $[O, b]$, $[b, I]$ all have infinite length.
   (i) Show that $[O, a]$ and $[O, b]$ are projective.
   (ii) Infer that $L(F^\omega)$ has no nontrivial valuation. (*Hint.* See Theorem 3.)

*4. Deduce a lattice identity valid in every metric lattice or subdirect product of such, but not in $L(F^\omega)$. (A. Lehman)

5. Show that any finite metric distributive lattice is isomorphic to a ring of sets, under an isomorphism which makes $v[x]$ equal to the measure of the set corresponding to x.

6. Show that a lattice L is distributive if and only if, given $a < b$ in L, there exists a *distributive* valuation v on L with $v[a] < v[b]$. (D. Vaida†)

*7. Find necessary and sufficient conditions for a metric space to be isometric to a metric lattice.‡

*8. Let A be a semigroup under $\vee$, which also has a unary operation $'$ and a real-valued function v, such that:
   (i) $v[a \vee b] \geqq \max (v[a], v[b])$, and
   (ii) $v[a \vee b] + v[a \vee b'] = v[a] + v[b \vee b']$.

Show that the free algebra for the above postulates is a Boolean algebra. (A. Lehman, SIAM Revs. **7** (1965), 253–73.)

5. Continuous Geometries

We now construct a remarkable class of complemented metric (modular) lattices, discovered by von Neumann [**1**], which are continuous-dimensional analogs of projective geometries.

For F be any field or division ring; consider the projective geometries $PG_{n-1}(F) = PG(F, n - 1)$ of length n over F.

† See also G. Trevisan, Rend. Mat. Univ. Padova **20** (1951), 386–400; J. Hasimoto, Proc. AMS **3** (1952), 1–2. To prove "only if", use Theorem VIII.15, Corollary 1.

‡ See L. M. Blumenthal, *Theory and applications of distance geometry*, Oxford, 1953, Chapter XV; L. M. Kelley, Duke Math. J. **19** (1952), 661–70.

LEMMA 1. *There is a monomorphism of* PG$(F, n-1)$ *to* PG$(F, 2n-1)$ *which maps* O *into* O, I *into* I, *and multiplies height by two.*

PROOF. If a and a' are any two complementary elements of height n in PG$(F, 2n-1)$, then the intervals $[O, a]$ and $[O, a']$ in PG$(F, 2n-1)$ are perspective and to each other (Theorem IV.11) under an isomorphism α, and isomorphic to PG$(F, n-1)$. Moreover, by Theorem III.16, the sublattice generated by these intervals in PG$(F, 2n-1)$ is isomorphic to their product. Hence the mapping $x \to [x, \alpha(x)]$ of $[O, a]$ into $[O, I]$ is the desired monomorphism.

We conclude that PG$(F, n-1)$ can be monomorphically embedded in PG$(F, 2n-1)$, so that the "normalized height function" $h[x]/h[I]$ is preserved. Iterating, we get a sequence of extensions of projective geometries:

$$\text{PG}(F, 1) \subset \text{PG}(F, 3) \subset \text{PG}(F, 7) \subset \cdots \subset \text{PG}(F, 2^m - 1) \subset \cdots,$$

in which both the lattice operations and the valuation $v[x] = h[x]/h[I]$ are preserved.

Taking the union of these extensions, we get an enveloping *metric lattice* which contains elements of every dyadic relative height $k/2^m$. By Theorem 5, this is a metrically dense sublattice of a complete metric lattice, which therefore contains elements of *every* dimension $d = \lim h[x_n]/h[I]$, $0 \leqq d \leqq 1$. This complete metric lattice is the *continuous geometry* CG(F) over F first constructed by von Neumann.

THEOREM 9. *The continuous geometry over any field or division ring* F *is a complete complemented modular lattice, in which any two equidimensional elements are perspective.*

SKETCH OF PROOF. We first show that the metric completion $\bar{M}$ of any metric complemented (modular) lattice M is complemented.

LEMMA 2. *Let* L *be any complemented metric lattice, and let* x' *be any complement of* x *in* L. *If* $d(x, y) < \epsilon$, *then* y *has a complement* y' *with* $d(x', y') < \epsilon$.

PROOF. Let $t = x \wedge y \leqq x$; for the complement t' of t, and any complement x' of x, the elements t, $t' \wedge x$, x' are independent elements with join I. Hence $t^* = (t' \wedge x) \vee x'$ is also a complement of t. But by (4), to replace x by t in $(t' \wedge x) \vee x$ will move it through a distance at most $d(t, x)$; the replacement gives $(t' \wedge t) \vee x' = x'$; hence $d(t^*, x') \leqq d(t, x)$. Similarly, we can find a complement y^* of y such that $d(y^*, t^*) \leqq d(y, t)$. The proof is completed by the triangle inequality (3), setting $y' = y^*$:

$$\begin{aligned} d(y^*, x') &\leqq d(x, x \wedge y) + d(y, x \wedge y) \\ &= v[x] - v[x \wedge y] + v[y] - v[x \wedge y] \quad \text{since } x, y \geqq x \wedge y, \\ &= v[x \vee y] - v[y] + v[y] - v[x \wedge y] \quad \text{by V1}, \\ &= d(x, y) \quad \text{by definition.} \end{aligned}$$

This proves Lemma 2.

It follows that to any Cauchy sequence a_n in M with limit a in $\bar{M}$ corresponds a Cauchy sequence of complements a_n' in M, whose limit a' in $\bar{M}$ will, by (4) satisfy $a \wedge a' = 0$, $a \vee a' = I$, establishing that $\bar{M}$ is complemented.

Completeness will be proved in Theorem 16, Corollary.

To prove that any two equidimensional elements are perspective, construct a convergent sequence of axes of perspectivity c_n in $\bar{M}$ for given Cauchy sequences $\{a_n\}$, $\{b_n\}$ in M. As in Lemma 2, we reduce to the case a_{n+1} and its dual. Suppose a, b are perspective by c in M, and that $a > a_1$, $b > b_1$, with $d(a, a_1) = d(b, b_1) = \delta_1$. Then, by hypothesis, $c \vee a_1$ and $c \vee b_1$ will have a common complement q with $d[q] = \delta_1$; hence a_1, b_1 will be perspective by $c \vee q$, where $d(c \vee q, c) = \delta_1$. Dually, suppose $a_2 > a$, $b_2 > b$, with $d(a, a_2) = d(b, b_2) = \delta_2$. Then $d[a_2 \wedge c] = d[b_2 \wedge c] = \delta_2$; and any common *relative* complement r of $a_2 \wedge c$ and $b_2 \wedge c$ in c will satisfy $d(c, r) = \delta_2$. Finally, a_2 and b_2 will be perspective by r.

Exercises for §5:

1. Let $M_1 \subset M_2 \subset M_3 \subset \cdots$ be a sequence of metric lattices. Show that $\bigcup M_k$ is a metric lattice.
2. (a) Show that a finite-dimensional projective geometry has essentially just one nontrivial valuation.
   (b) Show that, if v is any valuation on CG(F), then $v[x] = v[O] + \lambda d[x]$ for some constant λ.
3. Prove in detail that if any two equidimensional elements are perspective in a metric lattice M, the same is true in the metric completion $\bar{M}$ of M.
4. Show that every automorphism of any PG(F, $n - 1$) $\subset$ CG(F) can be extended to one of CG(F).

*5. Show that, if **Qu** is the ring of real quaternions and **R** the real field, CG(**Qu**) $\cong$ CG(**R**). (J. von Neumann)

*6. Let M be a projective geometry of finite length $n > 3$ which is also a compact topological lattice with $n + 1$ connected components under an extrinsic topology. Show that the coordinate-ring of M, if of characteristic ∞, is **R**, **C** (complex field), or **Qu**.

6. Jordan's Decomposition

Jordan's classic decomposition of functions of bounded variation into monotone summands was generalized by F. Riesz to additive set functions. It can be generalized to valuations on lattices.† A valuation $v[x]$ is said to be of *bounded variation* if and only if, for some finite K and all chains $x_0 < \cdots < x_n$,

$$\sum_{i=1}^{n} |v[x_i] - v[x_{i-1}]| < K.$$

We begin with some definitions. Consider the valuations $v[x]$ on a lattice L which satisfy $v[a] = 0$ for some fixed a. They evidently form a vector space, since any linear combination of such valuations is such a valuation. In this vector space, the valuations of bounded variation form a subspace.

We partially order this vector space by

† Jordan's decomposition is given in his *Cours d'Analyse*, Vol. 1, p. 54; the name is due to S. Saks. For the generalization of Riesz, see [**1**] and Verh. Zurich Congress, Vol. I, pp. 258–60. The present theorem was given in [**LT1**, p. 45].

(11) $$v \geqq v_1 \quad \text{means} \quad v[x] - v_1[x] \quad \text{is isotone.}$$

Then P1, P3 are obvious and P2 follows since we require $v[a] = 0$. Further, evidently

(12) $$v \geqq v_1 \text{ implies } v + v_2 \geqq v_1 + v_2 \text{ for all } v_2,$$
$$\text{and for positive } \lambda,\ \lambda v \geqq \lambda v_1 \text{ whenever } v \geqq v_1.$$

(*Remark.* These are just the conditions that the valuations form a partly ordered vector space; see §III.2, Example 4, and §XV.1.)

Next, we identify valuations with functions of intervals: if $x \leqq y$, we define $v[x, y] = v[y] - v[x]$. Condition V1 of §1 simply asserts that perspective and hence projective intervals have the same values. With any chain $\gamma: x = x_0 < \cdots < x_n = y$ subdividing $[x, y]$, we associate the *positive variation* of $v[x]$ on γ, defined as the sum of the positive increments of $v[x]$ along γ:

(13) $$v^+[x, y; \gamma] = \sum_{i=1}^{n} \max \{v[x_{i-1}, x_i], 0\}.$$

We define the limit

(14) $$v^+[x, y] = \sup_\gamma v^+[x, y; \gamma]$$

as the positive variation of v on $[x, y]$. We define the *negative variation* $v^-[x, y]$ of v on $[x, y]$ dually. Clearly $v[x]$ is of bounded variation if and only if every $v^+[x, y]$ and $v^-[x, y]$ is finite. We call these the positive and negative variations of $v[x]$ on $[x, y]$. Clearly, if γ'' is any subdivision of γ, then $v^+[x, y; \gamma''] \geqq v^+[x, y; \gamma]$. Again if γ' is any chain $x = x_0' < x_1' < \cdots < x_n' = y$ subdividing $[x, y]$, by modularity γ will have a subdivision γ'' whose intervals will be projective to corresponding intervals of some subdivision γ''' of γ', hence such that $v^+[x, y; \gamma''] = v^+[x, y; \gamma''']$ $\geqq v^+[x, y; \gamma']$. We infer that

(15) $$v^+[x, y] = \sup_{\gamma''} v^+[x, y; \gamma''] \text{ for refinements } \gamma'' \text{ of } \gamma.$$

If $v[x]$ is of bounded variation, since projective intervals have corresponding chains and (as remarked above) therefore have equal $v^+[x, y; \gamma]$ for corresponding chains, projective intervals have equal $v^+[x, y]$. It follows that

(16) $$v^+[x] = v^+[a, a \vee x] - v^+[a \wedge x, a]$$

is a valuation of L. Also, since substitution of $x_1 \geqq x$ for x expands $[a, a \vee x]$ and contracts $[a \wedge x, a]$, $v^+ \geqq 0$, by construction $v^+ \geqq v$. Conversely, if $v_1 \geqq v, 0$, then for every γ, $v_1[x, y] \geqq v^+[x, y; \gamma]$; hence $v_1[x, y] \geqq v^+[x, y]$ and so $v_1 \geqq v^+$. In summary,

THEOREM 10. *If $v[x]$ is of bounded variation on a lattice L, then $v \vee 0$ exists and is the v^+ defined by* (13)–(14).

It follows that the valuations of bounded variation on L with $v[a] = 0$ form a *vector lattice* (Chapter XV): the partial ordering which makes them a partly ordered vector space also makes them a lattice.

Exercises:

1. Define $v^-[x]$, the negative part of any valuation of bounded variation.
2. Show that, if $v[x]$ is of bounded variation, then $v[x] = v^+[x] + v^-[x]$ for all x.
3. Show that, if $f\colon \mathbf{R} \to \mathbf{R}$ is of bounded variation, then $f^+[x]$ and $f^-[x]$ are monotone functions (the "positive" and "negative" variations of f).
4. Show that, if $w[x, y]$ is any function of intervals in a lattice L which satisfies $w[x, y] + w[y, z] = w[x, z]$, $x \leqq y \leqq z$, and gives the same value to perspective intervals, then $v[x] = w[a, a \vee x] - w[a \wedge x, a]$ is a valuation.
5. Let v be any valuation on a relatively complemented lattice with O, I, such that $v[O] = 0$. Show that v is isotone if and only if $v[x] \geqq 0$ for all x, and of bounded variation if and only if $v[x]$ is bounded.

7. Intrinsic Topology of Chains

It is classic that the topology of the real line is determined by its order relation. We shall now see how far this can be generalized to arbitrary chains. In §§9–12, we shall study generalizations to arbitrary conditionally complete lattices and even posets. For chains, one may proceed as follows.†

DEFINITION. In a chain C, the *open intervals* are: (i) C itself, (ii) for any $a \in C$ the set $(a, +\infty]$ of all $x > a$, (iii) for any $a \in C$, the set $[-\infty, a)$ of all $x < a$, and (iv) for any $a < b$ in C the set (a, b) of all x satisfying $a < x < b$.

The *intrinsic neighborhood topology* of a chain C is obtained by taking the open intervals just defined as a neighborhood-basis (or basis of open sets). Hence we have the following result.

LEMMA 1. *A subset of a chain C is open in the intrinsic neighborhood topology if and only if it is a union of open intervals.*

The most general closed set in a chain is the intersection of the complements of open intervals. Moreover, in any chain with universal bounds $-\infty$ and $+\infty$, the complement of any open interval is the sum of one or two closed intervals. Hence the most general closed set is an intersection of finite sums of closed intervals. This proves

LEMMA 2. *In any chain C with universal bounds $-\infty$ and $+\infty$, the closed intervals form a subbasis for the family of closed sets.*

THEOREM 11. *Any chain is a normal Hausdorff space under its intrinsic topology, and the latter is invariant under automorphisms and dual automorphisms.*

PROOF. Clearly, any $p \in C$ has a neighborhood (i.e., C); again, the intersection of any two neighborhoods of p is a neighborhood of p; and third, any neighborhood of p is also a neighborhood of all its points, hence "open intervals" can be taken as a basis of open sets. Finally, if $p \neq q$ in C, then either p covers q, or $p > a > q$ for some a, or dually. In the first case $[-\infty, p)$ and $(q, +\infty]$ form a disjoint pair

† A. Haar and D. König, Crelle's J. **139** (1910), 16–28; see [**LT2**, p. 39] for other references.

of neighborhoods; in the second $[+\infty, a)$ and $(a, +\infty]$ do; dually for the last two cases. Hence C is a Hausdorff space.

Now let S and T be disjoint closed sets in C; then $S \cup T$ will be closed, and so its complement will be the union of open intervals I_α. In each I_α choose an x_α (this requires the Axiom of Choice). Each point $p \in S$ not interior to S must be separated from T on each side by some I_α, for T is closed and $S \cap T = \varnothing$. We adjoin to S the open interval (p, x_α) or (x_α, p) as the case may be. After augmenting T similarly, we have embedded S and T in disjoint open sets, completing the proof of Theorem 11.

We next consider *complete* chains. Every finite chain is complete; so is the real number system $\mathbf{R}$ if $-\infty$ and $=\infty$ (i.e., O and I) are adjoined. More generally, if X and Y are any two complete chains, then so are their ordinal sum $X \oplus Y$ and ordinal product $X \circ Y$. In this way many kinds of complete chains can be constructed.

Topologically, complete chains are distinguished from other chains by their compactness, as the following theorem shows.

THEOREM 12. *A chain C is complete if and only if it is compact in its neighborhood topology.*

PROOF. Let C be a complete chain. By Alexander's Theorem (§IX.8) C is compact if its subbasis of closed intervals (Lemma 2 above) has the compactness property. That is, it suffices to prove that, if $[a_\gamma, b_\gamma]$ $(\gamma \in \Gamma)$ is any collection of closed intervals such that $\bigcap_F [a_\phi, b_\phi]$ is nonvoid for any finite subset $F \subset \Gamma$, then $\bigcap_\Gamma [a_\gamma, b_\gamma]$ is nonvoid. But the first statement implies that

$$[a_\gamma, b_\gamma] \cap [a_\delta, b_\delta] = [a_\gamma \vee a_\delta, b_\gamma \wedge b_\delta]$$

is nonvoid for all $\gamma, \delta \in \Gamma$, whence $a_\gamma \leqq b_\delta$ for every $\gamma, \delta \in \Gamma$. Hence $c = \bigvee a_\gamma \in \bigcap_\Gamma [a_\gamma, b_\gamma]$, which is nonvoid.

Conversely, let X be any subset of a compact chain C. For each finite subset F of X, we define x_F as the least of the $x \in F$. Clearly none of the finite intersections $[-\infty, x_F]$ of the closed intervals $[-\infty, x](x \in X)$ is void; hence $\bigcap_{x \in X} [-\infty, x]$ is nonvoid, and X has a lower bound a. Now consider the set of all closed intervals $[a, x]$, where $x \in X$ and a is a variable lower bound to X. No finite intersection is void; hence $\bigcap [a, x]$ contains an element b. But any such b clearly is a lower bound of X which contains every lower bound of X; hence $b = \inf X$, which exists. Dually, $\sup X$ exists, completing the proof.

8*. Dense Subsets of Chains

We now digress, to show that the concepts of "denseness" introduced in §VIII.10, are equivalent to topological concepts.

THEOREM 13. *Let C be a chain with three or more elements. Then every order-dense subchain of C is topologically dense if and only if C has no isolated points.*

PROOF. Suppose C has an isolated point a. Then the complement S of a is order-dense in C; but $\{a\}$ being a neighborhood of a, a is not in the closure of S.

Thus S is not dense in C. Now suppose C has no isolated points, and let S be an order-dense subset of C. Let a be any element of C not in S, and let (b, c) be a neighborhood of a. Since a is not isolated, there exists $d \in (b, c)$ with $d \neq a$. If d is not in S then there exists $e \in S$ with e between a and d. In either case S intersects (b, c). Since a and the neighborhood (b, c) were arbitrary, this proves S is dense in C.

THEOREM 14. *A chain C is topologically connected if and only if it is conditionally complete and dense-in-itself.*

PROOF. Suppose C is connected. For any $a < b$ in C, $[-\infty, b)$ and $(a, +\infty]$ are open sets whose union is C; therefore, they have nonvoid intersection and there is an element between a and b. Thus C is dense-in-itself. To show that C is conditionally complete, let B be a subset of C with an upper bound x_0 but not having a l.u.b. Set $A = \{x \in C \mid x \geqq B\}$, $D = \{x \in C \mid x \notin A\}$. Given $x \in A$, there exists $y < x$ such that $y \in A$; clearly $(y, +\infty]$ is contained in A, and A is open. Given $x \in D$, there exists $y \in B$ such that $y > x$; clearly $[-\infty, y)$ is contained in D, and D is open. But A and D have void intersection and sum C, which contradicts the hypothesis that C is connected. Therefore, l.u.b. B must exist and dually, which proves that C is conditionally complete.

Now suppose that C is conditionally complete and dense-in-itself. Suppose $C = A \cup B$ with A, B open, disjoint, and not empty. We may assume that there exists $x_1 \in A$, $y_1 \in B$, with $x_1 < y_1$. Set $y_0 =$ g.l.b. $y \in B$ with $y > x_1$. Set $x_0 =$ l.u.b. $x \in A$ with $x < y_0$. Suppose $x_0 < y_0$. Then there exists a with $x_0 < a < y_0$; but $x_1 \leqq x_0 < a < y_0$ means a is not in B, and hence must be in A, which contradicts the definition of x_0. Thus we must have $x_0 = y_0$. Now suppose $x_0 \in A$, and let (a, b) be a neighborhood of x_0, which is contained in A. By the definition of y_0 we must have $y_0 \geqq b > x_1$, but this contradicts the fact that $x_0 = y_0$. We get a similar contradiction if we assume $x_0 \in B$. Therefore $C \neq A \cup B$, which proves that no decomposition of C with the assumed properties exists; hence C is connected.

Exercises for §§7–8:

1. Let C be any chain which is dense-in-itself.

(a) Show that a subset S of C is order-dense only if $\bar{S} = C$.

(b) Show that the intrinsic topology of any dense subset of C may be obtained from that of C by relativization, but that this is not true of all subsets of C.

2. Show that, if $x_\alpha \to a$ in a well-ordered set, then for some α, $x^\alpha_\beta \leqq a$ for all $\beta \geqq \alpha$.

3. A transfinite ordinal α is called a *limit*-ordinal if and only if $\alpha = \beta + 1$ for no β. Show that "limit-ordinals" are those which are topological limits of other ordinals.

4. For C a chain and $X \subset C$, show that $a \in \overline{X}$ if and only if a is the limit of a monotone well-ordered net $\{x_\alpha\} \subset X$.

5. Show that any isotone image of a complete chain is complete.

6. (a) Show that, in a complete chain, every open set can be expressed uniquely as the set-union of disjoint open intervals.

(b) Show that any chain C with a countable order-dense set has a countable basis of open sets.

7. Show that in $C = \mathbf{R} \circ \mathbf{2}$, the countable subset S of all pairs $(q, 1)$ and $(q, 2)$ with q rational, satisfies $\bar{S} = C$ but is not "order-dense". (Cf. Exercises 1a, 6b)

8. Show that any chain is a *completely* normal Hausdorff space in its intrinsic order topology.

9. Order and Star Convergence

It is a familiar fact that, in $\mathbf{R}$, the condition $x_n \to a$ (i.e., the convergence of a sequence $\{x_n\}$ to the limit a), is equivalent to the condition lim sup $\{x_n\}$ = lim inf $\{x_n\} = a$. It is this condition, and not the definition of open intervals as a basis of open sets, which defines an appropriate intrinsic order topology in lattices (and posets) generally. We can extend it from sequences to nets as follows.†

DEFINITION. Let $\{x_\alpha\}$ be any net of elements of a complete lattice. We define

$$\text{Lim inf}\,\{x_\alpha\} = \text{Sup}_\beta\,\{\text{Inf}_{\alpha \geqq \beta}\, x_\alpha\}, \tag{17}$$

$$\text{Lim sup}\,\{x_\alpha\} = \text{Inf}_\beta\,\{\text{Sup}_{\alpha \geqq \beta}\, x_\alpha\}. \tag{17'}$$

We say that $\{x_\alpha\}$ *order-converges* to a when

$$\text{Lim inf}\,\{x_\alpha\} = \text{Lim sup}\,\{x_\alpha\} = a. \tag{18}$$

Note that lim inf $\{x_\alpha\} \leqq$ lim sup $\{x_\alpha\}$ in any case.

COROLLARY. *If $x_\alpha \to a$ in a complete lattice, then there exist nets $t_\alpha \uparrow a$ and $u_\alpha \downarrow a$ with $t_\alpha \leqq x_\alpha \leqq u_\alpha$, and conversely.*

Explanation. By $t_\alpha \uparrow a$, we mean that the function $\alpha \to t_\alpha$ is isotone and that sup $t_\alpha = a$; the meaning of $u_\alpha \downarrow a$ is dual.

As usual, we define a subset X to be *closed* in the *order topology*, if and only if $\{x_\alpha\} \subset X$ and $x_\alpha \to a$ imply $a \in X$; in words, if and only if the limit of any order-convergent net of elements of X is itself in X.

THEOREM 15. *In any complete chain, the order topology coincides with the* (*open interval*) *neighborhood topology of* §7.

PROOF. $x_\alpha \to a$ if and only if every open interval (b, c) containing a contains for some β all x_α with $\alpha \geqq \beta$. But if this is true, $\text{Inf}_{\alpha \geqq \beta}\, x_\alpha \geqq b$; and since this is true for all $b < a$, Lim inf $\{x_\alpha\} \geqq a$. Dually, $x_\alpha \to a$ implies Lim sup $\{x_\alpha\} \leqq a$, and hence Lim sup $\{x_\alpha\}$ = Lim inf $\{x_\alpha\}$. Conversely, if Lim inf $\{x_\alpha\} = a =$ Lim sup $\{x_\alpha\}$, and $a \in (b, c)$, then $\text{Inf}_{\alpha \geqq \beta}\, x_\alpha > b$ and $\text{Sup}_{\alpha \geqq \gamma}\, x_\alpha < c$ for some β, γ; hence if $\delta \geqq \beta, \gamma$, then $x_\alpha \in (b, c)$ for all $\alpha \geqq \delta$; and so $x_\alpha \to a$.

Finally, by using the corollary stated above, we can consistently extend the definition of order-convergence given above from complete lattices to arbitrary posets.

DEFINITION. In any poset P, the net $\{x_\alpha\}$ is said to *order-converge* to the limit a (in symbols, $x_\alpha \to a$) when there exist nets $t_\alpha \uparrow a$ and $u_\alpha \downarrow a$ with $t_\alpha \leqq x_\alpha \leqq u_\alpha$.

By the corollary, $x_\alpha \to a$ in P if and only if $x_\alpha \to a$ in $\bar{P}$, *the completion by cuts* of P (§V.9). In other words, order-convergence in P is definable by *relativization*

† Order-convergence in lattices was first defined for sequences in G. Birkhoff [**3**, p. 453]. It and star-convergence (in vector lattices) were used in L. Kantorovich [**1**].

from order-convergence in the complete lattice $\bar{P}$. In all cases, it is evident that any poset P is a Fréchet L-space (§IX.6) with respect to order-convergence.

Star-convergence. In general, order-convergence need not have Urysohn's property F4 of §IX.6. However, as shown there, we can always obtain this property minimally by defining an appropriate star-convergence. In many important cases, as we shall see, *sequential* star-convergence suffices; hence we make the following definition.

DEFINITION. In a given poset P, a sequence $\{x_n\}$ *star-converges* to the limit a (in symbols, $x_n \to {*a}$) if and only if every subsequence $\{x_{n(k)}\}$ of $\{x_n\}$ contains a subsequence $\{x_{n(k(i))}\}$ which order-converges to a.

As shown in §IX.5, it follows that every poset is an L-space satisfying F4 under star-convergence.

Exercises:

1. Show that every σ-lattice is a Fréchet (sequential) L-space with respect to sequential order-convergence, and also an L-space with respect to order-convergence of nets.
2. Prove that every poset P is a T_1-space, if one defines the closure $\bar{X}$ of any set X to consist of the limits of order-convergent nets of elements of X.
3. Define star-convergence for nets, in any poset.
4. Show that, in the (vector) lattice of all square-integrable functions on $[0, 1]$, star-convergence is not equivalent to order-convergence.
5. Show that if $P = QR$ as a poset is a cardinal product of Q and R, then as a topological space (under the order topology) it is also the topological product of Q and R.
6. Show that any lattice is topologically dense in its completion by cuts in the order topology, but that this is not true for posets.
7. Show that a lattice which is compact in its order topology is necessarily complete.

*8. Construct a complete lattice which is not compact in its order topology.

9. Show that an element a of a complete lattice L is isolated in the order topology of L if and only if, for no chain $C \subset \{L - a\}$ is $a = \sup C$ or $a = \inf C$. (Kogan)
10. Show that a lattice need not be a Hausdorff space in its order topology. (E. S. Northam)

*11. Show that a Hausdorff space X is regular if and only if, for some complete lattice L, X is homeomorphic to a subset of L in the latter's order topology.†

10. Star Convergence in Metric Lattices

In this section, we shall study the relation between the *metric* topology in a metric lattice M, and the order topology of M. First, we establish the essential equivalence of metric completeness and order completeness.‡

THEOREM 16. *Any metrically complete lattice M is conditionally complete and satisfies*

$$x_n \uparrow x \quad \textit{implies} \quad v[x_n] \uparrow v[x] \quad \textit{and dually}; \tag{19}$$

conversely, any σ-complete metric lattice satisfying (19) *is metrically complete.*

† R. A. de Marr, Proc. AMS **16** (1965), 588–90.

‡ The results of §10 were obtained by J. von Neumann and the author, Ann. of Math. **38** (1937), 56, and some of them independently by L. Kantorovich [**1**]. See [**LT2**, p. 80].

PROOF. Let M be a complete metric lattice and let S be any bounded subset of M. Consider the joins $\sup X$ of the finite subsets X of S; since S is bounded and $v[X]$ isotone, the set of real numbers $v[\sup X]$ will be bounded; hence it will have a least upper bound v_S. Hence we can find $X_1, X_2, \cdots$, such that $v[\sup X_n] \geqq v_S - 2^{-n}$. Then letting U denote the (finite) set-union of X_n and X_m, we have by (3):

$$d(\sup X_m, \sup X_n) \leqq v[\sup U] - v[\sup X_m] + v[\sup U] - v[\sup X_n]$$
$$\leqq 2^{-m} + 2^{-n} \qquad \text{by construction.}$$

Hence by metrical completeness, the $\sup X_n$ converge metrically to some $s \in M$. Now let $x \in S$ be given; by (4)

$$v[x \vee s] - v[s] = \lim_{n \to \infty} \{v[x \vee \sup x_n] - v[\sup x_n]\} \leqq 2^{-m}$$

for all m. Hence $x \vee s = s$ and s is an upper bound for S; while if u is an upper bound for S then $u \geqq \sup x_n$ for all n and so $u \vee s = s$ by continuity (see (4)). Hence s is a least upper bound of S. The existence of $\inf S$ follows dually; hence M is conditionally complete.

To prove (19), note that if $x_n \uparrow x$, then $v[x_n] \uparrow$ and yet $v[x_n] \leqq v[x]$ for all n; hence $v[x_n] \uparrow c$ for some real number c. It follows that $d(x_m, x_n) = |v[x_n] - v[x_m]| \to 0$ as $m, n \to \infty$; hence $d(x_m, y) \to 0$ as $m \to \infty$ for some y, by metric completeness. Clearly $v[x_n] \uparrow v[y]$; moreover, $y \wedge x_m = (\lim x_n) \wedge x_m = \lim x_m = x_m$ (using metric limits); hence y is an upper bound to the x_n. But by definition, $x = \sup \{x_n\}$; hence $x \leqq y$ and $v[x] \leqq v[y]$. But we have already shown $v[x_n] \uparrow v[y]$ and $v[x_n] \leqq v[x]$ for all n; whence $v[x_n] \uparrow v[x]$, proving (19).

Conversely, let M be a σ-complete metric lattice assumed only to satisfy (19). From any Cauchy sequence, one can extract a subsequence $\{x_n\}$ satisfying $d(x_n, x_{n+1}) < 2^{-n}$ for $n = 1, 2, \cdots$. We shall show that for some y, $d(x_n, y) \to 0$, thereby proving metric completeness. Indeed, form $y_{n,r} = x_n \vee \cdots \vee x_{n+r}$. For fixed n, $y_{n,r} \uparrow$. By σ-completeness, $y_{n,r} \uparrow y_n$ where by (19) $d(y_{n,r}, y_n) \to 0$ as $r \to \infty$. Hence, using (4),

$$d(x_n, y_n) \leqq \sum_{r=0}^{\infty} d(y_{n,r}, y_{n,r+1}) = \sum_{r=0}^{\infty} d(y_{n,r} \vee x_{n+r}, y_{n,r} \vee x_{n+r+1})$$
$$< \sum_{r=0}^{\infty} 2^{-n-r} = 2 \cdot 2^{-n}.$$

Dually, define $z_{n,r} = x_n \wedge \cdots \wedge x_{n+r}$; then $z_{n,r} \downarrow z_n$ where $d(x_n, z_n) < 2 \cdot 2^{-n}$. Moreover, $y_n \geqq x_n \geqq z_n$; since $y_n \geqq y_{n,r+1} \geqq y_{n+1,r}$ for all r, $y_n \geqq y_{n+1}$; dually $z_n \leqq z_{n+1}$. By σ-completeness again, $x_n \downarrow y$ and $z_n \uparrow x$, where

$$d(y, z) = \lim_{n \to \infty} d(y_n, z_n) = \lim_{n \to \infty} [d(y_n, x_n) + d(x_n, z_n)] = 0$$

since $d(y_n, x_n) + d(x_n, z_n) < 2^{2-n}$ for all n. Hence $y = z$. Moreover,

$$d(x_n, y) = v[X_n \vee y] - v[x_n \wedge y] \leqq v[y_n] - v[z_n] < 2^{2-n}$$

so that $x_n \to y$ metrically. Q.E.D.

Since in any lattice with O and I, conditional completeness implies completeness, we obtain

COROLLARY. *In any metric lattice with universal bounds which satisfies* (19), *metric completeness,* (*order*) *completeness, and σ-completeness are equivalent.*

Now let M be a metrically complete metric lattice; and suppose that $\{x_n\}$ is a sequence which star-converges to x, that is, that every subsequence of $\{x_n\}$ has a subsequence order converging to x. If we can prove that any order-convergent subsequence is also metrically convergent to the same limit, then we will have proved that $\{x_n\}$ metrically converges to x. But suppose $u_n \downarrow y$, $w_n \uparrow y$ and $u_n \geqq y_n \geqq w_n$; then

$$d(y_n, y) = v[y_n \vee y] - v[y_n \wedge y] \leqq v[u_n] - v[w_n] \downarrow 0$$

by (19), proving $y_n \to y$ metrically, as desired.

Conversely, suppose $d(x_n, x) \to 0$. Then we can choose a subsequence $\{y_n\}$ such that $d(y_n, y_{n+1}) < 2^{-n}$, and hence such that $d(y_n, x) < 2^{-n+1}$. Now define $z_n = y_1 \vee \cdots \vee y_n$; then $z_1 \leqq z_2 \leqq \cdots$ and, by (4)

$$(20) \qquad d(z_n, z_{n+1}) = d(z_n \vee y_n, z_n \vee y_{n+1}) \leqq d(y_n, y_{n+1}) < 2^{-n}.$$

By metric completeness, there is a z such that $d(z_n, z) \to 0$; this z is then an upper bound (in fact a least upper bound) of $\{z_n\}$ and hence an upper bound for $\{y_n\}$. Similarly, y_n is bounded below. Now by the conditional completeness of metric complete lattices (Theorem 16), we can form $t_n = \bigwedge_{k \geqq n} y_k$ and $s_n = \bigvee_{k \geqq n} y_k$ which are bounded by the same bounds as $\{y_n\}$, and hence we have

$$(21) \qquad \bigvee_n t_n = \bigvee_n \left(\bigwedge_{k \geqq n} y_n\right) \geqq \bigwedge_n \left(\bigvee_{k \geqq n} y_k\right) = \bigwedge_n s_n,$$

where $s_n \downarrow$, $t_n \uparrow$. Furthermore, by (19) and (4):

$$d(t_n, x) = v[t_n] - v[x] = \sup_k v[y_n \vee \cdots \vee y_k] - v[x] \leqq \sup_k d(y_n \vee \cdots \vee y_k, x)$$

$$\leqq \sup_k \sum_{m=n}^{k} d(y_m, x) \leqq \sum_{m=n}^{\infty} d(y_m, x) = 2^{-m+1}.$$

Thus, $d(t_n, x) \to 0$. Similarly $d(s_n, x) \to 0$, so that

$$(22) \qquad \bigvee_n \left(\bigwedge_{k \geqq n} y_k\right) = x = \bigwedge_n \left(\bigvee_{k \leqq n} y_k\right).$$

We have therefore proved

THEOREM 17. *In a metrically complete metric lattice, metric convergence and star-convergence are equivalent.*

11. Topological Lattices

In general, a *topological algebra* is an algebra (as defined in Chapter VI) which is also a topological space (in the sense of Chapter IX), and whose operations are continuous in the topology specified. One can develop a theory of topological algebras in general ("universal topological algebra"),† analogous to that of "universal algebra" developed in Chapters VI–VII above, but we shall confine our attention here to the special case of topological lattices.

DEFINITION. A *topological lattice* is a lattice with a specified convergence topology, in which

(23) $$x_\alpha \to x \text{ and } y_\beta \to y \text{ imply } x_\alpha \wedge y_\beta \to x \wedge y,$$

and

(23′) $$x_\alpha \to x \text{ and } y_\beta \to y \text{ imply } x_\alpha \vee y_\beta \to x \vee y.$$

We have shown in (5) that *any metric lattice is a topological lattice in its metric topology*; in fact, the lattice operations are *uniformly* continuous, since they satisfy a Lipschitz condition with Lipschitz constant one. From this fact, it is easy to deduce the following important consequence.

THEOREM 18. *Any complete metric lattice is a topological lattice under order convergence.*

LEMMA. *For* (23) *and* (23′) *to hold in a complete lattice, it is sufficient that, respectively*:

(24) $$x_\alpha \uparrow x \textit{ implies } a \wedge x_\alpha \uparrow a \wedge x,$$

and

(24′) $$x_\alpha \downarrow x \textit{ implies } a \vee x_\alpha \downarrow a \vee x.$$

PROOF. (The condition is obviously necessary.) In (23), let u_α be the meet of the successors of x_α, and v_β the meet of the successors of y_β. We will prove that $u_\alpha \wedge v_\beta \uparrow x \wedge y$. Since $u_\alpha \wedge v_\beta$ is an isotone net, it suffices to show that $\bigvee (u_\alpha \wedge v_\beta) = x \wedge y$; This we now do. Since $x \geqq u_\alpha$ and $y \geqq v_\beta$, clearly

$$x \wedge y \geqq u_\alpha \wedge v_\beta \quad \text{for all } \alpha, \beta, \text{ and so } x \wedge y \geqq \bigvee (u_\alpha \wedge v_\beta).$$

Conversely, for each α, (24) implies

$$\bigvee (u_\alpha \wedge v_\beta) \geqq u_\alpha \wedge (\bigvee v_\beta) = u_\alpha \wedge y.$$

Hence, taking the join with respect to α,

$$\bigvee (u_\alpha \wedge v_\beta) \geqq \bigvee (u_\alpha \wedge y) = x \wedge y,$$

by L2 and (24) again. The proof is completed by duality.

PROOF OF THEOREM. Suppose that $x_\alpha \uparrow x$, and let $s = \sup_\alpha v[x_\alpha]$. We can select $t_n = x_{\alpha(n)}$ $(n = 1, 2, 3, \cdots)$ with $\alpha(n) \leqq \alpha(n+1)$ and $v[t_n] > s - 2^{-n}$. Since $\alpha(n) \leqq \alpha(n+1)$, $d(t_n, t_{n+1}) > 2^{-n}$; hence, by metric completeness, $t_n \to t$ metrically for some t, where $v[t] = s$. We shall now show that $t \geqq x_\alpha$ for all α. Otherwise, for some n, $v[t \vee x_\alpha] - v[t] > 2^{-n}$, implying $v[t_m \vee x_\alpha] - s < 2^{-n}$ for some m, n, whence $v[x_\beta] > s + 2^{-n}$ for some common successor of $x_{\alpha(m)} = t_m$ and

† For early ideas on this subject, see D. van Dantzig, Math. Ann. **107** (1933), 587–626, and G. Birkhoff [**3**, §26].

x_α, and so contradicting the definition of s. It follows that $t \geqq x$; but $v[t] = s \leqq v[x]$; hence $t = x$ and $d(x_\alpha, x) \downarrow 0$. By (5), however, $d(x_\alpha \wedge a, x \wedge a) \downarrow 0$ for all a. Hence $a \wedge x$, which is clearly an upper bound to the isotone net $\{a \wedge x_\alpha\}$, is its *least* upper bound and hence (order) limit. The proof is again completed by duality.

THEOREM 19. *A complete distributive lattice is a topological lattice under order-convergence if and only if it satisfies*

$$a \wedge (\bigvee x_\alpha) = \bigvee (a \wedge x_\alpha) \tag{25}$$

and, dually,

$$a \vee (\bigwedge x_\alpha) = \bigwedge (a \vee x_\alpha). \tag{25'}$$

These are the "infinite distributive laws" (1)–(1′) of §V.5. Comparing with Theorem V.16, we obtain the following corollary.

COROLLARY 1. *Any complete Boolean lattice is a topological lattice under order-convergence.*

Again, by definition (§V.10), Theorem 19 implies

COROLLARY 2. *Any complete topological distributive lattice is a complete Brouwerian lattice.*

PROOF OF THEOREM. For the isotone net $\{x_\alpha\}$, (25) is equivalent to (24); hence in any case (25) implies (24). Dually, (25′) implies (24′); and so, by the lemma, *any* complete lattice which satisfies the infinite distributive laws (25)–(25′) is topological.

Conversely, let a complete distributive lattice L be topological, and hence satisfy (24)–(24′). Then for any set A of elements $x_\alpha \in L$ we can form the isotone net of $y_F = \bigvee_F x_\alpha$, where F is any finite subset of A. By distributivity, $a \wedge y_F = a \wedge \bigvee_F x_\phi = \bigvee_F (a \wedge x_\phi)$, and the net $y_F \uparrow \bigvee_F x_\alpha = b$. Hence, by the lemma, $a \wedge y_F \uparrow a \wedge b$, which is to say:

$$a \wedge (\bigvee x_\alpha) = a \wedge (\bigvee y_F) = \bigvee (a \wedge y_F) = \bigvee \left(a \wedge \bigvee_F x_\phi\right) = \bigvee (a \wedge x_\alpha). \tag{26}$$

The proof of (25′) is dual, and gives the theorem.

Further results about lattices which are topological in their order topology are stated in the exercises below.

Exercises for §§10–11:

1. Show that (19) need not hold for positive valuations, even in chains.
2. Show that (19) holds in a metric lattice if and only if order-convergence of sequences implies metric convergence.
3. (a) Construct a metrically complete lattice which is not complete as a lattice.
   (b) Show that, in a metrically complete metric lattice, a sublattice is metrically bounded if and only if it is order-bounded.
4. Show that, for isotone nets $\{x_\alpha\}$, metric convergence is equivalent to order-convergence in complete metric lattices.

5. Show that a complete lattice is a topological lattice for order-convergence if and only if if it is one for star-convergence.

6. Show that, in the complete (Brouwerian) lattice of all closed subsets of the real line, $x_\alpha \downarrow x$ implies $(a \vee x_\alpha) \downarrow (a \vee x)$, but not dually: the lattice is not topological.

7. Same result for the additive subgroups of **Z**, the additive group of the integers. (Cf. §VIII.5, (3).)

*8. Show that the free lattice FL(n) is complete, but is not a topological lattice if $n > 2$.

9. Show that every "complete morphism" (i.e., one preserving arbitrary joins and meets) is continuous in the order topology.

10. Let L be any complete lattice in which every ascending well-ordered subset is countable, and dually. Show that if $s_\alpha \to a$ (order-convergence), then one can find a countable *sequence* $\{x_{\alpha(n)}\} \subset \{x_\alpha\}$ such that $x_{\alpha(n)} \to a$.

11. Let $P = \prod L_\alpha$ be a (cardinal) product of lattices L_α. Show that convergence in P is the Cartesian product of those of the L_α.

*12. (a) Let L be the Boolean lattice of all regular open subsets of $[0, 1]$. Show that there is no Hausdorff topology on L in which $x_n \uparrow a$ implies $x_n \to a$ topologically.

(b) Show that $x_{i,j} \to x_i$ for all i and $x_i \to a$ do *not* imply that $j(i)$ exists such that $x_{i,j(i)} \to a$ in L.†

13. (a) Show that a locally compact, connected topological lattice is a chain if and only if its topological dimension is 0 or 1.

(b) Show that any locally compact connected topological lattice has a basis of convex sets.‡

14. Show that the center of any compact topological lattice is totally disconnected.

15. (a) Show that, if a topological lattice L is homeomorphic with a connected, locally compact subset of the plane, then it is distributive.

(b) Show that this is not true of Euclidean three-space.

12. Interval Topology

The remarkable thing about the *order topology* defined in (complete) lattices by order- and star-convergence is that it is intrinsic—i.e., preserved under all (order) isomorphisms. Hence any isomorphism of lattices is necessarily a homeomorphism for the order topology.

However, one can define other interesting intrinsic topologies in lattices. One of the most interesting is the following modification of an interval topology invented by O. Frink.§

DEFINITION. Let P be any poset with universal bounds O and I. The *interval topology* on P is defined by taking the closed intervals $[a, b]$ as a subbasis of closed sets.

The main interest of the interval topology stems from the following general result.

THEOREM 20 (FRINK). *A lattice is compact in its interval topology if and only if it is complete.*

---

† Exercise 12 states results of E. E. Floyd, Pacific J. Math. **5** (1955), 687–9.

‡ Exercises 13–15 state results of L. W. Anderson and A. D. Wallace; see [**Symp**, pp. 195–7] for detailed references.

§ O. Frink [**1**]. Its modification for bidirected sets without universal bounds was suggested by the author, Revista Math. Tucuman **A14** (1962), 325–31.

The proof is based on Alexander's Theorem (Theorem IX.7, Corollary): the closed intervals $[a_\alpha, b_\alpha]$ of any complete lattice L have the compactness property. For, let us *direct* the finite intersections

$$(27) \qquad [a_F, b_F] = \bigcap_{\alpha \in F} [a_\alpha, b_\alpha] = \left[\bigvee_F a_\alpha, \bigwedge_F b_\alpha\right]$$

by dualizing set inclusions (in other words, we consider the *filter* of the $[a_F, b_F]$). If no $[a_F, b_F]$ is void, then the nets $\{a_F\}$ and $\{b_F\}$ exist, and $a_F \uparrow a$, $b_F \downarrow b$ for some $a = \bigvee a_F \leqq b = \bigwedge b_F$; hence

$$(28) \qquad \bigcap [a_\alpha, b_\alpha] = \bigcap [a_F, b_F] \supset [a, b]$$

is nonvoid. Since the $[a_\alpha, b_\alpha]$ have the compactness property and form a subbase of closed sets, L is compact.

Conversely, suppose that L is not complete, so that some set $S \subset L$ has no l.u.b. (or dually). Consider the intervals $[s, u]$, for arbitrary $s \in S$ and any upper bound† u of S. Then the $[s, u]$ have nonvoid finite intersections, since

$$\bigcap [s_\alpha, u] = \left[\bigvee_F s_\alpha, \bigwedge_F u_\alpha\right];$$

yet their intersection is void, since any element in it would have to be a l.u.b. of S.

The interval topology can also be defined in *bidirected* sets, possibly without universal bounds, as follows. (We recall that a poset D is bidirected when any $a, b \in D$ have an upper and a lower bound in D.)

DEFINITION. In a bidirected set D, let $\mathscr{C}$ be the $\bigcap$-ring of all intersections of finite unions of closed intervals $[a, b]$ of D. A set $S \subset D$ is *closed* in the *interval topology* of D if and only if $C \in \mathscr{C}$ implies $S \cap C \in \mathscr{C}$.

We now show that the interval topology is always *at least as weak* as the order topology in lattices. (In many important cases, the two are equivalent.)

THEOREM 21. *Every subset of a bidirected set which is closed in the interval topology is also closed in the order topology.*

PROOF. Suppose $x_\alpha \to c$, where $x_\alpha \in [a, b]$ for all α. Then $t_\alpha \leqq x_\alpha \leqq u_\alpha$, where $\bigvee t_\alpha = \bigwedge u_\alpha = c$. But for all α, $u_\alpha \geqq x_\alpha \geqq a$; hence $c = \bigcap u_\alpha \geqq a$. Dually, $c \leqq b$, whence $c \in [a, b]$ and closed intervals are closed. Since the closed intervals are a subbase of closed sets, the proof is complete.

One can extend Frink's theorem by showing that, in any *conditionally complete* lattice, a closed subset is *compact* if and only if it is *bounded*. For the proof, the reader is referred to the author's article (op. cit. supra).

Various other interesting intrinsic lattice topologies have also been proposed. For these, the reader is also referred to the periodical literature.‡

---

† If S has no upper bound, we consider the principal dual ideals of $[s, \infty)$.

‡ See O. Frink [**1**, §§12, 13]; B. C. Rennie, *The theory of lattices*, Foister and Jagg, Cambridge, 1952 (also Proc. London Math. Soc. **52** (1951), 386–400); A. J. Ward, Proc. Cambridge Philos. Soc. **51** (1955), 254–61; L. F. Papangelou, Math. Ann. **155** (1964), 81–107 and Pacific J. Math. **15** (1965), 1347–64.

Exercises:

1. (a) Show that, in the real field $\mathbf{R}$ $(-\infty, \infty)$, the (modified) interval topology is precisely the usual topology.
   (b) Extend this result to $\mathbf{R}^n$.
2. (a) Show that, in any chain, the interval topology is equivalent to the order topology.†
   (b) Show that the same is true in any lattice of finite breadth.
3. Show that, in the Cartesian product $\mathbf{R}^2$, the interval topology and the order topology are both equivalent to the usual topology.
4. Show that if $\mathbf{R}^2$ is completed by adjoining universal bounds $-\infty$ and $+\infty$, it ceases to be a Hausdorff space in its interval or order topology, which however remain equivalent.
5. Show that, in the complete atomic Boolean algebra $\mathbf{2}^{\omega}$, the order and interval topology are equivalent and give the Cantor ternary set.
6. Let $L(X)$ be the complete (dually Brouwerian) lattice of all closed sets of $[0, 1]$. Show that $L(X)$ is *not* compact in the order topology, which is therefore "stronger" than the interval topology.
7. Let L be any lattice. Show that, for ideals $J \subset L$, the following conditions are equivalent: (i) J is closed in the order topology, (ii) J is closed in the interval topology, (iii) J is "closed" in the sense of §V.9.

*8. Show that the direct product P of any family of lattices L_α with universal bounds is *homeomorphic* to their Cartesian product in the interval topology.

9. Show that, if L is an infinite lattice of length 2, then L has no compact Hausdorff topology which is *intrinsic* (i.e., invariant under all automorphisms and dual automorphisms).
10. Show that a Boolean lattice is a Hausdorff space in its interval topology if and only if it is atomic.‡

*11. Show that a lattice L with universal bounds is a Hausdorff space in its interval topology if and only if, given $a < b$ in L, the complement C of $[O, a] \cup [b, I]$ contains a finite subset F, such that every $c \in C$ is comparable to some $x \in F$.

*12. Construct a lattice L such that the square L^2 of L in its interval topology is not homeomorphic to the square of the interval topology on L.

*13. (a) Construct an infinite chain C which is compact and zero-dimensional in its interval topology, but admits no proper order-automorphisms.
   (b) Show that C is the Stone space of an infinite Boolean algebra with no proper automorphisms.§

PROBLEMS

81. Let M be a nontrivial modular lattice in which no quotient is projective to a proper part of itself. Does M necessarily have a nontrivial (real) valuation?

82. Does there exist a simple (modular) lattice with two linearly independent valuations?

*83. Prove that every chain is a normal Hausdorff space in its intrinsic topology without using the Axiom of Choice, or show that this is impossible.

84. When is a lattice a normal space in its intrinsic order topology?

---

† Exercises 2a, 3, 6, 8 are results of O. Frink [1]. For Exercise 2b, see T. Naito, Proc. AMS **11** (1960), 156–8; also Y. Matsushima, ibid., 233–5 and E. S. Wolk, ibid., 487–92.

‡ For Exercise 10, see M. Katetov, Colloq. Math. **2** (1951), 229–35. For Exercise 11 see E. S. Northam, Proc. AMS **4** (1953), 824–7, and S. A. Kogan, Uspehi Mat. Nauk **11** (1956), 185–90.

§ B. Jonsson, Proc. AMS **2** (1951), 766–70; L. Rieger, Fund. Math. **38** (1951), 209–16. Cf. Exercise 4 after §IX.10.

85. Is every complete morphism (i.e., for arbitrary joins and meets) of complete lattices continuous with respect to star-convergence? in the interval topology?

86. Find simple necessary and sufficient conditions for a bidirected set to be compact in its interval topology.

87. Can a lattice of infinite breadth be a Hausdorff space in its interval topology?

88. Describe geometric lattices having a locally Euclidean extrinsic topology, in which the elements of given height form a connected set.†

89. When and how can one extend a locally Desarguesian real hyperbolic geometry to an affine geometry?

90. If a continuous curve is a distributive lattice of breadth n, can it be embedded as a topological lattice in an n-cell.‡

91. Develop theories of (a) topological, (b) differentiable,§ and (c) analytic plane lattices, and of lattices of breadth two.

92. Develop general theories of differentiable and analytic lattices.

---

† See L. R. Wilcox, Duke J. Math. **8** (1941), 273–85; O. Haupt, G. Nobeling, and Chr. Pauc, Crelle's J. **181** (1940), 193–217. For the modular Desarguesian case, use A. Kolmogoroff, Ann. of Math. **33** (1932), 163–76.

‡ E. Dyer and A. Shields, Pacific J. Math. **9** (1959), 443–8; L. W. Anderson, J. Lond. Math. Soc. **37** (1962), 60–2.

§ Assuming, for example, that intervals $[a, b]$ are differentiable manifolds bounded by differentiable faces, edges, etc.

CHAPTER XI

BOREL ALGEBRAS AND VON NEUMANN LATTICES

1. Borel Algebras

Some of the deepest ideas of lattice theory are inspired by special problems of *real analysis*, such as those of measure and integration theory, operator theory, functional analysis, ergodic theory, and the like. This chapter will be devoted to such ideas.

Many of them involve the technical concept of a σ-lattice; hence we begin by defining it. A *σ-lattice* is a lattice in which every finite or countable subset $X = \{x_n\}$ has a meet $\inf X$ and a join $\sup X$. We shall call a Boolean σ-lattice a *Borel lattice*, and a σ-lattice which is also a Boolean algebra for finite joins and meets, a *Borel algebra*.

A subset S of a σ-lattice L is called a *σ-sublattice* of L when it contains with any finite or countable subset X of L also $\inf X$ and $\sup X$. A *Borel subalgebra* of a Borel algebra is a Boolean subalgebra which is also a σ-sublattice.

Example 1. Let X be any topological space, and $\mathbf{2}^X$ the (complete, atomic) Boolean algebra of *all* subsets of X; this is trivially a Borel algebra. Define a *Borel set* of X as any subset which belongs to the Borel subalgebra generated by the closed (or, equivalently, open) subsets of X. These Borel sets thus themselves form a Borel algebra, $\mathbf{2}^{[X]}$.

Evidently, a σ-sublattice is just a "subalgebra" of a σ-lattice, when the latter is considered as an infinitary algebra under countable join and meet. Using repeated entries, this includes finite joins and meets as special cases. The same is true of Borel subalgebras (and sublattices). Thus, any Borel subalgebra of a Borel algebra is itself a Borel algebra, and any intersection of Borel subalgebras of a Borel algebra A is itself a Borel subalgebra of A.

In general, a Borel subalgebra of $\mathbf{2}^X$ is called a *σ-field* of subsets of X, and a σ-sublattice of $\mathbf{2}^X$ is called a *σ-ring* of subsets of the set X. These notions play a basic role in the theories of measure and probability.

A *σ-morphism* from a σ-lattice L to a σ-lattice M is a function $h: L \to M$ such that

$$\bigwedge_n h(x_n) = h\Big(\bigwedge_n x_n\Big) \quad \text{and} \quad \bigvee_n h(x_n) = h\Big(\bigvee_n x_n\Big)$$

for every finite or countable subset $\{x_n\} \subset L$. Finally, a *σ-ideal* $J \subset L$ is defined by the two conditions:

(i) $a \in J$ and $x \leqq a$ imply $x \in J$, and

(ii) $X \subset J$ and X finite or countable implies $\bigvee X = \sup X \in J$.

THEOREM 1. *The kernel of any σ-morphism between σ-lattices is a σ-ideal. Conversely, let J be any σ-ideal of a Borel algebra A. Then A/J is a Borel algebra, and the Boolean epimorphism $A \to A/J$ is a σ-epimorphism.*

The proof is straightforward, and will be omitted.

THEOREM 2. *In any Borel algebra, all Boolean operations are continuous with respect to order-convergence.*

PROOF. In view of Corollary 1 of Theorem X.19, it suffices to prove that $x_n \to a$ implies $x'_n \to a'$. But since the order topology has a self-dual definition, this is trivial!

Exercises:

1. Prove Theorem 1.
2. Prove in detail that the Borel subalgebras of any Borel algebra form a complete lattice.
3. Prove that any σ-ideal of a σ-lattice is a σ-sublattice.
4. Show that any atomic Borel lattice with a countable number of points ("atoms") is complete.

*5. Show that the (atomic) Borel algebra of all Borel sets on the real line **R** has the cardinality of **R**.

*6. Show that the Borel algebra of all Borel sets on **R** is not complete. (*Hint.* See Theorem V.18.)

*7. Show that, in a Borel algebra A, every set of disjoint $a_i > 0$ is countable if and only if every well-ordered subset of A is countable ("σ-chain" condition)—but that this is not true in all Boolean algebras. [**Hal**, pp. 62–3].

2. Representations of Borel Algebras

In Corollary 3 of Theorem VIII.15, it was shown that any Boolean algebra is isomorphic with some field of sets (with preservation of complements and finite joins and meets). Again, by Theorem V.17, any complete Boolean algebra which satisfies the unrestricted distributive laws is isomorphic with the Boolean algebra of *all* subsets of its atoms, with preservation of *arbitrary* joins and meets. We now consider countable joins and meets. Theorems 3–4 below state two basic related results.

THEOREM 3 (LOOMIS-SIKORSKI†). *Any Borel algebra is a σ-epimorphic image of a σ-field of sets.*

PROOF. Let $\mathfrak{S}$ be the "representation space" of all those subsets $S \subset A$ which contain exactly one member (a or a') of each pair $\{a, a'\}$ of complementary elements of A. In the power set $P(\mathfrak{S})$ of all subsets of $\mathfrak{S}$, associate with each $a \in A$ the set $\tau(a) \subset \mathfrak{S}$ of all those $S \subset A$ with $a \in S$. Clearly $S \in \mathfrak{S}$ implies $S' \in \mathfrak{S}$, where S' signifies the complement of S in $P(A)$, the power set of A; moreover $a \in S$ if and only if $a' \in S'$; hence $[\tau(a)]' = \tau(a')$.

Now let $\Phi \subset P(\mathfrak{S})$ be the σ-field of subsets of $\mathfrak{S}$ generated by the $\tau(a)$ $(a \in A)$; since the $\tau(a)$ are closed under complementation, Φ is generated by the $\tau(a)$ under countable union and intersection. Let N be the family of all those countable intersections $\bigcap \tau(a_j)$ for which $\bigwedge a_j = O$ in A. Let J be the σ-ideal generated by

† L. Loomis, Bull. AMS **53** (1947), 757–60; R. Sikorski, Fund. Math. **35** (1948), 247–56.

N; it will consist of the $t \leqq \bigcup n_i$ for countable subsets of N. We shall prove that the correspondence $a \to \tau(a)$ is an isomorphism $\tau: \Phi/J \to A$. First, τ is a epimorphism of A onto Φ/J. For if $a = \bigvee a_i$, then $a' \wedge a_i \leqq a' \wedge a = 0$ for all i and $0 = a \wedge a' = a \wedge (\bigvee a_i)' = a \wedge \bigvee a_i'$. Hence, by definition of N and J, every $\tau(a') \cap \tau(a_i) \in N$, and $\tau(a) \cap \bigcup \tau(a_i') \in N$. By the former,

$$\tau(a') \cap [\bigcup \tau(a_i)] = \bigcup [\tau(a') \cap \tau(a_i)] \in J.$$

By the latter, $\tau(a) \cap [\bigcup \tau(a_i)]' = \tau(a) \cap \bigcap \tau(a_i') \in N$. Hence the symmetric difference between $\tau(a) = \tau(\bigvee a_i)$ and $\bigcup \tau(a_i)$ lies in J. A dual argument works for meets.

This σ-epimorphism will be an isomorphism if $\tau(a) \in J$ implies $a = O$. This we shall now show with the help of a "diagonal process", thereby completing the proof.

If $\tau(a) \in J$, then as above $\tau(a) \subset \bigcup_i (\bigcap_j \tau(a_{ij}))$, where $\bigwedge_j a_{ij} = O$ for all I. Hence, for any function $j(i)$, by the isotone law,

$$(1) \qquad \tau(a) \subset \bigcup_i \tau(a_{i,j(i)}), \quad \text{whence } \tau(a') \supset \bigcap_i \tau(a'_{i,j(i)}).$$

Now suppose $a \neq O$. Then $a' < I$, and so

$$I > a' = a' \vee O = a' \vee \bigwedge a_{ij} = \bigwedge (a' \vee a_{ij}).$$

Hence some $a' \vee a_{i,j(i)} < I$. Repeating the process

$$\begin{aligned} I > a' \vee a_{1,j} &= a' \vee a_{1,j(1)} \vee O \\ &= (a' \vee a_{1,j(1)}) \vee \bigwedge a_{2,j} = \bigwedge (a' \vee a_{1,j(1)} \vee a_{2,j}). \end{aligned}$$

By this "diagonal process", we get a function $j(i)$ such that, for all n,

$$I > a' \vee a_{1,j(a)} \vee \cdots \vee a_{n,j(n)}.$$

This sequence can contain neither a nor a complementary pair; hence we can find a "point" $S \subset \mathfrak{S}$ which contains a and every $a'_{i,j(i)}$. Clearly $S \in \tau(a)$, yet $S \notin \bigcup \tau(a_{i,j(i)})$ (since no $\tau(a_{i,j(i)})$ contains S and union in Φ is set-union). This contradicts (A), and completes the proof.

Using more sophisticated concepts of general topology, one can prove a sharper result.

THEOREM 3′ (SIKORSKI). *Let A be any Borel algebra, and $S(A)$ its Stone space. Let F be the σ-algebra of all Borel subsets of $S(A)$, and let Δ be the σ-ideal of all Borel sets of first category. Then A is isomorphic to F/Δ.*

More precisely, if θ is the isomorphism of A onto the subfield of open-and-closed subsets of $S(A)$ established in §IX.9, then the composite $\theta\phi\psi: A \to A\theta \to F \to F/\Delta$ is an isomorphism. For the proof, see [**Sik**, p. 117].

We shall now use a sharpened and extended form of Corollary 3 of Theorem VI.13, whose proof applies equally to infinitary algebras. This asserts that the free algebra with $\aleph$ generates in the family of all algebras generated by an algebra A, is isomorphic to the subalgebra of $A^{o(A^{\aleph})}$ generated by the "coordinate projections".

On the other hand, by the Loomis–Sikorski theorem, the free Borel algebra with $\aleph$ generators is isomorphic to the free Borel algebra FB ($\aleph$) with $\aleph$ generators generated by the Borel algebra† $A = \mathbf{2}$. By the result of the preceding paragraph, it follows that $FB(\aleph)$ is isomorphic to the Borel subalgebra of the (complete) Borel algebra $\mathbf{2}^{2^{\aleph}}$ of *all* subsets of the Cantor $\aleph$-space (Example 3 of §IX.8) which is generated by the clopen sets of points having one coordinate specified as 0 or 1. We conclude.

THEOREM 4. *The free Borel algebra with $\aleph$ generators is isomorphic with the σ-field (Borel subalgebra) of "Baire sets" of Cantor $\aleph$-space generated by the clopen sets.*

COROLLARY (RIEGER‡). *The free Borel algebra $FB(\omega)$ with countably infinite generators is isomorphic to the algebra of all Borel subsets of the Cantor ternary set.*

PROOF. Setting $\aleph = \aleph_0 = \omega$ in Theorem 4, we get the desired conclusion by observing that the Borel sets of the Cantor ternary set are generated by the clopen sets.

Unfortunately, the analog of Theorem 3 for Boolean $\aleph$-algebras with $\aleph > \aleph_0$ is false in general; cf. L. Rieger [**1**], and C.- C. Chang, Trans. AMS **85** (1957), 208–18.

3. Standard Borel Algebras

Although there presumably exists an enormous variety of nonisomorphic Borel algebras, many of the Borel algebras which arise most naturally in real analysis are isomorphic to one of a few "standard" Borel algebras.§ This principle is illustrated in a striking way by the following result.

THEOREM 5 (KURATOWSKI). *Let X be any uncountable Borel subset of a complete separable metric space. Then the Borel algebra of all Borel sets of X is the free Borel algebra with countably many generators.*

PROOF. By Remark 1 on page 358 of C. Kuratowski's *Topologie*, 2nd ed., Warsaw, 1948, the algebra in question is the same for all such X as for X the Cantor ternary set. The proof is completed by Theorem 4 above.

COROLLARY. *The free Borel algebra with countably many generators is "homogeneous": isomorphic as a Borel lattice with each of its interval sublattices which has the same cardinality.*

Because of this homogeneity, one cannot characterize nowhere dense subsets of X through their properties as elements of the Borel algebra of *all* Borel subsets of

† This neat interpretation of the Loomis–Sikorski theorem was pointed out to the author by Alfred Hales.

‡ L. Rieger [**1**]; see also R. Sikorski, Ann. Soc. Pol. Math. **23** (1950), 1–20; K. Takeuchi, J. Math. Tokyo **1** (1953), 77–9; [**LT2**, Problem 79].

§ In this connection, see the observations by R. S. Pierce in [**Symp**, pp. 129–40].

X. However, the study of nowhere dense sets leads to another interesting "standard" Borel quotient-algebra, closely related to that of Theorem 5.

We recall that a subset of a topological space X is of *first category* when it can be covered by (hence is) a countable union of nowhere dense sets. We also recall the classic

LEMMA (BAIRE). *Let X be any complete pseudo-metric or locally compact regular space. Then the sets of first category in X form a proper σ-ideal Z in the Borel algebra of all Borel subsets of X, which contains no open set.*

For the proof, see [**Kel**, p. 200, Theorem 34]; following Bourbaki, Kelley calls a set of the first category "meager".

COROLLARY. *In a complete pseudo-metric or locally compact regular space, each Borel set differs by a set of first category from one and only one regular open set.*

This shows that, in the Borel quotient-algebra of all Borel sets of X modulo sets of first category, the regular open sets form a complete set of representatives. We now refer to Theorem IX.3. After a little adaptation to extend Baire's Lemma, we obtain

THEOREM 6. *Let X be any Borel subset of Euclidean n-space with nonvoid interior. Then the Borel algebra B/Z of Borel sets of X, modulo sets of first category, is isomorphic to $\bar{B}_\omega$, the completion by cuts of the free Boolean algebra B_ω with countably many generators.*†

Are all definable sets Borel? Borel used to assert that only Borel sets were "definable",‡ an assertion which bears on the foundations of set theory (§VIII.16). In this connection, note that **R** has only $\mathbf{2}^{\aleph_0} = \mathbf{c}$ Borel subsets, and that only $\aleph_0$ sets can be defined by finite verbal statements! Also, that Borel's assertion implies the "nondefinability" of nonmeasurable sets, and hence that Theorem 13 below is meaningless.

A related question concerns the definability of non-Borel bijections of **R** onto other topological spaces. In this connection note that any Borel bijections $\beta: R \leftrightarrow X$ defines an isomorphism $\mathbf{2}^{[\mathbf{R}]} \leftrightarrow \mathbf{2}^{[X]}$; cf. Exercises 3–8 below.

Exercises for §§2–3:

1. Show that the Boolean algebra of countable sets modulo finite sets is not a Borel algebra.
2. Show that the Borel algebra of all subsets of **R** modulo countable sets is not isomorphic to any σ-field of sets.
3. Show that the algebra of Borel subsets of a T_1-space X is "standard" if and only if there is a Borel bijection from X to $\mathbf{2}^\omega$, the Cantor ternary set.
4. Show that $\mathbf{2}^\omega$ is homeomorphic to its Cartesian product with itself (its "square").

† This result was found by S. Ulam and the author around 1935; see [**LT1**, p. 103]. For Baire's Lemma and its extensions, see also C. Kuratowski, op. cit. supra, §11.

‡ L'ensemble des ensembles mesurables B a la puissance du continu, il existe donc d'autres ensembles mesurables que les ensembles mesurables B; mais cela ne veut pas dire qu'il soit possible $\cdots$ de prononcer un nombre fini de mots caracterisant un et un seul ensemble non mesurable B." (H. Lebesgue, *Lecons sur l'Integration*, Paris, 1904, p. 109.)

5. Construct a Borel bijection from $\mathbf{2}^\omega$ to $\mathbf{R}$, the real line.

6. Construct a Borel bijection from $\mathbf{R}$ to $\mathbf{R}^n$, for any positive integer n; also to $\mathbf{I}^\omega$, the Hilbert cube.

7. Let X and Y be T_1-spaces, and let f and g be Borel injections (one-one maps) of X into Y and vice-versa. Construct a Borel bijection from X to Y. (*Hint.* Use the Cantor–Bernstein construction.)

*8. Show that, if X and Y are any two nontrivial topological complexes, then there is a Borel bijection from X to Y. (*Hint.* Use Exercises 6 and 7.)

*9. For which T_1-spaces does there exist a σ-endomorphism from the Borel algebra of *all* Borel subsets of X onto the complete Boolean algebra of regular open subsets of X? (*Hint.* Consider "Baire" spaces.)

4. Boolean ($\aleph$, $\aleph'$)-algebras

Much research has been done on generalizing the preceding results from countable infinity $\aleph_0$ to other infinite cardinal numbers. But since Sikorski's book† contains a lucid and complete account of these generalizations, I shall only mention a few key results.

Definition. A Boolean algebra A is $\aleph$-*complete* (or a "Boolean $\aleph$-algebra") for a given cardinal number $\aleph$, when every subset of $\aleph$ (or fewer) elements of A has a meet and *a* join in A. A complete Boolean algebra is ($\aleph$, $\aleph'$)-*distributive* when the identity

$$\bigwedge_S \left[\bigvee_T a_{s,t}\right] = \bigvee_{S^T} \left[\bigwedge_S a_{s,\phi(s)}\right] \tag{2}$$

holds for arbitrary sets S and T with cardinalities at most $\aleph$ and $\aleph'$, respectively.

We have proved that $\aleph$-complete Boolean algebras are (m, $\aleph$)-distributive for every finite cardinal number m. However, a surprising theorem of Rieger [1] states that the Theorem of Loomis-Sikorski does not generalize to uncountable cardinal numbers. Namely, for $\aleph$ equal to the power of the continuum or greater, there exist $\aleph$-complete Boolean algebras which are *not* isomorphic to any quotient-algebra F/Δ of any $\aleph$-field of sets by any $\aleph$-ideal.

For any cardinal numbers $\aleph$ and $\aleph'$, one can construct a *free* $\aleph$-complete Boolean algebra with $\aleph'$ (free) generators. (Rieger [1]). The method of Chapter VI applies to infinitary algebras without essential change.

However, Gaifman and Hales‡ have shown that one *cannot* construct a free *complete* Boolean algebra with even countably many generators. To exist, such an algebra would have to have a cardinality exceeding any preassigned cardinal number! By similar arguments, Hales (op. cit.) using a technique invented by Crawley and Dean, has also shown that there is no free complete lattice with even three generators!

Exercises:

1. Show that there is a free complete Boolean algebra with n generators, for any finite n.

2. Show that a Boolean algebra can be embedded in a complete Boolean algebra with preservation of arbitrary joins and meets, if and only if it is (2, $\aleph$)-distributive for all $\aleph$.

† [**Sik**, Chapter II]. See also the articles by Pierce and Dwinger in [**Symp**].

‡ A. W. Hales, Fund. Math. **54** (1964), 45–66; H. Gaifman, ibid., 229–50.

3. Show that, for every infinite $\aleph$, there exists a $\aleph$-field of sets, whose completion by cuts is not $\aleph$-distributive. (R. S. Pierce†)

4. Show that a Boolean $\aleph$-algebra is isomorphic with a $\aleph$-field of sets if and only if, for any $a > 0$ in it, there exists a dual prime $\aleph$-ideal which contains a.

5. Show that, for any cardinal number $\aleph$, the free Boolean σ-algebra with $\aleph$ generators is isomorphic with a σ-field of sets.

*6. Show that the free $\aleph$-complete Boolean algebra with $\aleph$ free generators is not $\aleph$-isomorphic with a $\aleph$-field of sets, for $\aleph$ the power of the continuum. (Rieger [**1**])

7. Generalize Tarski's Theorem V.17 to Boolean $\aleph$-algebras, where $\aleph$ is any cardinal number.‡

*8. (a) Show that, for any countable poset, P, the lattice freely generated by P is a sublattice of the free lattice with three generators.

(b) Generalize to any cardinal number $\aleph$.

*9. If $\aleph$ is an infinite cardinal number with $\aleph^\omega = \aleph$, then there is a complete, homogeneous Boolean algebra of cardinality $\aleph$. (R. S. Pierce§)

5. Finite Measures; Measure Algebras

If B is a Borel algebra, a *finite measure* on B is a *nonnegative* valuation m on B which is *σ-additive* (or "countably additive") in the sense that

$$m\left[\bigvee_{n=1}^{\infty} x_n\right] = \sum_{n=1}^{\infty} m[x_n] \quad \text{if every } x_n \wedge \left(\bigvee_{i=1}^{n-1} x_i\right) = 0. \tag{4}$$

This trivially implies that $m[O] = 0$. General "measures" on B, assuming (positively) infinite values, will not be discussed.

A finite measure with $m[I] = 1$ is called a *probability* measure. From any nontrivial finite measure, one can construct a probability measure by setting $p[x] = m[x]/m[I]$; hence finite measures and "probabilities" are interchangeable.

In §§5–7, we shall survey the properties and construction of finite measures on Borel algebras, emphasizing lattice-theoretic aspects usually not emphasized in textbooks on measure and integration.¶ We begin by recalling two well-known facts.

LEMMA 1. *Let m be a finite measure on a Borel algebra B. Then the set of all $x \in B$ with $m[x] = 0$ is a σ-ideal N of B. Moreover m defines a finite measure on the Borel quotient-algebra B/N.*

We omit the proof, which is straightforward. Note that, in B/N, $x < y$ implies $m[x] < m[y]$: the measure defines a *positive* valuation on B/N.

DEFINITION. A *measure algebra* is a Borel algebra with a finite positive measure. A *probability algebra* is a measure algebra whose measure p satisfies $p[I] = 1$.

---

† Exercises 2–3 results of R. S. Pierce, Pacific J. Math. **8** (1958), 133–40. Exercises 4–5 state results of K. Takeuchi, Tokyo J. Math. **1** (1953), 77–9.

‡ For Example 7, see [**Sik**] or S. Enomoto, Osaka Math. J. **5** (1953), 99–115. For Example 8 see P. Crawley and R. A. Dean, Trans. AMS **92** (1959), 35–47, and **Symp**, pp. 31–42.

§ R. S. Pierce, Proc. AMS **9** (1958), 892–6; also [**Symp**, pp. 129–40].

¶ A notable exception is [**Ca**], where a very thorough discussion is given of measures on Borel algebras ("Somen").

LEMMA 2. *A measure algebra is a complete metric Boolean algebra with* $m[O] = 0$, *and conversely.*

This is a corollary of the results of §X.10; we omit the details.

It is easy to construct the most general measure algebra having a finite number of elements. Any finite Boolean algebra F consists of the joins of subsets S of its points p_k. By (4), the most general positive measure on F must be $m[\bigvee_S p_k] = \sum_S m_k$, where the $m_k = m[p_k]$ are positive constants. Conversely, any choice of such m_k gives a finite measure algebra.

In fact, we can let each p_k correspond to the half-closed interval $[m_1 + \cdots + m_{k-1}, m_1 + \cdots + m_k)$ of the x-axis. Using (4), this correspondence can be uniquely extended to an isometric isomorphism (measure-preserving isomorphism) from F to a field of elementary subsets of the x-axis. This shows that, by adding (4) to any set of postulates for Boolean algebra, we obtain a *complete set of postulates* for *finitary* measure theory, under the usual Boolean operations.†

Example 2. Let $B_\sigma = \mathbf{2}^{[\mathbf{I}]}$ be the (standard) Borel algebra of all Borel subsets of the interval $\mathbf{I} = [0, 1]$, and let m be the usual (Borel–Lebesgue) measure on B_σ. Then B_σ/N is a probability algebra, hence a complete metric Boolean lattice.

THEOREM 7. *The algebra* B_σ/N *of Example* 2 *is separable* (*i.e., it has a countable dense subset*).

PROOF. By Lemma 2, B_σ/N is a metric lattice. Moreover by Theorems 3 and 5, B_σ is generated by a countable set of generators. The proof is completed by the following simple lemma.

LEMMA 3. *Let C be a countable subset of a measure algebra M. Then the Borel subalgebra generated by C in M is a separable closed subset of M.*

PROOF. Let S be the Boolean subalgebra (*not* Borel subalgebra) of M generated by C; since S consists of Boolean polynomials $p(c_1, \cdots c_r)$ of *finite* subsets of C, S is countable. The metric *closure* $\bar{S}$ of S is also a Boolean subalgebra of M (since $\wedge$, $\vee$, $'$ are uniformly continuous); it is separable since S is countable; and it is a Borel subalgebra since every closed Boolean subalgebra of a measure algebra is a Borel subalgebra. Finally, since it is contained in any Borel subalgebra of M which contains C, and is a Borel subalgebra of M which contains C, it is the Borel subalgebra of Lemma 3.

We now prove the converse of Theorem 7, incidentally revealing the B_σ/N of Example 2 as a standard probability algebra, analogous to the B/Z of Theorem 6, for $X = (-\infty, \infty)$.

THEOREM 8. *Let M be a separable probability algebra without atoms* (*i.e., points of finite measure*). *Then M is isometrically isomorphic with the probability algebra B_σ/N of Example* 2.

† See G. Birkhoff, Proc. Cambridge Philos. Soc. **30** (1934), 115–22; H. P. Evans and S. C. Kleene, Amer. Math. Monthly **46** (1939), 141–8.

PROOF. Let $x_1, x_2, x_3, \cdots$ be a countable dense subset of M, and let S_n be the finite Boolean subalgebra of M generated by $x_1, \cdots, x_n$. As shown above, we can represent S_n by a field of subsets of $[0, 1] = \mathbf{I}$ so that m is measure. The set-union $\bigcup S_n = U$ is a Boolean subalgebra of $\mathbf{2}^{\mathbf{I}}$ with a positive valuation. Moreover, unless M contains an atom of positive measure, *every* interval $[0, a)$ of $\mathbf{I}$ is a metric limit of elements of U. Hence the image $\bar{U}$ of M contains every Borel subset of $\mathbf{I}$, modulo a set of measure zero. Since distinct elements of M are a positive distance apart, it follows that $M \cong B_\sigma/N$, as claimed.

Since the hypotheses of Theorem 8 are satisfied up to a factor $m[I]$ by any measure algebra, we have the

COROLLARY. *Let M be any nontrivial separable measure algebra without atoms. Then M is isometrically isomorphic to the measure algebra of measurable subsets of $[0, m[I]]$, modulo sets of measure zero. Moreover the isometric automorphisms of M are transitive on the elements with $m[x] = c$, for any finite c. Finally, M is homogeneous: isomorphic with all its nontrivial principal ideals.*

6. Outer Measure; Regular Measure

An *outer measure* on a Borel algebra A is defined as an isotone, extended† real-valued function m^* on A which satisfies $m^*[O] = 0$ and

$$m^*\left[\bigvee_{k=1} a_k\right] \leqq \sum_{k=1} m^*[a_k] \qquad \text{(subadditivity).} \tag{5}$$

We shall now show that one can always extend any measure on Borel sets to an outer measure defined for all subsets. (Trivially, any measure is an outer measure.)

Let B be any Borel subalgebra of any Borel algebra A, and let m be any measure on B. Define

$$m^*[a] = \inf_{b \geqq a, b \in B} m[b], \quad \text{for any } a \in A. \tag{6}$$

Define a *null* element as an $a \in A$ with $m^*[a] = 0$.

LEMMA 1. *The function m^* is an outer measure on A; the null elements form a σ-ideal of A. Hence the set $C \subset A$ of all $c \in A$ which differ from some $b \in B$ by a null element is a Borel subalgebra of A, and the restriction μ of m^* to C is a measure on C. Finally, $\mu^* = m^*$.*

We omit the proof of Lemma 1,‡ and call μ the *Lebesgue completion* of m, having the following example in mind. (If $C = B$, we call m *complete*, for the same reason.)

---

† That is, we allow the value ∞ to be assumed, whereas in §5, to bring out the connection with valuations, only finite measures were considered.

‡ See P. R. Halmos, *Measure theory*, van Nostrand, 1950, p. 55, Theorem B. Most of the proofs omitted in §§5–6 can be found in this excellent book.

Example 3. Let $A = \mathbf{2}^E$, where E is Euclidean n-space; let $B = \mathbf{2}^{[E]}$ be the Borel subalgebra of all Borel subsets of E; and let m be Borel measure. Then C is the class of Lebesgue measurable sets.

We now show how, conversely, from any outer measure m^* on a Borel algebra A, one can construct a measure on a Borel subalgebra $C \subset A$. We define $c \in C$ to mean that

$$m^*a = m^*[a \wedge c] + m^*[a \wedge c'] \quad \text{for all } a \in A. \tag{7}$$

We define $\nu = (m^*)^\dagger$ as the restriction of m^* to C.

LEMMA 2. *For any outer measure m^*, $(m^*)^\dagger = \nu$ is a complete measure. If m^* is the outer measure on A defined from a measure m on $A = B$ by* (6), *then $(m^*)^\dagger$ is the Lebesgue completion of m.*

COROLLARY. *If m^* is the outer measure defined from a measure m on A by* (6) *then $\nu = (m^*)^\dagger$ satisfies*

$$m^*A = \inf_{c \geqq a, c \in C} \nu[c], \quad \textit{for all } a \in A. \tag{8}$$

That is, in symbols, $((m^*)^\dagger)^* = m^*$.

Carathéodory† postulated this condition (8); we shall call an outer measure which satisfies (8) "closed". The preceding results essentially establish the following result.

THEOREM 9. *The correspondences $m \to m^*$ and $m^* \to (m^*)^\dagger$ are inverse bijections between the complete measures and the closed outer measures on any Borel algebra A.*

Regular measures. Now let X be any topological space; let $A = \mathbf{2}^X$ be the (complete) Borel algebra of all subsets of X, and let $B = \mathbf{2}^{[X]}$ be the (atomic) Borel subalgebra of all Borel subsets of X. A measure is called *regular on X* when it is the Lebesgue completion of its restriction to B. The following result is then a direct consequence of the facts stated above.

THEOREM 10. *Let X be a standard uncountable Borel space, so that $\mathbf{2}^{[X]}$ is the free Borel algebra with $\aleph_0 = \omega$ generators. Let m be any finite regular measure on X with $m[I] > 0$, but $m[p] = 0$ for every point $p \in X$. Then the measure algebra defined by m is isomorphic to the B_σ/N of Example 2, under an isomorphism which multiplies all measures by the constant factor $m[I]$.*

Jordan content and Lebesgue measure. Against the background of the preceding definitions, one can also construct Lebesgue measure from "Jordan content",‡ as follows. Jordan content is a nonnegative valuation with $v[O] = 0$,

† C. Carathéodory, *Vorlesungen über reelle Funktionen*, Teubner, Leipzig, 1927, p. 258, V. The concepts of outer measures and regular measures go back to C. Carathéodory, Gott. Nachr., Math.-Phys. Kl. (1914), 1–23.

‡ For an elementary discussion of Jordan content, see Tom M. Apostol, *Mathematical analysis*, Addison–Wesley, 1960, p. 225.

defined on the *Boolean subring* S of "elementary subsets" of finite volume in Euclidean n-space, of $A = \mathbf{2}^E$. We can define an *outer measure* m^* on A by changing (6) to

$$m^*[a] = \inf_{\vee x_i \geqq a} \sum_{i=1}^{\infty} v[x_i], \qquad i = 1, 2, 3, \cdots. \tag{9}$$

In words, $m^*[a]$ is the l.u.b. of the "total values" of *countable coverings* of a. The problem is to show that this outer measure *is* a closed outer measure, and that $m^*[a] = v[a]$ for all $a \in A$.

Exercises for §§5–6:

1. Prove Lemma 1.
2. Prove Lemma 2.
3. Show that a nonnegative *valuation* with $v[O] = 0$ on a Borel algebra is a (finite) *measure* if and only if it is *continuous* in the order topology of B.
4. Show that a Borel algebra with finite measure m cannot have an uncountable subset of disjoint elements of positive measure.
5. Prove Theorem 9 in detail.
6. Prove Theorem 10 in detail.
7. Show that, if (9) is applied to Jordan content on $\mathbf{R}$, then $m^*[a] = v[a]$ for any interval. (*Hint.* Use the Heine–Borel Theorem.)
8. Under the hypotheses of Exercise 7, show that the application of (7) to m^* on $\mathbf{2}^{\mathbf{R}}$ makes C contain every interval, and hence every Borel set.

*7. Existence of Measures

In §§5–6, we have assumed the existence of measures and outer measures as given, in all cases except that of finite Boolean algebras $\mathbf{2}^n$. Actually, the question of the *existence* of measures on prescribed Borel algebras is one of the deepest questions of measure theory. We present here a few relevant results.

One can prove the existence of nontrivial isotone *valuations* on any Boolean algebra by transfinite induction, the sharpest result being given by the following theorem of Tarski.†

THEOREM 11. *Any isotone valuation with $v[O] = 0$ defined on a subalgebra S of a Boolean algebra A can be extended to one defined on all of A.*

PROOF. Since our conditions are of finite character (Chapter VIII), it suffices to show that an extension to some subalgebra $T > S$ is always possible. But let $a \notin S$ be given; the elements $(a \wedge b) \vee (a' \wedge c)$ $[b, c \in S]$ form such a subalgebra. We define explicitly

$$\begin{aligned} m_1[a \wedge b] &= \operatorname{Inf} m[s] \quad \text{for } s \geqq a \wedge b \text{ in } S, \\ m_1[a' \wedge c] &= \operatorname{Sup} m[t] \quad \text{for } t \leqq a' \wedge c \text{ in } S. \\ m_1[(a \wedge b) \vee (a' \wedge c)] &= m_1[a \wedge b] + m_1[a' \wedge c]. \end{aligned} \tag{10}$$

† A. Tarski, Fund. Math. **15** (1930), 42–50; also ibid. **32** (1938), 45–63, and **33** (1945), 51–65.

We shall show that this is an extension, as desired.

$$m_1[a \wedge b] + m_1[a' \wedge b] = \text{Inf}_s\,\{m[s] + \text{Sup}_t\, m[t]\}.$$

But given $s \geqq a \wedge b$, $s' \wedge b \leqq (a' \vee b') \wedge b = a' \wedge b$ is a t of the desired kind. Hence, for any admissible s, we have

$$m[s] + \text{Sup}_t\, m[t] \geqq m[s] + m[s' \wedge b] \geqq m[s \vee (s' \wedge b)] = m[s \vee b] \geqq m[b].$$

Hence $m_1[a \wedge b] + m_1[a' \wedge b] \geqq m[b]$, for any $b \in S$. By duality, we get the reverse inequality, proving $m_1[b] = m[b]$.

It remains to prove that m_1 is additive; it is enough to show that a-components are additive—that for all $b, c \in S$, $(a \wedge b) \wedge (a \wedge c) = O$ implies

$$m_1[(a \wedge b) \vee (a \wedge c)] = m_1[a \wedge b] + m_1[a \wedge c].$$

For any $\epsilon > 0$, we can find $t \geqq a \wedge b$, $u \geqq a \wedge c$ in S, with $m[t] < m_1[a \wedge b] + \epsilon$, $m[u] < m_1[a \wedge c] + \epsilon$, hence

$$m[t \vee u] \leqq m[t] + m[u] \leqq m_1[a \wedge b] + m_1[a \wedge c] + 2\epsilon.$$

Since $t \vee u \geqq (a \wedge b) \vee (a \wedge c)$, $m_1[(a \wedge b) \vee (a \wedge c)] \leqq m_1[a \wedge b] + m_1[a \wedge c]$. Conversely, for any $v \geqq (a \wedge b) \vee (a \wedge c) = a \wedge (b \vee c)$ in S, we have $m[v] \geqq m[v \wedge b \wedge c'] + m[v \wedge c]$, since the two are disjoint. By straight Boolean algebra, $v \wedge b \wedge c' \geqq a \wedge b$; thus since $a \wedge c \wedge c' = 0 = a \wedge b \wedge c$,

$$\begin{aligned} v \wedge b \wedge c' &\geqq a \wedge (b \vee c) \wedge (b \wedge c') \\ &= (a \wedge b \wedge c') \vee (a \wedge b \wedge c) = a \wedge b. \end{aligned}$$

Hence $m[v \wedge b \wedge c'] \geqq m_1[a \wedge b]$; similarly, $m[v \wedge c] \geqq m_1(a \wedge c]$. Hence every $m[v] \geqq m_1[a \wedge b] + m_1[a \wedge c]$, and

$$m_1[(a \wedge b) \vee (a \wedge c)] = \text{Inf}\, m[v] \geqq m_1[a \wedge b] + m_1[a \wedge c). \qquad \text{Q.E.D.}$$

Not every Borel algebra can be made into a measure algebra, as the following result shows† (see also Exercise 4 above).

THEOREM 12. *Any measure algebra is weakly countably distributive, in the sense that if $a_{ij} \uparrow a_i$ for every fixed i ($i, j = 1, 2, 3, \cdots$), then there exists a sequence of functions $j_n(i) = j(n, i)$ such that*

$$\bigwedge_i \Bigl(\bigvee_j a_{ij}\Bigr) = \bigwedge_i a_i = \bigvee_n \,(\bigwedge a_{i,j(n,i)}). \tag{11}$$

PROOF. Since $a_{ij} \uparrow a_i$, we can choose $j(n, i)$ so that $m[a_i] - m[a_{i,j(n,i)}] \leqq 2^{-n-i}$. For any such $j(n, i)$, since joins and meets have Lipschitz modulus one,

$$m[\bigwedge a_i] - m\Bigl(\bigwedge_i a_{i,j(n,i)}\Bigr) \leqq \sum_{i=1}^{\infty} 2^{-n-i} = 2^{-n}.$$

† Theorems 12–13 are essentially due to S. Banach and C. Kuratowski, Fund. Math. **14** (1929), 127–31; Banach, ibid. **15** (1930), 97–101; S. Ulam, ibid. **16** (1931), 140–50. The general question of which Borel algebras can be made into measure algebras has been answered by D. Maharam, Ann. of Math. **48** (1947), 154–67, and J. L. Kelley, Pacific J. Math. **9** (1959), 1165–77.

Hence $m[\bigwedge a_i] - m[\bigvee_n (\bigwedge_i a_{i,j(n,i)})] \leqq 2^{-n}$ for all n; but the first expression includes the second by the isotonicity of lattice operations; hence the two are equal, proving (11).

THEOREM 13. *If the Continuum Hypothesis is true, then no nontrivial countably additive measure can be defined for all subsets of the continuum, such that every point has measure zero.*

PROOF. Form the set of all (single-valued) functions $\alpha: \alpha(i) = j$ from the positive integers to the positive integers. There are $2^{\aleph_0}$ of these; hence if the Continuum Hypothesis is true, we can well-order the set A of α so that each α has countably many predecessors. Now reject all α such that, for some $\beta < \alpha$ in A, $\alpha(i) < \beta(i)$ for some i. The residual set B of unrejected β has the property that to every $\alpha \in A$ corresponds a $\beta \in B$, with $\beta < \alpha$ and $\alpha(i) \leqq \beta(i)$ for all i (since the relation is transitive and A well ordered). Again, if B were countable, then we could enumerate its members $\beta_1, \beta_2, \beta_3 \cdots$; defining $\beta^*(k) = \beta_k(k) + 1$, we would get a new unrejected β^*, contrary to hypothesis. Hence B is uncountable. By the Continuum Hypothesis again, it is therefore in one-one correspondence with the continuum, so that points of the continuum can be denoted $p_\beta[\beta \in B]$. Now let X_k^i denote the set of all points p_β with $\beta(i) = k$; clearly $\bigvee_{k=1}^{\infty} X_k^i = I$, for all i, since $\beta(i)$ has some value. Hence, by (10), for some $r(i)$, $m[\bigwedge_i (\bigvee_{j=1}^{r(i)} X_j^i)] > 0$. But some $\beta_0 \in B$ satisfies $\beta_0(i) \geqq r(i)$ for all i. And $\beta(i) \leqq \beta_0(i)$ for all i will imply $\beta \leqq \beta_0$ in B; the set of these predecessors of β_0 is however always countable. Hence $\bigwedge_i (\bigvee_{j=1}^{r(i)} X_j^i)$ contains only *countably* many points p_β, and must have measure zero. This contradiction proves the theorem.

Exercises:

1. Let C be any chain which contains O and I in a distributive lattice L with universal bounds, and let v be any isotone (real) functional on C with $v[O] = 0$. Show that v can be extended to an isotone functional on L, and that the extension is unique if C is a maximal chain.

2. Show that, in $\mathbf{2^I}$, we can construct a nonnegative valuation with $v[O] = 0$ and $v[\mathbf{I}] = 1$.

3. Show that the metric completion of the metrized free Boolean algebra with countable generators is not isomorphic with its completion by cuts.

*4. In Example 2 of §5, show that $\mathbf{2}^{[X]}$ has a subalgebra† with just one representatative from every residue class of N.

5. (a) Show that, in an atomless measure algebra, for any sequence of positive x_i, $\inf_{0 < y_i \leqq x_i} (\bigvee_i y_i) = 0$.

(b) Show that the algebra B_σ/Z of Borel sets modulo sets of first category is not isomorphic with any measure algebra.

(c) Show that B_σ/Z admits no nontrivial measure.

6. Let C be a subset of the unit interval $I = [0, 1]$ which is homeomorphic to the Cantor ternary set $\mathbf{2}^\omega$. Show that there is a measure m on I with $m[p] = 0$ for any point p, yet $m[C] = m[I] = 1$. (Such a measure is called "singular" when C has Lebesgue measure zero.)

7. Show that $\mathbf{I} = [0, 1]$ contains the power of the continuum disjoint "Cantor ternary" subsets C_x.

---

† J. von Neumann, Crelle's J. **165** (1931), 109–15; D. Maharam, Proc. AMS **9** (1958), 987–94.

8. Von Neumann Lattices

A *von Neumann lattice* is (by definition) a complete, complemented modular topological lattice. Since any lattice of finite length is, trivially, a complete topological lattice, it follows that any modular geometric lattice (§IV.6) is a von Neumann lattice. Likewise, any complete metric lattice is a von Neumann lattice. Indeed, von Neumann's guiding idea was to free the theory of modular geometric and complete metric lattices from restrictions imposed by chain conditions. He had in mind many relevant examples, of which we mention two.

Example 4. Any (cardinal) product $\prod \Gamma_j$ of continuous geometries $\Gamma_j = \mathrm{CG}(D_j)$ is a von Neumann lattice.

Example 5. Any complete Boolean algebra is a von Neumann lattice. In particular, M/N is a von Neumann lattice.

We now recall the notion of the *center* of a lattice, as defined in §III.8. In Example 4, the center of the product $\prod \Gamma_j$ is the complete *atomic* Boolean algebra of all subsets of the set of indices j. More generally, Halperin† has shown how to construct from *any* complete Boolean algebra B and continuous geometry Γ a von Neumann lattice having B for center and whose "local components" contain copies of Γ as sublattices.

As in §III.9, an element of a (von Neumann) lattice L is in the center of L if and only if it is distributive. Moreover, by Theorem III.18, an element $a \in L$ is in the center of L if and only if its complement a' is *unique*. In this case, the complement a' is also in the center.

Further, by continuity, if the a_α are in the center of L and $a_\alpha \uparrow a$, then the a'_α are also in the center of L and $a'_\alpha \downarrow$. If L is complete, then $a'_\alpha \downarrow b$ for some $b \in L$, and by continuity

$$a \wedge b = (\sup a_\alpha) \wedge (\inf a'_\alpha) = \sup [a_\alpha \wedge (\inf a'_\alpha)] = O.$$

Dually, $a \vee b = I$, and we have proved

THEOREM 14. *The center of any von Neumann lattice L is a complete Boolean algebra.*

In fact, the center of L is the *structure lattice* of L, considered as an infinitary algebra (under unrestricted joins and meets).

DEFINITION. An *irreducible* von Neumann lattice is one whose center consists only of O and I.

We leave to the reader the proof of the important fact that, for any division ring D, the metric lattice CG(D) constructed in §X.5 is an irreducible von Neumann lattice.

One can rid Theorem 14 of the assumption of continuity, proving its conclusion in any complete, relatively complemented (not necessarily modular) lattice M, as follows.‡ Define (a, b) to be a *distributive pair* in a modular lattice M, when

† I. Halperin, Trans. AMS **107** (1963), 347–59. For the notion of "local component", see T. Iwamura, Japan J. Math. **19** (1944), 54–71.

‡ The author is indebted to Samuel Holland, Jr., for the exposition in the rest of this section.

every triple $(a_1 b, x)$ with $x \in M$ generates a distributive sublattice of M. Referring back to Janowitz' Lemma 2 of §III.10, we have

LEMMA 1. *In a modular lattice,* $a \nabla b$ *if and only if* $a \wedge b = O$ *and* $(a, b)D$.

The proof is straightforward, and will be omitted.

As an immediate corollary of the preceding lemma and Janowitz' Lemma 2, §III.10, we obtain

LEMMA 2. *In any complemented modular lattice,*

$$a \wedge b = O \quad \textit{and} \quad (a, b)D \tag{13}$$

if and only if

$$a_1 \leqq a, \quad b_1 \leqq b \quad \textit{and} \quad a_1 \sim b_1 \quad \textit{imply} \quad a_1 = b_1 = O. \tag{14}$$

PROOF. Compare conditions (i) and (iii) of the above-mentioned lemma, and use Lemma 1. (Recall that any complemented modular lattice is relatively complemented.)

Finally, using condition (ii) of Janowitz' Lemma, we obtain the following result as an immediate consequence.

LEMMA 3. *In any complete, relatively complemented lattice* L, $a \nabla b_\beta$ *for all* β *implies* $a \nabla (\bigvee b_\beta)$.

From this result there follows

THEOREM 14′ (M. F. JANOWITZ). *Let* L *be any relatively complemented lattice with* O *and* I. *Then the center of* L *is closed under the formation of arbitrary* sup *and* inf.

COROLLARY 1. *The center of any* σ*-complete relatively complemented lattice with* O *and* I *is a Borel algebra.*

COROLLARY 2. *The center of any complete relatively complemented lattice is a complete Boolean algebra.*

Exercises:

1. Let a complemented modular lattice contain a maximal subchain C. Show that it contains a maximal subchain dually isomorphic to C. (*Hint*. Use the Axiom of Choice.)
2. Show that, if a von Neumann lattice satisfies the Ascending Chain Condition, it is a modular geometric lattice.
3. Show that the center of the cardinal product $\prod \Gamma_j$ of Example 4 is $\mathbf{2}^J$, where J is the index set.
4. (a) Show that any interval sublattice of a von Neumann lattice L is itself a von Neumann lattice.

   (b) State and prove an analogous result for epimorphic images of L under *complete* lattice-morphisms (i.e., for arbitrary joins and meets).

*5. Show that every nontrivial interval sublattice of an irreducible von Neumann lattice is irreducible.†

† For the concrete case of CG(D), this "homogeneity" is easy to prove. For the general case, see [**CG**, p. 39].

6. Show that the center of any complete topological lattice is a complete Boolean algebra.

7. Show that CG(F) is *not* metrically compact. Infer that the metric topology and the interval topology on CG(F) differ.

*8. Show that an irreducible von Neumann lattice is compact in its metric topology if and only if it is finite. (A. Ramsay)

9. Perspectivity is Transitive

We recall (§III.11) that two elements a and b of a complemented modular lattice M are called *perspective* (in symbols, $a \sim b$) when they have a *common complement* d, such that:

$$a \wedge d = b \wedge d = O \quad \text{and} \quad a \vee d = b \vee d = I. \tag{12}$$

We recall also (§IV.6, Lemma 2) that perspectivity is transitive in any modular geometric lattice—i.e., in any von Neumann lattice of finite length. By brilliant combinatorial and continuity arguments (passages to the limit), von Neumann showed that this result was true *without* assuming any chain conditions; see his Theorem 18 below.

As his proof of this fact was long and technical, and even important simplifications by Halperin and F. Maeda leave it so, only an outline of the main steps involved will be given below. For many detailed proofs, the reader must consult books (notably [**CG**], [**KG**], and [**Sko**]) dealing specifically with continuous geometries and their generalizations.

For part of the proof, we again make no use of the assumption of continuity. We have

THEOREM 15. *If a, b, c are nonzero elements of a complemented modular lattice M, with $a \sim b \sim c$, then there exist $a_1 \sim c_1$ in M with $0 < a_1 \leqq a$, $0 < c_1 \leqq c$.*

PROOF. If $a \wedge c > 0$, set $a_1 = c_1 = a \wedge c$ and the result is trivial; hence we can suppose $a \wedge c = 0$.

But suppose $a \wedge c = 0$, yet that there do *not* exist a_1, c_1 satisfying the conditions of the theorem. Then condition (iii) of Janowitz' Lemma holds, and so $b = (b \vee a) \wedge (b \vee c)$. Let x be the common complement of b and c, and set $z = [x \wedge (a \vee b)] \vee [y \wedge (b \vee c)]$. Then, using L5:

$$\begin{aligned} a \vee z &= a \vee [x \wedge (a \vee b)] \vee [y \wedge (b \vee c)] \\ &= [(a \vee x) \wedge (a \vee b)] \vee [y \wedge (b \vee c)] \\ &= a \vee b \vee [y \wedge (b \vee c)] = a \vee [(b \vee y) \wedge (b \vee c)] = a \vee b \vee c, \end{aligned}$$

and similarly $b \vee z = a \vee b \vee c = a \vee z$. Continuing,

$$\begin{aligned} a \wedge z &= a \wedge \{[x \wedge (a \vee b)] \vee [y \wedge (b \vee c)]\} \\ &\leqq (a \vee b) \wedge \{[x \wedge (a \vee b)] \vee [y \wedge (b \vee c)]\} \\ &= [x \wedge (a \vee b)] \vee [(a \vee b) \wedge (b \vee c) \wedge y] \\ &= [x \wedge (a \vee b)] \vee (b \wedge y) = x \wedge (a \vee b). \end{aligned}$$

Thus $a \wedge z \leqq x \wedge (a \vee b)$, and so $a \wedge z \leqq a \wedge x \wedge (a \vee b) = a \wedge x = 0$. Similarly $c \wedge z = O$, and so $a \sim c$, a clear contradiction since $O < a \leqq a$ and $O < c \leqq c$.

We next define independence for arbitrary (not necessarily finite) sets. A subset S of elements of a complete modular lattice M is called *independent* as in §IV.4 when, for all σ, $a_\sigma \wedge \bigvee_{\tau \neq \sigma} a_\tau = O$. Three lemmas follow easily

LEMMA 1. *In any von Neumann lattice*: (i) *if S and T are independent subsets, and $\bigvee_S a_\sigma \wedge \bigvee_T a_\tau = 0$, then $S \cup T$ is independent*, (ii) *if every finite subset of S is independent, then S is independent.*

Clearly, (ii) states that independence is a property of "finite character" (Chapter VIII) in von Neumann lattices. It can be proved by simple passage to the limit, from the combinatorial results of §IV.4. This is also true of the next two lemmas, which state results of [**CG**, pp. 20–21].

LEMMA 2. *If $\{a, b, c\}$ is independent, and $a \sim b$, then $a \vee c \sim b \vee c$.*

LEMMA 3. *Let $\{a_\sigma\} = A$ and $\{b_\sigma\} = B$ $(\sigma \in J)$ be two subsets of a von Neumann lattice, with $a_\sigma \sim b_\sigma$ for all σ, and let $A \cup B$ be independent. Then $a_\sigma \sim b_\sigma$.*

We now come to one of the most technical parts of the proof; the details may be found in [**CG**, pp. 264–5]; they make use of Lemma 2 Theorem 15 and transfinite induction. The major results can be summarized as follows.

LEMMA 4. *Let $a \wedge b = 0$ in a von Neumann lattice L. Then there exist relative complements a', a'' and b', b'' in the interval sublattices* $[0, a]$ *resp.* $[0, b]$, *such that*: (i) $a' \sim b'$, (ii) $(a'', b'')D$.

THEOREM 16. *Given arbitrary $a, b \in L$, there exist relative complements a', a'' in $[O, a]$ and b', b'' in $[O, b]$ such that*: (i) $a' \sim b'$, (ii) $(a'', b'')D$, *and* (iii) $a'' \wedge b'' = O$.

Another very technical part of the proof concerns showing that, in a von Neumann lattice, *no interval can be projective with a proper part of itself:* the phenomenon arising in Example 3 of §III.12 and in Example 4 of §X.4 cannot occur in *topological* complete complemented modular lattices. Specifically, we have

THEOREM 17. *In a von Neumann lattice, $[O, a]$ projective to $[O, b]$ is impossible if $a < b$.*

A proof is given in [**CG**, Part I, Chapter IV]. The essential idea is to utilize *perspectivity by decomposition* (see §15 below): a triple (a, b, c) is said to be *perspective by decomposition* when there exist three *independent* sequences $\{a_i\}$, $\{b_i\}$, $\{c_i\}$ such that $a_i \sim b_i \sim c_i \sim a_i$ $(i = 1, 2, 3, \cdots)$ and

$$a = \bigvee a_i, \qquad b = \bigvee b_i, \qquad c = \bigvee c_i. \tag{15}$$

By using both Theorem 16 and Theorem 17, one can prove without much further difficulty†

† [**CG**, p. 265, Theorems 2.2–2.3], [**KG**, p. 91, Satz 2.2]; I. Halperin, Trans. AMS **44** (1938), 537–62. Halperin showed that only σ-completeness and σ-continuity need be assumed.

THEOREM 18 (VON NEUMANN). *In any complemented, modular, complete topological lattice, perspectivity is transitive: if $[O, a]$ is projective to $[O, b]$, then $a \sim b$.*

Perspectivity is also transitive in some important *non*modular orthomodular lattices, notably in the projection lattice of any w^*-algebra†. In this connection, we mention an unpublished result of D. M. Topping: the projection lattice of any von Neumann algebra is semimodular.

Exercises:

1. Show that, if e is in the center of a von Neumann lattice, then $e \sim e_1$ implies $e = e_1$.
2. Defining "independence" as in the text, show that in *any* complete lattice: "every subset of an independent set is independent".
3. Prove Lemma 1. (See [**CG**, p. 10, Theorem 2.3] for (ii).)
4. Prove Lemma 2.
5. Show that a subset R of elements of a von Neumann lattice is independent if and only if, for any subsets $S \subset R$ and $T \subset R$: $(\sup S) \wedge (\sup T) = \sup S \cap T$.
6. Show that if R is an independent subset of L, then $\bigwedge_\alpha (\sup S_\alpha) = \sup (\bigcap S_\alpha)$. (*Note.* Exercises 5–6 state Theorems 2.5–2.6 of [**CG**, pp. 11–12].)
7. Let $a_1, a_2, a_3, \cdots$ be an infinite independent sequence in a von Neumann lattice, such that $a_i \sim a_{i+1}$ for all i. Prove that every $a_i = 0$. ([**CG**, p. 21, Theorem 3.8])
8. Using Exercise 7 but not Theorem 17, show that if $a \sim c \sim b$, $b \wedge c = 0$, and $a \leqq b$, then $a = b$. ([**CG**, p. 22, Theorem 3.9])
9. Using Exercise 3 but not Theorem 17, show that if $a \sim b \sim c$, $a \wedge b = b \wedge c = 0$, and $b \leqq a \vee c$, then $a \sim c$.

10. Dimension Functions

The relation $a \sim b$ of perspectivity is an equivalence relation on any von Neumann lattice, by Theorem 18; hence there is a natural function from L to the set P of its equivalence classes $A = (a)$, $B = (b), \cdots$ mod $\sim$. We shall call this function $d: L \to P$ the *dimension function* of L, and show that its range P also has an interesting algebraic structure, under suitable complementation and (partial) addition operations, and an ordering relation. We first treat complementation.

LEMMA 1. *If $a_1 \sim a_2$, then $a_1' \sim a_2'$ for every pair of complements a_1' of a_1 and a_2' of a_2.*

PROOF. By hypothesis, $c \wedge a_j = a_j' \wedge a_j = O$ and $c \vee a_j = a_j' \vee a_j = I$ for $j = 1, 2$, and some $c \in L$. But this makes $a_1' \sim c \sim a_2'$, whence by Theorem 18 $a_1' \sim a_2'$ as claimed.

COROLLARY. *In P, the set A' of all complements of any $a \in A$ is a single equivalence class.*

DEFINITION. Let P be the range of the dimension function on a von Neumann lattice L. Then $A \prec B$ in P means that $a \leqq b$ for some $a \in A$ and $b \in B$, considered as subsets of L.

LEMMA 2. *$A \prec B$ implies $B' \prec A'$.*

† See P. A. Fillmore, Proc. AMS **16** (1965), 383–7. This paper contains some other important results.

PROOF. If $a \leqq b$ for some $a \in A$ and $b \in B$, and a' is any complement of a, then $a' \vee b \geqq a' \vee a = I$. Hence (§I.9 or §IV.6) if c is any relative complement of $a' \wedge b$ in a', $c \wedge b = O$ and $c \vee b = I$. Therefore $c \in B'$, but $c \leqq a' \in A'$, proving the result.

Since a' was *any* complement of a in Lemma 2, and the a_j' *any* complements of the a_j in Lemma 1, we can easily derive from Lemmas 1–2 the following

COROLLARY. *Each of the following conditions is necessary and sufficient for* $A \prec B$ *to hold*:

(i) $a' \vee b = I$ *for some* $b \in B$ *and some complement* a' *of some* $a \in A$,
(i′) $a \wedge b' = O$ *for some* $a \in A$ *and some complement* b' *of some* $b \in B$,
(ii) *for every* $b \in B$, $a' \vee b = I$ *for some complement* a' *of some* $a \in A$,
(ii′) *for every* $a \in A$, $a \wedge b' = O$ *for some complement* b' *of some* $b \in B$.

PROOF. Trivially, $A \prec B$ implies (i), since it implies $a' \vee b \geqq a' \vee a = I$. Conversely, (i) implies for any relative complement d of $a' \wedge b$ in b that $d \wedge a' = O$ and $d \vee a' = I$; hence a' has a complement $d \leqq b$, or $a \sim d$ for some $d \leqq b$. Since $a \sim d$ implies $d \in A$, this proves $A \prec B$. But, as in the proof of Lemma 2, this implies that *every* $a' \in A'$ satisfies $b' = c \in B'$ for some $b' \leqq a'$, and so $b' \wedge a = O$ and $b \vee a' = I$ for some complementary pair $b \in B$, $b' \in B'$, condition (ii′). The rest is obvious using duality (including Lemma 1).

LEMMA 3. *The relation* $\prec$ *quasi-orders* P.

PROOF. Trivially, $A \prec A$. As above, $A \prec B$ implies that $a \leqq b$ for every $a \in A$ and some $b \in B$, while $B \prec C$ implies this $b \leqq c$ for some $c \in C$. Hence, by definition, $A \prec B$ and $B \prec C$ imply $A \prec C$.

THEOREM 19. *Under the relation* $\prec$ *and complementation, the range* P *of the dimension function of any von Neumann lattice is an ortholattice.*

PROOF. By the Corollary of Lemma 2, $A \prec B$ and $B \prec A$ imply that $a \leqq b$ for *every* $a \in A$ and some $b \in B$, *and* that this $b \leqq a_1$ for *some* $a_1 \in A$. Hence $a \leqq b \leqq a_1$ for some a, $a_1 \in A$ and $b \in B$. But in a modular lattice, $a < a_1$ is incompatible with the condition that a, a' have a common complement ($a \sim a_1$); hence $a = b = a_1$, and $A = B$. Therefore P is a poset. For the fact that P is a lattice, see [**CG**, p. 269] or [**KG**, pp. 96–7].

THEOREM 20. *A von Neumann lattice is irreducible if and only if its dimension function has a simply ordered range.*

We refer the reader to [**CG**, pp. 39–40] and [**KG**, pp. 96–7] for proofs of this fact.

Disjoint sums. In any lattice, we define the *disjoint sum* $a \sqcup b$ as the join $a \vee b$ *if* $a \wedge b = O$. Disjoint sums can be used to define a *partial addition* in P, essentially because of the following lemma.

LEMMA 4 (MAEDA). *In any von Neumann lattice*:

(i) *if* $a_1 \sqcup a_2 = b_1 \sqcap b_2$ *and* $a_1 \sim b_1$, *then* $a_2 \sim b_2$,

(ii) *if* $a_1 \sqcup a_2 \sim b_1 \sqcup b_2$ *and* $a_1 \sim b_1$, *then* $a_2 \sim b_2$,

(iii) *if* $a_j \sim b_j$ $(j = 1, 2)$ *and* $a_1 \sqcup a_2$, $b_1 \sqcup b_2$ *exist, then* $a_1 \sqcup a_2 \sim b_1 \sqcup b_2$.

For the proof, see [**KG**, p. 91, Hilfsatz 2.2]. Using this result, and Theorem 19, one can prove

THEOREM 21. *Let P be the range of the dimension function on a von Neumann lattice L. When $A \prec B'$ (or, equivalently, $B \prec A'$), define $A + B$ to be the set of all $a + b$ ($a \in A, b \in B$). This partial addition is commutative and associative: for fixed A, the mapping $X \to X + A$ is an order-isomorphism of $[O, A]$ onto $[A', I]$ in P.*

Indeed, one can prove much more: one can prove that P is a *complete* lattice [**KG**, p. 117]; that P satisfies the σ-chain condition; and that the addition is *Archimedean.*† However, rather than discussing the implications of these facts, we shall discuss their generalization to *orthomodular lattices* (§II.14).

11. Dimension Ortholattices

Over the past decades, very significant extensions of von Neumann's ideas have been made, to orthomodular lattices. In this connection, we note that if one assumes both orthocomplements *and* modularity, then one need not assume continuity because of the following notable

THEOREM OF KAPLANSKY. *Any orthocomplemented complete modular lattice is a von Neumann lattice.*‡

However, our main purpose consists not so much in eliminating redundant postulates for von Neumann ortholattices, as in developing a theory which applies to the *non*modular orthomodular lattice of all subspaces of Hilbert space, already studied in §II.14. For this purpose, the most relevant available concept seems to be that of a "dimension lattice".§ In the theory of dimension lattices, one *postulates* the existence of an abstract dimension function, as well as that of (symmetric) orthocomplements. This contrasts strikingly with the theory of (modular) von Neumann lattices, in which *all* concepts can be uniquely defined in terms of the partial ordering relation alone.

DEFINITION. Let L be any *complete, orthomodular lattice.* A *dimensional equivalence relation* on L is an equivalence relation $\sim$ which satisfies:

D1. If $a \sim O$, then $a = O$.

D2. If $a \perp b$ (i.e., $a \leqq b^{\perp}$) and $c \sim a \vee b$, then $d \perp e$ exist such that $d \sim a$, $e \sim b$, $c = d \vee e$.

---

† I.e., that $A + A + \cdots + A \leqq I$ to any number of summands implies $A = O$. See Exercise 2 below.

‡ I. Kaplansky [1]. What we call a "von Neumann lattice" was called by Kaplansky a "continuous geometry".

§ This concept was developed independently by L. Loomis [2] and S. Maeda, J. Sci. Hiroshima Univ. **A19** (1955), 211–37. See also F. Maeda, ibid. **A23** (1959), 151–70.

D3. If $\{a_\alpha\}$ and $\{b_\alpha\}$ are families of pairwise orthogonal elements with the same indices α, and $a_\alpha \sim b_\alpha$ for all α, then $a_\alpha \sim b_\alpha$.

D4. If a and b are "perspective" (i.e., have a common complement in L), then $a \sim b$.

(Note that D4 is fulfilled trivially if L is a Boolean algebra, since then $a' = b'$ implies $a = b$.) A complete orthomodular lattice with a dimensional equivalence relation is called a *dimension ortholattice*.†

By the results of §§8–9, the continuous geometry CG($\mathbf{R}$) can be made into a dimension ortholattice, by:

(i) using Euclidean orthocomplements in all $\mathbf{R}^{2^n}$ and passing to the limit to define orthocomplements in CG($\mathbf{R}$), and

(ii) letting $\sim$ mean perspectivity (or, equivalently, $d[a] = d[b]$). Other examples of dimension ortholattices are the following.

Example 6. Let L be the lattice of all closed subspaces of Hilbert space invariant under a given self-adjoint linear operator A, and let $S \sim T$ mean that there is a unitary transformation U of S onto T such that $(\xi U)A = (\xi A)U$ for all $\xi \in S$.

Example 7. Let L be the lattice associated with a given measure algebra, and let $a \sim b$ mean that $[O, a]$ and $[O, b]$ are isomorphic both as lattices and metrically.‡

For some parts of the theory of dimension ortholattices, one can replace D4 by the following weaker assumption (which follows from D1–D4).

D4′. Unless $a \perp b$, there exist a_1, b, with $O < a_1 \leqq a$, $O < b_1 \leqq b$, and $a \sim b$.

The following example satisfies D1–D3 and D4′, but not D4 unless $A = cI$ is a scalar matrix.

Example 8. Let M be the orthomodular lattice of all subspaces of n-dimensional Euclidean space E_n. For a given positive definite symmetric matrix A, define the "relative trace" $d_A[S]$ of any subspace as the sum $\sum (\alpha_k A, \alpha_k)$ for any normal orthogonal basis of S. Let $S \sim T$ mean that $d_A[S] = d_A[T]$. Then D1–D3 and D4′ are satisfied.§

In a dimension ortholattice, an element e is said to be *invariant* if $a \leqq e$, $b \leqq e^\perp$ and $a \sim b$ imply $a = b = O$. It may be shown that the invariant elements of any dimension ortholattice L form a *complete Boolean algebra*, which lies in the center of L. (Loomis [**2**], Theorem 2). A *factor* is a dimension ortholattice in which the only invariant elements are O and I; in real von Neumann ortholattices

† Instead of D4, Loomis [**2**] postulated D4′ below, and used the term "dimension lattice" for the resulting system.

‡ See D. Maharam, Trans. AMS **65** (1949), 279–330, where this notion is used to classify measure algebras.

§ An interesting theorem on A. Gleason (J. Math. Mech. **6** (1957), 885–93) states that for $n > 2$, all real-valued dimensions on M (and on $L_c(\aleph)$) are such "relative traces".

(such as CG(**R**) and its powers), these factors are the irreducible ones (the "continuous geometries").

In a dimension ortholattice L, an element b is *finite* if $a \sim b$ and $a \leqq b$ imply $a = b$; it is *simple* if its only subelements are its intersections with invariant elements. A dimension ortholattice L is of Type I of $I \in L$ is a union of simple elements; L is of Type II if I is a union of finite elements but contains no nonzero simple elements; and L is of Type III if O is its only finite element.

Maeda and Loomis have shown (loc. cit.) that, in any dimension ortholattice of Type II_1, one can introduce a real-valued valuation ("dimension function") d such that $a \sim b$ holds if and only if $d[a] = d[b]$, which implies modularity. Any dimension ortholattice of Type I_n is also modular.

MacLaren† has defined an element a of a lattice L to be "modular" if and only if: (i) $[O, a]$ is modular, and (ii) (x, a) is a modular pair for all $x \in L$. He has proved that a complete symmetric ortholattice L is a dimension ortholattice in which every element is the join of finite elements if and (A. Ramsay) only if L is "nearly modular"—i.e., L is semimodular and every $a \in L$ is the join of modular elements.

Finally, A. Ramsay‡ has studied the question of which complete ortholattices can support nontrivial dimensional equivalence relations. He has shown that this is the case if and only if condition (ξ) of Chapter II holds: if a covers $a \wedge b$, then $a \vee b$ covers b.

Exercises (courtesy Dr. N. Zierler):

Let L be any orthomodular lattice in which, for any orthogonal sequence $a_1, a_2, a_3, \cdots, a_i$ exists.

1. Show that L is σ-complete.
2. Show that if L contains no uncountable set of disjoint elements, then L is complete.

Let P be an orthocomplemented poset (with O, I) in which, given $a \leqq b$, $a^\perp \wedge b$ exists and $b = (a^\perp \wedge b) \vee a$. Let $\bar{P}$ be the completion by cuts of P.

3. For $S \subset P$, let $S^\perp$ be the set of elements a such that $a \leqq s^\perp$ for all $s \in S$. Show that $S \to S^\perp$ is an orthocomplementation of $\bar{P}$.
4. For P a lattice, show that $\bar{P}$ is a symmetric ortholattice.
5. For P not a lattice, show that the orthomodular lattice $\bar{P}$ is not.
6. Show that if $S \to S^\perp$ is the only orthocomplementation of $\bar{P}$ which makes $\bar{P}$ an orthomodular lattice, and which satisfies $(a)^\perp = (a^\perp)$ for principal (closed) ideals.

PROBLEMS

93. Find necessary and sufficient conditions on a (separable, regular) T_1-space for its Borel sets to form the "standard" Borel algebra of Theorem 5.

94. Find necessary and sufficient conditions for an atomic Borel algebra to be the algebra of all Borel sets in a T_1-space.

95. Analyze any valuation on a metric lattice into its "discrete", "absolutely continuous", and "singular" parts.

† M. D. MacLaren, Trans. AMS **114** (1965), 401–16.

‡ A. Ramsay, Trans. AMS **116** (1965), 9–31.

96. If a Borel algebra B is generated by a subset G, does every $a > 0$ in A necessarily contain some finite or countable meet $\bigwedge g_i > 0$ of elements $g_i \in G$?

97. Characterize abstractly, as measure algebras: (a) the natural extension of Daniell measure to a torus of uncountably many dimensions, (b) the measure algebra of sets of finite p-dimensional Hausdorff measure in Euclidean q-dimensional space.†

*98. Without assuming the Continuum Hypothesis, prove that there exists no nontrivial measure on the Borel algebra of all subsets of the continuum, which gives all points measure zero.

99. *Construct* maximal ideals in the following Boolean quotient-algebras M/N: (i) countable sets modulo finite sets, (ii) $\mathbf{2}^{\aleph_1/\aleph_0}$, (iii) (measurable sets mod null sets), (iv) B/C (Borel sets mod first category). What about completely subdirectly irreducible ideals?

100. (G. Bruns) In a lattice L, Let $\mathscr{M}$ and $\mathscr{N}$ consist of all subsets of suitably bounded cardinalities, and let $\bigvee M$ and $\bigwedge N$ exist for all $M \in \mathscr{M}$, $N \in \mathscr{N}$. Call a monomorphism ϕ from L to a product $\prod C_k$ of complete chains an $(\mathscr{M}, \mathscr{N})$-representation of L when $\phi(\bigvee M) = \bigvee(\phi(M))$ and $\phi(\bigwedge N) = \bigwedge(\phi(N))$ for all $M \in \mathscr{M}$, $N \in \mathscr{N}$. When is the existence of such an $(\mathscr{M}, \mathscr{N})$-representation equivalent to some $(\mathbf{m}, \mathbf{n})$-distributive law

$$\bigvee_{i \in I} \left(\bigwedge_{j \in J_i} a_{ij} \right) = \bigwedge_{\alpha \in \Pi J_i} \left(\bigvee_{i \in I} a_{i\alpha(i)} \right),$$

and a dual $(\mathbf{m}', \mathbf{n}')$-distributive law?

101. Need a complete complemented modular lattice be a topological lattice? (S. Gorn)

102. Is perspectivity transitive in every modular ortholattice? In every orthomodular lattice?

103. Does the ortholattice of all closed subspaces of Hilbert space have a finite sub-ortholattice? (A. Ramsay)

104. If CG($\mathbf{R}$) is represented as a subdirect product of atomic projective geometries, are the latter all over $\mathbf{R}$? (O. Frink)

105. Find necessary and sufficient conditions on an irreducible von Neumann lattice, for it to be isomorphic to CG(D) for some division ring D.

106. Construct a von Neumann lattice having an arbitrary Borel algebra as center, and based on an arbitrary CG(D), as a suitable "direct integral". (A. Ramsay)

107. Find necessary and sufficient conditions on a von Neumann lattice for it to admit a positive valuation. Can an irreducible von Neumann lattice have two linearly independent positive valuations? (Cf. Problem 82.) (See also R. Sikorski, Colloq. Math. **11** (1963), 25-8, for nine unsolved problems on infinitary Boolean algebras.)

---

† F. Hausdorff, Math. Ann. **79** (1919), 157–79.

CHAPTER XII

APPLICATIONS TO LOGIC AND PROBABILITY

1. Boole's Isomorphism

Boolean algebra, and hence lattice theory, originated 125 years ago in an attempt "to investigate the fundamental laws of those operations of the mind by which reasoning is performed; to give expression to them in the symbolical language of a Calculus, and upon this foundation to establish the science of Logic and construct its method; to make that method itself the basis of a general method for the application of the mathematical doctrine of Probabilities . . ." [**Bo**, first paragraph].

Boole recognized that sets or "classes" form a Boolean algebra under the operations of intersection, set-union, and complement. He assumed [**Bo**, p. 42] that there was a universal class, so that (for example) the classes of "stones" and "not stones" were both well defined. He also recognized that, in ordinary speech, the notion of "class" is interchangeable with that of "quality" or "attribute", and in fact assumed that these could be used interchangeably [**Bo**, p. 49]. Thus we can summarize two of his basic assumptions as follows.

BOOLE'S FIRST LAW. Let attributes be designated by letters; the words "and", "or" and "not" by the symbols $\wedge$, $\vee$ $'$ respectively; and "x implies y" by $x \leqq y$. Then attributes form a Boolean algebra.†

BOOLE'S SECOND LAW. The correspondence $x \to \hat{x}$ from attributes to the class of objects having the given attribute is an *isomorphism* between the Boolean algebra of attributes and that of classes:

$$(1) \qquad \widehat{x \wedge y} = \hat{x} \cap \hat{y}, \qquad \widehat{x \vee y} = \hat{x} \cup \hat{y}, \qquad \widehat{(x')} = (\hat{x})'.$$

Boole showed that an important part of the combinatory logic of attributes and classes could be deduced algebraically from the preceding facts. A few applications are given below as Exercises. It is interesting to record that the converse result, that every Boolean algebra is isomorphic to an algebra of classes ("field of sets") was only proved in 1933 (Theorem VIII.10, Corollary 3), and that its proof requires the Axiom of Choice!

Actually, Boole did not consider the preceding "laws" as assumptions about

† In [**LT1**] and [**LT2**], a dual interpretation was given; by the Duality Principle, the two interpretations are polyisomorphic. The present interpretation is that of Boole; the dual interpretation expresses the fact the class of stones which are "red *or* round" is formed by taking the class of "red" stones *and* that of "round" stones. Perhaps for this reason, logicians often use & instead of $\wedge$.

logic, nor did he consider them as postulates for set theory or for a class of algebraic systems. Instead he regarded them as scientific *laws* of (human) *thought*, and he regarded Boolean algebra as a tool which could be *applied* to the solution of logical problems.

The question of their possible inconsistency apparently did not occur to him. However, it is now generally agreed that they involve the following paradox of Cantor,† discovered nearly fifty years after Boole's death. Let A be a supposed "universal class" of all attributes; with each set $S \subset A$ of attributes, we can associate the class of all (real or imaginary) objects possessing the attributes of S; this gives $2^{n(A)}$ classes, where $n(A)$ is the cardinality of A. Conversely, let U be a supposed "universal class" of all objects. With each subclass $X \subset U$ we can associate the attribute of membership in X. This gives $2^{n(U)}$ distinct attributes. There follows

$$(2) \qquad n(U) \geqq 2^{n(A)} \quad \text{and} \quad n(A) \geqq 2^{n(U)},$$

contradicting a well-known theorem of Cantor.

The preceding paradox illustrates once more the uncertainty which plagues the theory of infinite sets, an uncertainty which was discussed above in §VIII.16. The purpose of this chapter is not to discuss this uncertainty further, but to give a brief survey of some of the interesting applications of Boolean algebra and lattice theory which have been made since Boole's book was written.

2. **Propositional Calculus; Critique**

Boolean algebra applies also to propositions [**Bo**, Chapter XI]. Thus, for any two propositions P and Q, one can denote the propositions "P and Q", "P or Q", and "not P" by $P \wedge Q$, $P \vee Q$, and P', respectively. With respect to these interpretations, we have

BOOLE'S THIRD LAW. Propositions form a Boolean algebra. Applications of Boolean algebra may be made to the propositional calculus, similar to those described at the end of §1. In addition, in classical ("two-valued") logic, all propositions are either true or false, and never both. Moreover $P \wedge Q$ is true if and only if P and Q are both true; $P \vee Q$ is true when P or Q is true; of P and P', one is true and the other false. Hence we get

BOOLE'S FOURTH LAW. True propositions form a (proper) *dual prime ideal* in the Boolean algebra of all propositions: the complementary prime ideal consists of *false* propositions.

Under these assumptions, the compound proposition "P implies Q" (i.e., "if P, then Q"), which is denoted $P \rightarrow Q$, has a special meaning. It is true or false according as "Q or not-P" is true or false; hence one can interpret $P \rightarrow Q$ as meaning $P' \vee Q$, thus reducing the number of undefined operations in the algebra of logic.

† A. Fraenkel and Y. Bar-Hillel, *Foundations of set theory*, North-Holland Publishing Co., 1958, p. 8.

In two-valued logics, one can similarly symbolize the statement "P is equivalent to Q" by $P \sim Q$, and replace it by "P implies Q and Q implies P"—i.e., with the symmetric difference operation $(P' \vee Q) \wedge (Q' \vee P) = (P + Q)'$ of *complemented* Boolean *ring*-addition (§II.12).

One can also show that many compound propositions P are "tautologies", that is, true by virtue of their logical structure alone. This amounts algebraically to saying that $P \to I$. The simplest tautology is $P \vee P'$, ("P or not P"). It is a simple exercise in Boolean algebra to show that the following are also tautologies of two-valued logic:

$$\begin{gathered} O \to P, \quad P \to I, \quad P \to P, \quad P \sim P, \quad P \to (Q \to P), \\ (P \sim Q) \sim (Q \sim P), \quad (P \sim O) \vee (P \sim I), \quad (P \sim Q) \vee (Q \to P). \end{gathered} \tag{3}$$

The reader will find it easy to invent others.

On the other hand, one can raise many intuitive objections to the preceding logical rules, pushed to extremes. Thus the tautology $O \to P$ of (3) asserts that "a false proposition implies every proposition"; but what does this really mean?† Likewise, one may question the validity of the tautology $(P \to Q) \vee (Q \to P)$, which asserts that "of any two propositions P, and Q, either P implies Q or Q implies P".

Intuition also makes one skeptical of the validity of proofs by contradiction. Why should the disproof (or "reductio ad absurdum") of "not-P" imply the truth of P? Such negative proofs seem especially unsatisfactory when P asserts the existence of a number with given properties, and one can prove that "not-P" is self-contradictory, but one cannot *construct* an example of a number having the given properties.

Indeed, it is easy to construct logical systems in which there are undecidable propositions. In fact, Skolem and Gödel‡ have constructed a plausible and consistent logical system, in which there exist undecidable propositions concerning the natural integers. However, the existence proof requires the acceptance of a fixed logical system, including the Zermelo–Fraenkel axioms for set theory. The author's view is that imaginative mathematicians (like Cantor), drawing upon intuition and even physical analogies, may invent new logical algorithms which will "decide" questions previously considered to be undecidable. This remark applies also to the "uncertain hypotheses" discussed in §VIII.16. It seems especially relevant (to the author) since the standard logical systems have self-contained *countable* realizations, whereas the most difficult mathematical questions concern the *continuum*.

---

† In this connection, the following anecdote is appropriate. Russell is reputed to have been challenged to prove that the (false) hypothesis $2 + 2 = 5$ implied that he was the Pope. Russell replied as follows: "You admit $2 + 2 = 5$; but I can prove $2 + 2 = 4$; therefore $5 = 4$. Taking two away from both sides, we have $3 = 2$; taking one more, $2 = 1$. But you will admit that I and the Pope are two. Therefore, I and the Pope are one. Q.E.D".

‡ K. Gödel, Monatsh. Math. Phys. **38** (1931), 173–98.

Be this as it may, current opinion seems to favor the idea that there are many genuinely undecidable propositions, whose truth can be consistently asserted *or* denied. In the language of Chapter VI, each set of axioms "generates" a dual ideal of demonstrable propositions, and a (*not* necessarily complementary) ideal of false propositions. Typically (Richard paradox) both ideals are countable; hence in some sense "almost every" proposition is formally undecidable (or meaningless!).

Exercises:

1. (a) Using Boolean algebra, prove that the formulas of (3) are indeed tautologies, as well as $(x' \to O) \to x$.

 (b) Prove the same for $[(x \to y) \wedge (y \to z)] \to (x \to z)$.

2. Discuss the equivalence of "P unless Q", "P and/or Q", "if not-Q, then P". To what Boolean operation does "P or Q, but not both" correspond?

3. (a) Develop postulates for Boolean algebra in terms of x' and $x \to y$.

 (b) Show that one cannot express $\vee$ or $'$, in terms of $\to$ alone.

4. Show how, given any finite Boolean algebra A and prime ideal $P \subset A$, to construct a propositional calculus whose propositions are the different elements $x \in A$, and whose true propositions are precisely the $x \in P$.

5. Show that the epimorphisms from a Boolean algebra A of propositions onto a two-valued truth-table with the usual properties correspond one-one to the proper prime ideals of A.

3. Brouwerian and Modal Logics

As already stated, the Boole–Whitehead propositional calculus applies to *two-valued* deductive logic, in which every proposition is assumed to be either demonstrably true or demonstrably false. In algebraic language, it asserts the existence of an *epimorphism* of the Boolean algebra of "all" propositions to **2**. However, *many-valued* logics (so-called "modal logics") go back to ancient times. Thus Aristotelian logic recognized four modes:† necessary, contingent, possible, impossible.

Relatively modern algebraic systems for describing propositional calculi have been proposed by Lukasiewicz and Tarski, and by Post.‡

No attempt will be made to discuss these here; a major question concerns the extent to which there is a genuine (single-valued) *epimorphism* from the Boolean algebra of "all" propositions onto the proposed algebra of truth-values. (For this, Lukasiewicz and Tarski proposed the interval $[0, 1]$, while Post proposed the integers $0, \cdots, n - 1$ $(n > 2)$.) This difficulty becomes evident when one considers the properties of probability (see §5): the natural functions (probability measures) from propositions to $[0, 1]$ are *not* epimorphic for $\wedge$, $\vee$, or $\to$, though they *are* for $'$, since $p[x'] = 1 - p[x]$.

† See C. I. Lewis and C. H. Langford, *Symbolic logic*, New York, 1932; also the end of Gr. C. Moisil, Ann. Sci. de Jassy **22** (1936), 1–118, for further comments.

‡ J. Lukasiewicz and A. Tarski, C. R. Soc. Sci. Lettres Varsovie III **23** (1930), 1–21; a brief summary is in O. Frink, Amer. Math. Monthly **45** (1938), 210–19. See also E. Post, Amer. J. Math. **43** (1921); 163–85; P. C. Rosenbloom, Amer. J. Math. **64** (1942), 167–88; J. B. Rosser and A. R. Turquette, J. Symbolic Logic **10** (1945), 61–82.

A more fruitful modification of two-valued logic, in which the validity of proof by contradiction is *not* assumed, has been invented by L. E. J. Brouwer. In Brouwer's propositional calculus, although the implication $P \to (P')'$ is admitted, $(P')' \to P$ is not. Because of this, we shall denote "not-X" in Brouwerian logic by P^* instead of P'.

It was recognized by Stone [**3**] and Tarski† that there was a close analogy between Brouwerian logic and the distributive lattice of open sets of a topological space. However, the specific correlation with "Brouwerian lattices" (§II.11) seems to have been first accomplished in [**LT1**, §§161–2]. In it, the relative "pseudo-complement" or "residual" $A:P$ (see Chapter XIV) appears as the operation $P \to Q$ of "implication",‡ which is equivalent to $P' \vee Q$ in Boolean algebra. The connection is then made as follows.

DEFINITION. A *Brouwerian logic* is a propositional calculus which is a lattice with O and I, in which:

B1. $(P \to Q) = I$ if and only if $P \leqq Q$,

B2. $P \to (Q \to R) = (P \vee Q) \to R$ for all P, Q, R.

The proposition $P \to O$ will be denoted by P^* ("not-P"), and the relation $P = I$ ("P is true") by $\vdash P$. (Hence B1 is equivalent to $\vdash(P \to Q)$ if and only if $P \leqq Q$.)

THEOREM 1. *A Brouwerian logic L is a Brouwerian lattice, with $Q:P$ relabelled as $P \to Q$.*

PROOF. By B1, $t \leqq (x \to y)$ is equivalent to $\vdash t \to (x \to y)$; by B2, this is equivalent to $\vdash (t \to x) \to y$; and by B1 again, this is equivalent to $t \wedge x \leqq y$. Summarizing, $t \wedge x \leqq y$ if and only if $t \leqq (x \to y)$; that is, $x \to y = y:x$ in the sense of §V.10. Since $x \to y$ exists for all $x, y \in L$ by hypothesis, it follows that any Brouwerian logic is a Brouwerian lattice. Q.E.D.

It is interesting to see how B1–B2 imply that L is a distributive lattice. By Chapter I, (5)–(5′) and Theorem 9, it suffices to prove that $x \wedge (y \vee z) \leqq u$, where u stands for $(x \wedge y) \vee (x \wedge z)$. But trivially, $x \wedge y \leqq u$ and $x \wedge z \leqq u$; hence by hypothesis $y \leqq x \to u = u:x$ and $z \leqq x \to u$. It follows that $y \vee z \leqq x \to u$; that is, $x \wedge (y \vee z) \leqq u$. Q.E.D.

It is also interesting to consider "Brouwerian logics" as a family of algebras; this can be done by taking as basic a suitable system of postulates for $\to, \wedge, \vee$,—say those of Heyting.§ The most significant study in this direction is that of McKinsey and Tarski. Here as in [**LT1**] and [**LT2**], a notation dual to that above is used.

† A. Tarski, Fund. Math. **31** (1938), 103–34, esp. p. 132.

‡ For the notion of "implication algebra", see J. C. Abbott and P. R. Kleindorfer, Amer. Math. Monthly **68** (1961), 697–8.

§ A. Heyting, S.-B. Preuss. Akad. Wiss. (1930), 42–56; also *Mathematische Grundlagenforschung, Intuitionismus, Beweistheorie*, Berlin, 1935. For the study of McKinsey and Tarski, see Ann. of Math. **47** (1946), 122–62.

Cylindric and polyadic algebras. It should be emphasized that only a very small part of logic is lattice-theoretic. Even for two-valued logics, the discussion of *quantifiers* involves operations which are not definable in terms of inclusion†—hence not purely lattice-theoretic. These lead to "polyadic algebras" and "cylindric algebras", invented respectively by Halmos and by Henkin and Tarski.‡

4. Attributes in Classical Mechanics

The discussion of §§1–3 concerned ideas suggested primarily by pure logic: the analysis of linguistic traditions going back to antiquity. A quite different source of ideas is provided by mathematical models of the external universe. One of the most satisfactory of these is provided by the notion of a conservative *dynamical system* having a finite number of degrees of freedom, studied deeply in classical mechanics.§

Consider for example *n-body* systems, serving as models both for (Newtonian) celestial mechanics and for the (Maxwell–Boltzmann) kinetic theory of gases. The "state" of any such system Σ at any time t_0 is expressed by $6n$ real numbers: each body has three coordinates of position and three of velocity (or momentum). Hence $\mathbf{R}^{6n}$ ($6n$-dimensional Euclidean space) is the phase-space of all possible states of Σ; moreover from $\mathbf{x}(t_0)$ the state $\mathbf{x}(t)$ at all other times t is determined by appropriate force laws: the system is *deterministic*.

Each (instantaneously observable) "attribute" of Σ clearly defines a set in the given phase-space $\mathbf{R}^{6n}$: the set of all "states" in which Σ has the given attribute. According to classical Boolean logic (§1), the converse should also be true: with each subset S of $\mathbf{R}^{6n}$, one can associate the attribute that $\mathbf{x}(t_0) \in S$.

This is suspect even on purely logical grounds; as observed in §XI.3, it may be questioned whether any sets in $\mathbf{R}^{2n}$ other than Borel sets are effectively definable. Certainly, only a *countable* number of sets can be defined with a finite or countable vocabulary, by finite sentences. Moreover, by Theorem XI.13, it is unlikely that one could assign a nontrivial countably additive probability measure to every such attribute; we shall return to this point in §5.

The idea of associating an "attribute" of Σ with every subset of $\mathbf{R}^{2n}$ makes even less sense physically. Thus, since the accuracy of measurements is limited, one can never *observe* by direct experiment whether the kinetic energy of Σ at a given instant t_0 is a rational number or not: not even all Borel subsets of $\mathbf{R}^{2n}$ correspond to "observables" in a simple physical sense. However, all attributes which have an accepted physical meaning (e.g., for a gas, having temperature

† Quantifiers were already briefly considered by Boole [**Bo**, p. 63]; a few of their properties can be deduced from lattice-theoretic interpretations such as "$x > O$" (for some x).

‡ P. Halmos, Proc. Nat. Acad. Sci. **40** (1954), 296–301, and *Algebraic logic*, New York, 1962 (and references given there); see also [**Symp**, pp. 114–22]. For the ideas of Henkin and Tarski, see [**Symp**, pp. 83–113] and L. Henkin, *La structure algébrique des théories mathématiques*, Paris, 1956.

§ See for example E. T. Whittaker, *Analytical dynamics*, or any other good book on particle or statistical mechanics.

and pressure in a given interval) *do* correspond to Borel sets of the phase-space $\mathbf{R}^{2n}$.

Von Neumann† proposed that one should take the Borel algebra M/N at all Borel sets of $\mathbf{R}^{6n}$ module sets of measure zero as a model for the algebra of physical attributes. According to this model, one would say that the statement "$\Sigma m_i x_i^2$ is a rational number at time t_0" is almost certainly false (i.e., with probability one). This model has the great advantage that, by the Lebesgue density theorem, one can infer the presence or absence of any attribute with *arbitrary nearness to certainty* by making sufficiently accurate measurements.

Note that the preceding model differs from that of §1 in denying the validity of complete distributivity: the extended distributive laws of Chapter V, (4)–(4′). For, by Theorem V.17, this would imply that the algebra of (observable) attributes is isomorphic to the *atomic* Boolean algebra of *all* subsets of some class. Instead, von Neumann's model is only (weakly) countably distributive, in the sense of Chapter XI, (11). Thus, the algebra of attributes is a *topological lattice*, a property enjoyed by all Borel algebras (Theorem XI.2).

5. Classical Probability

Like logic, the theory of probability rests on uncertain philosophical foundations.‡ The basis most commonly accepted by mathematicians is the following.

DEFINITION. Probability is a probability measure on some Borel algebra of attributes.

By the Theorem of Loomis (§XI.2), any such Borel algebra can be realized by a σ-field of sets modulo a σ-ideal; the Lebesgue extension of the given probability measure can therefore be supposed defined on a σ-field of sets. In most texts, this is done.§

As regards *finite* probability algebras, one can realize the preceding definition very simply, using the representation theorem of §XI.5. Namely, one can represent any finite probability algebra by suitable elementary sets, representing each atomic (minimal proper) attribute α by a sector $S(\alpha)$ subtending $2\pi p[\alpha]$ radians on a *roulette wheel*. The probability that the wheel will stop spinning so that a fixed arrow will be opposite a point of $S(a)$ will then be $p[\alpha]$.

A basic philosophical problem concerns the consistency of the preceding definition with the simplest *statistical* definition of a probability as frequency.

Example 1. Let a repeatable experiment E consist in choosing a point at random on the unit circle $\mathbf{R}/(2\pi)$; and let it be performed a (countable) sequence of times. For a given Borel subset $S \subset E$, let $p_n[S]$ denote the fraction of the first n trials

---

† Ann. of Math. **33** (1932), 587–642, esp. pp. 595–8.

‡ This uncertainty is well illustrated in Math. Rev. **7** (1946), 186–93; see also J. M. Keynes, *A treatise on probabilities*, London, 1929.

§ A. Kolmogoroff, *Grundbegriffe der Wahrscheinlichkeitsrechnung*, Berlin, 1933; H. Cramer, *Mathematical methods of statistics*, Princeton, 1946; M. Loève, *Theory of probability*, 3rd ed. For the Borel algebra approach, see D. Kappos, *Strukturtheorie der Wahrscheinlichkeitsfelder*, Springer, 1960.

giving a point of E. Let $p_\infty = \mathrm{Lim}_{n\to\infty} p_n[S]$, when the limit exists. Then, for any *fixed* $S \subset E$, $p_\infty[S]$ exists and is equal to $m[S]/2\pi$ with *a priori* probability one, by the Law of Large Numbers.

Unfortunately, for any given sequence of trials, clearly $p_\infty[S]$ is *not* countably additive (only finitely additive), even when it is defined. Moreover, trivially, there exists a (countable) subset S_c with $m[S_c] = 0$ yet $p_\infty[S] = 1$.

Fortunately, these apparent inconsistencies can be resolved by using the concept of a separable *regular* probability measure, introduced in §XI.6. Namely, fix *a priori* a countable basis (or subbasis) of open sets—say the set of all closed intervals $[\alpha, \beta]$ with rational end points. Then with *a priori* probability one, $p_\infty[C] = m[C]/2\pi = (\beta - \alpha)/2\pi$ for all $C \in \mathscr{C}$. (This probability refers to the "product measure" on the space of all trial sequences.)

Moreover the Borel subalgebra generated by the $C \in \mathscr{C}$ includes *all* Borel subsets of the circle $\mathbf{R}/(2\pi)$. Further the *a posteriori* probability $p_\infty[C] = m[C]/2\pi$ can be extended to a *probability measure* on this Borel subalgebra in one and only one way, by Borel combination. The *Lebesgue completion* of this will give the usual regular probability measure on $\mathbf{R}/(2\pi)$. Finally, the preceding construction can be applied to any topological space X which has a countable generating family of Borel sets. The final conclusion then, is as follows.

THEOREM 2. *Let p be any regular probability measure on a topological space X with a standard Borel structure. Let $\mathscr{C}$ be any countable Boolean subalgebra of Borel sets $C_k \subset X$ which generates the Borel algebra $\mathscr{B}$ of all Borel sets of X. Then $p_\infty[C_k]$ is the a priori probability of C_k for almost every sequence of independent trials*—i.e., *with probability one in the product measure on the product space X.*

COROLLARY. *The Lebesgue completion of the Borel extension of p_∞ from $\mathscr{C}$ to $\mathscr{B}$ is possible and gives p, with a priori probability one.*

6. Logic of Quantum Mechanics

From the first, it was recognized that quantum mechanics involved departures from classical logical ideas. Among these departures, the "uncertainty principle" and the related principle of the noncommutativity of physical observations (on an atomic scale, especially) are most striking. Since the simplest verification of the distributive laws for the Boolean algebra of attributes is based on the permutability and repeatability of physical observations, it seems that distributivity (whose infinitary generalizations were criticized in §4) might be considered as especially suspect.

Moreover, the "observables" of quantum mechanics are usually† assumed to be Hermitian operators acting on a (complex) Hilbert space $\mathfrak{H}$. It follows that, if one defines an (observable) *attribute* of a quantum-mechanical "state" ψ as a statement of the form that the observation H on ψ will give a value λ in a Borel

† See J. von Neumann, *Mathematical foundations of quantum mechanics*, Princeton, 1955; also G. Mackey, Amer. Math. Monthly **64** (1957), 45–58.

set $S \subset \mathbf{R}$, then ψ has the attribute (H, S) with probability one if and only if ψ is in a *closed subspace* X of $\mathfrak{H}$. If S' is the complement of S, then the *negation* of (H, S) is certain if and only if $\psi \in X^{\perp}$, the orthocomplement of X.

If every such pair (H, S) is assumed to correspond to an observable attribute, then it follows that *every* closed subspace X of Hilbert space represents an attribute observable "with certainty" if and only if $\psi \in X$. This conjecture has frequently been made; it implies the following hypothesis as a corollary.†

HYPOTHESIS A. Observable quantum-mechanical attributes form a complete, atomic, orthomodular lattice which is isomorphic with the lattice of all closed subspaces of (separable) Hilbert space.

A weaker assumption is the hypothesis that the intersection $X \cap Y$ of any two closed subspaces X and Y which correspond to observable attributes, itself corresponds to such an attribute. This implies that $(X^{\perp} \cap Y^{\perp})^{\perp} = \overline{X + Y}$, the closure of the linear sum of two such subspaces, is again one. Therefore, it implies that observable attributes form an orthomodular lattice which is isomorphic with a subortholattice of the ortholattice of all closed subspaces of $\mathfrak{H}$.

Observe that $\psi \in X \cap Y$ means that successive measurements of a system in the state ψ will *certainly* verify the predictions (H, S) and (H_1, S_1) corresponding to X and Y, respectively. Thus, like $X^{\perp}$, it has a simple and direct physical meaning. The meaning of $\overline{X + Y}$ is somewhat more complicated, though, and *implication* does not appear as an operation at all (unlike in classical logic).

Motivated heuristically by the analogy with the lattice of all (closed) subspaces of finite-dimensional Hilbert spaces (i.e., Euclidean spaces), and by the plausibility of assuming the existence of positive valuations analogous to probability‡ (and to dimension in continuous geometries), von Neumann and the author conjectured that the lattice of (observable) attributes of a quantum-mechanical system might be *modular*. Like the lattice property (but unlike Hypothesis A), this conjecture can be considered a *scientific hypothesis* susceptible of direct experimental proof or disproof (see [**Symp**, pp. 160–3] for a more detailed discussion). Recently, however, Piron§ has shown that the sublattice of $L_c(\mathfrak{H})$ generated by the closed invariant subspaces for the free particle position and momentum operators is already nonmodular; hence the truth of the modularity hypothesis seems very unlikely.

On the other hand, $L_c(\mathfrak{H})$ is *nearly modular* in the sense defined in §XI.11. Namely, as already stated in §V.9, $L_c(\mathfrak{H})$ contains a dense modular sub-ortholattice $J \cup J^{\perp}$ consisting of the ideal J of elements of finite height and their orthocomplements (the elements of finite dual height). Furthermore, $L_c(\mathfrak{H})$ is precisely the completion by cuts of the ortholattice $J \cup J^{\perp}$. For details regarding this and

† See G. Birkhoff and J. von Neumann [**1**], where the ideas discussed here were first proposed.

‡ These can be interpreted as "a priori thermodynamic weights of states", for example. For related probabilistic ideas, see G. Bodiou, Publ. Inst. Stat. Univ. Paris **6** (1955), 11–25. Also D. MacLaren, Report ANL-7065, Argonne National Laboratory, 1965.

§ C. Piron, Helv. Physica Acta **37** (1964), 439–68.

related characterizations of $L_c(\mathfrak{H})$ and related lattices, the reader is referred to the original paper of Dr. MacLaren.†

PROBLEMS (Cf. those for Chapter VII.)

*108. Are the Axiom of Choice, Continuum Hypothesis, and Souslin Hypothesis "really" completely independent?

109. Is Problem 99 insoluble? In which cases is the assumption that such an ideal exists equivalent to the Axiom of Choice?

110. Develop an algebra of probability for quantum mechanics on a general class of symmetric ortholattices, without assuming an underlying Hilbert space or w^*-algebra.

† M. D. MacLaren, Pacific J. Math. **14** (1964), 597–612.

CHAPTER XIII

LATTICE-ORDERED GROUPS

1. *po*-groups

The rest of this book will be mainly concerned with *groups* (though also with semigroups and quasigroups) which are at the same time *lattices* (though sometimes only posets), in which *every group translation is isotone*. Such systems are called *lattice-ordered groups* or *l-groups*. The *additive* notation will be used for the group operation, so that group translations will be written $x \to a + x + b$, the group identity as 0, the inverse of x as $-x$, and $a + \cdots + a$ (to n summands) as na. Hence the assumption that group translations are isotone is equivalent to:

(1) If $x \leqq y$, then $a + x + b \leqq a + y + b$ for all a, b.

More generally, a *po-group* is defined as a group which is also a poset, and in which (1) holds. We first observe

LEMMA 1. *In any po-group G, every group translation is an order-automorphism.*

PROOF. Since the group translation $x \to (-a) + x + (-b)$ is isotone by (1), and is inverse to $x \to a + x + b$:

(1′) For all a, b, $a + x + b \leqq a + y + b$ implies $x \leqq y$.

COROLLARY. *Except in the trivial case $G = \{0\}$, a po-group cannot have universal bounds.*

For, by Lemma 1, if I is a universal upper bound for G, then $a + I = I$ for all $a \in G$ by Lemma 1, hence $a = 0$ for all a. A similar argument applies to universal lower bounds, and shows that the use of the symbol 0 for the group identity of a *po*-group leads to no confusion.

The preceding, almost trivial results have two basic consequences. First, x, x' can be used to denote any two elements of an l-group, since by what we have just shown complements cannot exist. Second, an l-group cannot be a "complete lattice" in the usual sense (unless it is $\{0\}$). For this reason, a *complete l-group* is defined to be an l-group which, as a lattice, is a *conditionally complete* lattice; this results in no ambiguity.

Otherwise, the terminology which describes different kinds of posets extends in an obvious way to *po*-groups. Thus, a *directed group* is a *po*-group which is a directed set when considered as a poset. A *po*-group which is simply ordered (i.e., a *chain* under its order relation) is called an *ordered group*.

In any *po*-group G, an element x is called *positive* when $x \geqq 0$; the set $P = G^+$ of all positive elements of G is called its *positive cone*. We have essentially assumed:

$$P \cap -P = 0, \qquad P + P \subset P, \qquad (-a) + P + a = P \quad \text{for all } a \in G.$$

The reason for this terminology becomes apparent when one visualizes the l-group (vector lattice; cf. Chapter XV) of all real three-vectors $\mathbf{x} = (x_1, x_2, x_3)$, and defines $\mathbf{x} \leqq \mathbf{y}$ to mean that $x_i \leqq y_i$ for $i = 1, 2, 3$. The following results are immediate.

LEMMA 2. *In any po-group*

$$x \leqq y \quad \text{and} \quad x_1 \leqq y_1 \quad \text{imply} \quad x + y \leqq x_1 + y_1. \tag{2}$$

Moreover P is invariant under all inner automorphisms of G.

Trivially, any subgroup of a *po*-group is itself a *po*-group, under the same group addition and order relation. Likewise, the dual of any *po*-group is a *po*-group, and the opposite of any *po*-group is a *po*-group. Other familiar *po*-groups are the following.

Example 1. Any ordered (integral) domain is an ordered group under its addition operation and order; P is the set of its nonnegative elements.

Thus, the additive group $(\mathbf{Z}, +, \leqq)$ of the integers is an ordered group; the same is true of the additive group $(\mathbf{Q}, +, \leqq)$ of rational numbers, and so on.

Example 2. Let $G = C[0, 1]$ be the set of all continuous real-valued functions $f(x)$ on $0 \leqq x \leqq 1$, and let P be the set of all nonnegative such functions. Then G is a (commutative) l-group.

THEOREM 1. *Any po-group G is determined to within isomorphism by its positive cone $P = G^+$, since*

$$a \leqq b, \quad b - a \in P, \quad \text{and } -a + b \in P \quad \text{are equivalent conditions.} \tag{3}$$

Moreover (i) $0 \in P$, (ii) *if* $x, y \in P$, *then* $x + y \in P$, (iii) *if* $x, y \in P$ *and* $x + y = 0$, *then* $x = y = 0$, (iv) *for all* $a \in G$, $a + P = P + a$. *Conversely, if G is any group, and P is a subset of G satisfying* (i)–(iv), *then condition* (3) *defines G as a po-group.*

PROOF. We get (3) easily from (1); thus $a \leqq b$ implies $0 = a - a \leqq b - a$. Condition (i) is trivial; moreover since $s \geqq 0$, $t \geqq 0$ imply

$$s + t \geqq s + 0 \geqq 0 + 0 = 0,$$

(ii) holds. Under the hypotheses of (iii), $0 = x + y \geqq x, y \geqq 0$ by (3), whence $x = y = 0$, proving (iii). Finally, by (3), $a + P$ and $P + a$ both are the set of all $x \geqq a$, proving (iv).

Conversely, let G be a group, and P any subset of G which fulfills the conditions listed; define $y \leqq x$ by (3). Then conditions (i), (ii), (iii) imply the rules P1, P3, P2, and, using (iv) in the form (iv'): $P = -a + P + a$. Therefore

$$\begin{aligned}(a + x + b) - (a + y + b) &= a + x + b - b - y - a \\ &= a + (x - y) - a \in P,\end{aligned}$$

hence (1) holds, so that G is a *po*-group.

Remark. Conditions (i)–(ii) assert that P is a submonoid of G, condition (iv), in the form (iv′) asserts that P is invariant under all inner automorphisms of G.

Example 3. Let Φ be any set of *po*-groups G_ϕ, and let Γ be the (unrestricted) *direct product* of the G_ϕ, as posets *and* as groups. Then Γ is a *po*-group.

For instance, if all G_ϕ are copies of $\mathbf{R}$, the real field in its natural order, then Γ is the po-group of all real functions on the set Φ, in their natural (partial) order. This is Example 3 of §I.1.

As a *po*-subgroup of the direct product $\Gamma = \Pi_\Phi G_\phi$ of Example 2, we have the *restricted direct product* of the G_ϕ, defined to consist of those elements f (functions $f: \phi \to f(\phi) = g_\phi \in G_\phi$) having only a *finite* number of nonzero components $g_\phi \neq 0$.

Definition. Let G and H be any two *po*-groups. The lexicographic *product* $G \circ H$ of G and H is the set of all (x, y) $(x \in G, y \in H)$, with: (i) $(x, y) + (x', y') = (x + x', y + y')$, and (ii) $(x, y) \geqq 0$ if and only if either $x > 0$, or $x = 0$ and $y \geqq 0$. This is always a *po*-group.

As a poset, $G \circ H$ is the *ordinal* product of G and H; as a group, it is their direct sum (just as in Example 3).

Lemma 3. *The lexicographic product $G \circ H$ of two nontrivial po-groups is an l-group if and only if G is a chain and H is a lattice.*

Sketch of Proof. Since $(x, y) \vee (x, y') = (x, y \vee y')$, H must be a lattice. Moreover if x, x' are incomparable in G, $(x, y) \vee (x', y) = (x \vee x', -\infty)$ where $-\infty$ is a universal *lower* bound of H, this implies $H = \{0\}$ by Lemma 1.

More generally, let N be any normal subgroup of a group G, where G/N and N are both *po*-groups with positive cones P_1 and P_2, respectively. Define $P \subset G$ to consist of those $g \in G$ which satisfy either: (i) $g \in P_2 \subset N$, or (ii) $(g + N)M \in P_1$. This makes G into a *po*-group, which may be called a *lexicographic extension* of N by G/N.

If G is any such lexicographic extension, then the mapping $G \to G/N$ is evidently on *o*-morphism, in the sense of the following definition.

Definition. An *isotone group-morphism* of the *po*-group G into a *po*-group H is an *o-morphism*.

Lemma 4 (Fuchs). *The kernel of any o-morphism θ: $G \to H$ between po-groups is a convex normal subgroup of G. Conversely, if N is any convex normal subgroup of the po-group G, and the positive cone in G/N is defined as the set of all $g + N$ with $g \in P$, then the mapping $g \to gN$ is an o-epimorphism of G onto G/N.*

For proofs of the preceding statements, see [**Fu**, p. 21]. Though the "kernel" N determines the congruence relation (cf. Chapter II, §§2–3), it does *not* determine the order relation in the image G/N. There exist bijective *o*-morphisms which are not isomorphisms, such as the natural mapping $\mathbf{R}^2 \to \mathbf{R} \circ \mathbf{R}$ of the direct product of $\mathbf{R}$ with itself onto its lexicographic product with itself.

2. Directed Groups

A *directed group* is a po-group G having the Moore–Smith property:

(4) Given $a, b \in G$, $c \in G$ exists such that $a, b \leqq c$.

LEMMA 1. *In any po-group, the Moore–Smith property is equivalent to the assertion that every element is the difference of positive elements.*

PROOF. Assuming (4) with $b = 0$, we get $a = c - (-a + c)$, where $0 \leqq c$ and $-a + c = -a + (c - a) + a \geqq -a + 0 + a = 0$. Conversely, if $a = a' - a''$ and $b = b' - b''$, where a', a'', b', b'' are positive, then $c = a' + b'$ is an upper bound to a and b.

LEMMA 2. *In any po-group G, the reflection $x \to -x$ is a dual automorphism of order in G.*

For, if $x \geqq 0$, then $0 = x + (-x) \geqq 0 + (-x) = -x$, and conversely $0 \geqq -x$ implies $x = x + 0 \geqq x + (-x) = 0$.

COROLLARY 1. *In any po-group G,*

$$(5) \qquad x \geqq y \quad \textit{implies} \quad a + (-x) + b \leqq a + (-y) + b,$$

for all $a, b \in G$.

COROLLARY 2. *Any directed group is bidirected.*

Given a *po*-group G (for example, one consisting of real functions under addition), we now ask: which subgroups D of G are directed? Evidently if $f \in D$, a directed subgroup of G, then the set $\{f, 0\}$ must have an upper bound. But if u is an upper bound of the set $\{f, 0\}$, and v an upper bound of the set $\{g, 0\}$, then $u + v$ is an upper bound of the set $\{f + g, f, g, 0\}$. Similar remarks apply to lower bounds, from which we conclude

LEMMA 3. *In any po-group G, the set $B(G)$ of all $f \in G$ such that $\{f, 0\}$ is bounded forms a directed subgroup of G, which contains every other directed subgroup of G.*

Example 4. Let L be any lattice, and let $a \in L$ be fixed. Let $V_a(L)$ be the commutative *po*-group of all (real) *valuations* $v[x]$ on L such that $v[a] = 0$. Let the positive cone of $V_a(L)$ consist of all *isotone* such valuations. Then $B(P_a)$ is the subset of those valuations which are of *bounded variation on finite intervals*. By Theorem X.10 and Lemma 1 of §3 below, $V_a(L)$ is even an l-group.

DEFINITION. An element a of a po-group G is called incomparably smaller than a given element $b \in G$ (in symbols, $a \ll b$) when $na \leqq b$ for every integer n. A *po*-group G is called *Archimedean* when $a \ll b$ implies $a = 0$. It is called *integrally closed* when

$$(6) \qquad na \leqq b \quad \text{for all } n = 1, 2, 3, \cdots \text{ implies} \quad a \leqq 0.$$

LEMMA 4. *Any integrally closed po-group G is Archimedean.*

PROOF. If $na \leqq b$ for all $n \in \mathbf{Z}$, then $na \leqq b$ for all $n \in \mathbf{Z}^+$, whence $a \leqq 0$; also $n(-a) = (-n)a \leqq b$ for all $n \in \mathbf{Z}^+$, whence $(-a) \leqq 0$ and $a \geqq 0$. Therefore $a = 0$.

Note that $G \circ H$ is *not* Archimedean (hence not integrally closed) except in the trivial cases in which $G = \{0\}$ and H is Archimedean, or $H = \{0\}$ and G is Archimedean. It will be proved in §3 that an l-group is integrally closed if and only if it is Archimedean. In §13, we will prove the much deeper result that *any integrally closed directed group is commutative.* For the present, we will content ourselves with much weaker results.

First, we recall from §1 that a *complete l-group* is an l-group in which every bounded set has a l.u.b. and a g.l.b.—i.e., it is an l-group which is *conditionally complete* as a lattice. We then prove

LEMMA 5. *Any subgroup of a complete l-group G is integrally closed (hence Archimedean).*

PROOF. By definition of completeness, $na \leqq b$ for all $n \in \mathbf{Z}^+$ implies that $c = \bigvee\{na\}$ exists. But this implies $c + a = \bigvee\{(n+1)a\} \leqq \bigvee\{na\} = c = c + 0$, since $\{(n+1)a\} \subset \{na\}$. And this implies $a \leqq 0$, by (1′).

We now describe an important example, due to Everett and Ulam [**1**], which illustrates various surprising possibilities in *po*-groups which are not-l-groups.

Example 5. Let G consist of all order-preserving homeomorphisms of the real line under substitution; and let $f \leqq g$ if and only if $f(x) \leqq g(x)$ for all x. Then P consists of those functions which satisfy $x \leqq f(x)$ for all x; moreover $f^+(x) = \max(f(x), x)$. In the usual notation of function theory, let $p(x)$ be any periodic function of period one on $(-\infty, +\infty)$ with $|p'(x)| \leqq 1/2$. If $h(x) = x + p(x)$ and $g(x) = x + 1/n$, then $-h + g + h = f$ satisfies $nf = ng$, yet $f \neq g$. Moreover the function ϕ defined on $[0, 1]$ by a broken line with two segments and vertex at $\phi(3/8) = 3/4$, and that ψ defined by one with vertices at $\psi(1/4) = 1/4$, $\psi(5/16) = 11/16$, and $\psi(3/4) = 3/4$, satisfies $0 < \psi^2 < \phi^2$ yet $\phi \nleqq \psi$.

Now consider the subgroup A of G containing all order-automorphisms of $[0, 1]$ which are defined by *algebraic* functions. These form a directed subgroup of G which is an Archimedean *po*-group but not integrally closed. This last can be shown by modifying the functions ϕ, ψ of Example 5 and their slopes by a sufficiently small amount, which can be done by the Weierstrass Approximation Theorem.

Exercises for §§1–2:

1. Prove that one can make any group into a *po*-group by letting the positive cone consist of 0 alone.
2. Prove that, in a group which is also a lattice, (1) is equivalent to the two distributive laws
$$a + (x \vee y) = (a + x) \vee (a + y) \quad \text{and} \quad (x + y) \vee b = (x + b) \vee (y + b).$$
3. Show that an l-group may be defined as a group, with a second binary operation which is idempotent, commutative, associative, and satisfies the distributive laws of Exercise 2.
4. Show that the following, though not in general lattices, are always *po*-groups or nearly so:

(a) G consists of the integers under addition; P consists of the integers $n > 2$.

(b) G is the multiplicative group of all nonzero rational numbers, P consists of all integers.

(c) G is the additive group of a field of characteristic ∞, which is "formally real" in the sense that $x_1^2 + x_2^2 + \cdots + x_n^2 = 0$ implies $x_1 = x_2 = \cdots = x_n = 0$. P is the subset of all sums of squares.

5. Prove in detail that Examples 1–2 define l-groups, and that Example 3 always defines a *po*-group.

6. Prove that the lexicographic product of any two *po*-groups is a *po*-group.

7. Prove that the lexicographic product $G \circ H$ of a directed group G and a *po*-group H is always a directed group.

8. (a) Prove that, in any *po*-group, the relation $\ll$ is anti-symmetric and transitive.

(b) Prove that any subgroup of an Archimedean *po*-group is itself an Archimedean *po*-group under the given order.

*9. Construct a directed group containing elements of order two. (*Hint.* Use Exercises 1 and 7.)

*10. Show that the subgroup A in Example 5 has the (2, 2) Riesz Interpolation Property. (Cf. [**LT2**, p. 52 and p. 53, Exercise 4]).

3. Properties of l-groups

By definition, an l-group is simply a *po*-group in which any two elements have a l.u.b. and a g.l.b. Obviously, any l-group is a directed group; hence all the results of §§1–2 apply to l-groups.

Distributive laws. Since group translations are isotone bijections with isotone inverses in any *po*-group, they are order-automorphisms. Hence addition is distributive on meets and joins in any l-group:

$$(7)\quad a + (x \vee y) = (a + x) \vee (a + y), \qquad (x \vee y) + b = (x + b) \vee (y + b),$$

$$(7')\quad a + (x \wedge y) = (a + x) \wedge (a + y), \qquad (x \wedge y) + b = (x + b) \wedge (y + b).$$

More generally, any group translation is an automorphism for unrestricted joins and meets, so that

$$(8)\qquad a + (\bigvee x_\sigma) + b = \bigvee (a + x_\sigma + b),$$

$$(8')\qquad a + (\bigwedge x_\sigma) + b = \bigwedge (a + x_\sigma + b),$$

the existence of the join (respectively meet) on the left side implying the existence of that on the right side of each equation.

Again, since $x \geqq 0$ implies $0 = x - x \geqq 0 - x = -x$ and conversely, the mapping $x \to -x$ is a dual order-automorphism of any *po*-group. Hence if $-a \vee -b$ exists, so does

$$(9)\qquad a \wedge b = -(-a \vee -b) = -(-b \vee -a).$$

Therefore the mapping $x \to a - x + b$ is a dual automorphism for any a, b in any *po*-group. Hence, in any l-group:

$$(10)\qquad a - (x \vee y) + b = (a - x + b) \wedge (a - y + b)$$

is an identity. Moreover, much as in (8)–(8′),

$$(11)\qquad a - (\bigvee x_\sigma) + b = \bigwedge (a - x_\sigma + b) \text{ and dually.}$$

The next two lemmas and subsequent theorems are essentially corollaries of the preceding results.

Lemma 1. *A po-group is an l-group if and only if, for all* $a \in G$, $a \vee 0 = a^+$ *exists in* G.

PROOF. The necessity is trivial. Conversely, let H be a *po*-group in which $c^+ \in H$ exists for all $c \in H$. Then, by (7),

$$(12) \qquad (a - b)^+ + b = [(a - b) \vee 0] + b = a \vee b \quad \text{for all } a, b \in H.$$

Since $(a - b)^+ + b$ exists by hypothesis, so does $a \vee b$. The existence of $a \wedge b$ in H now follows from (9).

LEMMA 2. *If the algebra $(A, +, \vee)$ is a group under $+$, a join-semilattice under $\vee$, and every group-translation is isotone, then $(A, +, \vee)$ is an l-group.*

PROOF. Again, meets exist by (9).

THEOREM 2. *The algebra $(A, +, \vee)$ is an l-group if and only if it is a group under $+$, a join-semilattice under $\vee$, and the distributive laws (7) hold.*

PROOF. Apply Lemma 2, and note that the distributive laws (7) already imply that group translations preserve order in the given semilattice. Now, referring back to the concepts of §VI.10, we obtain an important

COROLLARY 1. *The family of l-groups is equationally definable.*

Applying Theorem VIII.15 this gives

COROLLARY 2. *Every l-group is isomorphic with an l-subgroup of an unrestricted direct product (Example 3) of subdirectly irreducible l-groups.*

Setting $x = a$ and $y = b$ in (10), we get

LEMMA 3. *In any l-group, we have for all a, b,*

$$(13) \qquad a - (a \wedge b) + b = b \vee a.$$

COROLLARY. *In any commutative l-group, we have*

$$(14) \qquad a + b = (a \vee b) + (a \wedge b) \quad \textit{for all } a, b.$$

Example 6. Let G be the group of all positive rational numbers under multiplication (with 1 as the group identity); let G^+ consist of all positive integers. In this example, r^+ is the numerator of r when reduced to lowest terms. In Example 6, the *modular law* (14) specializes to the familiar identity $rs = (r, s)[r, s]$ of number theory.

DEFINITION. If a is an element of an l-group G, then $a^+ = a \vee 0$, and $a^- = a \wedge 0$; a^+ is called the *positive part* of a, and a^- the *negative part* of a.

Setting $b = 0$ in (13), we get $a - a^- + 0 = a^+$, whence

$$(15) \qquad a = a^+ + a^-, \quad \text{for all } a \in G.$$

In words, every element of an l-group is the sum of its positive and negative parts. In Example 4, this gives the Jordan decomposition of §X.6.

THEOREM 3. *If $na \geqq 0$ in an l-group, then $a \geqq 0$.*

PROOF. Expanding by (8′) n times, we get

$$n(a \wedge 0) = na \wedge (n - 1)a \wedge \cdots \wedge a \wedge 0.$$

But if $na \wedge 0 = 0$, this equals

$$(n-1)a \wedge (n-2)a \wedge \cdots \wedge a \wedge 0 = (n-1)(a \wedge 0).$$

Now cancelling, we get $a \wedge 0 = 0$, as claimed.

COROLLARY 1. *In any l-group, every element except* 0 *has infinite order.* (*Any l-group is "torsion-free".*)

PROOF. If $na = 0$, then $na \geqq 0$ and $na \leqq 0$. The proof follows from Theorem 3 and its dual. This result is not valid in all directed groups, as Exercise 9 of §2 shows.

COROLLARY 2. *In a commutative l-group,* $na \geqq nb$ *implies* $a \geqq b$.

PROOF. If $na \geqq nb$, then $na - nb = n(a-b) \geqq 0$. Hence, by Theorem 3, $a - b \geqq 0$. Adding b, $a \geqq b$.

This result is not valid in all noncommutative l-groups, as Example 5 shows.

4. Further Algebraic Properties

The algebraic theory of l-groups is quite unlike that of lattices generally. First, as we have already seen, there are no nontrivial complemented or relatively complemented l-groups. Second, as we now show, there are no nondistributive l-groups.

THEOREM 4. *Every l-group is a distributive lattice.*

PROOF. By Theorem I.10, it suffices to show that $a \wedge x = a \wedge y$ and $a \vee x = a \vee y$ imply $x = y$. But by (13), they imply

$$x = (a \wedge x) - a + (x \vee a) = (a \wedge y) - a + (y \vee a) = y.$$

Two important algebraic notions which apply to l-groups, but not to lattices generally, are those of disjointness and absolute value. The former is somewhat analogous to that of independence in modular lattices. We first prove

THEOREM 5. *In any l-group,*

$$(16) \qquad a \wedge b = 0 \quad \textit{and} \quad a \wedge c = 0 \quad \textit{imply} \quad a \wedge (b + c) = 0,$$

$$(16') \qquad a \vee b = 0 \quad \textit{and} \quad a \vee c = 0 \quad \textit{imply} \quad a \vee (b + c) = 0.$$

PROOF. Since a, b, c are positive, clearly $a \wedge (b + c) \geqq 0$. But by (7′), used twice,

$$\begin{aligned} 0 = 0 + 0 = (a \wedge b) + (a \wedge c) &= ((a \wedge b) + a) \wedge ((a \wedge b) + c) \\ &= (a + a) \wedge (b + a) \wedge (a + c) \wedge (b + c) \geqq a \wedge (b + c), \end{aligned}$$

since $a + a$, $b + a$, $a + c$ all contain a. Formula (16) follows; formula (16′) is just its dual.

DEFINITION. Two positive elements a and b are *disjoint*, (in symbols, $a \perp b$) if and only if $a \wedge b = 0$.

In Example 6, disjointness specializes to relative primeness. Theorem 5 asserts that the set of positive elements disjoint to any a is closed under addition. If $a \wedge b = 0$, we get further from (13), since $b \vee a = a \vee b$.

THEOREM 6. *Disjoint positive elements are permutable.*

$$\text{(17)} \qquad \textit{If } a \wedge b = 0, \textit{ then } a + b = b + a.$$

LEMMA 4. *If $b \wedge c = 0$, then $(b - c)^+ = b$ and $(b - c)^- = -c$.*

PROOF. By our preceding formulas,

$$(b - c) \vee 0 = (b \vee c) - c = b - (b \wedge c) + c - c = b,$$

and dually.

We now define the important and suggestive notion of the *absolute value* $|a|$ of an element a of an l-group, by the formula

$$\text{(18)} \qquad |a| = a \vee (-a).$$

THEOREM 7. *In any l-group, for all a:*

$$\text{(19)} \qquad |a| \geqq 0, \textit{ moreover } |a| > 0 \textit{ unless } a = 0,$$

$$\text{(20)} \qquad a^+ \wedge (-a)^+ = 0,$$

$$\text{(21)} \qquad |a| = a^+ - a^-.$$

PROOF. By definition and dualization:

$$|a| = a \vee (-a) \geqq a \wedge (-a) = -[(-a) \vee a] = -|a|,$$

whence $2|a| = |a| + |a| \geqq |a| - |a| = 0$. It follows by Theorem 3 that $|a| \geqq 0$, proving (19). Moreover if $|a| = 0$, then $0 = 2|a|$ and $a \vee -a = a \wedge -a$ in the proof, whence $a = -a$ and (by Theorem 3, Corollary 1), $a = 0$.

Next, dualizing (19), $a \wedge (-a) \leqq 0$, and so

$$0 = 0 \vee (a \wedge -a) = (a \vee 0) \wedge (-a \vee 0) \qquad \text{(distributive lattice)},$$

which is (20). Finally, by (19) again,

$$\begin{aligned} |a| &= 0 \vee |a| \vee 0 = [0 \vee a] \vee [(-a) \vee 0] \qquad \text{by L3} \\ &= a^+ \vee (-a)^+ \text{ (definition)} = a^+ + (-a)^+. \end{aligned}$$

by (20) and (17). Since $(-a)^+ = -a^-$, the proof is complete.

Setting $a - b$ in place of a in (21), we get

$$|a - b| = (a - b)^+ - (a - b)^- = [(a - b) \vee 0 + b] - [(a - b) \wedge 0 + b],$$

where the last equation is an elementary identity in groups. Hence, applying (7)–(7′) to the final expression:

$$\text{(22)} \qquad |a - b| = (a \vee b) - (a \wedge b).$$

THEOREM 8. *In any l-group, the absolute satisfies*

(23) $$|na| = |n| \cdot |a| \quad \text{for any integer } n.$$

(24) $$|(a \vee b) - (a^* \vee b)| \leqq |a - a^*| \quad \text{and dually}$$

(25) $$|(a \vee c) - (b \vee c)| + |(a \wedge c) - (b \wedge c)| = (a \vee b) - (a \wedge b).$$

PROOF. By (20), a^+ and $-a^- = (-a)^+$ are disjoint; hence by Theorem 6, they are permutable. Hence $na = na^+ + na^-$ (by elementary group theory); moreover na^+ and $-na^- = n(-a^-)$ are disjoint by Theorem 5 and induction. (This independently implies that na^+ and $-na^-$ are permutable.) But na^+ and $n(-a^-)$ are also positive; hence, by Lemma 4, $(na^+) = n(a^+)$ and $(na)^- = n(a^-)$. This implies (23) for positive n. The result for negative n now follows since $|-x| = |x|$.

Since (25) implies (24) obviously, it remains only to prove (25); this we now do.† First, we have

$$\begin{aligned}(a \wedge b) \vee (b \wedge c) \vee (c \wedge a) &= (a \wedge b) \vee [(a \vee b) \wedge c] \\ &= a \wedge b - (a \wedge b) \wedge c + (a \vee b) \wedge c \\ &= [a \wedge b - a \wedge b] + (a \wedge b) \vee c - c + (a \vee b) \wedge c.\end{aligned}$$

Dually, $(a \vee b) \wedge (b \vee c) \wedge (c \vee a) = (a \vee b) \wedge c - c + (a \wedge b) \vee c$. By Theorem 4 and Theorem II.8, these imply

(25′) $$(a \wedge b) \vee c - c + (a \vee b) \wedge c = (a \vee b) \wedge c - c + (a \wedge b) \vee c.$$

Using (25′), a direct calculation gives

$$\begin{aligned}|(a \vee c) - (b \vee c)| &+ |(a \wedge c) - (b \wedge c)| \\ &= (a \vee c) \vee (b \vee c) - (a \vee c) \wedge (b \vee c) \\ &\quad + (a \wedge c) \vee (b \wedge c) - (a \wedge c) \wedge (b \wedge c) \\ &= (a \vee b) \vee c - (a \wedge b) \vee c + (a \vee b) \wedge c - (a \wedge b) \wedge c \\ &= a \vee b - (a \vee b) \wedge c + c - (a \wedge b) \vee c + (a \vee b) \wedge c \\ &\quad - c + (a \wedge b) \vee c - a \wedge b \\ &= a \vee b - [(a \wedge b) \vee c - c + (a \vee b) \wedge c] \\ &\quad + [(a \vee b) \wedge c - c + (a \wedge b) \vee c] - a \wedge b \\ &= a \vee b - a \wedge b \qquad \text{by (25′)}.\end{aligned}$$

Notice finally that the following is true in any l-group:

(26) $$|b + c| \leqq |b| + |c| + |b|,$$

since $-|b| - |c| - |b| \leqq b + c + 0 = b + c \leqq |b| + |c| + |b|$.

Exercises for §§3–4:

1. Prove in detail that Examples 1–2 and 4–5 define l-groups.
2. Prove that the *po*-group of Example 3 is an l-group if and only if every factor G_ϕ is an l-group—and that the same is also true of restricted direct products.
3. Show that if $x, y \geqq 0$ in an l-group, then $a \wedge (x + y) \leqq (a \wedge x) + (a \wedge y)$.

† Following J. A. Kalman, Proc. AMS 7 (1956), 931–2.

4. Show that, in any commutative l-group A, for any positive integer n, the mapping $x \to nx$ is a monomorphism of A into itself (with respect to all l-group operations).

5. Prove in detail that $2a^+ = 0 \vee a \vee 2a$ in any l-group.

*6. Prove that $\mathbf{Z} \times \mathbf{Z}$, with generator $(1, -1)$, is the free l-group with one generator.

*7. Show that $2f = 2g$ (in multiplicative notation, $f^2 = g^2$) does not imply $f = g$ in the l-group of Example 5.

8. Prove that the center of any l-group is an l-subgroup, and that the centralizer $Z(S)$ of any l-subgroup of any l-group G is itself an l-subgroup of G.

9. Show that if $x \leqq a + b$ in an l-group, where a, b, x are all positive, then $x = s + t$, where $s \in [0, a]$ and $t \in [0, b]$.

*10. Show that (16)–(16′) hold in any po-group which satisfies the condition of Exercise 9.†

*11. In a group G, let a unary operation $x \to x^+$ satisfy $0^+ = 0$, $x = x^+ - (-x)^+$, $(x + y^+ - x) = (x + y - x)^+$, and $(x^+ - y)^+ + y = (y^+ - x)^+ + x$. Show that G is an l-group. (Vaida)

*5. Lattice-ordered Loops

Parts of the theory of l-groups do not depend on the associativity of addition. We therefore define a *po-quasigroup*‡ as a poset G which is also a quasigroup (§VIII.2), in which all quasigroup translations $x \to a + x$ and $x \to x + b$ are order-*automorphisms*. It is not enough to assume isotonicity; see Exercise 10 below. A po-quasigroup is called a *po-loop* when, as a quasigroup, it is a loop (i.e., has identity and inverses). A po-quasigroup (po-loop) which is a lattice as a poset is called an *l-quasigroup* (resp. *l-loop*).

Example 7. Let $F: \mathbf{R} \times \mathbf{R} \to \mathbf{R}$ be continuous, and satisfy $F(x, 0) = F(0, y) = 0$ and $-1 < \partial F/\partial x,\ \partial F/\partial y < 0$. Then $\mathbf{R}$ is an l-loop under the "addition" $x \circ y = x + y + F(x, y)$.

It can be proved that, if appropriate parentheses are inserted, the identities (7)–(11) hold in any l-group. We state without proof

THEOREM 9. *Any l-loop is a distributive lattice in which* (16)–(16′) *hold. Moreover* (14) *holds in any commutative l-loop.*

Various other properties of l-loops are stated in the exercises below.

Exercises:

1. Let ϕ be any order-automorphism of the real field $\mathbf{R}$. Show that $a \circ b = \phi^{-1}(\phi(a) + \phi(b))$ defines $\mathbf{R}$ as an l-loop.

2. Show that Example 7 defines an l-loop.

3. Prove converses to the results of Exercises 1–2 (assuming that loop addition is continuously differentiable, if necessary).

4. Show that any complete discrete (simply) ordered l-loop is isomorphic to $\mathbf{Z}$—hence an l-group.

5. Prove that any l-quasigroup satisfies variants of (7)–(11).

---

† Exercises 9–10 refer to the Riesz Interpolation Property; see F. Riesz [**1**] and G. Birkhoff [**6**, Theorems 49–50].

‡ The theory of l-quasigroups was originated by D. Zelinsky (Bull. AMS **54** (1948), 175–183) and I. Kaplansky; further results reported here were worked out by E. C. Ingraham, in consultation with the author.

6. Prove that any l-loop is a distributive lattice which satisfies (14) and (16)–(16′). Is this also the case in all l-quasigroups?

7. Prove that, in any l-loop, we have $a \vee b = [a - (a \wedge b)] + b$, and also $(a \vee b) - (a \wedge b) = (a - b) \vee (b - a)$.

8. Prove that, in any l-loop, disjoint positive elements are permutable, and their join equals their sum.

*9. Show that l-loops constitute an equationally definable family of algebras. What about l-quasigroups?

*10. Construct an "addition" $\oplus$ on the cardinal product $H \times K$ of two l-groups, of the form

$$(h, k) \oplus (h', k') = (h + h', k + k' + f(h, h')),$$

which makes $H \times K$ into a loop whose loop translations are isotone but *not* order-automorphisms.

6. Discrete l-groups

It is easy to determine the class of all "discrete" or "atomistic" l-groups, whose positive cone satisfies the descending chain condition (DCC) of §VIII.1:

(DCC) Every nonvoid set of positive elements contains a minimal member.

Indeed, for any cardinal number $\aleph$, the *restricted direct product* (§1) of $\aleph$ copies of the additive group $\mathbf{Z}$ of the integers is an l-group ($\mathbf{Z}^{\aleph}$) whose positive cone satisfies the DCC. We now show that there are no other discrete l-groups.

In any discrete l-group L, the minimal members of the set of all *strictly* positive elements will cover 0; we will call them *primes*. Since any two primes p and q are disjoint, they are permutable (Theorem 6). This yields

LEMMA 1. *The primes generate an Abelian subgroup, consisting of all elements which can be expressed as sums $n_1p_1 + \cdots + n_sp_s$ of a finite number of distinct primes, with integers n_i as coefficients.*

If $a > 0$, the set of differences $a - \sum n_ip_i$ which are positive contains a, and so is nonvoid. By the DCC the set has a minimal element m. Again by the DCC, every strictly positive element a contains a prime, namely, some minimal x such that $0 < x \leqq a$. If q were such a prime for m, then

$$0 \leqq m - q = a - \left(\sum n_ip_i + q\right) < m,$$

contrary to the minimality of m. Hence, m must be 0, so that $a = \sum n_ip_i$ for any $a > 0$ in L.

But every element can be expressed as a difference of positive elements: $c = c^+ - (-c)^+$ for all c. Hence we have

LEMMA 2. *Every element $a \in L$ belongs to the subgroup described in Lemma 1.*

(The case $a = 0$ is trivial.)

Now, for any $a \neq 0$ in L, we can write for *positive* m_i and n_j:

$$a = m_1p_1 + \cdots + m_rp_r - n_1q_1 - \cdots - n_sq_s, \tag{27}$$

the p_i and q_j being distinct primes. Moreover $a \geqq 0$ if and only if the set of q_i is empty, since

$$(m_1p_1 + \cdots + m_rp_r) \perp (n_1q_1 + \cdots + n_sq_s).$$

In conclusion, the order in L is identical with that in the restricted direct product of copies of $\mathbf{Z}$, one for each prime p of L (atom of L). This proves

THEOREM 10. *An l-group satisfies the chain condition* DCC *if and only if it is the restricted direct product* ($\mathbf{Z}^{\aleph}$) *of a set of copies of* $\mathbf{Z}$.

7. Ordered Groups

In any ordered group (§1), every element a is trivially either positive, negative, or 0. Moreover, by Corollary 1 of Theorem 4, in any l-group G, every element has infinite order. We now prove that a strong converse of this result holds in *commutative* l-groups

THEOREM 11 (F. LEVI). *Any torsion-free† Abelian group A can be made into an ordered group.*

PROOF. In such a group, the equation $nx = ma$ has at most one solution. For $nx = ny$ implies $n(x - y) = 0$, hence $x = y$. If $nx = ma$ has a solution, we denote it $(m/n)a$, and observe that all the laws of vector algebra hold for the multiplication by *rational* scalars so defined.

By a well-ordered *rational basis* for A, we mean a well-ordered (finite or infinite) subset of elements a_α of A such that every nonzero element of A is a finite rational combination

$$n_1a_{\alpha(1)} + \cdots + n_ra_{\alpha(r)} \qquad [\alpha(1) < \cdots < \alpha(r)]$$

of the a_α, while $\sum n_ia_{\alpha(i)} = 0$ implies that every $n_i = 0$—or equivalently, $\sum (m_i/n_i)a_{\alpha(i)} = 0$ implies that every $m_i/n_i = 0$. The existence of well-ordered rational basis can be proved directly, since any maximal well-ordered rationally independent subset is a basis. Finally, relative to such a basis, any element of A not 0 may be called positive or negative according as its first nonzero coefficient m_i/n_i is positive or negative. This "lexicographic" ordering of A clearly defines from it a simply ordered group.

COROLLARY. *A commutative group is the additive group of some l-group if and only if it is torsion-free.*

Note that the preceding construction yields an *Archimedean* ordered group (cf. §2) from A if and only if A is isomorphic with a subfield of the rational field $\mathbf{Q}$. However, there are many other Archimedean ordered groups, even uncountable ones. For example any subgroup of the real field $\mathbf{R}$ is an Archimedean ordered group. We will now prove that there are no other Archimedean ordered groups; the proof paraphrases the abstract theory of "magnitude".‡

† We recall that an Abelian group is called "torsion-free" when all its elements except 0 have infinite order.

‡ O. Hölder [1], 1–64, esp. pp. 13–14. For more recent references, see [**LT2**, p. 226, footnote 15].

LEMMA 1. *An l-group G is Archimedean if and only if it is integrally closed.*

PROOF. By Lemma 4 of §2, any integrally closed l-group G is Archimedean. Conversely, suppose G is Archimedean, and $na \leqq b$ for all $n = 1, 2, 3, \cdots$. Then $na^+ = (na)^+ = na \vee 0 \leqq b \vee 0$ as in the proof of Theorem 8, for $n = 1, 2, 3, \cdots$. But $na^+ \leqq 0 \leqq b^+$ for $n = 0, -1, -2, \cdots$; hence $na^+ \leqq b^+$ for $n = 0, \pm 1, \pm 2, \cdots$. Since G is Archimedean, we infer $a^+ = 0$ and $a = a^+ + a^- \leqq 0$. Q.E.D.

We now define a *strong unit* of a *po*-group G as an element e such that, for any $a \in G$, $ne > a$ for some positive integer $n = n(a)$,

LEMMA 2. *An ordered group G is Archimedean if and only if every $e > 0$ is a strong unit.*

PROOF. If $e > 0$ is a strong unit, then $a \neq 0$ implies $n|a| > b$ for all b and appropriate n; but $|a| = \pm a$; hence $(\pm n)a > b$ for some n, whence G is Archimedean. Conversely, if G is Archimedean, then, by Lemma 1, $ne \leqq b$ for all n is impossible; hence for any $e > 0$ some $ne > b$, and so e is a strong unit.

LEMMA 3. *In any ordered group, $na \geqq nb$ implies $a \geqq b$, and so $na = nb$ implies $a = b$.*

PROOF. Unless $a \geqq b$, we have $a < b$, whence $na < nb$ since addition is order-isomorphic. This is contrary to hypothesis.

We recall that $2a \geqq 2b$ did *not* imply $a \geqq b$ in Example 5 of §2. Since identical implications are preserved under formation of subdirect products (Theorem VI.18), it follows that the l-group of Example 5 *cannot* be represented as a subdirect product of ordered groups. This is noteworthy, since every commutative l-group is a subdirect product of ordered groups (§11 below).

THEOREM 12 (HÖLDER [**1**]). *Any ordered Archimedean group G is isomorphic to a subgroup of the additive group of all real numbers, and so is commutative.*

PROOF. Let $e > 0$ be arbitrary. For any $a \in G$, define $L(a)$ as the set of all rational fractions m/n $[n > 0]$ such that $me \leqq na$, and $U(a)$ as the set of all m/n with $me \geqq na$. We now prove

LEMMA 4. *For any $a \in G$, $L(a) \neq \varnothing$, $U(a) \neq \varnothing$, $L(a) \cup U(a) = \mathbf{Q}$ (the rational field), and $r \in L(a)$, $r' \in U(a)$ imply $r \leqq r'$. That is, $L(a)$ and $U(a)$ define a Dedekind cut.*

PROOF. Since G is ordered, either $me \leqq na$ or $me \geqq na$; hence every $m/n = r \in \mathbf{Q}$ is in $L(a)$ or $U(a)$. Hence, if $L(a) = \varnothing$, we would have $U(a) = \mathbf{Q}$ and $-me \leqq a$ for all $-m \in \mathbf{Z}$, which is impossible for $e \neq 0$ since G is Archimedean. Finally, $me \leqq na$ and $m'e \geqq n'a$ together imply

$$mn'e \leqq n'na = nn'a \leqq nm'e, \quad \text{or} \quad mn' \leqq nm' \quad \text{since } e > 0.$$

But this implies $m/n \leqq m'/n'$, since $nn' > 0$.

COROLLARY. *There is an isotone mapping $a \to x(a)$ from G into the additive group* $(\mathbf{R}, +)$, *in which $x(a)$ is the real number corresponding to the Dedekind cut of Lemma* 4.

LEMMA 5. *The mapping $a \to x(a)$ is a group monomorphism.*

PROOF. To show that $x(a + b) \geqq x(a) + x(b)$, it suffices to show that $m/n \in L(a)$ and $m'/n \in L(b)$ imply $(m + m')/n \in L(a + b)$. But if $me \leqq na$ and $m'e \leqq nb$, then

$$(*) \qquad (m + m')e \leqq na + nb, \quad \text{and} \quad (m' + m)e \leqq nb + na.$$

On the other hand, $a + b \leqq b + a$ implies that $na + nb \leqq n(a + b)$, while $b + a \leqq a + b$ implies that $nb + na \leqq n(a + b)$. Hence, in either event, $(m'm')e = (m' + m)e \leqq n(a + b)$, and so $(m + m')/n \in L(a + b)$.

Hence $x(a + b) \geqq x(a) + x(b)$. Dually, one can show that $x(a + b) \leqq x(a) + x(b)$ by proving that $U(a) + U(b) \subset U(a + b)$. The conclusion $x(a + b) = x(a) + x(b)$ of Lemma 5 follows.

To complete the proof of Theorem 12, it suffices to show that the kernel of the norphism is 0: that $x(a) = 0$ implies $a = 0$. But, if $x(a) = 0$, then for any positive integer n, $-e \leqq na \leqq e$, whence $a = 0$ since G is Archimedean.

Exercises for §§6–7:

1. Show that an l-group is simply ordered if and only if any two strictly positive elements have a strictly positive lower bound.

2. Show that an l-group is discrete if and only if it is "locally finite" (every interval $[a, b]$ is finite).

*3. Show that if G is an Archimedean l-group, and if the center of G contains the commutator subgroup of G, then G is commutative. (Everett and Ulam [**1**, p. 210].)

*4. Show that, in any ordered group, $-a - b + a + b \ll |a| + |b|$, but that the same is not true in Example 5 of §2.

*5. Show that any Abelian group which contains an element of infinite order can be made into a directed group. (Simbireva)

6. Prove that, in any ordered group, the conjugates of any subgroup which is *not* normal form an infinite chain.

*7. Prove that the automorphisms of any Archimedean ordered group form a group which is isomorphic to a subgroup of $(\mathbf{R}, +)$.

*8. Prove that, from every countable ordered group G, one can construct an Abelian ordered group having the same order-type (i.e., order-isomorphic to it).†

*9. A group G is said to be "cyclically ordered" when a ternary relation β is defined on triplets (a, b, c) of distinct elements, such that:

(i) exactly one of $(a, b, c)\beta$ and $(a, c, b)\beta$ holds,
(ii) $(a, b, c)\beta$ implies $(c, b, a)\beta$,
(iii) $(a, b, c)\beta$ and $(a, c, d)\beta$ implies $(a, b, d)\beta$,
(iv) the relation β is invariant under group-translation.

Prove that, in any cyclically ordered group G, the elements of finite order form a subgroup of the center of G which is isomorphic to a subgroup of the cyclically ordered group of rational numbers (mod 1). (L. Rieger)

---

† A. Malcev, Izv. Akad. Nauk **13** (1949), 473–82. For many of the above Exercises, answers can be found in [**Fu**].

*8. Orderable Groups

Ordered groups need not be commutative. For example, the group of all linear transformations $f: x \to ax + b$ $(a > 0;\ a, b \in \mathbf{R})$ under *composition* becomes an ordered group if we define its positive cone P by the condition

$$f \geqq 0 \quad \text{if and only if } a > 1, \text{ or } a = 1 \text{ and } b \geqq 0.$$

Another (non-Archimedean, by Theorem 12) noncommutative ordered group is the following.

Example 8. Let G be the nilpotent group with three generators a, b, c of infinite order, and the defining relations $a + c = c + a$, $b + c = c + b$, and $b + a = a + b + c$. Then every $x \in G$ can be uniquely written as $x = ma + nb + rc$ $(m, n, r \in \mathbf{Z})$. Let $x \geqq 0$ mean that either $m > 0$, or $m = 0$ and $n > 0$, or $m = n = 0$ and $r \geqq 0$. This choice of the positive cone makes G into an ordered group.

An interesting question, studied by F. Levi, B. H. Neumann, and L. Fuchs† is the following. Which abstract groups G can be made into (simply) ordered groups by a suitable choice of positive cone $G^+ = P$? Such groups are called *orderable* groups (or O-groups).

By Lemma 3 of §6, a group G cannot be orderable if it contains two distinct elements a and b such that $na = nb$ for some nonzero $n \in \mathbf{Z}$. Hence the group of Example 6 is not orderable. By Theorem 11, however, every torsion-free Abelian group is orderable. One can prove more: any torsion-free Abelian *po*-group can be made into an ordered group by strengthening the order.‡

Therefore, an Abelian group is orderable if and only if it satisfies the condition

$$na = nb \ (n \neq 0) \quad \text{implies } a = b. \tag{28}$$

However, this condition (though necessary) is not sufficient for a non-Abelian group to be orderable.

LEMMA 1 (B. H. NEUMANN). *In an orderable group*:

$$ma + nb = nb + ma \qquad (m, n \in \mathbf{Z} - \{0\}) \quad \textit{implies} \quad a + b = b + a. \tag{29}$$

PROOF. Suppose $a + b > b + a$ in a *po*-group. Then, by Lemma 1 of §1, substituting $a + b$ for $b + a$ in an expression *increases* its value. In particular, one can obtain $ma + nb$ from $nb + ma$ for any *positive* integers m, n by mn such substitutions; hence $a + b > b + a$ implies $ma + nb > nb + ma$ for any positive integers m, n. On the other hand, $a + b > b + a$ implies $b + (-a) > (-a) + b$, etc.; hence $a + b > b + a$ implies $ma + nb \neq nb + ma$ for any nonzero $m, n \in \mathbf{Z}$.

In an *ordered* group, $a + b \neq b + a$ implies that $a + b > b + a$ or that $b + a > a + b$. In either case, the argument of the preceding paragraph shows that $ma + nb \neq nb + ma$ for every nonzero $m, n \in \mathbf{Z}$.

---

† F. W. Levi, Proc. Ind. Acad. Sci. **16** (1942), 256–63 and **17** (1943), 199–201; B. H. Neumann, Amer. J. Math. **71** (1949), 1–18; L. Fuchs, Fund. Math. **46** (1958), 168–74.

‡ This theorem is due to P. Lorenzen, Math. Z. **45** (1939), 533–53. For a proof and references, see [**Fu**, §III.4].

COROLLARY. *Let G have two generators a, b of infinite order, and let $(-a)+b+a = -b$. Then G is not orderable.*

In the group G just defined, $nx = ny$ implies $x = y$ for any nonzero $n \in Z$; we omit the proof.

THEOREM 13 (SIMBIREVA†–NEUMANN). *If a group G has a well-ordered central series ending with the identity such that all quotient-groups $G_\alpha/G_{\alpha+1}$ are torsion-free, then G is orderable.*

We sketch the proof of Theorems 13–14; for the details, see [**Fu**, §IV.2].

SKETCH OF PROOF. The $G_\alpha/G_{\alpha+1}$ are Abelian and torsion-free; hence they are orderable. Choose a positive cone $P_\alpha/G_{\alpha+1}$ in each $G_\alpha/G_{\alpha+1}$. Then each nonzero $x \in G$ belongs to a first G_α, $\alpha = \alpha(x)$, since $\bigcap G_\alpha = 0$. Set $x > 0$ if $x \in P_\alpha$ for $\alpha = \alpha(x)$, and $x < 0$ otherwise (i.e., set $P = \bigcup (P_\alpha - G_{\alpha+1})$. Then this makes G into an ordered group.

THEOREM 14. *Every free group is orderable.*

SKETCH OF PROOF. In the free group G with $\aleph$ generators, let $[G, G] = G_1$, and let $G_{n+1} = [G, G_n]$, where $[H, K]$ stands for the subgroup generated by all commutators $x^{-1}y^{-1}xy$, $x \in H$, $y \in K$ (cf. §XIV.6). Then, by a theorem of Magnus and Witt, $\bigcap G_n = 0$, the group identity. The result now follows from Theorem 13.

Exercises:

1. Show that any ordered group G, whose structure lattice has finite length n, is isomorphic to an additive subgroup of the lexicographically ordered vector space ${}^n\mathbf{R}$.
2. Show that if an ordered group G is connected in the interval topology, then it is isomorphic to $(\mathbf{R}, +)$. (Iseki)
3. Show that every ordered group G is an order-epimorphic image of an ordered free group. (Iwasawa)
4. (a) Show that every countable ordered group can be embedded in an ordered group with two generators. (B. H. Neumann)‡

   (b) Show that every solvable countable ordered group whose commutator-subgroup series has length l can be embedded in a solvable ordered group with two generators, whose commutator-subgroup series has length $l + 2$ or less.

*5. (a) Construct an ordered group which coincides with its commutator-subgroup. (B. H. Neumann)

   (b) Construct a simple ordered group. (C. G. Chehata)

6. Show that the group Aut C of all order-automorphisms of a chain C is orderable if and only if Aut C is Abelian.§

† See E. P. Simbireva, Mat. Sb. **20** (1947), 145–78. This paper contains several other important results.

‡ Exercise 4 presents results of B. H. Neumann, J. London Math. Soc. **35** (1960), 503–12; for Exercise 5a, see Amer. J. Math. **71** (1949), 1–18; for Exercise 5b, see Proc. London Math. Soc. **2** (1952), 183–97.

§ P. M. Cohn, Mathematika **4** (1957), 41–50.

9. Congruence Relations; l-ideals

We now consider the *l-morphisms* $\phi: G \to H$ from an l-group G into an l-group H. As in Chapter II, §§2–3, any l-morphism is an o-morphism, but an o-morphism need not be an l-morphism. Since the family of l-groups is equationally definable, the image $\phi(G)$ of an l-group under any l-morphism is defined up to *isomorphism* (as a group *and* as a lattice) by G and the congruence relation induced by ϕ. Hence (by group theory) it is determined by its *kernel*: the inverse image $N = \phi^{-1}(0)$ of $0 \in H$ under ϕ; moreover N is necessarily a normal subgroup of G.

We first characterize the normal subgroups of a given l-group G which are kernels of l-morphisms of G.

Definition. An *l-ideal* of an l-group G is a normal subgroup of G which contains with any a, also all x such that $|x| \leqq |a|$.

Clearly G and 0 are l-ideals of G; they are called improper l-ideals; all other l-ideals of G are called proper l-ideals. Again let N be any l-ideal of G, and suppose that $a, b \in N$. If $a \wedge b \leqq x \leqq a \vee b$, then

$$\begin{aligned} |x| &= x \vee -x \leqq (a \vee b) \vee -(a \wedge b) = a \vee b \vee -b \vee -a \\ &= |a| \vee |b| \leqq |a| + |b|. \end{aligned}$$

Hence $x \in N$, and any l-ideal of an l-group is a *convex* l-subgroup. Conversely, since any normal l-subgroup of G contains with a also 0, $-a$, and $a \vee -a = |a|$, any convex normal l-subgroup is an l-ideal. Therefore, the l-ideals of an l-group G are its convex normal l-subgroups.

Theorem 15. *The congruence relations on any l-group G are the partitions of G into the cosets of its different l-ideals.*

Proof. If N is the set of elements congruent to 0 under a congruence relation, then $a \in N$ and $|x| \leqq |a|$ imply $a \wedge -a \leqq x \leqq a \vee -a$; hence $0 \wedge 0 \leqq x \leqq 0 \vee 0 \bmod N$, and so $x \in N$. Conversely, if N is an l-ideal, then $x \equiv x' \bmod N$ implies $|(x \vee y) - (x' \vee y)| \leqq |x - x'|$ by (24), and therefore $x \vee y \equiv x' \vee y \bmod N$. Using left–right symmetry and duality, we see that the partition of G into the cosets of N has the substitution property for both lattice operations, completing the proof.

Theorem 16. *The congruence relations on any l-group G form a complete algebraic Brouwerian lattice $\Theta(G)$.*

Proof. The preceding result is a corollary of Theorem 15 and §VIII.5, Lemma 3, Corollary. It can also be proved from the fact (Theorem VI.9) that the congruence relations on G as a lattice form a complete Brouwerian lattice $\Theta(G, \wedge, \vee)$. The congruence relations on L as an l-group form a closed sublattice $\Theta(G)$ of $\Theta(G, \wedge, \vee)$ by Theorem VI.8—hence they, too, form a Brouwerian lattice.†

Corollary 1. *The l-ideals of any l-group G form a complete algebraic Brouwerian lattice.*

† For another proof, see K. Lorenz, Acta Math. Acad. Sci. Hungar. **13** (1962), 55–67.

Note that, since $\Theta(G)$ is a sublattice of the lattice $\Theta(G, -)$ of all normal subgroups of G, also

$$(30) \qquad J \vee K = J + K = \{x + y\}, \qquad (x \in J, y \in K),$$

for any two l-ideals J and K of an l-group G.

COROLLARY 2. *The complemented l-ideals of any l-group G form a Boolean sublattice of the lattice of all l-ideals: the center of the lattice $\Theta(G)$.*

The following result implies various theorems of valuation theory;† since its hypothesis (*) holds trivially for coprime ideals (since then $N_i + N_j = G$), it also implies a generalized Chinese Remainder Theorem.

COROLLARY 3 (T. NAKANO). *Let $N_1, \cdots, N_r$ be l-ideals of an l-group G, and let $a_1, \cdots, a_r$ be elements of G such that*

$$(*) \qquad a_i \equiv a_j \pmod{N_i + N_j} \quad \textit{for every pair } i, j.$$

Then there exists an element $a \in G$ such that $a \equiv a_i \pmod{N_i}$ for $i = 1, \cdots, r$.

The proof is by induction; the case $r = 1$ is trivial. Assume that G contains an element b such that $b \equiv a_i \pmod{N_i}$ for $i = 1, 2, \cdots, r - 1$. Then, by (*),

$$b \equiv a_i \equiv a_r \pmod{N_i + N_r} \quad \text{for } i = 1, \cdots, r - 1;$$

that is, $a_r - b \in (N_i + N_r)$ for $i = 1, \cdots, r - 1$. Hence, using Theorem 16, $a_r - b \in \bigcap_{i=1}^{r-1} (N_i + N_r) = (\bigcap_{i=1}^{r-1} N_i) + N_r$. Therefore there exists $c \in \bigcap_{i=1}^{r-1} N_i$ such that $a_r - b - c \in N_r$. Setting $a = c + b$, clearly $a \equiv a_r \pmod{N_r}$ *and* $a \equiv b \equiv a_i \pmod{N_i}$ for $i = 1, \cdots, r - 1$. Q.E.D.

We now consider G primarily as a group $(G, -)$. As in Chapter VII, the representations of $(G, -)$ as a finite direct union of *groups* correspond one-one to finite Boolean algebras of normal subgroups under $\wedge = \cap$, $\vee = +$, with least element 0 and greatest element G. But $\Theta(G) = \Theta(G, -, \vee)$ is a *closed sublattice of* $\Theta(G, -)$, by Theorem VI.7 again. We conclude

THEOREM 17. *The representations of any l-group G as a finite cardinal product*

$$(31) \qquad G = G_1 \oplus \cdots \oplus G_r$$

correspond one-one to the finite (Boolean) subalgebras of the center of $\Theta(G)$.

COROLLARY. *Any two finite representations of an l-group G as a direct product* (31) *have a common refinement.*

Representations of the form (31) are also called "finite direct decompositions" of G. Regarding *infinite* direct decompositions, we state without proof the following

THEOREM OF SIMBIREVA. *Any two direct decompositions of a directed group G have a common refinement.*

† T. Nakano, Math. Z. **83** (1964), 140–6, and references given there.

For lexicographic decompositions (defined analogously) we have a similar result.

THEOREM OF L. FUCHS. *Any two lexicographic decompositions of a directed group have a common refinement.*

For discussions and proofs of these theorems, see [**Fu**, §§II.6 and II.7]. For the case of ordered groups, the second had been previously proved by A. Malcev (Izv. Acad. Nauk SSSR **13** (1949), 473–82).

10. Principal l-ideals

It is easy to characterize, in any l-group, those "principal" l-ideals which are generated by a single element a.

LEMMA 1. *In any l-group G, the smallest l-ideal containing a given element a is the set of all $x \in G$ such that*

$$|x| \leqq \sum_{i=1}^{n} [-g_i + |a| + g_i] \quad \text{for some } g_1, \cdots, g_n \in G. \tag{32}$$

SKETCH OF PROOF. If x satisfies (32), then it must be in any l-ideal of G which contains a. Conversely, the set of all x which satisfy (32) is an l-ideal of C, by (26) and the other inequalities of §4.

DEFINITION. The l-ideal defined in Lemma 1 is called the *principal* l-ideal generated by a, and denoted (a). By a *principal l-ideal* of an l-group G, is meant an l-ideal of the form (a), for some $a \in G$.

Clearly $(a) = (|a|)$ for any a. Moreover, in an ordered group, $(a) < (b)$ implies $|a| \ll |b|$. The converse is true in ordered Abelian groups, but not (for example) in the ordered group with generators b and a_k $(k \in \mathbf{Z})$, defining relations

$$a_i + a_j = a_j + a_i, \qquad -b + a_k + b = a_{k+1}, \quad \text{all } k \in \mathbf{Z}, \tag{33}$$

and ordering defined by the conditions

$$a_k \gg a_{k+1} > 0 \quad \text{and} \quad b \gg a_k, \quad \text{all } k \in \mathbf{Z}. \tag{33'}$$

LEMMA 2. *The join of any two principal l-ideals of an l-group G is a principal l-ideal of G:*

$$(a) + (b) = (|a| + |b|). \tag{34}$$

PROOF. Trivially, $|a| + |b| \in (a) + (b) = (a) \vee (b)$, whence $(|a| + |b|) \subset (a) \vee (b)$. Conversely, $(a) \subset (|a| + |b|)$ and $(b) \subset (|a| + |b|)$, whence $(a) \vee (b) \subset (|a| + |b|)$.

THEOREM 18. *The compact elements of $\Theta(G)$ are the principal l-ideals of G.*

PROOF. As in Theorem VIII.7, they are the *finitely generated* l-ideals† of G. But, by Lemma 2 above, these are precisely the principal l-ideals of G.

† Note that an l-ideal of G is a "subalgebra" of G under the unary operations $x \to |x|$, $x \to -x$, $x \to (-g) + x + g$ $(g \in G)$, and the binary operation $x + y$.

We now apply the results of §V.11 to the complete Brouwerian (algebraic) lattice $\Theta(G)$. We define two l-ideals J and K of G to be *independent* (in symbols, $J \perp K$) when $J \wedge K = 0$. We define J^* as the join $\bigvee K_\alpha$ of all l-ideals K_α such that $J \wedge K_\alpha = 0$; trivially, J^* is the set of all elements $x \in G$ such that $|a| \wedge |x| = 0$ for all $a \in J$. Then, by Glivenko's Theorem V.26, the mapping $J \to (J^*)^*$ is a *lattice-epimorphism of* $\Theta(G)$ onto the Moore family of all "closed" l-ideals of G. In other words, we have proved

THEOREM 19. *For any l-ideal J of an l-group G, let J^* be the set of all $x \in G$ such that $|a| \wedge |x| = 0$ for all $a \in J$. Then the correspondence $J \to J^*$ is an involutory Galois connection on $\Theta(G)$, and a dual lattice-morphism mapping $\Theta(G)$ onto the lattice of all "closed" l-ideals of G.*

Exercises for §§9–10:

1. (a) Prove that, if H and K are l-ideals of an l-group G, then $H \cap K$ and $H + K$ are also l-ideals of G.

(b) Justify the identities $H \cap K = H \wedge K$ and $H + K = H \vee K$.

2. Prove directly that, if H, J, K are any three l-ideals of the l-group G, then $H \cap (J + K) = (H \cap J) + (H \cap K)$. (*Hint.* Use Exercise 3 after §4.)

3. Show that the structure lattices of l-groups satisfy:

$$\Theta(GH) = \Theta(G)\Theta(H) \quad \text{and} \quad \Theta(G \circ H) = \Theta(G) \oplus \Theta(H).$$

(Here GH is the direct product of groups *and* lattices.)

4. Show that, in $\Theta(\mathbf{R} \circ \mathbf{R}^2)$, the set of closed l-ideals is not the center: it is a Boolean lattice not a sublattice.

5. Show that a subset of an l-group is an l-ideal if and only if it is a subalgebra under the binary operation $+$ and the unary operations $x \to |x|$, $x \to -x$, and (for all $g \in G$) $x \to T_g(x) = (-g) + x + g$.

6. Prove that a convex normal subgroup of an l-group is an l-ideal if and only if it is directed.

7. Show that the l-ideals of any atomistic l-group form an atomic Boolean algebra.

8. Show that every complemented l-ideal is closed.

9. Call a directed group *normal* when, for any $a > 0$, the set of all x such that $|x| \leqq na$ for some $n \in \mathbf{Z}^+$ is a normal subgroup. Show that an l-group is normal if and only if every $a > 0$ is the strong unit of some (principal) l-ideal.

*10. Construct a simple l-group which is not simply ordered.†

11. Commutative l-groups; Units

Many of the most important l-groups are commutative. Indeed (§14), *every* complete l-group is commutative. Commutative l-groups have a number of distinguishing special properties and simplifying features.

LEMMA 1 (KALMAN). *An l-group G is commutative if and only if*

$$|a + b| \leqq |a| + |b| \quad \textit{for all } a, b \in G. \tag{35}$$

PROOF. In any l-group, $a + b \leqq |a| + |b|$ by isotonicity, and $-(a + b) = (-b) + (-a) \leqq |b| + |a|$. If G is commutative, $|a| + |b| = |b| + |a|$ and (35) follows. Conversely, if (35) holds for all negative $a = -c$, $b = -d$ $(c, d \geqq 0)$, then $-(a + b) = d + c \leqq c + d$ for all $c, d \in G^+$. Likewise, $c + d \leqq d + c$,

† C. Holland, Proc. AMS **16** (1965), 326–9.

whence $c + d = d + c$: all positive elements are permutable. Consequently,

$$x + y = x^+ - (-x)^+ + y^+ - (-y)^+ = y^+ - (-y)^+ + x^+ - (-x)^+ = y + x$$

for all $x, y \in G$, and G is commutative.

LEMMA 2. *In a commutative l-group G, for any $a \in G$, the principal ideal (a) is the set of all $x \in G$ such that, for some positive integer n,*

$$|x| \leqq n|a| = |a| + \cdots + |a| \qquad (n \text{ summands}). \tag{36}$$

PROOF. If G is commutative, then for all $g_i \in G$, $-g_i + |a| + g_i = |a|$. Now use (32).

COROLLARY. *In a commutative l-group G, $(a) \wedge (b) = 0$ if and only if $|a| \wedge |b| = 0$.*

LEMMA 3. *In a commutative l-group G, for any $a \in G$, the set $\perp a$ of all x such that $|a| \wedge |x| = 0$ is an l-ideal.*

PROOF. By Theorem 4, $x \in \perp a$ and $y \in \perp a$ imply

$$0 \leqq |a| \wedge |x \pm y| \leqq |a| \wedge (|x| + |y|) = 0;$$

hence $\perp a$ is a subgroup of G. Also, $x \in \perp a$ and $|y| \leqq |x|$ trivially imply

$$0 \leqq |a| \wedge |y| \leqq |a| \wedge |x| = 0;$$

hence $\perp a$ is an l-ideal of G.

In any l-group, we define $a \perp b$ (in words, a and b are *disjoint*) to mean that $|a| \wedge |b| = 0$. We have just shown that, if G is *commutative*, the set $\perp a$ of all $x \perp a$ is an l-ideal. We will now show that $\perp a$ is a *closed* l-ideal, and relate the relation $a \perp b$ to the discussion of §10.

Indeed, the set $\perp(\perp a) = J(a)$ is just the pseudocomplement (Chapter V) of the l-ideal $\perp(a)$ in the sense of Theorem 19; hence it is a *closed* l-ideal. More generally, the *polar* (Chapter V) of any set $A \subset G$ under the symmetric relation $a \perp b$ is the intersection $A^\perp = \bigcap \perp a_\alpha$ of the l-ideals $\perp a_\alpha$ for variable $a_\alpha \in A$; hence it is an l-ideal, and $(A^\perp)^\perp$ is a complemented l-ideal. Hence so is $((A^\perp)^\perp)^\perp = A^\perp$, and so we have proved

THEOREM 20. *In any commutative l-group G, the sets "closed" under the polarity defined by the symmetric relation $a \perp b$ form a Boolean lattice.*

We have already defined a *strong unit* of a *po*-group G as an element $e \in G$ such that, for any $a \in G$, there is a positive integer n such that $a < ne$. We now define a *weak unit* of an l-group G as an element $c > O$ in G such that $c \wedge |x| = 0$ implies $x = 0$.

LEMMA 4. *Any strong unit is a weak unit.*

PROOF. By Theorem 6, any strong unit is positive. By Theorem 4, for any e, $e \wedge a = 0$ implies $ne \wedge a = 0$ for all n. But if e is a strong unit, $a \leqq ne$ for some n, and so $e \wedge a = 0$ implies $a = ne \wedge a = 0$, whence e is a weak unit.

Not all weak units are strong units. For example, the additive l-group of all continuous real functions on the domain $0 \leqq x < +\infty$ has the weak unit $f(x) = 1$, but no strong unit, since for no n and $f(x) \geqq 0$ is $(f(x) + x)^2 \leqq nf(x)$ for all x. On the other hand, in the l-group of all bounded real-valued functions on any domain, the function $f(x) = 1$ is a strong unit.

Clearly, Lemma 2 associates with each positive element a of a commutative l-group G the largest l-ideal (a) of which a is a *strong* unit, while Theorem 20 associates with a the largest l-ideal $J(a) = \perp(\perp a)$ of which a is a *weak* unit.

Now let A be any Abelian l-group. Unless A is simply ordered, by the invariance of order under group-translation A must contain an element a such that neither $a \geqq 0$ nor $0 \geqq a$. For this a, $a^+ > 0$ and $-a^- > 0$ must be disjoint, by (20). Therefore $(a^+) \cap (a^-) = 0$, since $c \leqq (a^+) \cap (a^-)$ implies $|c| \leqq na^+$ and $|c| \leqq n(-a^-)$ for some n, and so

$$|c| \leqq na^+ \wedge (-na^-) = n(a^+ \wedge (-a^-)) = n0 = 0.$$

By Theorem 15, since $(a^+) > 0$ and $(-a^-) > 0$, we infer

THEOREM 21. *A commutative l-group is either (simply) ordered or subdirectly reducible.*

From this result and Theorem VIII.15, we infer

THEOREM 22 (CLIFFORD). *Any commutative l-group is a subdirect product of ordered l-groups.*

Subdirect products of (simply) ordered l-groups are called *vector groups*; Clifford's Theorem states that every commutative l-group is a vector group. It has been proved recently by C. Holland† that every subdirectly irreducible *non*commutative l-group is isomorphic to a transitive l-subgroup of the l-group of all automorphisms of some chain.

Theorem of Hahn. Let A be any (simply) ordered Abelian group, and let $\Theta(A) = C$ be the "structure lattice" of all l-ideals of A.

We first show that C is a chain. Indeed, if J, J_1 are any two l-ideals of A with $J \not\subset J_1$, then J contains a positive element $a \notin J_1$. For any $b \in J_1$, it follows that a cannot be bounded by any $n|b|$, hence $b \ll a$ for all $b \in J_1$. Therefore, $J \not\subset J_1$ implies $J_1 < J$.

Further, by §VIII.5, C is a complete algebraic chain (note that C is both the structure lattice of A, and the subalgebra lattice of the l-group with a unary operator σ_s: $a \to |a| \wedge |s|$ for each $s \in A$). We next prove a special result.

LEMMA 1. *An l-ideal J of A is compact in the complete algebraic chain $C = \Theta(A)$ if and only if it is a principal l-ideal.*

PROOF. We know by Theorem VIII.7 that J is compact if and only if it is finitely generated. But a principal l-ideal is trivially finitely generated while conversely, a finitely generated l-ideal with generators $g_1, \ldots, g_n$ is principal with generator $|g_1| + \cdots + |g_n|$, by Theorem 18.

† C. Holland. Mich. Math. J. **10** (1963), 399–408.

COROLLARY. *If K is the chain of principal l-ideals of A, then $\Theta(A) \cong K$.*

LEMMA 2. *In a chain C, an element c is "compact" if and only if it covers another element c_1.*

We omit the proof; c_1 is clearly "subdirectly completely irreducible" and K is isomorphic to the lattice of *prime quotients* of C (if 0 is *not* called compact).

Further, if (a) is any principal l-ideal of A, and J is any maximal ideal not containing A, then (a) covers $(a) \wedge J$. It follows that the "gaps" (alias "jumps" or covering pairs) of A are dense in C; by definition, these gaps constitute the *skeleton* K of $C = (a)$. In a remarkable pioneer paper, Hahn [**1**] showed that any (simply) ordered Abelian group was monomorphic to an ordinal power of $\mathbf{R}$ with exponent K. A careful exposition of Hahn's main results, together with recent improvements and generalizations by M. Hausner, J. G. Wendel, P. Conrad, and others, may be found in [**Fu**, §IV.5]; see also §XV.4.

*12. Structure of l-groups

Theorem 21 provides the key to the structure of commutative l-groups with finite structure lattice $\Theta(G)$. We first observe

LEMMA 1. *In the distributive lattice of all l-ideals of any commutative l-group G, the l-ideals which contain any given meet-irreducible l-ideal J form a chain.*

PROOF. Since J is meet-irreducible, G/J is simply ordered by Theorem 21, and so $\Theta(G/J)$ is a chain. But by universal algebra (Chapter VI), the structure lattice $\Theta(G/J)$ is isomorphic to the interval $[J, G]$ of $\Theta(G)$. The conclusion now follows.

THEOREM 23. *Let G be any commutative l-group, the lattice of whose l-ideals has finite length. Then either G is a direct product (i.e., cardinal product), or it contains a maximal l-ideal M which contains every other proper l-ideal.*

PROOF. By Lemma 1, the sets of meet-irreducible l-ideals which are contained in the different maximal (proper) l-ideals of G are incomparable. Hence either (i) G has just one maximal proper l-ideal M, or (ii) the poset X of meet-irreducible l-ideals is a cardinal sum of unrelated components X_i. In the latter case, $\widetilde{\Theta(G)} = \mathbf{2}^{\check{X}}$ by Theorem III.3 (dualized), and so $\Theta(G) = \mathbf{2}^X = \mathbf{2}^{X_1 + \cdots + X_r} = \prod(\mathbf{2}^{X_i})$ has a nontrivial center. The conclusion now follows from Theorem 17.

Carriers. The result of Theorem 23 has been partially extended to non-commutative l-groups by P. Conrad, using the notion of a carrier ("filet") of P. Jaffard, which is defined as follows.

DEFINITION. In any l-group G, we define for any $a, b \in G^+$

$$a \sim b \quad \text{if and only if } a^* = b^*, \tag{37}$$

i.e., if and only if $a \perp x$ is equivalent to $b \perp x$. The equivalence classes for the relation (37) are called *carriers.*

If G is commutative, then (by §9) the "carriers" of G form a Boolean algebra isomorphic with the center of $\Theta(G)$. However, this is not always true in non-commutative l-groups, as the following example shows.

Example 9. Let the group G have three generators a, b, c of infinite order, and defining relations

$$a + b = c + a, \qquad a + c = b + a, \qquad b + c = c + b.$$

Let G^+ contain $ma + nb + n'c'$ if and only if $m > 0$, or $m = 0$ while $n \geqq 0$ and $n' \geqq 0$. In this example, the diagram of the carriers of G is

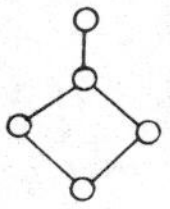

;

FIGURE 23

it is order-isomorphic to $\Theta(\mathbf{R} \circ \mathbf{R}^2)$.

We conclude this section by stating without proof two theorems concerning carriers.

THEOREM OF JAFFARD–R. S. PIERCE. *The mapping from the elements $a \in G^+$ to their "carriers" is a lattice-morphism with kernel* 0, *which satisfies*

$$\phi(a + b) = \phi(a) \vee \phi(b). \tag{38}$$

Moreover ϕ is the maximal lattice-morphism of G^+ whose kernel is 0.

For the proof, see [**Fu**, §V.3].

Developing the preceding concepts further, one can prove the following generalization of Theorem 24.

CONRAD'S THEOREM. *If the lattice of carriers of an l-group G has finite length, then either G is a cardinal product or it contains a unique maximal l-ideal M.*

In the second case, G is a *lexicographic extension* of M by the *ordered group* G/M. For the proof, see [**Fu**, §V.6]; the original result is due to P. Conrad, Michigan Math. J. **7** (1960), 171–80.

Exercises for §§11–12:

1. (a) Show that the l-ideals of any (simply) ordered group form a chain.
   (b) Show that, if the l-ideals of a commutative l-group G form a chain, then G is simply ordered.
2. Show that, in Example 7, the l-ideals form a chain.
3. (a) Show that, in Example 8, $J(c)$ is not an l-ideal.
   (b) Show that, in Example 8, the l-ideals form a chain even though the l-group is not simply ordered.
   (c) Infer that Theorem 22 cannot be extended to non-commutative l-groups.

*4. Let G be $C[0, +\infty)$, the additive group of all continuous real functions on $[0, \infty)$; let J be the l-ideal of those having compact support.
   (a) Show that G is Archimedean but J is not.
   (b) Show that $f \ll g$ in G/J if and only if $f = o(g)$.
   (c) Show that, given $f_1 \leqq f_2 \leqq f_3 \leqq \cdots$ countable, one can always construct g such that every $f_i \ll g$ in G/J. (Theorem of du Bois-Reymond)

*5. Show that, if G is an Archimedean l-group, and if the center of G contains the commutator subgroup of G, then G is commutative. (Everett and Ulam [**1**, p. 210])

6. (a) Show that any simple commutative l-group is isomorphic with an l-subgroup of the additive group of the real field **R**.

(b) Show that a torsion-free commutative group can be made into an Archimedean ordered group if and only if its cardinal number is bounded by that of **R**.

7. Prove that, in any commutative l-group G, for all a, b, c

$$2|a - b| \leqq |a - (b + c)| + |a - (b - c)|,$$
$$2(a \vee b) \leqq a \vee (b + c) + a \vee (b - c),$$
$$2(a \wedge b) \geqq a \wedge (b + c) + a \wedge (b - c).$$

(*Hint.* By considering separately the cases $a \geqq b$ and $a \leqq b$, prove that the inequalities hold in any ordered Abelian group.)

8. Show that, in Theorem 23, $a \notin M$ implies $a > 0$ or $a < 0$.

9. Show that every *finite* structure lattice of commutative l-groups is of the form $1 = \mathbf{2}^0$, $1 \circ L = \mathbf{2}^{1 \circ P}$, or $LM = \mathbf{2}^{P+Q}$, where $L = \mathbf{2}^P$ and $M = \mathbf{2}^Q$ are structure lattices of l-groups.

13. Complete l-groups

In §§13–16, we will study in greater depth the class of (conditionally) *complete* l-groups, that is, l-groups in which every bounded set has a l.u.b. and a g.l.b. The additive group $\mathbf{R} = (\mathbf{R}, +)$ of all real numbers is such a complete l-group. More generally:

LEMMA 1. *Any direct product $G = \prod G_\alpha$ of complete l-groups is itself a complete l-group.*

PROOF. In G, suppose that $\mathbf{a} \leqq \mathbf{x}_\sigma \leqq \mathbf{b}$ for all

$$\mathbf{x}_\sigma = (x_{\sigma 1}, x_{\sigma 2}, x_{\sigma 3}, \cdots) \in S \subset G.$$

Then $\mathbf{u} = (u_1, u_2, u_3, \cdots)$, where $u_\alpha = \bigvee_S x_{\sigma\alpha}$ for all α, exists and satisfies $\mathbf{u} = \sup_S \mathbf{x}_\sigma$ in G, and dually.

It follows that $\mathbf{R}^n$ is a complete l-group. The "vector lattices" $L_p[0, 1]$ and $L_p[-\infty, \infty]$, for $1 \leqq p \leqq \infty$, are other complete l-groups which will be studied more carefully in Chapter XV.

We now characterize the directed groups which can be embedded monomorphically (as *po*-groups) in complete l-groups. (We remark that, trivially, any (conditionally) complete directed group is a complete l-group.) By Lemma 5 of §2, such a directed group must be integrally closed, and hence Archimedean. It is a remarkable fact that this necessary condition is also sufficient.†

THEOREM 24. *The completion by nonvoid cuts of any integrally closed directed group is a complete l-group.*

PROOF. Let G be any directed group; for any nonvoid subset X of G which is bounded above, let $U(X)$ be the nonvoid set of its upper bounds; and let $X^\# = L(U(X)) \supset X$ be the nonvoid set of all lower bounds of $U(X)$ in G. Then the sets $X^\#$ form a conditionally complete lattice $\bar{G}$, the *conditional completion* of G by nonvoid cuts; cf. Chapter V.

† For the original references, see [**LT2**, p. 229, footnote 19], and [**Fu**, p. 95, footnote].

We now define an *addition* $+$ in $\bar{G}$ by the rule

$$(39) \qquad X^{\#} + Y^{\#} = (X^{\#} + Y^{\#})^{\#},$$

where $X^{\#} + Y^{\#}$ denotes as usual the set of all sums $x + y$ ($x \in X^{\#}$, $y + Y^{\#}$). Clearly, this addition is *isotone* and associative; moreover it agrees with addition in G under the monomorphic order-embedding $x \to L(U(x))$ of G in $\bar{G}$. In particular, the negative cone $0^{\#}$ is an *identity* (neutral element of $\bar{G}$), and so $\bar{G}$ is a *complete l-semigroup* (cf. §XIV.5) under $+$, $\subset$.

We next prove that

$$(40) \qquad X^{\#} + L(-X^{\#}) = X^{\#} + [-U(X^{\#})] \subset 0^{\#} \quad \text{for all } X^{\#} \in \bar{G}.$$

Indeed, since $x \to -x$ is an anti-automorphism, clearly

$$L(-X^{\#}) = L(U(L(-X^{\#}))) = -U(L(U(X^{\#}))) = -U(X^{\#}),$$

which proves the first equality of (40). Moreover, if $x \in X^{\#}$ and $y \in L(-X^{\#})$, then by hypothesis $y \leqq -x$ and so $x+y \leqq 0$. This shows that $0 \in U(X^{\#}+L(-X^{\#}))$, which implies

$$0^{\#} = L(0) \supset L(U(X^{\#} + L(-X^{\#}))) = (X^{\#} + L(-X^{\#}))^{\#},$$

and hence (40). Note that the preceding results hold in any directed group. We now prove

LEMMA 2. *If G is integrally closed, then*:

$$(40') \qquad U(X^{\#}) + L(-X^{\#}) \subset U(0).$$

This will show that $\subset$ can be replaced by $=$ in (40′), whence $\bar{G}$ is an l-group, completing the proof of Theorem 24.

PROOF OF LEMMA 2. Let $c \in U(X + L(-X))$, where X is bounded and nonvoid. For any fixed $y_j \in U(X) = L(-X)$, since $c \geqq x - y$; for all $x \in X$ and $y_j \in L(-X)$, clearly $c + y_j \geqq x$ for all $x \in X$; hence (by definition) $c + y_j \in U(X)$ also. It follows that $nc + y_j \in U(X)$ for every positive integer n. Since G is integrally closed, such a set can be bounded below only if $c \geqq 0$. This proves that $U(X + L(-X)) \subset U(0) = P$, the positive cone of G, which is (40′).

Fuchs [**Fu**, §V.10] proves slightly more. Following C. J. Everett (Duke Math. J. **11** (1944), 109–19), he characterizes the elements of $\bar{G}$ which have inverses, and observes that these form an *l-subgroup* G_D of $\bar{G}$. He calls this the *Dedekind extension* of G (cf. Chapter XIV); clearly, G is monomorphically embedded in G_D as an l-subgroup.

COROLLARY 1. *A directed group can be embedded monomorphically in a complete l-group (as a po-subgroup) if and only if it is integrally closed.*

By Lemma 1 of §7, we infer

COROLLARY 2. *An l-group can be embedded monomorphically in a complete l-group if and only if it is Archimedean.*

14. Infinite Distributivity; Closed *l*-ideals

We now show that the infinite distributive laws (10)–(10′) and (12) have far-reaching analogs, which can be interpreted as implying that all complete *l*-groups are topological lattices and topological groups.† We first prove

THEOREM 25. *The infinite distributive laws*

$$a \wedge \bigvee x_\alpha = \bigvee (a \wedge x_\alpha) \quad \text{and} \quad a \vee \bigwedge x_\alpha = \bigwedge (a \vee x_\alpha) \tag{41}$$

are valid in any complete l-group.

PROOF. Let $v = \bigvee x_\alpha$. Then, for all a

$$0 \leqq (a \wedge v) - (a \wedge x_\alpha) \leqq v - x_\alpha,$$

by (24). Hence $0 \leqq \bigwedge [(a \wedge v) - (a \wedge x_\alpha)] \leqq \bigwedge (v - x_\alpha)$, where, by (12), $\bigwedge (v - x_\alpha) = v - \bigvee x_\alpha = v - v = 0$. Therefore,

$$0 = \bigwedge [(a \wedge v) - (a \wedge x_\alpha)] = a \wedge v - \bigvee (a \wedge x_\alpha),$$

which proves the first identity of (41). The second follows by duality.

To give topological interpretations to our results, we recall from Chapter X that a net $\{x_\alpha\}$ of elements of a conditionally complete lattice is said to *order-converge* to the limit x, when

$$\bigvee_\beta \left[\bigwedge_{\alpha \geqq \beta} x_\alpha \right] = \bigwedge_\beta \left[\bigvee_{\alpha \geqq \beta} x_\alpha \right] = x. \tag{42}$$

In another notation, the condition is that

$$\liminf \{x_\alpha\} = \limsup \{x_\alpha\} = x. \tag{42′}$$

THEOREM 26. *In any complete l-group, $x_\alpha \to x$ and $y_\beta \to y$ (in the sense of order-convergence) imply*

$$x_\alpha \wedge y_\beta \to x \wedge y, \qquad x_\alpha \vee y_\beta \to x \vee y, \qquad x_\alpha + y_\beta \to x + y. \tag{43}$$

PROOF. In (43), the Cartesian product of nets (Chapter IX) is understood. By hypothesis, $u'_\alpha \leqq x_\alpha \leqq u_\alpha$ and $v'_\beta \leqq y_\beta \leqq v_\beta$, where $u'_\alpha \uparrow x$, $u_\alpha \downarrow x$, $v'_\beta \uparrow y$, $v_\beta \downarrow y$. Moreover, by isotonicity:

$$u'_\alpha \circ v'_\beta \leqq x_\alpha \circ y_\beta \leqq u_\alpha \circ v_\beta \qquad (\circ = \wedge, \vee, \text{or } +). \tag{44}$$

Hence it suffices to prove that $u_\alpha \circ v_\beta \downarrow x \circ y$, and dually. Moreover, again by isotonicity, $u_\alpha \circ v_\beta \geqq x \circ y$ for all α, β, where $u_\alpha \circ v_\beta$ is antitone in α, β. Hence it suffices to show that $\bigwedge_{\alpha,\beta} (u_\alpha \circ v_\beta) = x \circ y$ for $\circ = \wedge, \vee$, and $+$.

But trivially, by the general associative law:

$$\bigwedge_{\alpha,\beta} (u_\alpha \wedge v_\beta) = (\bigwedge u_\alpha) \wedge (\bigwedge v_\beta) = x \wedge y.$$

† Though not in the usual sense; cf. §XV.8.

Again, using (41) twice, we have

$$\bigwedge_{\alpha,\beta} (u_\alpha \vee v_\beta) = \bigwedge_\alpha \left[\bigwedge_\beta (u_\alpha \vee v_\beta)\right] = \bigwedge_\alpha \left[u_\alpha \vee \bigwedge_\beta v_\beta\right]$$
$$= \bigwedge_\alpha [u_\alpha \vee y] = \left(\bigwedge_\alpha u_\alpha\right) \vee y = x \vee y.$$

Finally, using (10′) similarly twice, we see

$$\bigwedge_{\alpha,\beta} (u_\alpha + v_\beta) = \bigwedge_\alpha \left[\bigwedge_\beta (u_\alpha + v_\beta)\right] = \bigwedge_\alpha \left[u_\alpha + \bigwedge_\beta v_\beta\right]$$
$$= \bigwedge_\alpha [u_\alpha + y] = \left(\bigwedge_\alpha u_\alpha\right) + y = x + y.$$

This completes the proof.

Theorem 26 shows that the operations $\wedge$, $\vee$, $+$ are continuous with respect to order-convergence. We shall now study the l-ideals which are closed in the order topology.

Closed l-ideals. We know (Theorem 17) that the complemented l-ideals of any l-group G correspond to its direct factors. We also know (Theorems 19–20) that if G is commutative, then its complemented l-ideals are "closed" under the polarity defined by disjointness.

We now show that, if G is a *complete* l-group, then an l-ideal $J \subset G$ is "closed" in the preceding sense (that $\perp(\perp J) = J$) if and only if it is "closed" in the following, topological sense, *and* if and only if it is complemented.

DEFINITION. An l-ideal J of a complete l-group G is *closed* if and only if it contains, with any bounded subset $\{x_\alpha\}$, also $\bigvee x_\alpha$.

Remark. Since the mapping $x \to -x$ of G leaves J setwise invariant and inverts order, it follows that J also contains $\bigwedge x_\alpha$. We now complete the proof of

THEOREM 27. *In a complete l-group G, the following are equivalent conditions on an l-ideal J:*

(i) *J is complemented,* (ii) $J = \perp(\perp J)$, (iii) *J is closed.*

PROOF. The implications (i) $\to$ (ii) and (i) $\to$ (iii) are obvious from Theorem 19, with $J^* = \perp J$. Also, the implication (ii) $\to$ (iii) is obvious, since $\perp(\perp J)$ is "closed" in the sense of the definition just given, by (41). To complete the proof, we now show that if J is "closed" in a complete l-group, then $J + \perp J = G$ (the condition $J \cap \perp J = 0$ is obvious), whence (iii) $\to$ (i). But, for any $a \geqq 0$ of G, define the J-component of a as $b = \sup_{x \in J} a \wedge x$; by hypothesis, $b \in J$. Moreover clearly $0 \leqq b \leqq a$; define $c = -b + a$; clearly $0 \leqq c \leqq a$. Moreover for any positive $y \in J$, since $b + y \in J$, we have

$$b \geqq a \wedge (b + y) = (b + c) \wedge (b + y) = b + (c \wedge y).$$

Hence $c \wedge y = 0$ for all positive $y \in J$, and so $c \in \perp J$. This shows that $J + \perp J$ contains all *positive* elements of G. Moreover since the elements of J are permutable with those of $\perp J$, both J and $\perp J$ are *normal* subgroups of G, and $J + \perp J$ contains all elements $d = d^+ + d^-$ of G. This proves that $J + \perp J = G$.

Since any intersection of closed l-ideals is closed, we have the following corollary.

COROLLARY. *The complemented (i.e., closed) l-ideals of any complete l-group form a complete Boolean algebra.*

Historical remark. Theorem 27 was first proved by F. Riesz [**1**] assuming commutativity. The author [**LT2**, Theorem 19] avoided this assumption.

Exercises for §§13–14:

1. Show that a directed group is a σ-complete l-group if and only if every countable set of positive elements has a g.l.b.
2. Show that the group of linear functions $f(x) = ax + b$ $(a > 0)$, under composition, can be made into an l-group but not into an integrally closed one.
3. Let S be any subset of a complete l-group G. Show that $G \cong S^{\perp} \oplus S^{\perp\perp}$. Define a *Cauchy sequence* in an l-group G as a sequence $\{x_n\}$ such that $|x_n - x_m| \leqq u_n$ for some *null sequence* u_n which order-converges to 0, and all $m \geqq 0$.

*4. Show that, if G is any l-group, then its Cauchy sequences modulo null sequences form an l-group $\bar{G}$ to which G is monomorphic.†

5. Let G be an l-group such that, given a double sequence $\{a_{kn}\}$ such that for fixed k $a_{kn} \downarrow 0$, there exists a "diagonal" sequence $a_{k,n(k)} \to 0$, then $\bar{\bar{G}} \cong \bar{G}$ in Exercise 4.
6. Exhibit an Archimedean commutative l-group not a subdirect product of l-subgroups of **R**. (Kaplansky‡)

15. Theorem of Iwasawa

The proof of Theorem 27 can be applied further, to show that the assumption that J is a *normal* subgroup is redundant, for closed l-ideal in a complete l-group.

LEMMA 1. *In a complete l-group G, let S be a subgroup which:*
(i) *contains with any a also all $x \in G$ such that $|x| \leqq |a|$, and*
(ii) *contains with any bounded set of elements a_α, also $\bigvee a_\alpha$.*
Then S is a normal subgroup of G, and hence an l-ideal.

PROOF. For any $a \geqq 0$ of G, define b and c as in paragraph two of the proof of Theorem 27. As shown there, both J and $\perp J$ are normal subgroups of G.

Using this fact, we can prove the remarkable result that *any complete l-group is commutative.* The idea of the proof is to decompose such a group G, for any element c, into the subdirect factors $G/(c^+)^{\perp}$ and $G/(c^+)^{\perp\perp}$, in which $c \leqq 0$ and $c \geqq 0$ respectively. To simplify the proof, we also note

LEMMA 2. *If all positive elements of an l-group G are permutable, then G is commutative.*

PROOF. $a + b = a^+ + a^- + b^+ + b^-$, $b + a = b^+ + b^- + a^+ + a^-$. If positive elements are permutable, then a^+ and a^- are permutable with b^+ and b^-; hence $a^+ + a^- + b^+ + b^- = b^+ + a^+ + a^- + b^- = b^+ + b^- + a^+ + a^-$, completing the proof.

† Exercises 4–5 state results of C. J. Everett, Duke Math. J. **11** (1944), 109–19; cf. [**Fu**, §V.11].

‡ P. Conrad, J. Harvey and C. Holland, Trans. AMS **108** (1963), 143–69; see p. 166, Example **4**.

THEOREM 28 (IWASAWA). *Any complete l-group is commutative.*†

PROOF. By Lemma 2, it suffices to show that if $a > 0$ and $b > 0$, then $a + b = b + a$. Again, by the remark above about decomposing G, we can confine attention to the cases $a - b \geqq 0$ and $b - a \geqq 0$, and to the cases $-a - b + a + b \geqq 0$ and $-a - b + a + b \leqq 0$. This gives us four cases, of which $a \geqq b \geqq 0$ and (by left–right symmetry) $b + a \geqq a + b$ is typical. We define t by $-a + b + a = b + t$, so that $t \geqq 0$. We further define $b_n = -na + b + na$, $t_n = -na + t + na$, for every integer n; thus the b_n, t_n are the transforms of b, t under the group of inner automorphisms generated by $x \to -a + x + a$. Observe that since $0 \leqq b \leqq a$, the interval $[0, a]$ being invariant under all these inner automorphisms, we have $0 \leqq b_n \leqq a$ for all a. Moreover we can prove $b_{n+1} = b + t + t_1 + \cdots + t_n$ by induction, since

$$\begin{aligned} -a + (b + t + t_1 + \cdots + t_{n-1}) + a \\ &= -a + b + a + (-a + t + a) \sum_{i=1}^{n-1} (-a + t_i + a) \\ &= (b + t) + t_1 + t_2 + \cdots + t_n. \end{aligned}$$

Now suppose $t_1 = -a + t + a \geqq t$. Then, for any integer n, $t_{n+1} - t_n = -na + (t_1 - t) + na \geqq 0$, since inner automorphisms preserve order. Hence $0 \leqq t \leqq t_1 \leqq t_2 \leqq t_3 \leqq \cdots$, and, for any positive integer n, we have

$$nt \leqq t + t_1 + t_2 + \cdots + t_{n-1} = -b + b_{n+1} \leqq a.$$

Hence, by Lemma 5 of §2, $t \leqq 0$; but $t \geqq 0$ by hypothesis; hence $t = 0$.

Similarly, consider the case $t_1 \leqq t$. Then we have as above $0 \leqq t \leqq t_{-1} \leqq t_{-2} \leqq t_{-3} \leqq \cdots$. Also,

$$b = a + (b + t) - a = a + b - a + a + t - a = a + b - a + t_{-1},$$

and so $a + b - a = b - t_{-1}$. By induction on n, we can show that

$$b_{-n-1} = a + (b - t_{-1} - t_{-2} - \cdots - t_{-n}) - a = b - t_{-1} - t_{-2} - \cdots - t_{n-1}.$$

Hence, for any positive integer n,

$$nt \leqq t_{-n} + t_{-n-1} + \cdots + t_{-2} + t_{-1} = -a - b_{-n-1} + a + b = -b_n + b \leqq b.$$

Hence if $t_1 \leqq t$, $t \leqq 0$ and so $t = 0$ as before.

But, by use of components, we can reduce the general case to the two cases $t_1 \geqq t$ and $t_1 \leqq t$, projecting onto $[(t_1 - t)^+]^\perp$ and $[(t_1 - t)^+]^{\perp\perp}$. Hence, in any case, $t = 0$, and $a + b = b + a$, completing the proof.

COROLLARY. *Any Archimedean directed group is commutative.*

Exercises:

1. Prove Lemma 1 in detail.
2. (a) A *component* of a positive element e of an l-group A is an element f such that $f \wedge (e - f) = 0$. Prove that the components of any such e form a Boolean algebra.

(b) Show that, if A is a complete l-group with weak unit e, then the Boolean algebra of (a) is isomorphic with center of $\Theta(A)$.

† Japan. J. Math. 18 (1943), 777-89.

3. Show that e is a weak unit of a complete l-group A if and only if $(x \wedge ne) \to x$ for every $x \in A$.

4. Show that a complete l-group not isomorphic to **R** or **Z** is directly decomposable.

5. Show that a complete l-group whose structure lattice has length n is isomorphic to $\mathbf{R}^k\mathbf{Z}^{n-k}$ for some $k = 0, 1, \cdots, n$.

*6. Show that a complete l-group which is not atomistic (§5) has at least the cardinality of the continuum.

PROBLEMS (See also [**Fu**, pp. 209–13].)

111. Find conditions for a group, lattice, resp. l-group to be isomorphic to the l-group of all order-automorphisms of an appropriate chain. (*Hint.* The automorphisms of any chain C yield automorphisms of its completion $\bar{C}$; consider their fixpoints.)

112. Find necessary and sufficient conditions for an abstract group G to be group-isomorphic with an l-group.†

113. Determine all ways in which the free group with n generators can be made into an l-group (an ordered group).‡

114. Which directed groups are topological groups and topological lattices in the ("new") interval topology?

115. Develop a common abstraction which includes Boolean algebras (rings) and l-groups as special cases.

116. Develop postulates for real affine spaces in terms of the ternary operation $a - b + c$.§

117. Are there incomplete l-groups in which: (a) every "closed" l-ideal is complemented, or (b) the correspondence $K \to (K^*)^*$ of Theorem 19 is a lattice endomorphism? What are they? What does this mean for $\Theta(G)$?

118. Develop a general theory of commutative l-groups with "operators" (e.g., of l-modules).

119. Generalize the results of Theorem 10 etc. to l-loops.

120. What are the possible (finite) structure lattices of l-groups?

121. Can the pathological behavior found in Example 5 of the text occur in an l-group of finite breadth? In a Lie tl-group?¶

---

† Conditions for G to be isomorphic with an ordered group were found by K. Iwasawa, J. Math. Soc. Japan **1** (1948), 1–9.

‡ See [**Fu**, p. 49] for relevant literature.

§ See Certaine, Bull. AMS **49** (1943), 869–77 and "betweenness" as defined by Behrend-Greve: (s, y, z) if $y = \tau x + (1 - \tau)z$, where $0 \leqq \tau \leqq 1$, Math. Z. **78** (1962), 298–318.

¶ For the concepts of Lie l-group and tl-group, see G. Birkhoff, Comment. Math. Helv., Speiser Festchrift, pp. 209–17.

Chapter XIV

LATTICE-ORDERED MONOIDS

1. *po*-groupoids

The concept of a lattice-ordered monoid (or "l-monoid") arises naturally in ideal theory, where it has roots in the work of Dedekind, and was studied carefully by Krull.† The modern theory of l-monoids began with basic papers by Ward and Dilworth [**1**] and Dilworth [**1**], later extended and sharpened by J. Certaine [**1**]. It applies not only to ideal lattices, but also to Brouwerian lattices, relation algebras, and l-groups.‡ We shall begin this chapter by considering an even more general concept than that of l-monoid.

Definition. A *po-groupoid* (or *m-poset*) is a poset M with a binary multiplication which satisfies the *isotonicity* condition

$$(1) \qquad a \leqq b \quad \text{implies} \quad xa \leqq xb \quad \text{and} \quad ax \leqq bx,$$

for all $a, b, x \in M$. When multiplication is commutative or associative, M is called a *commutative po*-groupoid or *po-semigroup*, respectively. A *po*-semigroup with identity (or "neutral element") 1, such that

$$(1') \qquad x1 = 1x = x \quad \text{for all } x,$$

is called a partly ordered monoid, or *po-monoid*. A *zero* of an m-poset M is an element $0 \in M$ such that

$$(2) \qquad \text{for all } x \in M, \quad 0 \leqq x \quad \text{and} \quad 0x = x0 = 0.$$

Trivially, a *po*-groupoid can have at most one zero.

Example 1. Any *po*-group G is a *po*-monoid; moreover the positive cone $P = G^+$ and negative cone $-P = G^-$ of any *po*-group are also *po*-monoids.

It is understood that, when considered as *po*-monoids, the additive *po*-groups of Chapter XIII are to be rewritten as multiplicative *po*-groups. Thus, in Example 1, we shall usually write the group operation in G, G^+, and G^- as multiplication, the identity as 1, and we will write x^n in place of nx, used in Ch. XIII. This turns

† R. Dedekind, Ges. Werke, Vol. III, pp. 62–71; W. Krull, S.-B. Phys. Med. Soc. zu Erlangen **56** (1924), 47–63.

‡ For the connections between the concept of l-monoid and l-groups, Brouwerian lattices, and the algebra of relations, see G. Birkhoff [**6**, §27]; also [**LT2**, Chapter XIII], [**Tr**, Part II], and [**Fu**, Part III]. For other early papers by Ward and Dilworth, see these and the references of the text.

out to be more convenient for various reasons. Note that G^+ has a *zero*, whereas G and G^- do not. The "identity" of the negative cone G^- is also its universal upper bound; hence to write it as 0 would be very confusing.

In §2, we will characterize those *po*-monoids which arise as the positive (or negative) cones of directed groups.

Example 2. In any semigroup† S, the "power set" $P(S) = \mathbf{2}^S$ of all its subsets $X, Y, Z, \cdots$ forms a *po*-semigroup with zero $\varnothing$, when (partially) ordered by inclusion and multiplied as in the "calculus of complexes"—so that XY is the set of all products xy $(x \in X, y \in Y)$. If S is a monoid with identity 1, then so is $P(S)$, with identity $\{1\}$.

Example 3. In any ring R, the additive subgroups $X, Y, Z, \cdots$ form an *m*-poset with zero under inclusion, if XY is defined as the set of all finite sums $\sum x_i y_i$ $(x_i \in X, y_i \in Y)$. If R is an associative ring with unity 1, then this *po*-semigroup is a *po*-monoid, whose identity is the additive subgroup generated by 1.

2. Divisibility Monoids

We now define an important class of *po*-monoids in which the order is given by the multiplication operation; such *po*-monoids are said to be *naturally* ordered.

DEFINITION. A *divisibility monoid* is a *po*-monoid M in which $a \leqq b$ is equivalent to $b \in Ma$, and also to $b \in aM$. The *dual* of a divisibility monoid is one in which the same is true of the relation $a \geqq b$.

Note that any join-semilattice with O is trivially a *commutative* divisibility monoid under the "multiplication" $\vee$; dually, any meet-semilattice with I is the dual of a commutative divisibility monoid, with I as identity. In these *po*-monoids, the multiplication is defined by the order:

LEMMA 1. *In any divisibility monoid, $ab = 1$ implies $a = b = 1$.*

PROOF. Trivially, $1 \leqq a, b$—while $ab = 1$ implies $a, b \geqq 1$. Hence (by P2) $ab = 1$ implies $a = b = 1$. Hence, in any divisibility monoid, the monoid identity is a universal lower bound.

LEMMA 2. *Let G be any (multiplicative) po-group, with positive cone P. Then:*

(i) *P is a divisibility monoid.*

(ii) *P is cancellative: $ax = ay$ or $xb = yb$ implies $x = y$.*

If G is a directed group, then also

(iii) *Any $a, b \in P$ have an upper bound in P,*

(iv) *Any $g \in G$ can be written $g = ab^{-1}$ $(a, b \in P)$,*

(v) *In G, $ab^{-1} \leqq cd^{-1}$ $(a, b, c, d \in P)$ if and only if, for some $x, y, z \in P$, $ax = cy$, $bx = dyz$.*

The proofs of (i)–(iv) are trivial.

† For the general theory of semigroups, see A. H. Clifford and G. B. Preston, *Algebraic theory of semigroups*, AMS, 1961. This will be referred to below as [**C-P**].

To prove (v), let $ab^{-1} \leqq cd^{-1}$ in G, let u be any upper bound to a, b, c, d and set $ax = cy = u$, and $b' = bx$, $d' = dy$. Then

$$ub'^{-1} = ab^{-1} \leqq cd^{-1} = ud'^{-1}$$

whence $d' \leqq b'$, and so $b' = d'z$ for some $z \in P$. Substituting, $bx = dyz$, as claimed. Conversely, if the conditions of (v) are satisfied, then

$$ab^{-1} = (ax)(bx)^{-1} = (cy)(dyz)^{-1} \leqq (cy)(dy)^{-1} = cd^{-1}.$$

COROLLARY. *Let G and H be any two directed groups, with positive cones G^+ and H^+. Then any semigroup isomorphism between G^+ and H^+ can be uniquely extended to a po-group isomorphism of G with H.*

We now show that conditions of Lemma 1 and (i)–(iii) completely characterize the class of positive cones of directed groups, as *po*-semigroups.

THEOREM 1. *A semigroup S is the positive cone of a directed group G if and only if: (α) it is cancellative, (β) $Sa = aS$ for all $a \in S$, (γ) $ab = 1$ implies $a = b = 1$.*

SKETCH OF PROOF (cf. [**LT2**, p. 218], [**Fu**, p. 14]). Given S, it is easy to verify that the equation $db_d = bd$ defines a morphism $\phi_d: b \to b_d$ from S to the group of its automorphisms. We then let G consist of the couples (a, b) with $a, b \in S$. We get a group if we define

$$(a, b) = (c, d) \quad \text{means} \quad ad = cb_d \tag{3}$$

$$(a, b)(c, d) = (ac_b, db); \tag{4}$$

we leave the details as an exercise.†

COROLLARY 1. *A po-monoid P is the positive cone of a directed group G if and only if it is a divisibility monoid with cancellation, in which any two elements have a common multiple.*

COROLLARY 2 (VON NEUMANN). *The monoid M of Theorem 1 is the positive cone of an l-group G if and only if any $a, b \in S$ have a least common multiple in M.*

Exercises for §§1–2:

1. (a) Let G be any *po*-groupoid with 1. Show that the set G^+ of $x \geqq 1$ in G is a sub-*po*-groupoid, and that so is the set G^- of $x \leqq 1$.

   (b) Show that $a \in G^+$ is equivalent to $ay \geqq y$ for all $y \in G$, and also to $ya \geqq y$ for all $y \in G$.

2. (a) Define the direct (cardinal) product PQ of two *po*-groupoids P and Q, and prove it always a *po*-groupoid.

   (b) Does the natural definition of lexicographic product $P \circ Q$ of P and Q always give a *po*-groupoid?

3. Prove in detail that Example 2 defines a *po*-monoid.
4. Same question for Example 3.
5. Prove statements (i)–(iv) of Lemma 2, §2.
6. Prove Theorem 1 of §2.

---

† Note that, more generally, any "right reversible" cancellative semigroup can be embedded in a group [**C-P**, p. 35].

7. Prove that an additive *po*-monoid P consists of the positive elements of some *po*-group if and only if:†

(a) $x < y$ implies $a + x < a + y$ and $x + b < y + b$ for all $a, b \in P$, and

(b) $x \leqq y$ if and only if $s + x = y$ for some $s \in P$ and if and only if $x + t = y$ for some $t \in P$ (P is naturally ordered).

3. Axioms for Magnitude

The earliest studies of divisibility monoids concerned the *simply ordered* case: they were made in an attempt to derive suitable axioms for *magnitude*. The *Archimedean* case, when $a^n \equiv b$ for all positive integers n implies $a = 0$, is easily treated. We first prove

LEMMA 1. *Any simply ordered Archimedean divisibility monoid is cancellative.*

PROOF. Suppose $ab = ac$, where $b \neq c$. Either $b < c$ or $c < b$ (simple ordering); suppose the former. Then $c = bx$ for some $x \neq 1$, whence $ab = abx$. Iterating, $ab = abx^n = x^n$ for all n. By the Archimedean hypothesis, $x = 1$ follows, implying $c = b$ contrary to hypothesis.

THEOREM 2 (HÖLDER [**1**]). *Any simply ordered Archimedean divisibility monoid is monomorphic to* $(\mathbf{R}, +)$, *and hence commutative.*

SKETCH OF PROOF. We use the ancient method of Eudoxus. Fix $a > 1$; then for any x consider the set L of rational numbers m/n such that $x^n \geqq a^m$, and the set U of rational numbers such that $x^n \leqq a^m$. Just as in Theorem XIII.12, L and U are the two halves of a Dedekind cut; moreover by the cancellative law, one can show that the correspondence $x \to (L, U)$ is an order-monomorphism which carries products into sums. Since $a \leqq b$ implies $ax = b$, the image of the monomorphism so constructed even consists of the positive elements ("cone") of a *subgroup* of the additive group of $\mathbf{R}$.

Ingenious results have been obtained, which extend the preceding results to *weakly Archimedean* simply ordered divisibility monoids and semigroups, in which $a^n < b$ for all positive integers n implies $a \leqq 1$; these are not necessarily cancellative. As examples, we define for any $n \in \mathbf{N}$ the *cyclic* divisibility monoid $\mathbf{N}_n$ of order $n + 1$, consisting of the powers a^k ($k \in \mathbf{N}$) of some a, with $a^k a^l = a^m$ for $m = \min(k + l, n)$.

LEMMA 2. *Any simply ordered weakly Archimedean divisibility monoid M with minimal element $a > 1$ is isomorphic to* $(\mathbf{N}, +, \leqq)$ *or some* $\mathbf{N}_n$, *hence commutative.*

PROOF. Let $b \neq 1$ and a be given. Since M is Archimedean, there will be a largest positive integer k such that $a^k < b \leqq a^{k+1}$; since M is a divisibility semigroup, $b = a^k x$ for some $x \in M$. Because a is minimal, this implies $a^k a \leqq a^k x = b$. But $b \leqq a^{k+1}$; hence $b = a^{k+1}$. The rest of the proof is trivial.

† O. Nakada, J. Fac. Sci. Hokkaido Univ. **11** (1951), 181–9, and **12** (1952), 73–86; [**Fu**, p. 154].

Lemma 2 is due to Hölder [1] and Fuchs. In [**Fu**, Chapter XI], many other properties of weakly Archimedean simply ordered divisibility monoids are derived, including the fact that they are necessarily commutative. (*Caution.* What we call "weakly Archimedean" is called "Archimedean" in [**Fu**].)

The modern theory of "magnitude" has developed in other directions, and refers especially to *partially* ordered divisibility monoids. One noteworthy approach is that of Tarski [**Ta**], based on the concept of a *cardinal algebra.* Others concern measure and dimension lattices, and have been already described in Chapter XI. Unfortunately, these last are *not* in general closed under "addition": they only define "partial algebras" in the terminology of Chapter VI.

4. Examples of l-groupoids and l-monoids

An l-groupoid is not just a *po*-groupoid which is a lattice under its partial ordering relation: products must also be distributive on joins. Though this follows automatically from isotonicity in *po*-groups, it does not hold in *po*-groupoids—not even in *po*-monoids. Our basis definition is thus as follows.

Definition. A multiplicative semilattice, or *m-semilattice*† is a semilattice M under $\vee$ with a multiplication such that

$$a(b \vee c) = ab \vee ac \quad \text{and} \quad (a \vee b)c = ac \vee bc, \tag{5}$$

for all $a, b, c \in M$. If M is a lattice with a multiplication and (5) holds, then M is called an "m-lattice" or *l-groupoid.* An l-groupoid which is a semigroup (monoid) under multiplication is called an *l-semigroup* (resp. *l-monoid*, short for lattice-ordered monoid).

Note that (1) follows trivially from (5): if $b \leqq c$, then $ac = a(b \vee c) = ab \vee ac$, whence $ab \leqq ac$. In other words, any m-semilattice is trivially a *po*-groupoid (i.e., m-poset).

Note also that any join-semilattice with O is not only a divisibility monoid, if $\vee$ is taken as multiplication; it is an l-monoid, since $a \vee (b \vee c) = (a \vee b) \vee (a \vee c)$. The dual is not true: a meet-semilattice with I, though always the dual of a divisibility monoid under the multiplication $\wedge$, is not an l-monoid under the multiplication $\wedge$ unless the lattice is distributive, because otherwise (5) fails.

Note also that any *po-group* which is a lattice satisfies (5) and its dual

$$a(b \wedge c) = ab \wedge ac \quad \text{and} \quad (a \wedge b)c = ac \wedge bc; \tag{5'}$$

it remains an l-group under dualization. However, in general, the dual of an l-semigroup is *not* an l-semigroup.

One verifies easily that the subsets of any monoid M (Example 2) form an l-monoid with zero $\varnothing$, which is a *Boolean algebra* as a lattice. This shows that l-monoids (unlike l-groups) can be complemented lattices with respect to their order; on the other hand, they need not be distributive or even modular (as lattices).

† First studied in [**Tr**]; the French word is "gerbier".

Example 4. The join-endomorphism $\alpha, \beta, \gamma, \cdots$ of any semilattice S form an m-semilattice, under the definitions

$$a(\alpha\beta) = (a\alpha)\beta \quad \text{and} \quad a(\alpha \vee \beta) = a\alpha \vee a\beta. \tag{6}$$

As another typical example, the additive subgroups of any ring R (Example 3 of §1) form an l-groupoid, which is modular as a lattice. If R is an associative ring with unity 1, then this l-groupoid is an l-monoid.

Observe that the class of l-groupoids is *equationally definable* in the sense of Chapter VI. Hence the general algebraic concepts of subalgebra, epimorphic image, and direct product apply to l-groupoids. In particular, any m-sublattice, epimorphic image, or direct product of l-groupoid(s) is itself an l-groupoid. A similar observation applies to l-semilattices, l-semigroups, and l-monoids.

The preceding observation permits one to construct various interesting additive m-semilattices of *real functions*, under their usual partial order. The *upper semicontinuous* functions of n variables form one interesting commutative l-semigroup; the *subharmonic* functions form another.

Because of the enormous generality of the concepts of po-groupoid and even l-monoid, most of the deeper theorems about them are applicable only to special kinds of po-groupoids (such as l-groups). One such kind of po-groupoid consists of integral po-groupoids, defined as follows.

DEFINITION. An element a of a po-groupoid with identity 1 is *integral* if and only if $a \leqq 1$. An *integral po*-groupoid is one in which $a \leqq 1$ for all a.

Thus, the negative cone of any l-group is an *integral po*-groupoid. The two-sided ideals of any ring form another (cf. Example 3), for which the following theorem is more significant.

THEOREM 3. *In any l-groupoid M*,

$$(a \wedge b)(a \vee b) \leqq ba \vee ab \quad \text{for all } a, b. \tag{7}$$

If M is an integral l-groupoid, then

$$a \vee b = 1 \quad \text{implies} \quad a \wedge b = ba \vee ab, \tag{8}$$

and

$$a \vee b = a \vee c = 1 \quad \text{implies} \quad a \vee bc = a \vee (b \wedge c) = 1. \tag{9}$$

If M has an element $z \leqq 1$ *satisfying* $zx = xz = z$ *for all* $x \in M$, *then this z is a zero of M as an l-groupoid.*

PROOF. Ad (7), $(a \wedge b)(a \vee b) = [(a \wedge b)a] \vee [(a \wedge b)b] \leqq ba \vee ab$, by (7) Ad (8), if $a \vee b = 1$, then $a \wedge b \leqq ba \vee ab$, by (7). But by (1), $ba \leqq 1a = a$ in an integral m-lattice; likewise $ba \leqq b1 = b$; hence $ba \leqq a \wedge b$. Similarly, $ab \leqq a \wedge b$, whence (8) follows by P2. Ad (9), clearly $a, b, c \leqq 1$ in any integral m-lattice. Hence $a \vee bc \geqq a \geqq aa, ba, ac$, and

$$1 = 1 \vee 11 \geqq a \vee bc \geqq aa \vee ba \vee ac \vee bc = (a \vee b)(a \vee c) = 11 = 1.$$

Likewise, since $b \geqq bc$, $c \geqq bc$, we have $b \wedge c \geqq bc$, so $1 \geqq a \vee (b \wedge c) \geqq a \vee bc = 1$, proving (9). Finally, the assertion about z is immediate: z is a universal lower bound for M since, for all $x \in M$, $z = zx \leqq 1x = x$, whence $z = z \wedge x = x \wedge z$.

Exercises:

1. Show that if G is any l-monoid, then $ax \geqq x$ for all $x \in G$ if and only if $a \in G^+$, while $ax \leqq x$ for all $x \in G$ if and only if $a \in G^-$.
2. Prove that the join-endomorphisms of any chain C form a distributive l-monoid under composition and the usual natural ordering (namely, $\phi \leqq \psi$ means $x\phi \leqq x\psi$ for all $x \in C$).
3. Show that, in any cancellative l-monoid, $a^n \geqq 1$ implies $a \geqq 1$ and $a^n = 1$ implies $a = 1$.†
4. If, in a commutative cancellative l-semigroup, $a^n \leqq b^n$ and b is invertible, then $a \leqq b$. (*Hint.* Use Exercise 3.)
5. (a) In the lattice **2**, let $OO = O$ and $OI = IO = II = I$. Show that this defines an l-monoid.
   (b) Show that $0, 1, \infty$ form an l-monoid under multiplication, regardless of whether $0\infty = \infty 0$ is defined as 0, 1, or ∞.
6. Show that any finite m-semilattice with 0 is an l-groupoid.

*7. Construct an l-monoid with 10 or fewer elements in which $a \vee b = 1$, yet $ab \neq ba$.

*8. Construct a commutative l-monoid of 5 elements in which $(a \wedge b)(a \vee b) < ab$.‡

9. In Exercise 2 after §2, when are PQ and $P \circ Q$ l-groupoids, respectively?

5. Residuation

One of the most important concepts in the theory of l-groupoids is that of residual, defined as follows.

Definition. Let L be any *po*-groupoid. The *right-residual* $a .\!\cdot b$ of a by b is the largest x (if it exists) such that $bx \leqq a$; the *left-residual* $a \cdot\!. b$ of a by b is the largest y such that $yb \leqq a$. A *residuated lattice* is an l-groupoid L in which $a .\!\cdot b$ and $a \cdot\!. b$ exist for any $a, b \in L$; a *residuated l-monoid* (*semigroup*) is a residuated lattice which is a monoid (semigroup).

Example 5. In any ring R, the two-sided ideals form a residuated lattice, under the multiplication of Example 3 (§1).

Corollary. *A lattice L is a residuated lattice, when xy is defined as $x \wedge y$, if and only if it is a Brouwerian lattice. In this case, L is an integral commutative l-semigroup.*

The proof is almost trivial since, if $xy = x \wedge y$, the definition of relative pseudocomplement $b{:}a$ given in Chapter V coincides with that of $b \cdot\!. a = b .\!\cdot a$ given above, and that of pseudocomplement a^* coincides with that of $0{:}a$. (In *commutative* residuated lattices, we write $a{:}b$ for both $a .\!\cdot b$ and $a \cdot\!. b$.)

Any *po*-loop is residuated; moreover $x .\!\cdot y$ is the same as the x/y defined in Chapter VII, while $x \cdot\!. y = x\backslash y$. Hence, in any *po*-group, we have $x .\!\cdot y = y^{-1}x$

† [**Tr**, p. 139]; apply the proof of Theorem XIII.3 above.

‡ Exercises 7–8 state results of Certaine [**1**].

and $x^{\cdot}.y = xy^{-1}$, and $x(z.^{\cdot} y) = xy^{-1}z = (x^{\cdot}.y)z$. It is a corollary that, for *po*-loops and *po*-groups, residuals are definable in terms of the multiplication operation alone: their study therefore belongs to pure loop (resp. group) theory, and will not be pursued here. The major point is to have a simple criterion for deciding when a given *po*-semigroup is a *po*-group; one such criterion is the following

LEMMA 1. *A po-monoid is a po-group if and only if it is residuated, and satisfies* $x(1.^{\cdot} x) = (1^{\cdot}.x)x = 1$.

The proof is trivial: if the preceding identity holds, then every $x \in G$ has a right- and a left-inverse.

THEOREM 4. *In any residuated lattice, we have*

$$(a \wedge b).^{\cdot} c = (a.^{\cdot} c) \wedge (b.^{\cdot} c) \quad \textit{and symmetrically,} \tag{10}$$

$$a.^{\cdot} (b \vee c) = (a.^{\cdot} b) \wedge (a.^{\cdot} c) \quad \textit{and symmetrically,} \tag{11}$$

$$ab \leqq c, \qquad b \leqq c.^{\cdot} a, \qquad \textit{and} \quad a \leqq c^{\cdot}.b \quad \textit{are equivalent,} \tag{12}$$

$$(ab).^{\cdot} a \geqq b \quad \textit{and} \quad (ab)^{\cdot}.b \geqq a. \tag{13}$$

In any residuated semigroup

$$(a.^{\cdot} b)^{\cdot}.c = (a^{\cdot}.c).^{\cdot} b \quad \textit{is the largest } x \textit{ such that } bxc = a, \tag{14}$$

and

$$a.^{\cdot} (bc) = (a.^{\cdot} b).^{\cdot} c \quad \textit{and} \quad a^{\cdot}.(bc) = (a^{\cdot}.c)^{\cdot}.b. \tag{15}$$

Proofs of these results are easy, and are left as exercises (see Dilworth [**1**] or [**Tr**, pp. 153–4]). In equations (10) and (11), the existence of the left side implies that of the right side.

Almost trivially, we have the following result.

LEMMA 2. *In any po-groupoid, the functions $a.^{\cdot} b$ and $a^{\cdot}.b$ are isotone in a and antitone in b.*

This result implies $a.^{\cdot} (b \wedge c) \geqq a.^{\cdot} b$ and $a.^{\cdot} (b \wedge c) \geqq a.^{\cdot} c$. By the definition of $\vee$ as least upper bound, there follows the inequality (16) of

LEMMA 3. *In any residuated lattice, we have*

$$a.^{\cdot} (b \wedge c) \geqq (a.^{\cdot} b) \vee (a.^{\cdot} c) \quad \textit{and symmetrically,} \tag{16}$$

$$b \leqq a.^{\cdot} (a^{\cdot}.b) \quad \textit{and} \quad b \leqq a^{\cdot}.(a.^{\cdot} b). \tag{17}$$

Since $(a^{\cdot}.b)b \leqq a$ (by definition of $a^{\cdot}.b$), the first inequality of (17) is a corollary of the definition of $a.^{\cdot} (a^{\cdot}.b)$. The second inequality follows by left-right symmetry.

COROLLARY. *Any integral element a of a residuated lattice with unity satisfies*

$$a \leqq 1.^{\cdot} (1^{\cdot}.a) \leqq 1 \quad \textit{and} \quad a \leqq 1^{\cdot}.(1.^{\cdot} a) \leqq 1. \tag{18}$$

PROOF. Since $a \leqq 1$,

$$1^{\cdot}.a \geqq 1^{\cdot}.1 = 1;$$

hence by Lemma 2,

$$1.\!\cdot(1\cdot\!.a) \leqq 1.\!\cdot 1 = 1.$$

On the other hand, since

$$(1\cdot\!.a)a \leqq 1, \qquad 1.\!\cdot(1\cdot\!.a) \geqq a,$$

completing the proof of the first inequality. The second follows by symmetry.

Complete l-groupoids. Most residuated lattices arising in applications are complete, and satisfy the infinite distributive laws

$$(19) \qquad a(\bigvee b_\beta) = \bigvee (ab_\beta) \quad \text{and} \quad (\bigvee a_\alpha)b = \bigvee (a_\alpha b).$$

This leads us to make the following definitions.

DEFINITION. A *complete* l-groupoid, or *cl-groupoid*, is a complete lattice with a binary multiplication satisfying (19). A cl-groupoid with associative multiplication is called a cl-semigroup, if it has a 1, it is called a *cl-monoid*.

The modules of a ring (Example 3, §1) constitute a typical cl-groupoid; (19) follows from the fact that the operations involved are finitary (binary); we omit the verification. We next show that most cm-lattices are residuated, generalizing Theorem V.24 concerning complete Brouwerian lattices.

THEOREM 5. *Any cl-groupoid, $a.\!\cdot b$ exists if $bx \leqq a$ for some x, and $a\cdot\!.b$ exists if $yb \leqq a$ for some y.*

PROOF. Let u be the join of all x_α such that $bx_\alpha \leqq a$. Then

$$bu = b(\bigvee x_\alpha) = \bigvee bx_\alpha \leqq a$$

by (19); hence

$$u = a.\!\cdot b.$$

The existence of $v = b\cdot\!.a$ under the stated assumptions can be proved similarly.

COROLLARY 1. *Any cl-groupoid with zero is residuated.*

Using Theorem 3, we obtain:

COROLLARY 2. *If R is any associative ring, then the cl-semigroups of all modules of R and of all two-sided ideals of R are residuated.*

In fact, if H and K are subrings of R, then $H.\!\cdot K$ and $H\cdot\!.K$ are the right- and left-quotients of H by K, in the sense of ideal theory. The case $H = 0$ is of especial importance; $0.\!\cdot K$ is called the right-annihilator of K. We will discuss these ideas further below.

Exercises:

1. Show that any chain is an l-groupoid under any isotone multiplication.
2. (a) Show that the positive integers form a residuated lattice under ordinary multiplication and $m|n$.
   (b) Prove the same result for the nonnegative integers.
3. Show that the nonnegative integers form a residuated lattice under their usual ordering, and either the addition or the multiplication operation.
4. Show that the "positive cone" of $x \geqq 1$ in any l-monoid is a sub-l-semigroup which is not residuated, except in trivial cases.
5. Show that, in any commutative l-monoid, $b \leqq a\!:\!(a\!:\!b)$.

6. Show that any l-groupoid which satisfies the ascending chain condition is residuated, provided the inequalities $xa \leqq b$ and $ay \leqq b$ have solutions x, y for all a, b.

7. Construct a three-element residuated po-groupoid which is not a lattice. **[Tr,** p. 152]

8. Let G be a commutative cl-monoid. Show that the largest residuated lattice contained in G consists of all $a \in G$ such that $xa \leqq 1$ for some $x \in G$.

9. Let $\bar{G}$ be the completion by nonvoid cuts of a directed group G, constructed as in Theorem XIII.24. Show that $\bar{G}$ is a residuated lattice and an l-monoid.

10. Let M be a commutative monoid under multiplication, in which the cancellation law holds and $ab = 1$ implies $a = b = 1$. Let G be the usual extension of M to a group, and define a "v-ideal" of G as a subset V which contains, with any $g \in G$, all gx $(x \in M)$.

(a) Show that the v-ideals of G form a complete l-group if and only if, in M, $a^n | b^n c$ for fixed a, b, c and every positive integer n implies $a|b$.

(b) Extend to the noncommutative case.

6. Elementary Applications

Some elementary applications of the preceding ideas are easily made. For example, the l-monoids of ideals (left-, right-, and two-sided) in any associative ring are easily constructed from the l-groupoid of its additive subgroups ("modules"), by the following definition.

DEFINITION. In a po-groupoid M, an element a is called *subidempotent* if $aa \leqq a$; it is called a *left-ideal* element if $xa \leqq a$ for all $x \in M$, and a *right-ideal* element if $ax \leqq a$ for all $x \in M$. An element which is both a left- and a right-ideal element is called an *ideal* element.

THEOREM 6. *Let L be any l-semigroup. Then the right-ideal elements of L, the left-ideal elements of L, and the* (*two-sided*)*-ideal elements of L are sub-l-semigroups* (*subsemigroups and sublattices*).

PROOF. The case of right-ideal elements is typical. If a and b are right-ideal elements, then by (1)

$$(a \wedge b)x \leqq ax \leqq a \quad \text{and} \quad (a \wedge b)x \leqq bx \leqq b$$

for all $x \in L$; hence $(a \wedge b)x \leqq a \wedge b$. Also, by (5),

$$(a \vee b)x = ax \vee bx \leqq a \vee b.$$

Finally, by associativity,

$$(ab)x = a(bx) \leqq ab.$$

Again, the binary operations of left- and right-residuation define a class of Galois connections having various applications.† We have

THEOREM 7. *For any fixed element c of any residuated lattice L, the correspondences* $x \rightarrow c.\cdot x = x^*$ *and* $y \rightarrow c\cdot.y = y^\dagger$ *define a Galois connection on L.*

PROOF. By definition of Galois connection (Chapter V), this means that the correspondences in question are antitone and that

$$x \leqq c.\cdot(c\cdot.x) \quad \text{and} \quad x \leqq c\cdot.(c.\cdot x) \quad \text{for all } x.$$

These results were proved above.

† P. Dubreil and R. Croisot, Collectanea Mathematica 7 (1954), 193–203.

DEFINITION. For any $c \in L$, L a residuated lattice, an element $x \in L$ is right c-closed if and only if $x = c.\cdot(c\cdot.x)$, and left c-closed if and only if $x = c\cdot.(c.\cdot x)$.

COROLLARY 1. *An element $x \in L$ is right c-closed if and only if $x = c.\cdot y$ for some y, and left c-closed if and only if $x = c\cdot.y$ for some y.*

The fact that the meet of any two right c-closed elements is right c-closed follows since $(a \vee b)^* = a^* \wedge b^*$. On the other hand, the join of two right c-closed elements need not be right c-closed.

Note also the conditions of P. Dubreil:† the c-closures of x are defined by the equations $\bar{x} = cx.\cdot c$ and $\bar{x} = cx\cdot.c$, respectively.

The preceding definitions can be applied to the l-semigroup L of all modules $A, B, C, \cdots$ of an associative ring R (Example 3 of §1). In this example, the Galois connection defined by $X \to C.\cdot X$ and $Y \to C\cdot.Y$ can also be derived concretely from the *polarity* (Chapter V) defined by the binary relation $xy \in C$. If C is a right-ideal, then so is $C.\cdot X$; if C is a left-ideal, then so is $C\cdot.Y$. Hence we have

COROLLARY 2. *If C is a (two-sided)-ideal of an associative ring R, then the right C-closed modules are right-ideals of R, and the left C-closed modules are left-ideals.*

The concept of c-closure can also be applied to the additive residuated lattice of bounded subharmonic functions (see §3) on a region R. The 0-closed functions $0{:}x$ are just the *harmonic* functions on R.

7. Integral l-groupoids

We have defined an *integral* l-groupoid as an l-groupoid L having an identity e which is also a universal upper bound for L, so that $x \leqq e$ for all $x \in L$. Obviously, the negative cone of any l-group is such an integral m-lattice. A more typical example is the following.

Example 6. Let R be any ring with unity 1 (we do not assume associativity).‡ Then the (two-sided) ideals of R form an integral m-lattice with identity $e = R$ and "zero" $z = 0$.

The following results generalize some well-known definitions and theorems about ideals. (Note that, in the integral l-groupoid defined by the negative cone of an l-group, two elements are "coprime" if and only if they are *disjoint* in the sense of Chapter XIV.)

DEFINITION. A *maximal* element of an integral l-groupoid L is an element covered by e; a *prime* element is an element p such that $xy \leqq p$ implies $x \leqq p$ or $y \leqq p$. Two elements $a, b \in L$ are *coprime* if $a \vee b = e$.

LEMMA 1. *If a, b are coprime, then*

$$x = xa \vee xb = (x \wedge a) \vee (x \wedge b) \quad \textit{for all } x. \tag{20}$$

† Bull. Soc. Math. France **81** (1953), 289–306; L. Lesueur, ibid. **83** (1955), 166.

‡ For the associative case, see W. Krull, Math. Z. 28 (1928), 481–503.

PROOF. We begin by noting that, since $xy \leqq xe = x$ and $xy \leqq ey \leqq y$, we have

$$xy \leqq x \wedge y, \quad \text{for all } x, y. \tag{21}$$

We then infer (20) from the sequence of inequalities

$$x = xe = x(a \vee b) = xa \vee xb \leqq (x \wedge a) \vee (x \wedge b) \leqq x.$$

LEMMA 2. *If a, b are coprime and $a \wedge b \leqq x$, then*

$$x = (x \vee a) \wedge (x \vee b) = (x \wedge a) \vee (x \wedge b). \tag{22}$$

PROOF. In any lattice, $(x \wedge a) \vee (x \wedge b) \leqq x \leqq (x \vee a) \wedge (x \vee b)$. Under the present hypotheses, by (20) and (5),

$$\begin{aligned}(x \vee a) \wedge (x \vee b) &= a[(x \vee a) \wedge (x \vee b)] \vee b[(x \vee a) \wedge (x \vee b)]\\ &\leqq a(x \vee b) \vee b(x \vee a) = (ax \vee ab) \vee (bx \vee ba).\end{aligned}$$

Again, since $ax \leqq a \wedge x$ and $ab \leqq a \wedge b = a \wedge (a \wedge b) \leqq a \wedge x$, $ax \vee ab \leqq a \wedge x$. Similarly, $bx \vee ba \leqq b \wedge x$. Substituting, $(x \vee a) \wedge (x \vee b) \leqq (a \wedge x) \vee (b \wedge x)$, completing the proof.

LEMMA 3. *If x and y are both coprime to a, then so are $x \wedge y$ and xy.*

This simply restates the fact that $a \vee b = a \vee c = e$ implies $a \vee bc = a \vee (b \wedge c) = e$.

THEOREM 8. *If a, b are coprime, then the interval $[a \wedge b, e]$ is lattice-isomorphic with the cardinal product $[a, e] \times [b, e]$.*

PROOF. Given $a \wedge b \leqq x \leqq e$, define $t = x \vee a$ and $u = x \vee b$; clearly $t \in [a, e]$ and $u \in [b, e]$. Conversely, given $t \in [a, e]$ and $u \in [b, e]$, form $\phi(t, u) = t \wedge u$; these are single-valued correspondences from $[a \wedge b, e]$ to $[a, e] \times [b, e]$ and conversely. By Lemma 2, $\phi(x \vee a, x \vee b) = x$; conversely, since $a \leqq t$,

$$t \leqq (t \wedge u) \vee a \leqq (t \vee a) \wedge (u \vee a) = t \vee a \quad (\text{since } u \vee a = e) = t;$$

similarly, $u = (t \wedge u) \vee b$; hence the correspondences are inverse. But they are obviously isotone; hence they are both isomorphisms.

THEOREM 9. *Every complemented integral l-groupoid L is a Boolean algebra, with $xy = x \wedge y$.*

PROOF. Let $a \in L$ be arbitrary, and a' any complement of a. Then by Theorem 8, a and a' are in the center of L; hence L is a Boolean algebra. Further, by (20) and (5),

$$x \wedge a = (xa \vee xa') \wedge (xa \vee x'a) = xa \vee (x \wedge x')a \vee x(a' \wedge a) \vee (xa' \wedge x'a).$$

All terms but the first are 0, whence $x \wedge a = xa$. Q.E.D.

Ascending chain condition. We now assume that L is an integral l-groupoid which satisfies the ascending chain condition of §VIII.1. In Example 6, this holds if R is a finite extension of a commutative (associative) ring satisfying the ascending chain condition, or if R is a polynomial ring over such a ring.

With each maximal element m_i of L, we associate the set M_i of all elements x which are coprime to all maximal $m_j \neq m_i$. By Lemma 3, this is a dual ideal of L (in the lattice-theoretic sense), which is closed under multiplication. Hence it is an integral l-groupoid. In ideal theory, the ideals in $\bigcup M_i$ have been called "einartig"; moreover in algebraic number theory, the einartig ideals are the prime-power ideals. Hence the following result may be regarded as a partial generalization of the Fundamental Theorem of Ideal Theory.

THEOREM 10. *Let L be any integral l-groupoid satisfying the ascending chain condition. The sublattice S generated by the sets M_i of einartig elements is the restricted cardinal product $M_1 \times M_2 \times M_3 \times \cdots$ of the M_i.*

PROOF. By (9), each M_i is an interval sublattice (dual ideal) closed under multiplication. Again, if $x \in M_i$ and $y \in M_j$ $[i \neq j]$, then $x \vee y$ is not contained in any maximal element, and so $x \vee y = e$. The same remark holds, by (9), if $x = x_1 \cdots x_n$ $[x_i \in M_i]$ and $y \in M_{n+1}$. Hence, by Theorem 8 and induction, the interval sublattice generated by the M_i is their cardinal product. In the presence of the ascending chain condition, we can extend this result to an infinite number of factors, provided all but a finite number of "components" are e.

Caution. Though the M_i are closed under multiplication, their cardinal product need not be.

Exercises for §§6–7:

1. Let G be an l-groupoid with universal lower bound O. Show that the distributive law $a \bigvee_{\varnothing} x_\alpha = \bigvee_{\varnothing} (ax_\alpha)$ for the void set $\varnothing$ implies that O is a multiplicative zero.

2. Show that a residuated groupoid is a residuated quasigroup if and only if it satisfies $x(y.\cdot x) = (y\cdot.x)x = x$ for all x, y.

3. Show that, in a residuated groupoid G with 1, the residuals $a.\cdot a$ and $a\cdot.a$ are the non-negative idempotents of G.

4. Call a *po*-groupoid with unity 1 *integrally closed* when $a.\cdot a = a\cdot.a = 1$ for all a. Show that any residuated cancellative monoid is integrally closed.†

5. Show that the ideals $K > \{0\}$ in the rational field $\mathbf{Q}$ form a commutative cl-monoid, in which, if H denotes the ideal of all fractions whose denominators are powers of a fixed prime p, $E:H$ does not exist.

*6. Show that the free modular lattice with three generators cannot be made into an integral l-groupoid. (Ward-Dilworth)

*7. Show that the following conditions on an integral l-groupoid are all equivalent, and imply distributivity: $a:b \vee b:a = 1$, $a:(b \wedge c) = (a:b) \vee (a:c)$, $(b \vee c):a = (b:a) \vee (c:a)$.

*8. Show that if $1:(1:x) = x$ and $1:(xy) = (1:x)(1:y)$ in an integral residuated groupoid L, then L is a commutative l-group. (Certaine)

*9. Show that, in a conditionally complete residuated semilattice, $a.\cdot a = 1$ is equivalent to $a\cdot.a = 1$.

*10. (a) Show that $aa' \equiv 0$ in any complemented integral l-groupoid G with 1 and universal lower bound 0.

(b) Infer that G is a Newman algebra (§II.13), and a Boolean lattice if an l-monoid.

† Exercises 3–4 state results proved in [**Tr**, pp. 161–2]. Exercise 9 is [**Tr**, Theorem 10, p. 165].

8*. Commutation Lattices

Before trying to generalize the deeper properties of ideals of algebraic numbers, we sketch some features of the algebra of commutation in groups and Lie algebras, which have natural lattice-theoretic generalizations.

DEFINITION. A *commutation lattice* is a complete algebraic lattice L, which is a *po-groupoid* with zero 0 under a *commutative* (but not necessarily associative) *join-continuous multiplication* (ab) such that

$$(ab) \leqq a \vee b, \tag{23}$$

$$(a(b \vee c)) \leqq (ab) \vee (ac) \vee ((ab)c), \tag{24}$$

$$(ab) \leqq b \quad \text{and} \quad (ac) \leqq c \quad \text{imply} \quad (a(bc)) \leqq bc, \tag{25}$$

for all $a, b, c \in L$. By "join-continuous", we mean that

$$b_\delta \uparrow b \text{ (as a directed set)} \quad \text{implies} \quad (ab_\delta) \uparrow (ab). \tag{26}$$

Example 7. Let $L(G)$ be the lattice of all normal subgroups of any group G; and for any $S \in L(G)$, $T \in L(G)$ let (S, T) be the subgroup of G generated by all commutators $(s, t) = s^{-1}t^{-1}st$, $s \in S$, $t \in T$. Then $L(G)$ is a commutation lattice under inclusion and the "multiplication" (S, T).

SKETCH OF PROOF. Since $(t, s) = (s, t)^{-1}$, (23) holds. Since $(a, bc) = (a, c)(c, (a, b))$, (24) follows from this. Finally, (25) holds since $(S, T) \leqq T$ means that T is invariant under all inner automorphisms induced by S. (Join-continuity (26) holds since all operations involved are algebraic—i.e., dependence relations of finite character.)

We now show that various properties of commutation in groups hold in any commutation lattice—e.g., in any such lattice associated with the subalgebras of a Lie ring or algebra.†

LEMMA 1. *In any commutation lattice:*

$$ab \leqq b \quad \textit{and} \quad ac \leqq c \quad \textit{imply} \quad a(b \vee c) \leqq b \vee c, \tag{27}$$

$$ab \leqq a \quad \textit{and} \quad ac \leqq a \quad \textit{imply} \quad a(b \vee c) \leqq a. \tag{27'}$$

PROOF. Substituting into (24) and using the hypotheses:

$$a(b \vee c) \leqq ab \vee ac \vee ((ab)c) \leqq b \vee c \vee (bc) \leqq b \vee c \quad \text{by (23)},$$

$$a(b \vee c) \leqq ab \vee ac \vee ((ab)c) \leqq a \vee a \vee (ac) \leqq a \vee a = a \quad \text{by P1}.$$

(Though (25) was not used here, it seems unlikely that it can be proved similarly from (23) and (24).)

LEMMA 2. *The ideal elements of any commutation lattice L form a sublattice and subgroupoid, which is an l-groupoid in which* (21) *holds.*

† The discussion extends that on pp. 204–5 of [**LT2**], from commutators of normal subgroups to commutators of subgroups.

PROOF. By definition (§6) the ideal elements of L are the $a \in L$ such that $Ia = aI \leqq a$. In Example 7, they are the normal subgroups of G, and the properties stated are well known. In any *po*-groupoid, $Ib \leqq b$ and $Ic \leqq c$ imply $I(b \wedge c) \leqq Ib \wedge Ic$ by isotonicity; hence the ideal elements are a meet-semilattice. Because of (24) and (23), they imply in any commutation lattice also

$$I(b \vee c) \leqq (Ib) \vee (Ic) \vee ((Ib)c) \leqq b \vee c \vee bc \leqq b \vee c;$$

hence the ideal elements are a sublattice. By (24), they are also a subgroupoid (closed under multiplication). Finally, since $ab \leqq aI$ (by isotonicity), and $ab \leqq Ib$, $ab \leqq aI \wedge Ib \leqq a \wedge b$ for ideal elements, this proves (21).

LEMMA 3. *In any commutation lattice, $a:a$ and $O:a$ exist for all a; moreover the ideal elements form a residuated lattice in which:*

$$(ab) \leqq a \wedge b. \tag{28}$$

PROOF. By (27′), $ab \leqq a$ and $ac \leqq a$ imply $a(b \vee c) \leqq a$; hence the set of $x \in L$ such that $ax \leqq a$ is an ideal. By (26), the join of all the elements x in this ideal has the same property; hence it is $a:a$. A similar argument gives $O:a$. Again, (28) holds in this lattice since, by isotonicity, $(ab) \leqq (aI) \leqq a$ and $(ab) \leqq (Ib) \leqq b$. Finally, the lattice is residuated since, in it, $(ab) \leqq d$ and $(ac) \leqq d$ imply

$$a(b \vee c) \leqq ab \vee ac \vee ((ab)c) \leqq d \vee d \vee (ac) \leqq d \vee d = d,$$

and we can apply (26) to the ideal of elements $b \in L$ such that $ab = d$, to construct $d:a$ for any $a, d \in L$.

In any commutation lattice, we now define $I^1 = I$ and $I^{n+1} = I^nI$ recursively, and ${}^0I = O$ and ${}^{n+1}I = ({}^nI):I$ recursively. We say that L is *nilpotent* if some $I^n = O$. We call the sequence of I^n the *lower* central series of L, and the dual sequence of ${}^mI$ the *upper* central series of L.

THEOREM 11. *In any nilpotent commutation lattice, the upper and lower central series have the same length r. If $I = a_0 > a_1 > a_2 > \cdots > a_r = O$ is any nilpotent chain, with $a_iI \leqq a_{i+1}$, then*

$$I^{i+1} \leqq a_i \leqq {}^{r-i}I \quad \textit{for all } i = 1, \cdots, r. \tag{29}$$

Exercises:

1. Prove in detail that, in Example 7, $L(G)$ always defines a commutation lattice.
2. Show that the additive subgroup of any Lie ring forms a commutation lattice under the multiplication of Example 3.
3. Prove that, in Example 7, $S:S$ exists and is the normalizer of S, and $1:S$ exists and is the centralizer of S, for any subgroup S.
4. Prove that $L(G)$ in Example 7 is *not* an l-groupoid, for G the free group with two generators.
5. Let L be any commutation lattice of finite length. Show that $O:a$ and $a:a$ exist for all $a \in L$.
6. Show that the ideal elements of any commutation lattice satisfy the identity $(a(bc)) = ((ab)c) \vee ((ac)b)$ of P. Hall.

7. In a commutation lattice, define $c^1 = II$, and recursively $c^{m+1} = c^m c^m$. Prove that $c^n = O$ implies $I^{2^n} = O$. (P. Hall)

(For other Exercises on commutation lattices, see D. H. McLain, Proc. Glasgow Math. Assn. **3** (1956), 38–44; I. V. Stelleckii, Dokl. Akad. Nauk SSSR **128** (1959), 680–3; E. Schenkman, Proc. AMS **9** (1958), 375–81.)

Define a *ringoid* as an algebra R with a binary multiplication and (possibly) other binary operations f, such that $af_\alpha(b, c) = f_\alpha(ab, ac)$ and $f_\alpha(b, c)a = f_\alpha(ba, ca)$ for all $a, b, c \in R$. Define a *module* of a ringoid R as a subset $H \subset R$ such that $b, c \in H$ imply $f_\alpha(b, c) \in H$, and an *ideal* of R as a subset K such that $b \in K$ and $a \in R$ imply $\{ab, ba\} \subset K$.

8. Prove that if $Oa = aO = O$ for all a in a ringoid R, then $f_\alpha(O, O) = O$ for all f_α.

9. Show that "multiplicative ideals" can be considered as ringoid ideals.

10. Show that the modules of any ringoid form a residuated cm-lattice in which the ideals form a residuated cm-sublattice.

11. Extend Exercise 10 to S-modules, or modules H which for a given subset $S \subset R$, contain all products sx and xs ($x \in H$, $s \in S$).

9. Maximal and Prime Elements

Much of commutative ideal theory generalizes readily to integral l-groupoids. In §§9–10, we shall illustrate this principle in the context of the theory of *divisibility* (prime factorization). To this end, we now define three related concepts, which are equivalent in any divisibility semigroup in which the unique factorization theorem holds.

DEFINITION. Let M be any integral po-groupoid. An element $m \in M$ is called *maximal* when it is covered by 1; an element $p < 1$ such that $ab \leqq p$ implies $a \leqq p$ or $b \leqq p$ is called *prime*; an element $p < 1$ such that $ab = p$ implies $a = p$ or $b = p$ is called *indecomposable*.

LEMMA 1. *In any integral po-groupoid, any maximal element is prime and indecomposable.*

PROOF. Let m be maximal, and let $xy = m$. Then $x = m$ or $x = 1$ since $m = xy \leqq x1 = x$ and m is maximal. Likewise, $y = m$ or $y = 1$. Since $x = y = 1$ would imply $m = xy = 1$, either $x = m$ or $y = m$, and so m is indecomposable.

Again, unless $x \leqq m$, $x \vee m = 1$. Hence if $xy \leqq m$ but $x \nleqq m$, then

$$y = 1y = (x \vee m)y \leqq xy \vee my = m \vee m1 = m,$$

so that m is a prime.

We now apply the preceding definitions to cancellative commutative monoids, including especially integral domains.

LEMMA 2. *Let G be any commutative cancellative monoid. Then the relation $a|b$ defines an integral po-monoid $P(G)$ on the sets of associative elements. For each $a \in P(G)$, the mapping $x \to ax$ is an order-isomorphism of $P(G)$ onto the lattice ideal A of elements $c \leqq a$.*

To prove that $P(G)$ is a po-monoid is to show that $ax \sim ay$ implies $x \sim y$. But $ax|ay$ implies $ay = ya = xaz = axz$ for some $z \in G$, whence $y = xz$ by the cancellation law in $P(G)$ follows.

Now let G be the multiplicative monoid of the nonzero elements of an *integral domain* D. Then $a \sim b$ is equivalent to the statement that a and b generate the same nonzero principal ideal $(a) = (b)$ of D. We have the

COROLLARY 1. *The nonzero principal ideals of any integral domain D form a commutative, cancellative divisibility monoid $S(D)$.*

Since $(ab) \subset (a)(b)$, Corollary 1 of Theorem 1 applies, and we have

COROLLARY 2. *The nonzero principal ideals of any integral domain form a po-monoid which is isomorphic to the negative cone of a directed group.*

LEMMA 3. *In any integral po-monoid satisfying the ascending chain condition, each element $c \neq 1$ is a product of indecomposable factors.*

PROOF. If the conclusion fails, then the nonvoid set of all elements *not* so decomposable must contain a maximal member c. This c cannot be indecomposable (i.e., covered by 1), or the conclusion would hold trivially. But if c is decomposable, then $c = ab$ where $a > c$ and $b > c$. Since c was maximal among elements not products of indecomposable factors, we have $a = p_1 \cdots p_r$ and $b = q_1 \cdots q_s$, whence $c = p_1 \cdots p_r q_1 \cdots q_s$, giving a contradiction.

The following example shows that one cannot prove more, without making a further assumption.

Example 8. Let G be the additive monoid of pairs (m, n) of nonpositive integers whose sum $m + n$ is even. Then $(-2, 0) + (0, -2) \leqq (-1, -1)$, and so $(-1, -1)$ is maximal but not prime. Also, since $(-2, -2) = (-2, 0) + (0, -2) = (-1, -1) + (-1, -1)$, one need not have a unique factorization theorem.

Actually, the relevant assumption is precisely the *lattice* hypothesis, as we now show.

THEOREM 12. *Let L be an l-monoid which is isomorphic to the negative cone of a directed group, and satisfies the ascending chain condition. Then every $c \neq 1$ in L can be uniquely factored into prime factors.*

The proof is a corollary of §XIII.5.

COROLLARY (JAFFARD†). *The nonzero integers of an algebraic number field $F = \mathbf{Q}(\theta)$ satisfy the unique factorization theorem if and only if their divisibility po-semigroups is a lattice.*

PROOF. The integers form a subdomain of F [**B-M**, Chapter XIV]. Moreover since the lattice of all ideals of F satisfies the ascending chain condition (Chapter VIII), so does the subset of the principal ideals. We now invoke Corollary 2 of Lemma 2.

It is actually sufficient that their divisibility *po*-semigroup be a *semilattice* under g.c.d. For a necessary and sufficient test for this, see H. Pollard, *The theory of algebraic numbers*, Wiley, 1950, Theorem 9.5.

† P. Jaffard, *Les systèmes d'idéaux*, Dunod, Paris, 1960, p. 81, Theorem 4.

Artin equivalence. We next consider an interesting congruence relation introduced by Artin into commutative residuated l-monoids with unity (e.g., into the l-monoid of ideals of a commutative ring R with unity 1). This is closely related to the equation $x(1:x) = 1$, which holds in a residuated l-monoid if and only if it is an l-group!

Two elements a, b of such an l-monoid L are *Artin equivalent* when $1:a = 1:b$. By Theorem 7, this is equivalent to saying that $a^{**} = b^{**}$, where we define $x^* = 1:x$; thus it is analogous to the equivalence relation studied in Theorem V.26 of Glivenko. It is related to the identity $x(1:x) = 1$ since the latter implies the identity

$$a = a1 = a(1:a)(1:(1:a)) = 1(1:(1:a)) = 1:(1:a).$$

If Artin equivalence were a congruence relation for all operations, then the Artin-equivalent elements would define an l-group. With the ascending chain condition, one could then prove unique factorization into primes (cf. §XIII.5).

By Glivenko's Theorem, Artin equivalence is a congruence relation for joins. It is also a congruence relation for multiplication, since if $a^* = b^*$, then

$$1:ax = (1:a):x = (1:b):x = 1:bx.$$

For it to be a congruence relation for meets, it is necessary and sufficient that L be *integrally closed*, in the sense that $a:a = 1$ for all $a \in L$; this is proved in [**Tr**, p. 243].

10. Abstract Ideal Theory

The general theory of ideals in Noetherian rings centers around the concepts of primary and irreducible ideal, and of the radical of an ideal. As was first shown by Ward and Dilworth [**1**, Part IV], much of this theory is true in general Noetherian l-monoids. The present section develops this idea.†

DEFINITIONS. An l-monoid is *Noetherian* when it is integral, commutative, and Noetherian as a lattice. In an l-monoid L, and element q is (right-)*primary* when $ab \leqq q$ implies that either $a \leqq q$ or $b^n \leqq q$ for some finite integer n. The *radical*‡ $\sqrt{a}$ of an element $a \in L$ is the join of all $x \in L$ such that $x^n \leqq a$ for some $n = n(x, a)$.

LEMMA 1. *In any integral commutative l-monoid L, $x^m \leqq a$ and $y^n \leqq a$ imply $(x \vee y)^{m+n} \leqq a$.*

PROOF. By distributivity,

$$(x \vee y)^{m+n} = \bigvee_{k=0}^{m+n} x^k y^{m+n-k}.$$

† This idea was implicit in W. Krull, S.-B. Phys.-Med. Soc. zu Erlangen **56** (1924), 47–63, and developed by him in many later papers.

‡ For other lattice-theoretic approaches to the radical, see S. A. Amitsur, Amer. J. Math. **74** (1952), 775–86 and **76** (1954), 100–36; E.-A. Behrens, Math. Z. **64** (1956), 169–82.

For $k \leqq m$, $y^{m+n-k} \leqq ay^{m-k} \leqq a$, while for $k > m$, $x^k \leqq a$ similarly; hence $x^k y^{m+n-k} \leqq a$ for all $k = 0, 1, \cdots, m + n$. The conclusion follows.

COROLLARY 1. *In any integral commutative l-monoid L, the set $R(a)$ of elements $x \in L$ such that $x^m \leqq a$ for some $m \in \mathbf{Z}^+$ is a (lattice) ideal.*

In a Noetherian lattice, every ideal is principal, and hence contains the join of its elements. We conclude

COROLLARY 2. *In any Noetherian l-monoid $(\sqrt{a})^n \leqq a$ for some least positive integer n, the "exponent" of a. Hence $\sqrt{a}$ is its own radical.*

LEMMA 2. *In a Noetherian l-monoid, q is primary if and only if:*

$$ab \leqq q \quad \text{and} \quad a \nleqq q \quad \text{imply} \quad b \leqq \sqrt{q}. \tag{30}$$

PROOF. If q is primary, then (30) is immediate since $b^n \leqq q$ is equivalent to $b \leqq \sqrt{q}$. Conversely, (30) implies $b^n \leqq (\sqrt{q})^n \leqq q$ for n the exponent of q, and hence that q is primary, by definition.

LEMMA 3. *The radical $\sqrt{q}$ of any primary element of a Noetherian l-monoid is a "prime".*

PROOF. If $ab \leqq \sqrt{q}$, then $(ab)^n = a^n b^n \leqq q$ for some n. Hence either $a^n \leqq q$ or $(b^n)^k = b^{nk} \leqq q$ for some finite k. In the first case, $a \leqq \sqrt{q}$; in the second, $b \leqq \sqrt{q}$; hence by definition (§8), $\sqrt{q}$ is prime.

In general, the converse of Lemma 3 is false, even for ideals: powers of prime ideals need not be primary. For a counterexample, see Zariski–Samuel, *Commutative rings*, Vol. 1, p. 154. However, we have:

LEMMA 4. *If m is a maximal element of a Noetherian l-monoid, then any power $m^k = q$ of m is primary.*

PROOF. Suppose $ab \leqq q$ but $b \nleqq m = \sqrt{q}$. Then $m \vee b = 1$ since m is maximal; hence

$$1 = (m \vee b)^k = m^k \vee cb, \quad \text{where } c = \bigvee_{j=1}^{k} m^{j-1} b^{k-j}.$$

Therefore $a = a1 = am^k \vee c(ab) \leqq m^k \vee m^k = q$.

LEMMA 5. *Any finite meet of primary elements q_j having the same radical $p = \sqrt{q_j}$ is primary with radical p.*

PROOF. Let $c = q_1 \wedge \cdots \wedge q_r$; clearly $\sqrt{c} = \bigwedge \sqrt{q_j} = p$. Suppose $ab \leqq c$ but $a \nleqq c$. Then $a \nleqq q_j$ for some j; yet $ab \leqq c \leqq q_j$; hence by Lemma 2, $b \leqq \sqrt{q_j} = p = \sqrt{c}$. Q.E.D.

LEMMA 6. *Let $c = q_1 \wedge \cdots \wedge q_r$ be an irredundant meet of primary elements q_j with different radicals p_j. Then c is not primary.*

PROOF. Some radical $p_j = \sqrt{q_j}$ will be minimal, say p_1. Then $p_j \nleqq p_1$ for $j > 1$; let n be the maximum of the exponents of the q_j. Since p_1 is prime, $(p_2 \cdots p_r)^n \nleqq p_1$; yet $(p_2 \cdots p_r)^n \leqq q_2 \wedge \cdots \wedge q_r$. Hence

$$q_1(p_2 \cdots p_r)^n \leqq q_1 \wedge q_2 \wedge \cdots \wedge q_r = c,$$

where $q_1 \nleqq c$ and $((p_2 \cdots p_r)^n)^m \nleqq p_1$ for *any* m. Now set $b = (p_2 \cdots p_r)^n$; we have shown that $q_1 b \leqq c$, yet $b^m \leqq q_1$ for no m, since this would imply $b^m \leqq p_1$, disproved above. This implies that c is *not* primary.

Ideals in Noetherian rings. Although some of the theory of ideals in Noetherian rings is valid in (commutative) Noetherian integral l-monoids generally, this is not true of the classic result that *every irreducible ideal is primary*. This is related to other special properties of ideals in Noetherian rings. Define an element q of a commutative l-monoid M to be "principal" when $b \leqq q$ implies that $qc = b$ for some $c \in M$. Then the l-monoid $M = J(R)$ of all ideals in any (commutative) Noetherian ring R contains a *submonoid* S of principal elements (the principal ideals), such that every $a \in M$ is the join of a finite number of $x_i \in S$. Ward and Dilworth [**1**, Theorem 11.2] have shown that "every irreducible is primary" in any Noetherian, modular l-monoid with such a submonoid S; see also Exercise 4 below.

In general, there is suspisingly little interaction between residuation and modularity or distributivity in lattices. However, it is interesting to find out for which rings the lattice of all ideals is distributive. In "orders" of algebraic numbers generated by a single algebraic integer, it is known† that this is equivalent to being integrally closed.

It may also be noted that properties of the distributive lattice of all *algebraic varieties* in an affine or projective space, defined in §VIII.3, imply properties of radical ideals in the associated polynomial ring since the "polar" of any algebraic variety is a radical ideal. (The converse is true over the complex field.)

11. Fundamental Theorem of Ideal Theory

We will now consider how far the fundamental theorem of ideal theory, as first proved by Dedekind, can be deduced from the theory of l-semigroups. This result concerns "prime factorization" is any finite extension $F = Q(\theta)$ of the rational field Q, an *algebraic number field*. If E is the subring of all integers of F, a *Dedekind ideal* of F is an E-module $A > 0$ of F such that:

(i) $nA \subset E$ for some rational integer $n \in Z$.

Elementary arguments suffice to show that the Dedekind ideals of F form a commutative residuated monoid $S(F)$ with identity E and zero 0. Moreover the "negative cone" of Dedekind ideals $J \subset E$ consists precisely of "ideals" of E in the usual sense, and these satisfy the ascending chain condition by Chapter VIII. To summarize these statements simply, we now make two related definitions.

DEFINITION. A *Dedekind* l-monoid is a commutative residuated l-monoid whose negative cone satisfies the ascending chain condition. Any Noetherian l-monoid is an integral Dedekind l-monoid.

† G. Birkhoff, Proc. Nat. Acad. Sci. U.S.A. **20** (1934), 571–3.

LEMMA 1. *The Dedekind ideals of any algebraic number field F form a Dedekind l-monoid.*

Lemma 1 summarizes the statements made two paragraphs earlier. On the other hand, restating Lemma 1 of §5, we have

LEMMA 2. *A Dedekind l-monoid S is an atomistic l-group if and only if $x(1:x) = 1$ for all $x \in S$.*

By Chapter XIII, the integral elements of any atomistic l-group can be uniquely factored into "primes". Hence, to prove the unique factorization theorem for ideals in any algebraic number field, it suffices to show that $A(E:A) = E$ for any Dedekind ideal A. We shall now investigate how much of this proof is supplied by the theory of Dedekind l-monoids. We first prove

THEOREM 13. *In a Dedekind l-monoid L, let:*

(I) $0 < a < 1$ *imply that all chains between a and 1 are finite, and*

(II) $p(1:p) = 1$ *for every maximal proper integral element p.*

Then every element $a \in L$ with $0 < a < 1$ has a unique factorization into maximal $p_i < 1$.

We approach the proof through a series of three lemmas.

LEMMA 1. *If $1 > p$ and $p(1:p) = 1$, then*

$$1 > p > p^2 > p^3 > \cdots \quad \textit{and} \quad 1 < (1:p) < (1:p)^2 < \cdots. \tag{31}$$

PROOF. Since $p \leqq 1$, $p^{r+2} \leqq p^r$ for $r = 1, 2, 3, \cdots$. If $p^{r+1} = p^r$, then

$$p = p^{r+1}(1:p)^r = p^r(1:p)^r = 1,$$

giving a contradiction. The proof that $1 < (1:p) < (1:p)^2 < \cdots$ is similar.

LEMMA 2. *If $1 > p > a > 0$, then $1 > (1:p)a > a$.*

PROOF. Since $p > a$, $1 = (1:p)p \geqq (1:p)a$. Moreover $1 = (1:p)a$ would imply $p = p1 = p(1:p)a = 1a = a$, contrary to hypothesis. Likewise, since $(1:p) > 1$, $(1:p)a \geqq 1a = a$. Moreover $(1:p)a = a$ would imply $a = 1a = p(1:p)a = pa$, and hence

$$a = pa = p(pa) = p^2a = p^2(pa) = p^3a = \cdots.$$

Since $a \leqq 1$, this would imply $a = p^ra \leqq p^r1 = p^r$ for all r. Hence, by (31), we would have an *infinite* chain of elements $\{p^r\}$ between 1 and a, contrary to hypothesis.

COROLLARY. *Under the hypotheses of Theorem* 13, *every prime nonzero integral element is maximal.*

PROOF. Let a be any nonzero nonmaximal integral element. Then $a < p < 1$ for some maximal $p < 1$. Hence $q = (1:p)a > a$ by Lemma 2, so that $p > a$. But $pq = p(1:p)a = a$ since $p(1:p) = 1$; hence a is nonprime.

LEMMA 3. *If $0 < a < 1$, then a is a product of elements p_i covered by 1 (primes).*

PROOF. By the finite chain condition, either the conclusion holds or there is a maximal element a for which it fails. This maximal element cannot be covered by 1, because the conclusion holds trivially for elements covered by 1. Hence, by the Corollary of Lemma 2, $a = pr$ where $1 > p > a$ and $1 > r > a$. By induction, $p = p_1 \cdots p_r$ and $q = q_1 \cdots q_s$, where the p_i and q_j are covered by 1. Hence $a = p_1 \cdots p_r q_1 \cdots q_s$, which we wanted to prove.

To prove Theorem 13, it only remains to prove that the factorization is unique (up to rearrangement of factors). This can be proved by the usual argument. If $a = p_1 \cdots p_r$ and $a = q_1 \cdots q_s$, where the p_i and q_i are covered by 1, then $p_1 \geqq a = q_1 \cdots q_s$. Hence, by the Corollary of Lemma 2 and induction, $p_1 \geqq q_j$ for some j. Since p_1 and q_j are both maximal, $p_1 = q_j$. Hence

$$\begin{aligned} p_2 \cdots p_r &= (1:p_1)p_1p_2 \cdots p_r = (1:p_1)a = (1:q_j)a \\ &= q_1q_2 \cdots q_{j-1}(1:q_j)q_j \cdots q_r = q_1 \cdots q_{j-1}q_{j+1} \cdots q_s, \end{aligned}$$

whence uniqueness follows by induction on r.

Discussion. Although Theorem 12, Corollary, is much more elegant in appearance, Theorem 13 is more useful for algebraic number theory. This is because one can prove in about five pages† that assumptions I and II hold in the Dedekind l-monoid of all Dedekind ideals of any algebraic number field F, viz.:

I. If A is a nonzero integral E-module, then all chains between A and E have finite length (finite chain condition).

II. If P is a maximal proper integral E-module of F, then $P(1:P) = 1$.

Actually, one can prove much more than assumption I very directly. Since the lattice of E-modules is modular, all connected chains between a and 1 have the same length.

In conclusion, one needs only about five pages of technical reasoning about algebraic numbers *per se* to establish the fundamental theorem of ideal theory, *if* one is willing to assume the elements of the theory of l-monoids. Possibly one can do better, though I doubt if one can replace assumption I by the ascending chain condition.

Exercises for §§9–11:

1. Show that, in a divisibility po-monoid, any prime element is maximal, and an element is maximal if and only if it is indecomposable.
2. Show that, in any divisibility l-monoid the concepts of prime element, indecomposable element, and maximal element are mutually equivalent.
3. Show that the ideals $K > 0$ in the rational field $\mathbf{Q}$ form a cl-monoid in which, if H denotes the ideal of all fractions whose denominators are powers of a fixed prime p, $E:H$ does not exist.
4. Let M be a Noetherian, modular l-monoid in which $a, b \in M$ imply $a^k \wedge b \leqq ab$ for some k; show that every meet-irreducible element is primary. (Ward-Dilworth [**1**, Theorem 11.1])
5. Construct a Noetherian, modular l-monoid M in which not every meet-irreducible element is primary.

† One must prove Lemmas (8.21)–(8.23) of Pollard, op. cit. in §9. The proof of these uses his Theorem 8.7 ("every prime ideal is maximal"), which must also be proved since to assume the Corollary to Lemma 3 of §8 would involve circular reasoning.

6. Let G be an integral commutative po-groupoid in which, given $a \leqq b$, one can find c such that $a = bc$. Then every indecomposable element of G is maximal. [**Tr**, p. 219]

7. Let G be a commutative integral po-groupoid in which, given $a \leqq b < 1$, there exists $c > a$ such that $a = bc$. Show that, in G, every prime element is maximal. [**Tr**, p. 221]

8. Prove that, in any Noetherian l-monoid, the radical of any element a is the meet of the minimal prime elements $p_i \geqq a$. (Lesieur [**1**, Theorem 2.2])

9. Show that, given a finite distributive lattice L, there exists a linear algebra having L for its ideal lattice.†

12. Frobenius l-monoids

We now turn our attention to residuation in *noncommutative* l-monoids. Consider the *full matrix algebra* $M_n(F)$ of all $n \times n$ matrices with entries in a given field F. The linear associative algebra $M_n(F)$ is isomorphic to the ring of all endomorphisms of the n-dimensional vector space $V_n(F)$ over F. Moreover, by classic theorems of Wedderburn,‡ $M_n(F)$ is a regular ring (§VIII.6) in which every left- or right-ideal is principal.

Specifically, if a right-ideal B of $M_n(F)$ contains one endomorphism with null-space $S \in V_n(F)$, then it contains every endomorphism whose null-space includes S, and the set of all such endomorphisms of $V_n(F)$ is a right-ideal $J(S)$ of $M_n(F)$. Dually, if a left-ideal C of $M_n(F)$ contains one endomorphism with range S, then C contains every endomorphism whose range is in S; moreover the set of all such endomorphisms is a left-ideal $K(S)$ of $M_n(F)$, for any subspace S. Finally, $0 .\!\cdot K(S) = J(S)$ and $0 \cdot\!. J(S) = K(S)$. Hence $M_n(F)$ is a Frobenius ring in the sense of the following definition.

Definition. A ring in which $0 .\!\cdot (0 \cdot\!. J) = J$ for all right-ideals J and $0 \cdot\!. (0 .\!\cdot K) = K$ for all left-ideals K is a *Frobenius ring*. A Frobenius algebra§ is a Frobenius ring which is a linear associative algebra.

Lemma. *Any semisimple linear associative algebra A of finite order over a field F is a Frobenius algebra.*

The preceding lemma follows directly from Wedderburn's results, which state that A is the direct sum of full matrix algebras $A_i = M_{n(i)}(D_i)$, where the D_i are division algebras over F. The right-ideals of A are the direct sums of those of the A_i, and likewise for the left-ideals; annihilators $0 \cdot\!. J$ and $0 .\!\cdot K$ are also the direct sums of their components. Finally, the statements about $M_n(F)$ made earlier apply also to any $M_n(D)$, provided one specifies left- and right-subspaces (D-modules) appropriately.

Definition. A residuated po-groupoid in which $0 .\!\cdot (0 \cdot\!. h) = h$ for every right-ideal element h and $0 \cdot\!. (0 .\!\cdot k) = k$ for every left-ideal element k is a *Frobenius po-groupoid.*

---

† E.-A. Behrens, Math. Ann. **133** (1957), 79–90.

‡ J. H. M. Wedderburn, Proc. London Math. Soc. **6** (1907), 77–118.

§ The theory of Frobenius algebras was developed by T. Nakayama, Ann. of Math. **40** (1939), 611–33 and **42** (1941), 1–21.

THEOREM 14. *The linear subspaces of any Frobenius algebra form a Frobenius l-monoid $M(A)$, which is a projective geometry considered as a lattice.*

PROOF. The first statement is a corollary of the definition; the second was already noted in §I.9.

THEOREM 15. *In any Frobenius l-monoid M, the left- and right-ideal elements form dually isomorphic sublattices.*

PROOF. By hypothesis, they are dually isomorphic subsets; by Theorem 6, they are sublattices.

Remark. In semisimple linear associative algebra, they are *complemented* (modular) sublattices, but the proof of this fact seems to require ring theory.

Exercises:

1. Let A_1, A_2 be two Frobenius rings in which

$$A_j .\!\cdot A_j = A_j \cdot\!. A_j = 0 \qquad (j = 1, 2).$$

Show that $A_1 \oplus A_2$ is a Frobenius ring.

2. Show that any regular principal ideal ring is a Frobenius ring.
3. Prove that no nilpotent linear associative algebra is a Frobenius algebra.†
4. Show that, for any field F, the quotient-ring $F[x]/(x^n)$ is a Frobenius ring, with simply ordered ideal lattice.
5. Do Frobenius rings form a "family" of algebras?
6. Relate residuals in Frobenius rings to the discussion of polarity in Example 2 of §V.7.
7. Let the diagram below define an integral l-monoid, with:

$$a_i b_i = b_i, \qquad c^2 = d, \qquad a_1 b_2 = b_2 a_1 = cd = 0.$$

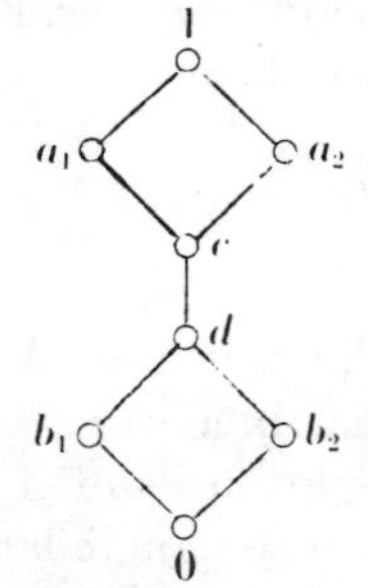

Show that this defines a Frobenius l-monoid. (P. Norman)

13. Algebra of Relations

The algebra of all binary relations on an arbitrary set J of elements $\alpha, \beta, \gamma, \cdots$ provides a final application of the theory of l-monoids. A binary relation on J may be defined by its *relation matrix* $\|r_{\alpha\beta}\|$, by letting $r_{\alpha\beta} = 1$ if the relation holds between α and β, and letting $r_{\alpha\beta} = 0$ if it does not. One easily verifies

† See M. Hall, Ann. of Math. **39** (1938), 220–34 and **40** (1939), 360–9, where various other related results are proved.

LEMMA 1. *Relation matrices form a Boolean lattice if* $\|r_{\alpha\beta}\| \leqq \|s_{\alpha\beta}\|$ *is defined to mean* $r_{\alpha\beta} \leqq s_{\alpha\beta}$ *for all* α, β. *When multiplied by the rule*

$$rs = t \quad \text{means that} \quad r_{\alpha\beta} = \bigvee_{\gamma} r_{\alpha\gamma} s_{\gamma\beta}, \tag{32}$$

they form an l-monoid $\mathcal{R}_n$, *where n is the cardinality of J.*

The l-monoid $\mathcal{R}_n$ is isomorphic with the l-monoid of all *join-endomorphisms* of the Boolean algebra $\mathbf{2}^n$. It is also related to Example 2 of §1. Let G be the monoid of elements e_{ij} and 0, with the multiplication rules for *matrix units*:

$$\begin{aligned} e_{ij}e_{kl} &= e_{il} \quad \text{if } j = k \\ &= 0 \quad \text{if } j \neq k, \end{aligned} \tag{33}$$

and $e_{ij}0 = 0e_{ij} = 0$. Then $\mathcal{R}_n$ is isomorphic with the l-monoid of all subsets of G which contain 0; this brings out the analogy with full matrix algebra. Finally, $\mathcal{R}_n$ can be defined as the l-monoid of all l-modules in the l-ring (Chapter XVII) of all real $n \times n$ matrices.

Nonnegative matrices. With any $n \times n$ nonnegative matrix A with entries $a_{ij} \geqq 0$, we can associate a relation matrix $R = r(A)$, by letting $r_{ij} = 0$ if $a_{ij} = 0$ and $r_{ij} = 1$ otherwise. We have

THEOREM 16. *The mapping* $A \to r(A)$ *is a morphism from the l-semiring (Chapter* XVII) *of nonnegative* $n \times n$ *matrices to the relation algebra* $\mathcal{R}_n$:

$$\begin{gathered} r(AB) = r(A)r(B), \qquad r(A \wedge B) = r(A) \wedge r(B), \\ r(A \vee B) = r(A + B) = r(A) \vee r(B). \end{gathered} \tag{34}$$

We leave the proof of (34) to the reader. By definition, A is "positive" when $r(A) = I$, the relation matrix consisting entirely of ones; it is "irreducible" when $\bigvee_n r(A^n) = I$. If we define a "cycle of length m" of a relation matrix as a function $f\colon \mathbf{Z}_m \to i(m)$ such that $r_{i(m-1),i(m)} = 1$, and the "index" $i(R)$ of $\|r_{ij}\|$ as the g.c.d. of the lengths of its cycles, then an irreducible matrix is "primitive" if and only if $i(R) = 1$. Otherwise, R is *cyclic*, and R^n is ultimately periodic with period $i(R)$.

Frobenius has made a classic study of the asymptotic behavior of $r(A^n) = [r(A)]^n$ for large n. He has shown that $r(A^n) = I$ for all sufficiently large n if and only if A is "irreducible" and "primitive".

14. Postulates for Relation Algebras

Though the algebra of relations was worked out in the nineteenth century, as a part of the algebra of logic, the systematic study of postulate systems for relation algebras dates only from 1940.† We will now characterize relation algebras as special l-monoids, as in [**LT2**, pp. 209–13].

The l-monoid $\mathcal{R}_n$ has a $0 = \|0\|$ and a unity $e = \|\delta_{ij}\|$, as well as a $I = \|1\|$. Again, each $r = \|r_{ij}\| \in \mathcal{R}_n$ has a *converse* $\breve{r} = \|r_{ji}\|$. We now show that the unary

† See C. S. Peirce, Mem. Acad. Arts Sci. **9** (1870), 317–78; [**Sch**, Vol. 3]; J. C. C. McKinsey, J. Symbolic Logic **5** (1940), 85–97; A. Tarski, ibid **6** (1941), 73–89.

operation of conversion can be defined in terms of residuation and Boolean operations.

LEMMA. *In the relation algebra* $\mathcal{R}_n$, *conversion and residuation can be defined from each other*:

$$e' .\!\cdot r' = e' \cdot\!. r' = \breve{r} \qquad (\textit{all } r), \tag{35}$$

$$r .\!\cdot s = (\breve{s}r')' \quad \textit{and} \quad r \cdot\!. s = (r'\breve{s})' \qquad (\textit{all } r, s). \tag{36}$$

PROOF OF (35). The inequality $r's \leqq e'$ states that $\bigvee_k (1 - r_{ik})s_{kj} = 0$ if $i = j$, that is (when valid) that $r_{ik} = 0$ implies $s_{ki} = 0$. But this statement is equivalent to $s \leqq \breve{r}$; hence $\breve{r}$ is the largest s such that $r's \leqq e'$; in other words, $e' .\!\cdot r' = \breve{r}$. The other equation of (35) now follows by symmetry.

We omit the proof of (36), which is an unpublished result of W. C. Davidson.

The preceding lemma justifies defining a *relation algebra* as a residuated l-monoid L with zero 0 and unity e, which is a Boolean algebra when considered as a lattice, in which

$$e' .\!\cdot s = e' \cdot\!. s \quad \text{for all } s, \tag{37}$$

and, denoting $e' .\!\cdot r' = e' \cdot\!. r'$ by $\breve{r}$, in which

$$\breve{\breve{r}} = r \quad \text{and} \quad \breve{(rs)} = \breve{s}\breve{r}. \tag{38}$$

Various other rules follow from the above postulates. Since $e' .\!\cdot s$ is an antitone unary operation in any residuated l-monoid, $e' .\!\cdot r'$ is isotone in any relation algebra. Moreover, since $\breve{\breve{r}} = r$, it is an isotone *bijection*. As a corollary,

$$\breve{(r \vee s)} = \breve{r} \vee \breve{s}, \qquad \breve{(r \wedge s)} = \breve{r} \wedge \breve{s}, \qquad (\breve{r})' = \breve{(r')}. \tag{39}$$

To apply the concepts of universal algebra (Chapter VI) to relation algebras as defined above, one must consider both complementation *and* residuation (or, equivalently, conversion) as basic operations, in addition to the usual basic operations for l-monoids. The concepts of relation subalgebra and of morphisms for relation algebras must be altered correspondingly.

There are other, equivalent sets of postulates for relation algebras. Thus, the identities $\breve{(rs)} = \breve{s}\breve{r}$ and $\breve{\breve{r}} = r$ imply $\breve{e} = e$ in any l-monoid. Using this fact, and the fact that $\breve{(r \vee s)} = \breve{r} \vee \breve{s}$ implies the other two identities of (39) as above, we obtain the following result.

THEOREM 17. *A relation algebra is a c$\mathcal{L}$-monoid L with unary operations of complementation and conversion, such that complementation makes L into a Boolean algebra, in which*

$$\breve{(rs)} = \breve{s}\breve{r}, \qquad \breve{\breve{r}} = r, \qquad \breve{(r \vee s)} = \breve{r} \vee \breve{s}, \tag{40}$$

and in which (36) *defines residuals.*

Again, Tarski has shown that the single equation

$$(41) \qquad [\breve{r}(rs)'] \vee s' = s'$$

is sufficient to imply the properties of residuation. In view of (40), (41) is equivalent (writing $r = \breve{x}$ and $s = y'$) to $[x(\breve{x}y')'] \vee y = y$, or (using (36)) to $x(x.\cdot y) \leqq y$. Hence we have:

THEOREM 18 (TARSKI). *A relation algebra is a c$\mathcal{L}$-monoid which is a Boolean algebra with a unary conversion operation satisfying* (40) *and* (41).

Concrete relation algebras. By a "concrete" or "proper" relation algebra, is meant a *subalgebra* of the "full" relation algebra $R_\aleph$ ($\aleph$ an arbitrary cardinal number), with respect to all the operations defined above. Lyndon† has shown that not every relation algebra as defined above is a "concrete" relation algebra (i.e., monomorphic to some $R_\aleph$). Specifically, he has shown that the "atoms" p_{ij} of any concrete relation algebra must satisfy

$$(42) \qquad p_{13}p_{32} \wedge p_{14}p_{42} \wedge p_{15}p_{52} > 0 \text{ implies}$$
$$(p_{41}p_{15} \wedge p_{42}p_{25}) \wedge (p_{41}p_{13} \wedge p_{42}p_{23})(p_{31}p_{15} \wedge p_{32}p_{25}) > 0.$$

And he has exhibited a relation algebra of order 2^{56} whose "atoms" fail to satisfy (42).

Exercises for §§13–14:

1. Show that a binary relation r on any set is:
   (a) Symmetric precisely when $\breve{r} = r$, and transitive when $r^2 \leqq r$.
   (b) A partial ordering when $r \wedge \breve{r} = e$ and $r^2 = r$.
   (c) A strict inclusion relation when $r \wedge \breve{r} = 0$ and $r^2 \leqq r$.
   (d) An equivalence relation when $r^2 = r = \breve{r} = e$.
2. For a relation r, define: $F(r) = r \vee e$, $S(r) = r \vee \breve{r}$, and $T(r) = \bigvee_n r^n$. Show that, under composition and order, F, S and T generate a nine-element *po*-semigroup.‡
3. (a) Show that $\breve{r} = r'$ is impossible in a nontrivial relation algebra (i.e., one having more than one element).
   (b) Show that $ax = xa = e$ in a relation algebra implies $x = \breve{a}$.
4. Show that, in $\mathscr{R}_n$, $rs \wedge t = 0$ if and only if $rst \leqq e'$. Infer that it implies $r \wedge st = 0$.
5. Prove that the results of Exercise 4 hold in any relation algebra.
6. Prove that, in $\mathscr{R}_n$: (i) $rs \leqq e'$ implies $sr \leqq e'$, (ii) $r > 0$ implies $IrI = I$, and (iii) $r > 0$ implies $r\breve{r} > 0$.
7. Show that the (direct) product of two nontrivial relation algebras is a relation algebra in which $IrI < I$ for some $r > 0$.
8. Show that the relation algebra $\mathscr{R}_n$ is "simple" (i.e., that it has no proper congruence relations).

A *Boolean l-monoid* is an l-monoid which, as a lattice, is a Boolean lattice.

*9. Show that any Boolean l-monoid can be represents by sets.

† R. C. Lyndon, Ann. of Math. **51** (1950), 707–29, and **63** (1956), 294–307; see also Michigan Math. J. **9** (1961), 21–8.

‡ T. Tamura, Bull. AMS **70** (1964), 113–20.

*10. Extend the preceding result to Boolean algebras with any given set of join-endomorphisms.†

PROBLEMS

122. Develop a theory of subdirectly irreducible integral l-groupoids.

123. Develop a theory of l-groupoids in which also $a(x \wedge y) = ax \wedge ay$ and $(x \wedge y)a = xa \wedge ya$.‡

124. Construct the free commutative l-monoid with n generators. Does there exist a free *complete* commutative l-monoid with n generators?

125. Construct the free l-loop with one generator.

126. For which groups do the subgroups form an l-groupoid under inclusion and commutation?

127. Develop a theory of continuous relation algebras, applicable to the continuous Boolean algebras B/Z and B_0/N of Chapter XI.

128. Show how to construct the tensor product of any two given relation algebras, so that $\mathscr{R}_m \otimes \mathscr{R}_n = \mathscr{R}_{mn}$.

---

† For the results of Exercises 9–10, see [**LT2**, pp. 211–13] and B. Jonsson and A. Tarski, Amer. J. Math. **73** (1951), 891–939 and **74** (1952), 127–62.

‡ See C. Choudhury, Bull. Calcutta Math. Soc. **49** (1957), 71–4.

CHAPTER XV

VECTOR LATTICES

1. Basic Definitions

Real vector spaces, considered as additive (Abelian) groups, are often *po*-groups and even *l*-groups. By definition, a *partly ordered vector space* V is such a *po*-group, in which $x \geqq 0$ in V and $\lambda \geqq 0$ in $\mathbf{R}$ imply $\lambda x \geqq 0$ in V. By this definition, $x \to \lambda x$ is an order-preserving group-automorphism for any $\lambda > 0$, with order-preserving inverse $y \to \lambda^{-1}y$; therefore $x \to \lambda x$ is an *automorphism* of V as a *po*-group, for any positive scalar λ.

The theory of Abelian *po*-groups, developed in Chapter XIII, clearly applies to any partly ordered (real) vector space V. Thus V is called a *directed vector space* when it is directed as a *po*-group—i.e., when $V = P - P$, where P is the "positive cone" of all $x \geqq 0$ in V. When V (considered as an additive *po*-group) is an *l*-group, it is called a *vector lattice*. When V is (conditionally) complete or σ-complete as an *l*-group, it is likewise called a *complete* (respectively σ-complete) *vector lattice*.

The theory of vector lattices was initiated by F. Riesz [**1**], L. Kantorovich [**1**], and H. Freudenthal [**1**]; because of its importance for functional analysis, it was rapidly developed by other authors, whose main results are summarized below. However, no attempt will be made to summarize what is known about partly ordered *topological* vector spaces, having an (extrinsic) topology not defined in terms of addition and order; in particular, many results of H. Nakano, L. Nachbin, F. F. Bonsall, I. Namioka, and H. Schaefer will not be discussed.

This is because, as emphasized in [**LT1**, §§126–7], so many basic properties of the most important real function spaces can be developed in terms of addition and order alone as undefined concepts. In particular, this is true of many properties of boundedness, dual spaces, of many natural (intrinsic) topologies, and of much of the theory of positive linear operators (to be developed in Chapter XVI).

Example 1. For any cardinal number $\aleph$, $\mathbf{R}^{\aleph}$ is a vector lattice, provided one defines $\mathbf{x} \geqq \mathbf{0}$ to mean that all $x_i \geqq 0$,

In Example 1, for $\aleph = 3$, the "positive cone" is the first octant: an infinite trihedron. This observation will now be generalized.

LEMMA 1. *In any partly ordered vector space V, the "positive cone" P is a convex cone satisfying $P \cap -P = 0$. Conversely, if P is any convex cone in the real vector space V, and $P \cap -P = 0$, then V is partly ordered by the relation*

$$(1) \qquad x \leqq y \quad \textit{if and only if } (y - x) \in P.$$

PROOF. If $x \in P$ and $y \in P$, then $\lambda x \in P$ for any $\lambda > 0$ by hypothesis and $x + y \in P$ by Chapter XIII, (2). These conditions characterize real convex cones; in particular, they make $\{x, y\} \subset P$ imply $\alpha x + (1 - \alpha)y \in P$ if $0 < \alpha < 1$.

Conversely, if P is a convex cone in V, then by hypothesis λx and $x + y = 2[(x + y)/2]$ are in P if x and y are, and $\lambda > 0$.

Example 2. For any positive integer n, the cone in $\mathbf{R}^n$ defined by

$$\sum x_i \geqq 0 \quad \text{and} \quad \sum_{i<j} x_i x_j \geqq 0 \tag{2}$$

is the positive cone of a partly ordered vector space which is *not* a vector lattice for $n > 2$. (Its positive cone has a "circular" cross-section, being defined equivalently by the inequalities $(\sum x_i)^2 \geqq \sum (x_i^2)$ and $\sum x_i \geqq 0$.)

LEMMA 2. *Any direct sum (union) of vector lattices is a vector lattice.*

We omit the (trivial) proof; cf. Chapter XIII, Example 3.

LEMMA 3. *If V is any vector lattice, then the lexicographic union $\mathbf{R} \circ V$ is also a vector lattice.*

We again omit the proof, which is a trivial extension of that of §XIII.1, Lemma 3. It is a corollary that $\mathbf{R}^3$, $\mathbf{R} \circ \mathbf{R}^2$, $\mathbf{R}(\mathbf{R} \circ \mathbf{R})$, and ${}^3\mathbf{R} = \mathbf{R} \circ (\mathbf{R} \circ \mathbf{R})$ are all three-dimensional vector lattices; we shall show in §2 that there are no others.

Vector lattices have important applications to functional analysis, stemming from the fact that *most of the standard real function spaces are vector lattices*, and in a very natural way.

Example 3. Let X be any topological space, and let $C(X)$ be the set of all continuous real functions on X. Then $C(X)$ is a vector lattice if one defines P as the set of all nonnegative functions.

Example 4. Let Ω be any measure algebra (§XI.5), and let $M^p(\Omega)$ be the set of measurable real functions on Ω for which $\int |f(x)|^p d\mu < +\infty$. Then $M^p(\Omega)$ is a vector lattice if one chooses for P the set of everywhere nonnegative $f(x)$.

The usual space $L^p(\Omega)$ is the quotient-algebra $M^p(\Omega)/N(\Omega)$ where $N(\Omega)$ is the l-ideal of functions which vanish except on a set of measure zero (cf. §7).

Since any vector lattice is an Abelian l-group, most of the formulas of Chapter XIII apply automatically to any vector lattice. For convenient reference, we restate some of these formulas here.

THEOREM 1. *In any vector lattice V, we have:*

$$a + (x \vee y) = (a + x) \vee (a + y), \qquad a + (x \wedge y) = (a + x) \wedge (a + y). \tag{3}$$

$$a + b = a \wedge b + a \vee b. \tag{4}$$

$$a \wedge b = a \wedge c = 0 \quad \textit{implies} \quad a \wedge (b + c) = 0. \tag{5}$$

$$|a + b| \leqq |a| + |b|, \quad \textit{where } |a| = a^+ - a^- > 0 \quad \textit{unless} \quad a = 0. \tag{6}$$

$$|\lambda a| = |\lambda| \cdot |a| \quad \textit{for any } \lambda \in R,\ a \in V. \tag{7}$$

$$\lambda(x \wedge y) = \lambda x \wedge \lambda y \quad \textit{and} \quad \lambda(x \vee y) = \lambda x \vee \lambda y \quad \textit{if } \lambda \geqq 0. \tag{8}$$

PROOF. See Chapter XIII, formulas (9), (14), (16), and Theorem 9, for derivations of (3)–(6), respectively. As regards (7), we have $|\lambda a| = |\lambda| \cdot |a|$ for $\lambda = |\lambda| > 0$ by definition, and for $\lambda = 0$ trivially. For $\lambda < 0$,

$$|\lambda| \cdot (a) = -\lambda(a^+ - a^-) = \lambda a^- - \lambda a^+ = (\lambda a)^+ - (\lambda a)^- = |\lambda a|,$$

since for $\lambda < 0$ the correspondence $a \to \lambda a$ is a *dual* automorphism, interchanging $\vee$ and $\wedge$.

Exercises:

1. Prove the following identities in any vector lattice:

$$-(-x \vee -y) = x \wedge y, \; -(x \wedge y) = -x \vee -y.$$

2. Prove that any vector lattice is distributive as a lattice.
3. Prove the following identities in any vector lattice:
   (a) $f \vee g = f + (g - f)^+, f \wedge g = f - (g - f)^-$.
   (b) $f \vee g = 1/2(f + g + |f - g|)$.
4. Prove the following inequalities in any vector lattice:
   (a) $(f + g)^+ \leqq f^+ + g^+$,
   (b) $||f| - |g|| \leqq |f - g| = |f^+ - g^+| + |f^- - g^-|$.
5. In a real vector space V, let P be a subset satisfying: (i) $P \cap -P = 0$, (ii) $P + P \subset P$, (iii) $\lambda P = P$ for any scalar $\lambda > 0$. Show that P is the "positive cone" for a suitable partial ordering of V, which makes V into a directed vector space if and only if $P - P = V$.
6. In the plane $\mathbf{R}^2$, let P be the set of all (x, y) such that $x \geqq 0$ and $y \geqq x^{1/2}$. Which of the postulates for a vector lattice fails?
7. (a) Let the real vector space V be an l-group under addition and some ordering. Show that, for any positive rational scalar λ, $x \to \lambda x$ is a lattice-automorphism.
   (b) Show that this need not be true for irrational λ, even when V is complete as an l-group.

2. *l*-ideals

As was shown in Chapter XIII, §§9–12, the notion of an l-ideal is the key to the understanding of the structure of l-groups. The same is true for vector lattices; as a first step towards establishing this, we prove

LEMMA 1. *In a vector lattice V, the l-ideals are precisely the subspaces which contain with any x all y such that $|y| \leqq |x|$, and so are the congruence modules for the vector lattice operations.*

PROOF. Let J be any l-ideal, i.e., a subgroup of V containing with x all y such that $|y| \leqq |x|$. If $a \in J$, then $na \in J$ for each positive integer n. If λ is any real number, pick a positive integer n exceeding $|\lambda|$; then $|\lambda a| = |\lambda| \cdot |a| \leqq n|a|$, and so $\lambda a \in J$. Hence J is a linear subspace. The converse is obvious.

COROLLARY. *Any morphism of l-groups ϕ: $V \to W$ between vector lattices is linear.*

Real function theory abounds in examples of l-ideals of vector lattices. For example, let $\mathbf{R}^{\aleph}$ be the set of all real functions on an arbitrary set X of cardinality $\aleph$. Then the set $B(X)$ of all bounded real functions on X is an l-ideal of $\mathbf{R}^{\aleph}$. Also, for any subset $S \subset X$, the set of all real functions having "support" in S (i.e., vanishing outside of S) is an l-ideal of $\mathbf{R}^{\aleph}$.

Again, let $M(\Omega)$ be the set of all (real) functions measurable on a given measure space Ω. Then $M^p(\Omega)$, as defined in Example 4, is an l-ideal of M. Similarly, the set $N(\Omega)$ of so-called "null functions", vanishing except on a set of measure zero, is an l-ideal of $M(\Omega)$.

THEOREM 2. *Any "simple" vector lattice (i.e., one having no proper congruence relations) is isomorphic to* $\mathbf{R}$.

PROOF. For any element $a \neq 0$ in a simple vector lattice V, the principal l-ideal (a) (§XIII.10) generated by a must be V; hence V must be Archimedean. Likewise, being simple, V is subdirectly irreducible, and hence by Theorem XIII.21 simply ordered.

Being an Archimedean (simply) ordered group, V is isomorphic with a subgroup of the real field $\mathbf{R}$, by Theorem XIII.12. By the Corollary to Lemma 1 above, this order-isomorphism must be linear; hence $V = \mathbf{R}$. Q.E.D.

Theorem 2 is a strengthened analogue of Theorem XIII.12; it has the following corollary.

COROLLARY. *The structure lattice $\Theta(L)$ of any n-dimensional vector lattice L has length n.*

We now exploit again Theorem XIII.21. With Lemma 1, it yields as a corollary the following analogue of Theorem XIII.22.

THEOREM 3. *Any vector lattice is a subdirect product of simply ordered vector lattices.*

Next, we apply Lemma 1 to Theorem XIII.23. In combination with Theorem 2, this yields as a corollary

LEMMA 2. *Let V be any n-dimensional vector lattice. Then either V is a direct sum of vector lattices of lower dimension, or V contains a largest proper l-ideal M such that V/M is simple.*

By Theorem 2, we know that $V/M \cong \mathbf{R}$, whence

COROLLARY. *In Lemma 2, M is an $(n - 1)$-dimensional subspace of V.*

Further, since M contains every proper l-ideal of V, $a \notin M$ implies that the principal l-ideal $(a) = V$. Moreover unless $(a^+) = V$ or $(a^-) = V$, $(a^+) \subset M$ and $(a^-) \subset M$, which is impossible since $(a) = (a^+) \vee (a^-)$ and $(a) = V > M$. Suppose $(a^+) = V$; then since $(a^+) \wedge (a^-) = 0$ for any a, $(a^-) = 0$ and $a = a^+ \geqq 0$. Likewise, $(a^-) = V$ implies $a \leqq 0$. Since $a = 0$ is impossible, we conclude

LEMMA 3. *In Lemma 2, $a \notin M$ implies that $a > 0$ or $-a > 0$.*

In either case, choosing $b \notin M$ to be positive, it will follow that the elements $\lambda b + m = x$ $(\lambda \in \mathbf{R},\ m \in M)$ exhaust V, and that $x > 0$ if and only if $\lambda > 0$ or $\lambda = 0$ and $m > 0$. This proves†

† For another approach see [**LT2**, p. 240, Exercise 2].

THEOREM 4. *Every n-dimensional vector lattice V is either a direct sum of vector lattices of lower dimension, or the lexicographic union $V = \mathbf{R} \circ M$ of $\mathbf{R}$ with the largest proper l-ideal M of V, which is $(n - 1)$-dimensional.*

COROLLARY 1. *Every finite-dimensional vector lattice can be built up from copies of $\mathbf{R}$ by direct and lexicographic union.*

COROLLARY 2. *Every Archimedean vector lattice of linear dimension n (n finite) is isomorphic to $\mathbf{R}^n$.*

3. Function Lattices

The most important vector lattices represent sets of real *functions*, isomorphically or epimorphically. By definition, a *function lattice* is a vector lattice which is *isomorphic* with a vector lattice of functions on some set under pointwise $\pm$, $\vee$, $\wedge$,—i.e., which is a subdirect copy of copies of $\mathbf{R}$. In other words, a function lattice is a vector lattice which is isomorphic to a vector sublattice of $\mathbf{R}^{\aleph}$ for some cardinal number $\aleph$ or, more briefly, is monomorphic to some $\mathbf{R}^{\aleph}$ (as a vector lattice).

The question of whether or not a given vector lattice V is a function lattice can be answered when its maximal l-ideals are known. This is because of the following corollary to Theorem 2.

LEMMA 4. *The epimorphisms of a given vector lattice V onto copies of $\mathbf{R}$ are given by the totality of maps $V \to V/M_\alpha$ onto quotients of maximal (proper) l-ideals M_α of V, uniquely up to a positive scalar factor.*

PROOF. If M is a maximal proper l-ideal in V, then V/M is simple and so is isomorphic to $\mathbf{R}$.

COROLLARY (NAKAYAMA). *A vector lattice V is isomorphic to a subdirect product of copies of $\mathbf{R}$ if and only if the intersection of its maximal proper l-ideals is 0.*

PROOF. Suppose V is a vector sublattice of $\mathbf{R}^I$ for some index set I. For each element of I, the kernel of the corresponding projection $\mathbf{R}^I \to \mathbf{R}$ is a maximal proper l-ideal in V, and evidently the intersection of these kernels is 0. Conversely, suppose the intersection of the maximal proper l-ideals in V is 0. By general algebraic theory, V is isomorphic to a subdirect product of simple vector lattices, so by Theorem 2, V is isomorphic to a subdirect product of copies of R.

Kaplansky has noted that the real functions on $0 \leqq x \leqq 1$, defined and continuous except at a finite number of simple poles of second order, $\alpha_i/(x - a_i)^2$, form an Archimedean vector lattice, the intersection of whose maximal l-ideals consists of all continuous functions and so is not 0.

Maximal l-ideals of $C(K)$. It is easy to determine the maximal l-ideals of $C(K)$, the vector lattice of all real continuous functions on a compact (hence normal) Hausdorff space K. For any point $x \in K$, the set M_x of all $f \in C(K)$ such that $f(x) = 0$ is clearly $C(K)$ or a maximal l-ideal of $C(K)$. We now show that $C(K)$ has no other maximal (proper) l-ideals.

Let J be any proper l-ideal of $C(K)$ not contained in any M_x. For any $x \in K$, there will then be some $f \in J$ with $f(x) \neq 0$, and hence such that $f(y) \neq 0$ throughout an open neighborhood $U(x)$ of x. Since K is compact, there must be a finite set of such open neighborhoods $U_1, \cdots, U_n$ which cover K; let $f_1, \cdots, f_n$ be the corresponding functions $f_i \in J$, with $f_i(y) \neq 0$ for all $y \in U_i$. Then $g = |f_1| + \cdots + |f_n| \in J$ will be a function which is identically positive on $K = \bigcup U_i$, and hence bounded away from zero there. For *any* $h \in C(K)$, therefore, $|h| \leqq \lambda g$ for all sufficiently large real λ, and so $J = C(K)$. We conclude that every proper l-ideal is contained in some M_x, and so the M_x are the only maximal l-ideals.

Closed l-ideals. More generally, we can describe "closed" l-ideals (Chapter XIII, §§9–10) of the vector lattice $C(X)$ of all continuous functions on any normal Hausdorff space X: they form a Boolean lattice isomorphic with the lattice of all regular open sets of X. This isomorphism associates with each regular open set A of X, the set $A'^{\dagger}$ of all $f \in C(X)$ which vanish identically on its closed complement A' (such sets may be called "regular closed sets").

To show this, consider the polarity (Chapter V) associated with the binary relation $f(x) = 0$ between $C(X)$ and X. For any l-ideal (indeed, any subset!) J of $C(X)$, the "polar" $J^* = S$ of J is a closed subset of X (by definition, J^* is the set of all $x \in X$ with $f(x) = 0$ for all $f \in J$). Moreover, by definition, $g \in J^{\perp}$ if and only if $g(y) = 0$ for all $y \in S'$—or equivalently, g being continuous, for all $y \in S'^{-}$. Also, S'^{-}, being the closure of an open set, is *regular*. This proves, that, for any l-ideal J of $C(X)$, $J^{\perp} = (S'^{-})^* = ((J^*)'^{-})^*$, where $S = J^*$ is a regular closed subset of X.

Conversely, let T be any regular closed set, with regular open complement T'; let $T'^{-\prime} = K: T' = U$ be the (regular, open) *pseudocomplement* of T in the Brouwerian lattice of all open sets of X. Then the sets $T^{\dagger} = F$ and $(T'^{-})^{\dagger} = G$ are closed l-ideals of $C(X)$ with $F^{\perp} = G$ and $G^{\perp} = F$, as we now show. Indeed, $G \subset F^{\perp}$ is obvious. Now suppose $h \in F^{\perp}$. Since X is normal, there exists for any $y \in T'$ an $f \in C(K)$ with $f(y) = 1$ and $f \in F$—i.e., $f(t) = 0$ for all $t \in T$; but $h \wedge f = 0$, hence $h(y) = 0$ for all $y \in T'$ follows. This proves $F^{\perp} \subset G$, and so $F^{\perp} = G$. Similarly, $G^{\perp} = F$. In summary, for any regular closed set T of X, $F = T^* = (F^{\perp})^{\perp}$ is a closed l-ideal. Since the converse was proved in the last paragraph, the proof is complete.

Since any compact Hausdorff space is normal and hence regular, and since the regular open sets form a neighborhood-basis in any regular topological space [**Kel**, p. 113], we derive as a corollary of the preceding results the fact that any compact Hausdorff space K is determined up to homeomorphism by the vector lattice $C(K)$ of its continuous functions: the points p_α of K can be identified as the maximal ideals $M_\alpha \subset C(K)$, and the neighborhoods of any point as the closed l-ideals F of $C(K)$: a point p_α is in the neighborhood U_F if and only if $F \not\subset M_\alpha$.

Exercises for §§2–3:

1. Show that the l-ideals of $\mathbf{R}^n$ form a Boolean lattice isomorphic to $\mathbf{2}^n$ and identify each l-ideal with the set of all vectors whose support is contained in some subset of $\mathbf{n}$.
2. Show that the l-ideals of any vector lattice form an algebraic Brouwerian lattice.

3. Show that every three-dimensional vector lattice is isomorphic to one and only one of the following, $\mathbf{R}^3$, ${}^3\mathbf{R}$, $\mathbf{R} \circ \mathbf{R}^2$, $\mathbf{R}(\mathbf{R} \circ \mathbf{R})$.

4. Draw the diagram of the structure lattice of each of the four vector lattices of Exercise 3.

5. Let $\phi(n)$ be the number of all nonisomorphic vector lattices of dimension n, and $\psi(n)$ the number of those which are directly indecomposable. Show that $\phi(n) = \psi(n + 1)$, and compute the following table:

$n =$	1	2	3	4	5	6
$\psi(n) =$	1	1	2	4	9	19
$\phi(n) =$	1	2	4	9	19	47.

6. In a directed vector space V, define an *l-ideal* as a subspace S which contains: (i) with any $a > 0$, also $[0, a]$, and (ii) with any b, also some $c > 0$ and $d < 0$ with $b = c + d$.

(a) Show that this definition is equivalent to that of the text when V is a vector lattice.

(b) What can you say about V/S? about the poset of all l-ideals of V (ordered by set-inclusion).

*7. Show that the vector lattice $C(X)$ of all continuous functions on a normal topological space X is σ-complete if and only if X is the Stone space of a Borel algebra.†

*8. (a) Let J be any ideal of any Abelian l-group A. Show that the set of all translates of J is closed under finite intersection.

(b) Prove that the converse is true if A is a real vector space.‡

4. Simply Ordered Vector Lattices

Theorem 3 makes it interesting to analyze the structure of *simply* ordered vector lattices. It follows from Theorem 4 that every finite-dimensional simply ordered vector lattice is isomorphic with some ordinal power ${}^n\mathbf{R}$ of the real field. In trying to extend this result, a first important step is

LEMMA 1. *Let $a \neq 0$ and b generate the same principal l-ideal $(a) = (b)$ in a simply ordered vector lattice V. Then there exists a unique real number λ such that $|\lambda a - b| \ll |a|$.*

PROOF. Suppose $a > 0$; let L and U be the sets of real numbers β such that $\beta a \leqq b$ and $\beta a \geqq b$, respectively. Since V is ordered, $L \cup U = \mathbf{R}$; since $a > 0$, $\beta \in L$ and $\gamma \in U$ imply $\beta \leqq \gamma$. Moreover $L \neq \varnothing$ and $U \neq \varnothing$ because $(a) = (b)$. Let λ be the *cut* defined in $\mathbf{R}$ by (L, U), and let $c = \lambda a - b$. Then, for any $n \in \mathbf{Z}^+$, $-a/n < c < a/n$. Hence $n|c| \leqq a$, and so $|c| = |\lambda a - b| \ll |a|$.

COROLLARY. *A (simply) ordered vector space V is either isomorphic with $\mathbf{R}$ or non-Archimedean.*

In the infinite-dimensional case, the analysis of the structure of simply ordered vector lattices requires sophisticated use of transfinite induction. An easy result is

LEMMA 2. *Let V be any ordered vector space. Then the principal l-ideals of V form a chain which is dense in the (simply ordered) structure lattice $\Theta(V)$ of V.*

---

† M. H. Stone, Canad. J. Math. **1** (1949), 176–86. For extensions, see R. S. Pierce, ibid. **5** (1953), 95–100.

‡ For Exercise 8, see J. G. MacCarthy, Trans. AMS **100** (1961), 241–51.

PROOF. Trivially, $(a) < (b)$ if and only if $|a| \ll |b|$, and every l-ideal J of V is the join of the principal l-ideals which are contained in J. What is more special, there is a maximal proper sub-l-ideal contained in every principal l-ideal (a), $a \neq 0$: this consists of all $x \in V$ such that $|x| \ll |a|$. This shows that there is a dense set of "gaps" (discrete jumps) in $\Theta(V)$.

THEOREM 5. *If $\Theta(V) = W$ is well-ordered, then V is isomorphic to the restricted ordinal power* $({}^W\mathbf{R})$.

PROOF. Using transfinite induction, we can select a well-ordered set of positive elements $e_1 \ll e_2 \ll e_3 \ll \cdots$, such that each principal ideal $(e_{\alpha+1})$ is the smallest l-ideal properly containing (e_α) and, for any limit-ordinal ω, (e_ω) is the least l-ideal properly containing $\bigcup_{\alpha<\omega}(e_\alpha)$. By induction, using Lemma 1, one can show that the e_α so constructed form a *basis* for V considered purely as a vector space. Moreover a finite linear combination $\Sigma\, \lambda_\alpha e_\alpha > 0$ if and only if the nonzero λ_α with "largest" subscript is positive. But this defines $({}^W\mathbf{R})$.

If V has a *countable* basis, then one can prove without further assumption that $V \cong ({}^W\mathbf{R})$, where W is the "skeleton" of principal ideals.† But the most important theorem about simply ordered vector spaces is the following classic result of Hahn [**1**].

HAHN-ERDÖS EMBEDDING THEOREM. *Every simply ordered vector space V is isomorphic to a subspace of a lexicographic union of functions over the "skeleton" consisting of the gaps in $\Theta(V)$.*

For the precise meaning of this statement, and its proof, see [**Fu**, IV.4–IV.5]; it involves Hahn's ordinal power (§VIII.14).

5. Free Vector Lattices

In this section we will construct the free vector lattice $FVL(n)$ on n generators, whose existence follows from the general considerations of Chapter VI.‡ By Theorem 3, we know that $FVL(n)$ is a subdirect product of simply ordered vector lattices S_α, each of which (being an epimorphic image of $FVL(n)$) must itself have n generators $x_1, \cdots, x_n$. We next prove.

LEMMA 1. *Any simply ordered vector space with n generators $x_1, \cdots, x_n$ is an epimorphic image of* ${}^n\mathbf{R}$.

PROOF. Without loss of generality, we can assume that $x_1 > x_2 > \cdots > x_n > 0$. Next by Lemma 1 of §3, we can replace the x_i by $y_1 = x_1$, $y_2 < x_2 - \alpha_{21}x_1 \ll y_1$, $y_3 = x_3 - \alpha_{31}x_1 - \alpha_{32}x_2 \ll x_2$, and so on. But these y_i generate ${}^n\mathbf{R}$ if all nonzero, and an epimorphic image of ${}^n\mathbf{R}$ in any case.

† [**Fu**, pp. 60–1]; the result is due to J. Erdös, Publ. Math. Debrecen **4** (1956), 334–43. See also P. Conrad, J. Indian Math. Soc **22** (1958), 1–25 and 27–32.

‡ See E. C. Weinberg, Math. Ann. **151** (1963), 187–99 and **159** (1965), 217–22; D. Topping, Canad. J. Math. **17** (1965), 411–28; K. Baker, ibid. (to appear).

LEMMA 2. *The vector lattice* ${}^n\mathbf{R}$ *is an epimorphic image of a vector sublattice of* $\mathbf{R}^\omega$.

PROOF. In $\mathbf{R}^\omega$, let $\mathbf{f}^j = (1^j, 2^j, 3^j, \cdots)$, and let T be the vector sublattice of $\mathbf{R}^\omega$ spanned by $\mathbf{f}^1, \mathbf{f}^2, \cdots, \mathbf{f}^n$. In T, define

$$\mathbf{f} \equiv \mathbf{g}\ (\theta) \quad \text{if and only if} \quad \lim_{r\to\infty} f_r/g_r = 1. \tag{9}$$

Since, for all $\mathbf{f} \in T$, it is true that

$$f_r = \lambda_0 + \lambda_1 r + \cdots + \lambda_n r^n \text{ for some finite set of } \lambda_i \in \mathbf{R} \text{ and all sufficiently large } r, \tag{10}$$

the equivalence relation (9) is a *congruence relation* on T. In T/θ, evidently $0 < \mathbf{f}^1 \ll \mathbf{f}^2 \ll \cdots \ll \mathbf{f}^n$; hence T/θ (which is generated by $\mathbf{f}^1, \cdots, \mathbf{f}^n$ since T is) is isomorphic with ${}^n\mathbf{R}$. This completes the proof.

To summarize, we have shown that $FVL(n)$ is a subdirect product of simply ordered vector lattices S_α with n generators, each of which is an epimorphic image of a vector sublattice of $\mathbf{R}^\omega$. But this proves

THEOREM 6. *The free vector lattice $FVL(n)$ with n generators is a member of the family of vector lattices generated by the vector lattice* $\mathbf{R}$.

COROLLARY. *Any identity of vector lattice operations which is valid in* $\mathbf{R}$ *is valid in every vector lattice.*

As a further corollary, $FVL(n)$ can be *constructed* by the construction of §VI.7, as the sublattice of real functions on the domain $\mathbf{R}^n$ of all real vectors $\mathbf{x} = (x_1, \cdots, x_n)$ which is generated by the coordinate projections $x_i: \mathbf{x} \to x_i$, hence as a vector sublattice of $\mathbf{R}^{\mathbf{R}^n}$. But these are all continuous and piecewise linear. This proves the following result.

THEOREM 7. *The functions $x_1, \cdots, x_n$ in the vector space $\mathbf{R}^n$ generate the free vector lattice with n generators, which consists exclusively of continuous functions, which are piecewise homogeneous linear in conical polyhedral sectors of $\mathbf{R}^n$, with common vertex* $\mathbf{0}$.

These functions are determined by their value on the *unit sphere*, and are thus piecewise linear on spherical polyhedra. For $n = 1$, the unit sphere has two points, and we have the

COROLLARY 1. $FVL(1) \cong \mathbf{R}^2$.

COROLLARY 2. *Every free vector lattice is a subdirect product of copies of* $\mathbf{R}$.

Exercises for §§4–5:

1. Show that ${}^n\mathbf{R}$ has no set of $m < n$ generators.
2. Show that the ordinal product (lexicographic union) $V \circ W$ of two vector lattices V and W is a vector lattice if and only if V is simply ordered.
3. Prove in detail that $FVL(1) \cong \mathbf{R}^2$.

4. Establish an isomorphism between $FVL(2)$ and the vector lattice of all real continuous functions on the unit circle having the form $f(\theta) = a_i \cos \theta + b_i \sin \theta$ on each of a finite number of segments of the circle.

*5. Show that any polyhedral complex is determined up to homeomorphism by the vector lattice of all continuous functions which are piecewise linear on some simplicial subdivision. (A. Gleason)

*6. Let V be the set of all real continuous "piecewise polynomial" functions f on $(0, \infty)$—that is, of continuous functions equal to some polynomials on each of a finite number of subintervals (x_{i-1}, x_i) with $0 = x_0 < x_1 < \cdots < x_{n(f)} = \infty$.

(a) Show that V is a vector lattice.

(b) Define $f\theta g$ on V to mean that $\mathrm{Lim}_{x\to\infty} f(x)/g(x) = 1$. Prove that $V/\theta \cong {}^{\omega}\mathbf{R}$, where ω is the first countable ordinal.

*7. Prove that there exists a function $\alpha \to f_\alpha$ from the set of all countable ordinals α to real functions on $(0, \infty)$, such that $\alpha < \beta$ implies $f_\alpha \gg f_\beta$.†

*8. Show that every vector lattice is an epimorphic image of a subdirect product of copies of $\mathbf{R}$.

*9. (D. Topping) Show that the complete vector lattice L_1/N of (integrable functions)/(null functions) has no maximal l-ideal, and so is not a subdirect product of copies of $\mathbf{R}$.

*10. (a) Show that every free Abelian l-group is a subdirect product of copies of $\mathbf{Z}$.

(b) Show that the identities true in all commutative l-groups are precisely those true in $\mathbf{Z}$. (E. C. Weinberg)

6. Integrally Closed Directed Vector Spaces

For some applications, it is more essential that a partly ordered vector space be integrally closed and directed, than that it be a lattice. For example, the partly ordered vector space of Example 2 has many important applications but is not a lattice, and the same is true of the following example.

Example 5. Let V be the vector space of all $n \times n$ symmetric matrices, and let P be the cone defined by

$$\sum a_{ij}x_ix_j \geqq 0 \quad \text{for all } \mathbf{x} = (x_1, \cdots, x_n). \tag{11}$$

The fact that Example 2 and Example 5 (like Examples 1, 3, and 4) are integrally closed (hence Archimedean) directed vector spaces is a corollary of the following result.

THEOREM 8. *A finite-dimensional partly ordered vector space V is directed if and only if its positive cone P has a nonvoid interior;‡ if directed, it is integrally closed if and only if P is also topologically closed.*

PROOF. If P has a nonvoid interior with element a, then $P - P$ contains $(a + U) - (a + U) = 2U$ for some ball U: $\|a\| < \epsilon, \epsilon > 0$. Hence $P - P \supset \lambda U$ for all $\lambda > 0$ (P being a cone), and so $P - P = V$. Conversely if the interior of P is void, then (by an elementary theorem on convex sets§), P is contained in an

† This generalizes an old theorem of du Bois-Reymond; see G. H. Hardy, *Orders of infinity*, 2d ed., Cambridge Univ. Press, 1924, p. 8.

‡ We are here using the "natural" Euclidean topology for V, which is independent of the choice of basis.

§ See F. A. Valentine, *Convex sets*, McGraw-Hill, Chapter VI.

$(n - 1)$-dimensional affine subspace which includes the origin; hence $P = P - P$, contradicting $P - P = V$. The proof of the first statement of Theorem 8 is completed by reference to Lemma 1 of §XIII.2.

Now let V be an n-dimensional directed vector space with positive cone P; by the preceding paragraph, P is also n-dimensional. Moreover, almost by definition (Chapter XIII, (6)), (V, P) is integrally closed if and only if

$$\left(a - \frac{1}{n} b\right) \in P \quad \text{for all } n \in \mathbf{Z}^+ \quad \text{implies} \quad a \in P. \tag{12}$$

Evidently, (4′) holds whenever P is topologically closed. We now recall the "principle of linear accessibility" for convex sets,† which states that if c is any interior point and a any boundary point of a convex set S, then the segment (a, c) of $\lambda a + (1 - \lambda)c$, $0 \leqq \lambda < 1$, lies in S. It follows, setting $c = a - b$ in (12), that if (V, P) is integrally closed and directed, then P is a topologically closed convex cone. Q.E.D.

Corollary 1. *If the positive cone P of an Archimedean vector lattice has a nonvoid interior, then P is closed.*

Corollary 2. *In an integrally closed directed vector space, let $f > 0$, $g > 0$, $\alpha_n f \geqq g$ for $n = 1, 2, 3, \cdots$ and $\alpha_n \to \alpha$. Then $\alpha f \geqq g$.*‡

We have seen (Theorem XIII.24) that the completion by nonvoid cuts of any integrally closed directed group is a complete l-group. In any integrally closed directed vector space, the mappings $x \to \lambda x$ are order-automorphisms and dual automorphisms for $\lambda > 0$ and $\lambda < 0$, respectively, as well as group automorphisms. These automorphisms have a natural extension to the completion by cuts of such a space, which shows that

Theorem 9. *The completion by nonvoid cuts of any integrally closed directed vector space is a (conditionally) complete vector lattice.*

In general, it is not easy to visualize just what these completions by cuts give. However, A. G. Waterman has recently proved the following theorem. The completions by cuts of: (i) the free vector lattice with n generators, (ii) the directed vector space of Example 2, and (iii) the directed vector space of Example 5 are all isomorphic to the lattice of normal upper semicontinuous functions on the $(n - 1)$-sphere.

Strong unit. M. Krein§ has shown that much of Theorem 8 is valid in any integrally closed directed vector space with a strong unit e. In any such space V, let M_e be the set of "extreme", convex, vector subspaces of V, and let $\bar{M}_e$ be the "W^*-closure" (by weak completion) of M_e. Then, applying the Krein–Milman

† V. C. Klee, Duke Math. J. **18** (1951), 443–66 and 875–83.

‡ See G. Birkhoff [**7**, p. 38, Lemma 1]; the subspace spanned by f and g is two-dimensional.

§ M. Krein, Doklady URSS **28** (1940), 13–17; M. Krein and S. Krein, Mat. Sb. **13** (1943), 1–38. For the Krein–Milman Theorem, see M. Krein and D. Milman, Studia Math. **9** (1940), 133–7.

Theorem to the closed convex cone V^+ (which has a nonvoid interior), one gets the following result.

KREIN REPRESENTATION THEOREM. *There is a (natural) monomorphism from any integrally closed directed vector space V with strong unit, into the space $C(\bar{M}_e)$ of all continuous real functions on $\bar{M}_e$.*

Note that, relative to the strong unit e chosen, the infimum $\|x\|_e$ of all λ with $|x| \leqq \lambda_e$ defines a "uniform norm" on V. If V is an (Archimedean) *vector lattice* (with strong unit e), then $x \wedge y = 0$ imply $\|x \vee y\|_e = \|x\|_e \vee \|y\|_e$; hence V is an (M)-space in the sense of §15 below.

7. Dual Spaces

The concept of the *dual* (or "conjugate") of a directed vector space has many applications; it can be defined in general terms as follows ([**LT1**, §146], [**K-N**, p. 225]).

Let (V, P) be a directed vector space, and let P^* be the set of linear functionals on V which are nonnegative on P. Then P^* is a convex cone in the set V^* of all *differences* $f - g$ of nonnegative linear functionals $f, g \in P^*$. Hence (V^*, P^*) is a directed vector space; it is by definition the *dual* of (V, P), and its elements are called *bounded* linear functionals on V.

If (V, P) is integrally closed and *finite-dimensional*, then (V^*, P^*) has the same properties (and linear dimension). Moreover there is a natural isomorphism $(V, P) \cong (V^{**}, P^{**})$ between (V, P) and its second dual, considered as directed vector spaces. This duality plays an important role in the theory of linear programming. However, we shall concentrate our attention below on the case of *vector lattices* of unrestricted dimension.

On vector lattices, the modular identity $x + y = (x \wedge y) + (x \vee y)$ makes any linear functional $v[x]$ satisfy the identity

$$v[x] + v[y] = v[x + y] = v[(x \wedge y) + (x \vee y)] = v[x \wedge y] + v[x \vee y]. \tag{13}$$

as well as $v[0] = 0$. This proves the first statement of

LEMMA 1. *On any vector lattice V, every linear functional is a valuation satisfying $v[0] = 0$, and every isotone valuation is a positive functional.*

The converse is not true: on any simply ordered vector space, *every* functional is a valuation, yet very few functionals are linear.

Some isotone linear functionals are lattice-morphisms from V to $\mathbf{R}$, but such functionals are exceptional. For instance, among the isotone linear functionals $ax + by$ with nonnegative a, b on $\mathbf{R}^2$, only those with $a = 0$ or $b = 0$ are morphisms of lattices.

It follows from Lemma 1 that the *Jordan decomposition* of §X.6 can be applied to linear functionals on any vector lattice V. Hence one can construct the positive and negative parts v^+ and v^- of any *bounded* linear functional. This proves

THEOREM 10. *The bounded linear functionals on any vector lattice V form another vector lattice* V^*.

For any bounded linear functional ϕ on V, let $Z(\phi)$ denote the subspace of all $f \in V$ with $\phi(f) = 0$. Define the *absolute null-space* of any vector lattice V as the intersection $Z = \bigcap Z(\phi)$. This is analogous to the subspace $\bigcap K_\alpha$ of the intersection of the maximal proper l-ideals of V (the "radical" of V), discussed at the start of §3. It is the set of vectors in the null-space of every bounded linear functional.

We now define a vector lattice to be **R**-*separable* when its absolute null-space is 0. Namioka [**K-N**, p. 226] has shown that if P is closed in the strongest locally convex topology for V, then (V, P) is **R**-separable. In §12, we will show that any Banach lattice is **R**-separable. We now prove

THEOREM 11. *If V is any* **R**-*separable vector lattice, then V is naturally isomorphic to a vector sublattice of* $(V^*)^*$.

PROOF. Let $f \in V$ and define $\psi_f(\phi) = \phi(f)$ for all $\phi \in V^*$. ψ_f is clearly a linear functional on V^*. We shall prove that ψ_f is a bounded linear functional, by showing that it is the difference of two positive linear functionals. (By the Jordan decomposition, a valuation is bounded if and only if it is the difference of two isotone valuations.) But $\psi_f(\phi) = \phi(f) = \phi(f^+) - \phi(-f^-) = \psi_{f^+}(\phi) - \psi_{-f^-}(\phi)$. Since $f^+ \geqq 0$, $\phi \geqq 0$ implies $\phi(f^+) \geqq 0$ so that ψ_{f^+} is positive. Similarly, ψ_{-f^-} is positive so that ψ_f is bounded for each $f \in V$.

Since $f \geqq 0$ implies $\psi_f \geqq 0$, ψ_f is isotone. It is one-one because if $\psi_f(\phi) = 0$ for all $\phi \in V^*$, then $\phi(f) = 0$ for all ϕ and so $f \in \bigcap Z(\phi) = \{0\}$. To prove that the linear mapping $\gamma^2: f \to \psi_f$ is a *monomorphism of l-groups*, as claimed in the theorem, it remains by §XIII.3 to show that $\psi_{f^+} = (\psi_f)^+$. Moreover, since γ^2 is isotone (proved above), it is obvious that $\psi_{f^+} \geqq \psi_f$ and $\psi_f \geqq 0$, and so that $\psi_{f^+} \geqq \psi_f \vee 0 = (\psi_f)^+$. Therefore it remains to prove that $(\psi_f)^+ \geqq \psi_{f^+}$; this we now do.

By Theorem XIII.20, f^+ and f^- belong to complementary l-ideals J and $J^\perp$ of V. Hence from any $\phi \geqq 0$ in V^*, we can construct $\phi_J \geqq 0$ by

$$\phi_J(g) = \begin{cases} \phi(g) & \text{for all } g \in J, \\ 0 & \text{for all } g \in J^\perp. \end{cases} \tag{14}$$

In particular, $\phi_J(f^+) = \phi(f^+)$, $\phi_J(f^-) = 0$, and $\phi \geqq \phi_J \geqq 0$. Now suppose $\psi \geqq \psi_f$ and $\psi \geqq 0$. Then, for any $\phi \geqq 0$, we will have

$$\psi(\phi) \geqq \psi(\phi_J) \geqq \psi_f(\phi_J) = \phi_J(f) = \phi_J(f^+) = \phi(f^+) = \psi_{f^+}(\phi).$$

Therefore, $(\psi_f) + (\phi) \geqq \psi(\phi) \geqq \psi_{f^+}(\phi)$ for all $\phi \geqq 0$, which is to say $(\psi_f)^+ \geqq \psi_{f^+}$. Q.E.D.

Thus $V \subset (V^*)^*$ for any **R**-separable vector lattice, and so any **R**-separable V can be represented as a vector lattice of bounded linear functionals. Vector lattices whose absolute null-space is 0 and which satisfy $V = (V^*)^*$ are called *reflexive*.

Exercises for §§6–7:

1. Show that a vector lattice is Archimedean if and only if it is integrally closed.
2. Show that, if $(x, y) > (0, 0)$ is defined to mean that $x > 0$ and $y > 0$, the real plane is a directed Archimedean vector space which is not integrally closed.
3. Let V be any partly ordered topological vector space in which the positive cone P is topologically closed. Show that (V, P) is integrally closed.
4. Show that any subspace of an Archimedean partly ordered vector space is Archimedean.
5. Show that any subdirect product of Archimedean directed vector spaces is an Archimedean directed vector space.

*6. Let $C = C(0, 1]$ be the vector lattice of all continuous functions on $0 < x \leqq 1$. Show that the set B of all bounded continuous functions is an l-ideal of C. Show that C is integrally closed, whereas C/J is not Archimedean.

7. Using a "Hamel basis" with respect to rational scalars, construct an additive function $\mathbf{R} \to \mathbf{R}$ which is not bounded.
8. Without assuming §X.6, show how to decompose any bounded (real) linear functional on a vector lattice into its positive and negative components.

8. Complete Vector Lattices

Many of the most important vector lattices are (conditionally) complete. Thus

LEMMA 1. *Any direct sum (cardinal product) of complete (σ-complete) vector lattices is complete (resp. σ-complete).*

We omit the proof. Further, since l-ideals are convex, and any convex subset of a conditionally complete lattice is itself conditionally complete, we have

THEOREM 12. *Any l-ideal of a complete (σ-complete) vector lattice is itself complete (resp. σ-complete).*

It is a corollary that the space $B(X)$ of all functions bounded on any domain X is a complete vector lattice. Thus the Banach space (b) of all bounded sequences is a complete vector lattice, as an l-ideal of $\mathbf{R}^d$.

The vector lattice $C[0, 1]$ of all continuous functions on $0 \leqq x \leqq 1$ is however not complete; neither is the vector lattice $M[0, 1]$ of all functions measurable on $[0, 1]$; the same is moreover true if $[0, 1]$ is replaced by any other nontrivial, nonatomic measure space Ω. On the other hand, by Lemma 1, the set $N[0, 1]$ of all functions vanishing except on a set of measure zero, being an l-ideal of the complete vector lattice $\mathbf{R}^c$ of *all* real functions on $[0, 1]$ (and hence of $M[0, 1]$) is a complete vector lattice.

What is more important, the quotient-space (and vector lattice) $V(\Omega) = M(\Omega)/N(\Omega)$ of all measurable functions modulo null functions is complete. We now prove this result for the special case that Ω is the measure algebra of Borel-measurable subsets of the line $(-\infty, +\infty)$, and we let $\bar{M} = M(-\infty, \infty)/N(-\infty, \infty)$ denote the complete measure algebra of such sets modulo null sets. Let $X(f, a)$ denote the set on which $f(x) \leqq a$. Then $X(f, a)$ is an order-preserving function from the infinite interval $\mathbf{R}$: $(-\infty, +\infty)$ to the complete Boolean algebra B_σ/N of Theorem XI.10. Moreover if $X(f, a) = X(g, a)$ for all a, then the set on which $|f(x) - g(x)| < 1/n$ for all n is null; hence $f(x) - g(x)$ is a null function and $\bar{f} = \bar{g}$ in B_σ/N.

Again, for any $f \in V(-\infty, \infty)$, $\inf_a X(f, a) = \phi$ and $\sup_a X(f, a) = \mathbf{R}$. Conversely, any $X(a)$ having the properties just described is an $X(f, a)$—and $f(x)$ can be constructed through approximating step-functions assuming countably many distinct values. Since, finally $f \geqq g$ is equivalent to $X(f, a) \leqq X(g, a)$ for all a, we conclude

Lemma 2. *The space* $V(-\infty, \infty) \cong M(-\infty, \infty)/N(-\infty, \infty)$ *is lattice-isomorphic with the set of functions from* $\mathbf{R}$ *to* B_σ/N *which are isotone and satisfy* $\inf_a X(a) = 0$ *and* $\sup_a X(a) = I$.

But this is a convex subset of the lattice $(B_\sigma/N)^{\mathbf{R}}$, in the notation of §III.1, and the latter is complete since B_σ/N is. Hence

Theorem 13. *The space of measurable functions modulo null functions is a complete vector lattice.*

It is a corollary that the vector lattices L^p and $M \cap B/N \cap B$ are complete, since they are l-ideals (convex subsets) of the space $M(\Omega)/N(\Omega)$.

Exercises:

1. Prove Lemma 1 from first principles.
2. Prove that any convex sublattice of a conditionally σ-complete lattice is itself a conditionally σ-complete lattice.
3. Show that, in any σ-complete vector lattice, the operation of multiplication by a real scalar λ may be defined in terms of addition and order, and so need not be introduced as an undefined primitive concept.
4. Show that, for a finite-dimensional vector lattice V, the following conditions are equivalent: (i) V is complete, (ii) V is σ-complete, (iii) V is Archimedean.
5. Define l-ideals l^p of $\mathbf{R}^\omega$ analogous to $L^p(0, 1)$. (*Hint.* Construct a suitable atomic measure space.)
6. Show that the vector lattice $C(X)$ of Example 3 is not σ-complete unless X is discrete.
7. (*a) Show that the vector lattice $M(0, 1)$ is (conditionally) σ-complete but not complete.
   (b) Show that $M(0, 1)/N(0, 1)$ has maximal l-ideals, but no maximal (proper) *closed* l-ideal.
8. Show that $L^2(0, 1)$ is isomorphic as a vector lattice to $L^2(-\infty, \infty)$.
9. Show that $L^2(0, 1)$ and l^2, though isomorphic as Euclidean vector spaces (inner product spaces), are not isomorphic as vector lattices. (*Hint.* Look for minimal l-ideals and maximal closed l-ideals.)

9. Order-convergence; Components of Weak Units

We next extend to vector lattices the results of Theorem XIII.26 concerning the continuity of all l-group operations with respect to order-convergence. Since the topology of $\mathbf{R}$ is sequential, we consider only sequences, proving

Lemma 1. *In any* σ-*complete vector lattice* V:

$$\text{(15)} \qquad \textit{if } \lambda_n \to \lambda \quad \textit{and} \quad f_n \to f, \quad \textit{then} \quad \lambda_n f_n \to \lambda f.$$

Proof. $|\lambda_n f_n - \lambda f| \leqq (\sup |\lambda_n|)\cdot|f_n - f| + |\lambda_n - \lambda|\cdot|f|$. The first summand tends to zero since multiplication by the constant scalar $\sigma = \sup |\lambda_n|$ is an automorphism of V for $\sigma > 0$, preserving all intrinsic topologies; the case $\sigma = 0$ is trivial. By Theorem XIII.26, it therefore suffices to show that the second summand tends to zero. Hence, since the case $|f| = 0$ is trivial, it suffices to prove

(16) If $g > 0$ and $\mu_n \to 0$, then $\mu_n g \to 0$.

But if $h \leqq \mu_n g$ for a sequence of μ_n tending to zero, then $h \ll g$; hence $h = 0$ since V (being σ-complete) is Archimedean. Hence $\mu_n g \to 0$ in (16), completing the proof.

In summary, we have proved

THEOREM 14. *In any σ-complete vector lattice, all (finitary) vector lattice operations are continuous with respect to order-convergence.*

Caution. Although the preceding result also implies continuity with respect to star-convergence, this need *not* imply continuity in an equivalent neighborhood topology. It is a corollary, by Edgar's Theorem (p. 220, below Urysohn's Theorem) that all vector lattice operations are also continuous with respect to star-convergence.

We will now apply some results from §XIII.11 to any *σ-complete vector lattice V with weak unit e*, constructing a Borel algebra of "components" of e. If S is any closed l-ideal in V, then $V \cong S \oplus S^*$, where $S^* = \perp S$ is the complement of S in $\Theta(V)$. For the given weak unit e of V, we now let e_S be the component of e in S for the direct decomposition $V \cong S \oplus S^*$. In particular, we have

(17) $$e_S = \sup \{e \wedge s | x \in S\} \in S.$$

It is also evident from the decomposition that $e_{S^*} = e - e_S$.

THEOREM 15. *In any σ-complete vector lattice, the mapping $S \to e_S$ is a lattice isomorphism from the Borel algebra of closed l-ideals onto the Boolean lattice of all "components" of e.*

PROOF. From the decomposition, it is evident that $e_S \wedge e_{S^*} = 0$ and $e_S + e_{S^*} = e_S \vee e_{S^*} = e$. Also, by Theorem I.10, the complemented elements of the distributive lattice defined by the interval $[0, e]$ form a Boolean lattice. Further, it is obvious from (17) that the mapping $S \to e_S$ is isotone from the Borel algebra of closed l-ideals into the Boolean lattice of components of e. To prove Theorem 15, it therefore suffices to show that the mapping (17) is one-one onto. This is true because, for any component f of e such that $f \wedge (e - f) = 0$, the set of all elements $x \in V$ such that

(18) $$\lim_{n \to \infty} nf \wedge |x| = |x|$$

is a closed l-ideal $S(f)$ of V with complement $S(e - f)$; we omit the details. Finally, $e_{S(f)} = f$ and $S(e_S) = S$.

10. Stieltjes Integral Representation

Let V be a σ-complete vector lattice with weak unit e, as in §8. We define a *resolution* of e as a family of components e_λ of e, such that

(19) $$\lambda \leqq \mu \text{ implies } e_\lambda \leqq e_\mu,$$

† See E. E. Floyd, Pacific J. Math. 5 (1955), 687–9; also §X.11.

and, in the intrinsic order topology,

$$(20) \qquad \operatorname*{Lim}_{\lambda \to -\infty} e_\lambda = \bigwedge e_\lambda = 0, \qquad \operatorname*{Lim}_{\lambda \to \infty} e_\lambda = \bigvee e_\lambda = e.$$

By §8, this amounts to choosing a closed l-ideal E_λ for each real number λ, such that

$$(21) \qquad \lambda < \mu \text{ implies } E_\lambda \subset E_\mu, \qquad \bigcap E_\lambda = 0, \qquad \bigcup E_\lambda = V.$$

Relative to any such resolution of e, we can define a *Stieltjes integral* $\int \lambda de_\lambda$ as follows. For any finite real number M and any partition π of the interval $[-M, M]$ into subintervals $[\lambda_{i-1}, \lambda_i]$, we define

$$u_\pi = \sum_i \lambda_{i-1}(e_{\lambda_i} - e_{\lambda_{i-1}}), \qquad v_\pi = \sum_i \lambda_i(e_{\lambda_i} - e_{\lambda_{i-1}}),$$

thinking of them as lower and upper bounds to the integral, respectively. One easily shows that for any π, π',

$$u_\pi \leqq u_{\pi \wedge \pi'} \leqq v_{\pi \wedge \pi'} \leqq v_{\pi'};$$

moreover $0 = v_{\pi'} - u_\pi \leqq 2\epsilon e$, where ϵ is the length of the largest subinterval of π or π'. Hence, since $2\epsilon e \downarrow 0$ as $\epsilon \downarrow 0$, the u_π and v_π approach the same limit g, which we *define* as the meaning of the symbol $\int_{-M}^{M} \lambda de_\lambda$. We then define

$$(22) \qquad \int_\lambda^{-\lambda} \lambda de_\lambda = \operatorname*{Lim}_{n \to \infty} \int_n^{-n} \lambda de_\lambda$$

if this limit exists (in the sense of order convergence).

We now show that any $f \in V$ has a Stieltjes integral representation of the preceding type. This was first shown by Freudenthal [**1**], making various redundant assumptions. The construction is analogous to that of self-adjoint operators on Hilbert space, in terms of an appropriate "resolution of the identity".

Given $f \in V$, define $E_\lambda^\perp(f)$ as the closed l-ideal of V having $(\lambda e - f)^+$ for weak unit, and let $E_\lambda(f)$ be its complement. It is easy to prove (21) (from the fact that e is a weak unit); hence the $E_\lambda(f)$ form a resolution of the identity. Further,

$$(23) \qquad \lambda e \leqq f \leqq (\lambda + \Delta\lambda)e \quad \text{on } E_{\lambda+\Delta\lambda}(f) \cap E'(f).$$

Now set $\lambda_k = k/2^n$ with n a fixed positive integer; $J_k = E_{\lambda_{k-1}} \cap E_{\lambda_k}^\perp$; and $\Delta e_k(f) = e_{\lambda_k} - e_{\lambda_{k-1}}$ as the corresponding components of e. We will then have on the component J_k, since $e = \Delta e_k$ there,

$$(24) \qquad f - 2^{-n}e \leqq \lambda_{k-1}\Delta e_k(f) \leqq f \leqq \lambda_k \Delta e_k(f) = f + 2^{-n}e.$$

Now taking the countable linear sum over the disjoint components J_k of V, which exists since the extreme sums are bounded, we obtain for the partition π_n, $f - 2^{-n}e \leqq u_{\pi_n} \leqq v_{\pi_n} \leqq f + 2^{-n}e$. Therefore,

$$(25) \qquad \int_{-\infty}^{\infty} \lambda de_\lambda(f) = \operatorname*{Lim}_{n \to \infty} \sum \lambda_k \Delta e_k(f) = f, \quad \text{any } f \in V.$$

In summary, we have proved

THEOREM 16. *For any weak unit e in a σ-complete vector lattice V, each element f has a Stieltjes integral representation of the form* (25).

Stone representation. Since the closed l-ideals of V form a Boolean algebra, they can also be identified with the open-and-closed subsets of a (zero-dimensional, compact) space, the *Stone space* $S(V)$ of V. It is natural to try to interpret the representation (25) in terms of suitable functions on this space. The components of e appear as continuous functions on V; if e is identified with the function $e(x) \equiv 1$, the convergence in (25) becomes uniform.

One might think, therefore, that each f would also be represented as a continuous function. However, this cannot be since continuous functions on compact spaces must be bounded. Moreover the "uniform" convergence above is *relative* to the particular choice of weak unit e (cf. §12); hence there is no unique natural representation of σ-complete vector lattices by continuous functions.

The situation becomes much simpler, however, if V has a *strong unit.* "Uniform" convergence has the same meaning relative to every strong unit, and every function is bounded under the representation (25) if the unit e is identified with $e(x) \equiv 1$ on $S(V)$. One then obtains various representation theorems (see §§15–17 below).

Exercises for §§9–10:

1. Show that if e is a weak (strong) unit of a partly ordered vector space, then so are all its positive scalar multiples.
2. (a) Show that $L^p(-\infty, \infty)$ has a weak unit for all p, $1 \leqq p \leqq \infty$.
   (b) Prove that $L^p(0, 1)$ has a strong unit if and only if $p = \infty$.
3. Show that if $V_1, V_2, V_3, \cdots$ are vector lattices with weak units, then the same is true of their unrestricted product ($=$ direct sum as a vector space).
4. Show that the restricted cardinal product $(\mathbf{R}^{\omega})$ of a countable number of copies of $\mathbf{R}$ has no weak unit.
5. Prove that any σ-complete vector lattice satisfies the following generalized Cauchy condition: $\{f_n\}$ is order-convergent if and only if $\operatorname{Lim}_{m,n} |f_m - f_n| = 0$.

*6. Show how to construct a *Cantor* completion of any vector lattice, consisting of its Cauchy sequences modulo its null sequences.†

7. Call a vector lattice, "σ-full" if, whenever $a_i \wedge a_j = 0$ for all $i \neq j$, $\bigvee a_i$ exists. Show that the vector lattice of all Lebesgue-measurable function $(-\infty, \infty)$ is σ-full.
8. In $L^p(0, 1)$, what is the meaning of (25) for $e(x) \equiv 1$? Discuss the effect of varying p.

11. Bounded Linear Functions

A basic theorem of algebra states that the set of *all* linear functions $\phi: V \to W$ from the (real) vector space V to a second vector space W is itself a vector space, often denoted $\operatorname{Hom}_{\mathbf{R}}(V, W)$. If V and W are partly ordered vector spaces, then the definition

$$(26) \qquad \phi \geqq 0 \text{ means that } \phi \text{ is isotone,}$$

† See C. J. Everett, Duke Math. J. **11** (1944), 109–19; [**Fu**, V.11]; B. Banaschewski, Math. Nachr. **16** (1957), 51–71.

makes $\mathrm{Hom}_{\mathbf{R}}(V, W)$ into a partly ordered vector space; the proof is straightforward (cf. [**LT2**, p. 244, Theorem 6]), and will be omitted.

For applications to real analysis, the preceding vector space is much too large; a more appropriate set of functions in most cases seems to consist of those which are *bounded* in the following sense.

DEFINITION. A linear function $\phi: V \to W$ between directed vector spaces is *bounded* when the set $\{\phi, 0\}$ is bounded in $\mathrm{Hom}_{\mathbf{R}}(V, W)$.

LEMMA 1. *Any bounded linear function maps bounded sets into bounded sets.*

PROOF. Let $S \subset [a, b] \subset V$ be given, and let ψ, ψ' be upper and lower bounds to $\{\phi, 0\}$ in $\mathrm{Hom}_{\mathbf{R}}(V, W)$. Then, for all $x \in [a, b]$:

$$\phi(x) = \phi(a) + \phi(x - a) \leqq \phi(a) + \psi(x - a) \leqq \phi(a) + \psi(b - a),$$

and dually $\phi(x) \geqq \phi(b) + \psi'(b - a)$. Hence the image $\phi(S)$ of S under $\phi: V \to W$ is a subset of the interval $[\phi(b) + \psi'(b - a), \phi(a) + \psi(b - a)]$ and so is bounded.

We now restrict attention to the special case of bounded linear functions $\phi: V \to W$, where V is a vector lattice and W a *complete* vector lattice. (The case $W = \mathbf{R}$ of the dual space of a vector lattice (§6) is included.)

DEFINITION. For given ϕ as above, define ϕ^+ by:

(27) $\qquad \phi^+(f) = \sup_{x \in [0, f]} \phi(x) \qquad$ if $f \geqq 0$, and

(27′) $\qquad \phi^+(g) = \phi^+(g^+) - \phi^+(-g^-) \qquad$ for any $g \in V$.

We will now show that ϕ^+ as defined by (27)–(27′) is $\phi \vee 0$. But if ψ is any additive upper bound to $\{\phi, 0\}$, then trivially $\psi \geqq \phi^+$ as defined by (27)–(27′). Conversely, $f \geqq 0$ in V clearly implies $(\phi^+ - \phi)(f) \geqq \phi(f) - \phi(f) = 0$ and $(\phi^+ - 0)(f) \geqq \phi(0) - 0(f) = 0$; hence $\phi^+ = \phi \vee 0$ in the ordering defined by (26) on $\mathrm{Hom}_{\mathbf{R}}(V, W)$, if ϕ^+ as defined by (27)–(27′) is linear.

It thus remains to show that ϕ^+ is linear. Now, it is trivial that $(\lambda\phi)^+ = \lambda\phi^+$ for any $\lambda \geqq 0$. We also observe (again leaving the proof to the reader)

LEMMA 2. *Any function $\psi: V \to W$ which is linear on the positive cone P of a vector lattice V can be extended by* (27′) *to one which is linear on V.*

It therefore remains to show that ϕ^+ is additive on positive elements. But by Exercise 9 after §XIII.4, the set of $x \in [0, g + h]$ is the same as the set of $y + z$ ($y \in [0,g]$, $z \in [0,h]$). Therefore $\sup_{[0,g+h]} \phi(x) = \sup_{y \in [0,g],\, z \in [0,h]} \phi(y + z)$ and, ϕ being linear and addition continuous in $\mathbf{R}$,

$$\sup_{[0,g+h]} \phi(x) = \sup_{[0,g]} \phi(y) + \sup_{[0,h]} \phi(z).$$

Substituting back into (26), this proves $\phi^+(y + z) = \phi^+(y) + \phi^+(z)$ for any positive y, z, as required. This completes the proof that $\phi^+ = \phi \vee 0$.

By Lemma 1 of §XIII.3, the following theorem is essentially a corollary of the existence of $\phi \vee 0 = \phi^+$ for all bounded linear $\phi: V \to W$.

THEOREM 17. *The bounded linear functions from a vector lattice to any complete vector lattice themselves form a vector lattice under the partial ordering* (26).

COROLLARY. *For linear functions $\phi: V \to W$, where V is a vector lattice and W a complete vector lattice, boundedness is equivalent to each of the following conditions:* (i) *ϕ carries bounded sets into bounded sets, and* (ii) *ϕ is the difference of nonnegative functions.*

PROOF. By Lemma 1, boundedness implies (1); by Theorem 17 above and Chapter XIII, (15), (i) implies (ii). Finally, if $\phi \geqq 0$ and $\psi \geqq 0$, then

$$-\phi - \psi = \phi - \psi \leqq \phi + \psi,$$

hence (ii) implies boundedness.

Exercises:

1. (a) Denote the partly ordered vector space of all linear functions $\phi: X \to Y$ (X, Y p.o. vector spaces) by Hom (X, Y). Discuss the distributivity of Hom with respect to formation of direct products.

(b) Same question if X and Y are complete vector lattices, and Hom (X, Y) signifies the vector lattice of all bounded linear functions $\phi: X \to Y$.

2. Define a "generalized valuation" on a lattice L as a function $v: L \to X$ (X a commutative l-group) such that $v[x] + v[y] = v[x \wedge y] + v[x \vee y]$.

(a) Show that, if L has a strictly isotone generalized valuation, then it is a modular lattice.

(b) For v any strictly isotone generalized valuation, define the "generalized distance" $|x - y| = v[x \vee y] - v[x \wedge y]$, and discuss its properties.

(c) Define $x_n \to a$ to mean that $|x_n - a| \to 0$ in X. Discuss the resulting sequential topology.†

3. In Exercise 2, let X be a complete lattice, and call v *bounded* when $v = v' - v''$, where v', v'' are isotone. Construct a "generalized Jordan decomposition" $v = v^+ + v^-$, analogous to that of §X.6.

12. Banach Lattices

A real Banach space is a real vector space V with a real norm function $\|f\|$ such that $\|\lambda f\| = |\lambda| \cdot \|f\|$ and such that the "distance" $\|f - g\|$ makes V into a complete metric space. Every familiar Banach space is also a vector lattice in a natural sense. Moreover, in all familiar examples:

$$(28) \qquad |f| \leqq |g| \quad \text{implies} \quad \|f\| \leqq \|g\|.$$

These facts motivate the following definition (Kantorovich [**1**, §9]).

DEFINITION. A *Banach lattice* is a real Banach space which is also a vector lattice, in which the order and the norm are related by the implication (28).

The relation (28) trivially implies

$$(29) \qquad \|f\| = \| |f| \| \quad \text{for all } f.$$

We shall now prove a few facts about Banach lattices.

LEMMA 1. *No vector lattice V which contains a sequence $\{f_n\}$ such that all sequences $\{\lambda_n f_n\}$ are order bounded can be made into a Banach lattice.*

PROOF. Suppose V were made into a Banach lattice, and set $\lambda_n = n/\|f_n\|$.

---

† For further results, see L. M. Blumenthal, Palermo Rend. **1** (1952), 343–60 and **10** (1961), 175–92.

(We may assume $f_n \neq 0$ without loss of generality.) If $|\lambda_n f_n| \leqq u$ for all n, then $n = |\lambda_n| \cdot \|f_n\| = \|\lambda_n f_n\| = \| |\lambda_n f_n| \| \leqq \|u\|$ for all n, which is impossible.

COROLLARY. *Neither of the vector lattices* $\mathbf{R}^\omega$ *or* $\mathbf{R}^c$ *can be made into a Banach lattice.*

LEMMA 2. *The operations* $f + g, f \wedge g, f \vee g$ *are Lipschitz continuous in any Banach lattice.*

PROOF. From commutativity and Chapter XIII, (25), we have

$$|(f \circ h) - (g \circ h)_1| \leqq |f - g|,$$

where $\circ$ denotes $+$, $\wedge$, or $\vee$. By (28), this implies the corresponding metric inequality; hence all three operations are Lipschitz continuous, with modulus of continuity one, in the usual metric $\|f - g\|$.

COROLLARY. *In any Banach lattice* V, *the positive cone* P *and all intervals* $[a, b]$ *are metrically closed sets. Moreover* $a_n \to a$, $b_n \to b$, $x_n \to x$, *and* $a_n \leqq x_n \leqq b_n$ *for all* n *imply* $a \leqq x \leqq b$. *(Here metric convergence is intended.)*

LEMMA 3. *Any separable Banach lattice has a weak unit* e.

PROOF. V separable means there is a countable metrically dense subset $\{f_n\}$ of V. Define

$$e = \sum_{n=1}^{\infty} \frac{|f_n|}{2^n \|f_n\|} \quad \text{summed over } f_n \neq 0.$$

This is a well-defined element of V because the sequence of partial sums is a Cauchy sequence, and V is a complete metric space. Let g be an element > 0. Since $\{f_n\}$ is dense in V, there is an n such that $\|g - f_n\| < \|g\|$. If $g \wedge e = 0$, then $g \wedge ke = 0$ for any real number λ. Hence $0 \leqq g \wedge |f_n| \leqq g \wedge (2^n\|f_n\|)e = 0$, and so $g \wedge f_n^+ = 0$. But this contradicts the fact that, by Lemma 2:

$$\|g\| = \|(g \wedge g) - (g \wedge f_n^+)\| \leqq (g \vee 0) - (f_n \vee 0)\| \leqq \|g - f_n\| < \|g\|.$$

That is, $g > 0$ implies $g \wedge e > 0$, and so e is a weak unit.

We now consider the relation between metric boundedness and order boundedness, in Banach lattices.

LEMMA 4. *In any Banach lattice* L, *every subset which is order-bounded is metrically bounded.*

PROOF. If x and y are in $[a, b]$, then so are $x \wedge y$ and $x \vee y$. Hence $|x - y| \leqq |b - a|$ and $\|x - y\| \leqq \|b - a\|$. This shows that the diameter of any interval $[a, b]$ is finite, being in fact $\|b - a\|$.

LEMMA 5. *In a Banach lattice* L, *a positive element* e *is a strong unit if and only if some multiple* $[-ne, ne] = n[-e, e]$ *of* $[-e, e]$ *contains the unit ball* U.

PROOF. If $U \subset n[-e, e]$, then $|f| \leqq n\|f\|e$ for all $f \in L$, since $|f|/\|f\| \in U$. Hence e is a strong unit.

Conversely, suppose $n[-e, e]$ fails to contain U for any n. Then we can find a sequence of elements $f_k \in U$ such that $2^{-2k}|f_k|$, fails to be bounded by e. Form $g = \Sigma 2^{-k}|f_k|$; since L is complete, this will exist and contain every $|f_k|$. Hence $2^{-k}g \geqq 2^{-2k}|f_k|$ cannot be contained in e for any k, and so e is not a strong unit.

COROLLARY 1. *A Banach lattice L has a strong unit if and only if its unit ball U is order-bounded.*

PROOF. If $U \subset [a, b]$, then $|a| + |b|$ is a strong unit of L by Lemma 2. Conversely, if U has a strong unit e, then $U \subset [-ne, ne]$.

COROLLARY 2. *If order-boundedness is equivalent to metric boundedness in a Banach lattice L, then L has a strong unit.*

Conjugate spaces. A linear functional ϕ on a Banach space B is called (metrically) *bounded* when there is a constant C such that $|\phi(f)| \leqq C\|f\|$ for all $f \in B$. It is classic that the set of all such bounded linear functionals on a given Banach space forms a *conjugate* Banach space B^*, and that $B \subset (B^*)^*$ (Hahn–Banach Theorem). We next show that, if B is a Banach lattice,† this conjugate B^* is the same as the order conjugate B^* already constructed in Theorem 10.

THEOREM 18. *Metric boundedness and order-boundedness are equivalent for linear functionals ϕ on a Banach lattice.*

PROOF. If ϕ is metrically bounded, then for any $a > 0$ in the given Banach lattice B, $|x| \leqq a$ implies by (28) $\|x\| \leqq \|a\|$, and hence $|\phi(x)| \leqq C\|x\| \leqq C\|a\|$. This shows that ϕ carries order-bounded sets in B into order-bounded sets in **R**; hence that ϕ is order-bounded. Conversely, if ϕ is a metrically unbounded linear functional, then there is a sequence $\{x_n\}$ such that $\|x_n\| \leqq 2^{-n}$ yet $\|\phi(x_n)\| \to +\infty$. By the corollary of Lemma 2 the elements $a = \sum_{n=1}^{\infty} x_n^-$ and $b = \sum_{n=1}^{\infty} x_n^+$ bound a set of y on which $\phi(y)$ is not metrically bounded and hence order-unbounded. Thus ϕ is not order-bounded. Q.E.D.

This proves, as stated above, that the (conjugate space)B^* of B is the same for the vector lattice as for the Banach space of B. But any Banach space is **R**-separable, because given any $x \neq 0$ one knows by the Hahn–Banach Theorem that there exists a bounded linear functional ϕ such that $\phi(x) \neq 0$. We infer:

COROLLARY 1. *Any Banach lattice is* **R***-separable. Any reflexive Banach space which is a Banach lattice is a reflexive vector lattice.*

Combining Theorems 18 and 10, we have also

COROLLARY 2. *The conjugate of any Banach lattice is again a Banach lattice.*

Exercises:

1. Prove in detail the Corollary of Lemma 1.
2. Show that the vector lattice $M(-\infty, \infty)/N(-\infty, \infty)$ of all measurable functions modulo null functions cannot be made into a Banach lattice.

† The assumption of metric completeness is redundant; see [**K-N**, Theorem 24.3]. However, for simplicity, we shall not in general try to dispense with this assumption below.

3. Show that no "full" vector lattice with a countable set of disjoint positive elements a_i can be made into a Banach lattice.

4. In a vector lattice V with strong unit e, define

$$(*) \qquad \|f\|_e = \sup \{\lambda | f \in [-\lambda e, \lambda e]\}.$$

Show that this makes V into a Banach lattice.†

5. (a) Show that if f_n is a decreasing sequence of elements of a Banach lattice which converges metrically to 0, then it order-converges to 0.

(b) Show that, in any Banach lattice, a metrically convergent sequence star-converges to the same limit.

6. Let V be any Banach lattice in which $u_n \downarrow 0$ implies that some sequence $\{ku_{n(k)}\}$ is order-bounded. Show that order-convergence in V implies metric convergence, and that star-convergence is equivalent to metric convergence.

*7. Prove that, for a complete Banach lattice, the following three conditions are equivalent:‡ (i) every bounded linear functional is order-continuous, (ii) $f_n \downarrow 0$ implies $\|f_n\| \to 0$, (iii) for nets $\{f_\alpha\}$ $f_\alpha \downarrow 0$ implies $\|f_\alpha\| \to 0$.

13. Relative Uniform Convergence

Order-convergence in vector lattices was discussed in §9; the extension of order-convergence to star-convergence in general lattices had been previously studied in Chapter X; and the extension of the latter to a neighborhood topology had been considered in §IX.6. In §X.12, we also constructed an intrinsic interval topology in an arbitrary bidirected set. This provides another interesting intrinsic topology in any vector lattice V, which is generally weaker than that defined by order-convergence, and indeed seems analogous to the "weak" topology of Banach.§

We now define a third intrinsic topology for Archimedean directed vector spaces, following a basic idea of E. H. Moore. We shall say that a sequence $\{f_n\}$ of elements of a directed vector space converges *relatively uniformly* to an element f if and only if, for some u and $\lambda_n \downarrow 0$, $|f_n - f| \leqq \lambda_n u$.

LEMMA 1. *In any Archimedean directed vector space, $f_n \to f$ and $g_n \to g$ imply $f_n \circ g_n \to f \circ g$ for relative uniform convergence, whether $\circ$ is $\wedge$, $\vee$, or $+$.*

PROOF. Let $|f_n - f| \leqq \lambda_n u$ and $|g_n - g| \leqq \mu_n v$; then $(\lambda_n + \mu_n)(u + v)$ bounds the relevant differences.

Next, we shall correlate relative uniform convergence with order-convergence.

LEMMA 2. *In a σ-complete vector lattice, a sequence $\{f_n\}$ order-converges to f if and only if $|f_n - f| \leqq w_n$ for some $w_n \downarrow 0$.*

PROOF. Since $x \to x - f$ is a lattice automorphism which leaves absolutes and differences unchanged, we can assume that $f = 0$. In this case $w_n = \bigvee_{k \geqq n} |f_k|$ provides the desired sequence. The converse is immediate since $\limsup |f_n| \leqq \lim w_n = 0$ and dually.

---

† For a generalization, see M. H. Stone, Ann. of Math. **48** (1947), 851–6.

‡ The result of Exercise 7 is due to Ogasawara; see also M. Nakamura, Tôhoku Math. J. **1** (1949), 100–8.

§ In this topology, $f_\alpha \to f$ if and only if $\phi(f_\alpha) \to \phi(f)$ for every $\phi \in V^*$.

THEOREM 19. *In any Archimedean vector lattice V, relative uniform convergence implies order-convergence. The two are equivalent if $u_n \to 0$ implies that some sequence $\{ku_{n(k)}\}$ is order-bounded in V.*

PROOF. If $|f_n - f| \leqq \lambda_n u$ and $\lambda_n \downarrow 0$, then $\{f_n\}$ order-converges to f. Conversely if $|f_n - f| \leqq u_n$ and if u is an upper bound to the $ku_{n(k)}$, then $|f_n - f| \leqq u/k$ for all $n > n(k)$.

Relative uniform star-convergence is defined (as in §IX.6) to be relative uniform convergence of some subsequence of every subsequence.

THEOREM 20. *In any Banach lattice, metric convergence is equivalent to relative uniform star-convergence.*

PROOF. By the homogeneity of both topologies, we need only consider convergence to 0. But if $|f_n| \leqq \lambda_n u$ and $\lambda_n \downarrow 0$ then $\|f_n\| \leqq \lambda_n \cdot \|u\| \downarrow 0$. Thus relative uniform star convergence implies metric star-convergence and so metric convergence (because if a sequence of real numbers x_n fails to converge to y then there is a subsequence $x_{n(k)}$ bounded away from y and so every $x_{n(k(i))}$ is bounded away from y). Conversely if $\|f_n\| \to 0$ we can choose $n(k)$ such that $k^3\|f_{n(k)}\| \to 0$ and then construct $V = \sum_{k=1}^{\infty} kf_{n(k)}$ with $|f_{n(k)}| \leqq |V|/k$.

It follows that, up to linear topological equivalence, the norm in any Banach lattice X is determined by its vector lattice properties alone! Therefore one can use order as a *substitute* for the norm!

Relative uniform topology. As has been shown by High Gordon [1], relative uniform *convergence* is closely associated with the following relative uniform *topology* first introduced by I. Namioka [I].

DEFINITION. In any integrally closed directed vector space, define the *relative uniform topology* as the finest locally convex topology in which every order-bounded set is topologically bounded. Gordon [1, p. 421, foot] has shown that this relative uniform topology is the finest *locally convex* topology such that, if $f_n \to f$ relatively uniformly, then $f_n \to f$ topologically. It is also the Mackey topology associated with the set V^* of bounded linear functionals on V. (Note that, if local convexity were not required, this would be just the "polar" of relative uniform convergence, in the sense of §IX.6.) In this topology, a neighborhood-basis for 0 consists of those convex symmetric sets which contain some positive scalar multiple $[-\epsilon a, \epsilon a]$ of every *order* interval $[-a, a]$.

Namioka [1, Theorem 8.5] has shown that in Archimedean vector lattices the operations $\wedge$, $\vee$, $+$ are automatically continuous in the relative uniform topology; (for Banach lattices this has been proved above).

When (V, P) is a σ-complete $\mathbf{R}$-separable vector lattice, then it is "tonnelé" (a t-space) in the relative uniform topology. In this case, Gordon [1, Theorem 5.14] has shown that the relative uniform topology on (V, P) is the topology induced on (V, P) by $(V, P)^{**}$.

14. Uniformly Monotone Norms

The remaining three sections of this chapter will deal with special classes of Banach lattices which play a central role in applications. Among these should be mentioned the Banach lattices $L^p(\Omega)$ ($1 \leqq p \leqq \infty$), where Ω is a measure algebra, and $C(X)$, where X is a general topological space. The case $p = 2$ and $\Omega = B_\sigma/N$ (§XI.5), the algebra of measurable sets modulo null sets, giving the commonest *Hilbert space* $L^2(-\infty, \infty) \cong L^2[0, 1]$, is perhaps the most important, but it is the cases $p = 1, \infty$ of the spaces $L(\Omega)$ and their duals ("conjugates") which will be mainly considered here.

For $1 \leqq p < \infty$, the norms in $L^p(\Omega)$ and $l^p(\mathbf{N})$, where $\mathbf{N}$ is the sequence of nonnegative integers, satisfy the following condition, analogous to uniform convexity (cf. Exercise 4 below).

DEFINITION. The norm in a Banach lattice $B = (V, P, \|\,\|)$ is *uniformly monotone* when, given $\epsilon > 0$, one can find $\delta > 0$ so small that

$$(30) \qquad f, g \in P, \quad \|f\| = 1, \quad \text{and} \quad \|f + g\| \leqq \|f\| + \delta \quad \text{implies} \quad \|g\| \leqq \epsilon.$$

A Banach lattice whose norm is uniformly monotone is called a *UMB-lattice*.

Almost trivially, any metrically closed vector sublattice of a UMB-lattice is again a UMB-lattice. UMB-lattices also have various other less trivial nice properties, such as the following.

THEOREM 21. *In a UMB-lattice, any metrically bounded net of elements which is directed in the lattice order (or its dual) converges metrically.*

PROOF. We can confine attention to the successors of a fixed element $f_\alpha = a$; moreover since $f \to f - a$ is an isometric lattice-automorphism, we can assume without loss of generality that all $f_\beta (\beta \geqq \alpha)$ are nonnegative. Also, by changing the scale, we can assume that $\sup_{\beta \geqq \alpha} \|f_\beta\| = 1$. But in this case, choosing β so that $\|f_\beta\| > 1 - \delta$, we will have $\|f_\gamma - f_\beta\| < \epsilon$ for all $\gamma \geqq \beta$, proving the result.

COROLLARY. *Any UMB-lattice V is (conditionally) complete as a vector lattice.*

For, if a set $S \subset V$ has an upper bound, then so do the finite joins $\bigvee_F x_i (F \subset S)$, and these form a metrically bounded directed set.

THEOREM 22. *In any UMB-lattice, order convergence is equivalent to relative uniform convergence.*

PROOF. Let $U_n \downarrow 0$. By Theorem 21, $\{u_n\}$ converges metrically to some element a. By Lemma 2 of §12, $a \wedge 0 = 0$ and $a \vee u_n = u_n$ for all n; hence $0 \leqq a \leqq u_n$ for all n, and so $a = 0$. It follows that

$$(31) \qquad u_n \downarrow 0 \quad \text{implies} \quad \|u_n\| \downarrow 0$$

in any *UMB*-lattice. But (31) implies that, for some $n(k)$, $\|u_{n(k)}\| \leqq 1/k2^k$; hence the infinite linear sum $\sum_{i=1}^{\infty} k u_{n(k)}$ exists and is (§12, Lemma 2) an upper bound to the sequence $\{k u_{n(k)}\}$. By Theorem 19, this implies the statement to be proved.

COROLLARY. *In any UMB-lattice, star-convergence is equivalent to metric convergence.*

Now let V be any UMB-lattice and ϕ any bounded additive functional on V. We shall show that ϕ decomposes V into complementary closed l-ideals ("components") on which ϕ is positive, negative and zero respectively.

LEMMA 1. *Let N^+, N^-, and N^0 be defined as the sets of f such that $0 < x \leqq |f|$ implies $\phi(x) > 0$, $\phi(x) < 0$, and $\phi(x) = 0$, respectively. Then N^+, N^-, and N^0 are independent l-ideals.*

PROOF. Let $f, g \in N^+$. Then $0 < x \leqq |f + g|$ implies $0 < x \leqq |f| + |g|$, whence $x = y + z$, where $0 \leqq y \leqq |f|$, $0 \leqq z \leqq |g|$, and $y > 0$ or $z > 0$; hence $\phi(x) = \phi(y) + \phi(z) > 0$. Therefore N^+ is closed under addition. Again, if $f \in N^+$ and $|h| \leqq |f|$, then $0 < x \leqq |h|$ implies $0 < x \leqq |f|$, whence $\phi(x) > 0$ and $h \in N^+$. We conclude that N^+ is an l-ideal. Similarly, N^- and N^0 are l-ideals. They are trivially independent since if $f \neq 0$, then $f \in N^+$ implies $\phi(|f|) > 0$, $f \in N^-$ implies $\phi(|f|) < 0$, and $f \in N^0$ implies $\phi(|f|) = 0$.

LEMMA 2. *The functional $\phi(x)$ attains its supremum M on any interval $0 \leqq x \leqq f$.*

PROOF. Choose $x_m \in [0, f]$ so that $\phi(x_m) > M - 3^{-m}$; we shall show that $g = \lim\sup \{x_m\}$, which exists by the Corollary of Theorem 21, and satisfies $g \in [0, f]$, also satisfies $\phi(g) = M$. Indeed,

$$\phi(x_m \vee x_n) = \phi(x_m) + \phi(x_n) - \phi(x_m \wedge x_n) > M - 3^m - 3^n,$$

whence

$$\phi\left(\bigvee_{i=m}^{m+r} x_i\right) > M - (3^{-m} + \cdots + 3^{-m-r}) > M - 2\cdot 3^{-m}.$$

Passing to the limit,

$$M \geqq \phi\left(\bigvee_{k \geqq m} x_k\right) \geqq M - 2\cdot 3^{-m}.$$

Passing to the limit again as $m \to \infty$, by continuity in the metric and hence (Theorem 22, Corollary) order-topology, $\phi(g) = M$.

LEMMA 3. *In Lemma 2, let u be the infimum of the x between 0 and f such that $\phi(x) = M$; define v dually; let $w = f - u - v$. Then $u \in N^+$, $v \in N^-$, $w \in N^0$, and $f = u + v + w$.*

PROOF. The existence of u and v (and hence of w) follows from the completeness of V. Again, if $\phi(x) = \phi(y) = M$, then $\phi(x \wedge y) + \phi(x \vee y) = 2M$. But $\phi(x \wedge y) \leqq M$ and $\phi(x \vee y) \leqq M$ by hypothesis; hence $\phi(x \wedge y) = \phi(x \vee y) = M$. It follows by Theorem 19 and continuity that $\phi(u) = \phi(\bigvee_X x) = M$, where X is the set of $x_\alpha \in [0, f]$ with $\phi(x_\alpha) = M$. Moreover $0 < x \leqq u$ implies $\phi(u - x) < \phi(u)$ and so $\phi(x) = \phi(u) - \phi(u - x) > 0$; hence $u \in N^+$. Dually, $v \in N^-$. Hence $u \wedge v = 0$ by Lemma 1, whence $u + v = u \vee v \leqq f$ and $0 \leqq w \leqq f$. Finally, $0 < x \leqq w$ implies $\phi(x) + \phi(u) = \phi(x + u) \leqq \phi(u)$, whence $\phi(x) \leqq 0$; dually, it implies $\phi(x) \geqq 0$, and so $\phi(x) = 0$. Hence $w \in N^0$. But $f = u + v + w$ is obvious, completing the proof.

THEOREM 23. *If ϕ is any bounded additive functional on a UMB-lattice V, then V is the direct product of the closed l-ideals N^+, N^-, and N^0 of Lemma 1 above.*

PROOF. By Lemma 1, the l-ideals in question are independent; by Lemma 3, their sum is V.

Exercises for §§13–14:

1. Show that Lemma 1 of §14 holds in any vector lattice, whereas Lemmas 2–3 are not true in $C[0, 1]$.

2. Show that (31) fails in the Banach lattice (b) of all bounded sequences $f = (f_1, f_2, f_3, \cdots)$, with norm $\|f\| = \sup |f_i|$.

3. (a) Prove in detail that $L^p(-\infty, \infty)$ is a UMB-lattice, if $1 \leqq p < +\infty$.
   (b) Show that (30) fails in the spaces $C[0, 1]$ and in (b).

*4. A Banach space is "uniformly convex" when there exists for any $\epsilon > 0$ a $\delta > 0$ such that, if $\|f\| = \|g\| = 1$ and $\|f - g\| \geqq \epsilon$, then $\|(f + g)/2\| \leqq 1 - \delta$. Show that any uniformly convex Banach lattice is a UMB-lattice.

5. Let ϕ be any linear functional on a complete Banach lattice V such that $x_\alpha \downarrow 0$ implies $\phi(x_\alpha) \rightarrow 0$. Show that V is decomposed by ϕ as in Theorem 23.†

*6. Let V be a directed vector space with positive cone P. Show that the following conditions are equivalent: (i) C is closed in the finest locally convex topology on V, (ii) C is closed for every topology on V which makes every positive linear functional continuous, (iii) $f \in V$ is positive if (and only if) $\phi(f) \geqq 0$ for every positive linear functional ϕ on V.

*7. Prove that, for a convex symmetric subset $S = -S$ of an integrally closed directed vector space V, the following statements are equivalent:
   (i) S is a neighborhood of 0 in the relative uniform topology of V,
   (ii) If $f_n \rightarrow 0$ relatively uniformly, all but a finite number of terms of $\{f_n\}$ are in S.

*8. Prove that a linear functional ϕ on V in Exercise 6 is continuous in the relative uniform topology if and only if $f_n \rightarrow 0$ (relatively uniformly) implies that $\phi(f_n) \rightarrow 0$.‡

*9. (a) Show that any normed linear space can be so partially ordered that metric convergence of nets is equivalent to order-convergence.
   (b) Show that if a topological linear space can be partially ordered so that convergence of nets is equivalent to order-convergence, then it is normable.
   (c) Show that (b) fails if only sequences are considered.

15. (*L*)-spaces

It is easy to verify that the Banach lattices $L^1(0, 1)$, $L^1(-\infty, \infty)$ and (l) satisfy the following condition

(32) If $f > 0$ and $g > 0$, then $\|f + g\| = \|f\| + \|g\|$.

Banach lattices which satisfy (32) are called (abstract) (L)-spaces; this class of Banach lattices has been given a penetrating analysis by Kakutani [1].§

Evidently, (32) implies that the norm is uniformly monotone, with $\delta = \epsilon$. Therefore any (L)-space is a *UMB*-lattice, and hence (Theorem 21, Corollary) it is a complete vector lattice.

---

† See M. Nakamura, Proc. Japan Acad. **26** (1950), 9–10.

‡ Exercises 6–8 state results of H. Gordon [**1**, pp. 420–2]. Exercise 9 states results of R. E. de Marr, Pacific J. Math. **14** (1964), 17–20.

§ The concept was introduced by the author, Proc. Nat. Acad. Sci. **24** (1938), 154–9.

Remark. In any vector lattice which is also a Banach space, let there be given a norm satisfying (32). Then the *derived* norm, defined as

$$(33) \qquad \|f\|' = \|f^+\| + \|f^-\| = \| |f| \| \qquad \text{(by (32))}$$

also satisfies (32). Moreover, as was shown by Kakutani [**1**, Theorem 1], it gives an equivalent topology. Hence the postulate (27), that $|f| \leqq |g|$ implies $\|f\| \leqq \|g\|$ is nearly redundant for (L)-spaces.

We now construct a large class of "concrete" (L)-spaces. Let A be any Boolean algebra, and consider the set $V(A)$ of all bounded valuations on A which satisfy $v[O] = 0$ (i.e., such that $x \wedge y = O$ implies $v[x \vee y] = v[x] + v[y]$). Under any isomorphic representation of A as a field Φ of sets, $V(A)$ is just the set of all *bounded, additive set-functions* on A.

THEOREM 24. *Under the definition*

$$(34) \qquad \|v\| = \sup_{x,y} \{v[x] - v[y]\} = \sup_x \{v[x] - v[x']\},$$

those bounded valuations on any Boolean algebra A which satisfy $v[O] = 0$ form an L-space $V(A)$.

PROOF. We recall from §X.6 that valuations of *bounded variation* on any lattice L form a vector lattice (Chapter X, (11)–(12) and Theorem 10). If L is a Boolean algebra, then for any finite chain $\gamma: 0 = x_0 < x_1 < \cdots < x_n = I$, letting $a = \bigvee_P (x_i \wedge x'_{i-1})$ and $a' = \bigvee_{P'} (x_i \wedge x'_{i-1})$, where P is the set of all i with $v[x_i] > v[x_{i-1}]$ and P' is the complementary set of i with $v[x_i] \leqq v[x_{i-1}]$, we have

$$v^+[O, I] = \sup v[a] \quad \text{and} \quad v^-[O, I] = \inf v[a'].$$

Hence, for valuations on L, having bounded variation is equivalent to being *bounded*, and $\|v\|$ as defined by (34) is just the total variation of v on $[O, I]$. Therefore, for $v > 0$, $\|v\| = v[I]$, from which (32) is obvious.

We next consider the subset $V_\sigma(B)$ of all σ-additive valuations which satisfy $v[O] = 0$ on a Borel algebra B. We observe without proof† the

LEMMA. *Each of the following conditions on a bounded valuation v on a Borel algebra B is equivalent to the σ-additivity of v:* (i) *$x_n \downarrow O$ implies $v[x_n] \to 0$, and* (ii) *$x_n \to x$ in the order-topology implies $v[x_n] \to v[x]$.*

THEOREM 25. *If B is a Borel algebra, then $V_\sigma(B)$ is a metrically closed l-ideal of $V(B)$.*

COROLLARY 1. *For any Borel algebra B, the set $V_\sigma(B)$ of all σ-additive valuations with $v[O] = 0$ is an (L)-space.*

COROLLARY 2. *The probability distributions on B form a metrically closed, convex subset of $V(B)$.*

† For proofs, see Kakutani [**1**], or [**LT2**, p. 255].

Clearly, a nonnegative σ-additive valuation on B is just a "finite measure" in the sense of §XI.5; general σ-additive valuations are often called "signed measures". Applied to the present example, Theorem 16 yields as a corollary the following classic result.

LEBESGUE–RADON–NIKODYM THEOREM. *Let e and f be any finite measures on a Borel algebra B, and let $e[a_n] \downarrow 0$ imply $f[a_n] \downarrow 0$. Then f can be written in the form* (25).

Conversely, let V be any (L)-space, and let A be the complete Boolean algebra of its closed l-ideals $J, K, \cdots$ (cf. §XIII.13). Given $f \in \mathbf{L}$, its components $f_J, f_K, \cdots$ satisfy $f_0 = 0$ (whence $\|f_0\| = 0$ and, by (33)

$$(35) \qquad J \wedge K = 0 \quad \text{implies} \quad \|f_{J \vee K}\| = \|f_J\| + \|f_K\|.$$

That is, the function $f(J) = \|f_J\|$ is a *valuation* on A with $f(0) = 0$. One easily verifies that the correspondence $f \to f(J)$ is a norm-preserving *l-monomorphism* from V into $V(A)$. This proves

THEOREM 26. *Any (L)-space V can be represented l-monomorphically by valuations on the complete Boolean algebra of its closed l-ideals.*

Of especial interest for applications to probability theory, including ergodic theory and the theory of multiplicative processes (Chapter XVI), is the special case of the lattice $V_\sigma(A)$ of all those σ-*additive* valuations on a Boolean σ-algebra A which satisfy $f[0] = 0$. This lattice has been studied in Chapter XI. We now quote without proof the following important result of Kakutani [**1**].

THEOREM 27 (KAKUTANI). *Any separable (L)-space V is l-monomorphic and isometric with a closed vector sublattice of the concrete (L)-space $L^1(0, 1)$.*

We will only sketch the proof; for further details see Kakutani [**1**] or [**LT2**, pp. 255–6].

First, construct a weak unit e as in Lemma 3 of §12. Then (cf. §10, end) construct the complete Boolean algebra $L_c(V)$ of all closed l ideals of V. For any such closed l-ideal J and the chosen e, we can use the *valuation* $\|f\|$ on V^+ to construct the *component* $e_J \in V$ of e in J, and $e'_J = e_{J\perp} = e - e_J$, establishing an isomorphism between $L_c(V)$ and the components of e, where we can suppose $\|e\| = 1$ without loss of generality. Under the *measure* $\|e_J\|$, $L_c(V)$ is by a straightforward extension of Theorem XI.8 a subalgebra of the measure algebra of all measurable subsets of the (open or closed) unit interval, modulo sets of measure zero, with $m(S) = \|e_S\|$.

The proof can now be completed by reconsidering the Stieltjes integral representation (25) of Theorem 16. For any $f \in V$, we have $\|f\| = \int_{-\infty}^{\infty} |\lambda| \, de_\lambda(f) = \|f^+\| + \|f^-\|$; hence the representation is isometric. In particular, f^+ has the "right" meaning; hence the correspondence is l-monomorphic. Finally, since norm and lattice operations are preserved, and since infinite joins in metric lattices are just metric limits of finite joins, and since any abstract (L)-space is a complete lattice, the representation (25) has for image a *closed* sublattice of $L^1(0, 1)$, which is thus a universal separable (L)-space.

16. (M)-spaces

The Banach lattices B and (b) of all bounded (real) functions, and of all bounded sequences, both in the uniform norm, have properties which are dual to those of (L)-spaces. The same is true of the Banach lattice $C^*(X)$ of all bounded continuous functions on any topological space. The properties referred to characterize these spaces as (M)-spaces, in the sense of the following definition.

DEFINITION. An *(M)-space* is a Banach lattice $\mathbf{M}$ whose norm satisfies

$$\text{(36)} \qquad \text{If } f > 0 \text{ and } g > 0, \text{ then } \|f \vee g\| = \|f\| \vee \|g\|.$$

If the unit ball of $\mathbf{M}$ has a largest element u, this u is called the *unity* of $\mathbf{M}$; it is evidently a strong unit of $\mathbf{M}$.

(M)-spaces with such a unity satisfy a remarkable representation theorem, due to M. and H. Krein and Kakutani.†

THEOREM 28. *Any (M)-space with unity is isomorphic with the lattice of* all *continuous functions on a suitable compact space $K(\mathbf{M})$. If $\mathbf{M} = C(X)$ for some completely regular X, then $K(\mathbf{M})$ is just the Stone-Cech compactification of X.*

We now sketch a proof of this representation theorem.

LEMMA 1. *If J is a proper l-ideal of the (M)-space $\mathbf{M}$ with unity u, and $x \in J$, then $\|u - x\| \geqq 1$.*

PROOF. Suppose $\|u - x\| < 1$. Then $u - x \leqq \alpha u$ for some $\alpha < 1$, whence $x \geqq \beta u$ for $\beta = (1 - \alpha) > 0$. This would imply $\mathbf{M} = (u) \subset (x) \subset J$, a contradiction.

LEMMA 2. *Let $\mathbf{M}$ be an (M)-space with unity u, and let $a > 0$ be given in $\mathbf{M}$. Then $\mathbf{M}$ contains a proper (metrically) closed l-ideal J such that $\|a - x\| \geqq \|a\|$ for all $x \in J$, unless $\mathbf{M} = \mathbf{R}$ is one-dimensional.*

PROOF. The case $a = \lambda u$ is covered by Lemma 1. Otherwise, the inequality $\alpha u \leqq a \leqq \beta u$ will be satisfied by a largest α and smallest $\beta = \|a\| > \alpha$; set $\gamma = (\alpha + \beta)/2$. Then the metric closure of the principal l-ideal $(\alpha - \alpha u)$ will have the stated property.

COROLLARY. *In Lemma 2, $\mathbf{M}$ will contain a maximal (metrically) closed proper l-ideal J such that*

$$\text{(37)} \qquad \operatorname{Inf}_{x \in J} \|a - x\| = \|a\|;$$

moreover $\mathbf{M}/J \cong \mathbf{R}$ will be one-dimensional.

---

† M. and S. Krein, Doklady URSS **27** (1940), 427–30; S. Kakutani [**2**]. For an alternative approach, see [**K-N**, Theorem 24.5]. Namioka's "spectrum" refers to maximal (proper) metrically closed l-ideals; his "real" lattice homomorphisms" (p. 238) are l-morphisms onto $\mathbf{R}$.

For, the closure of the set-union of any chain of closed l-ideals whose members satisfy (37), itself satisfies (37).

Under the norm (37), $\mathbf{M}/J$ is clearly an (M)-space with unit $(u + J)/J$ for any (metrically) closed l-ideal J.

Hence $\mathbf{M}$ is isomorphic with a Banach sublattice of the (M)-space of *all* bounded real functions on the set of its maximal closed l-ideals. It remains to construct the appropriate (compact) topology on this set. This can be best done by using the weak topology; see §17.

LEMMA 3. *If a Banach lattice has a unity u, then it is an (M)-space.*

PROOF. Under the stated hypothesis, $f > 0$ and $g > 0$ imply $f \leqq \|f\|u$, $g \leqq \|g\|u$, and hence

$$\|f \vee g\| \leqq \|(\|f\|u \vee g\|g\|u)\| \leqq \|(\|f\| \vee \|g\|)u\| = \|f\| \vee \|g\|.$$

Since the reverse inequality is trivial, the lemma is proved.

17. Duality Between (L)- and (M)-spaces

The duality between (L)-spaces and (M)-spaces is foreshadowed by the (Riemann–)Stieltjes integral. Let A be the Boolean algebra of "elementary subsets" ([**LT2**, §XI.1]) of the unit hypercube K, let $\mathbf{M} = C(K)$ be the (M)-space of all continuous functions $f(\mathbf{x})$ on K; and let $\mathbf{L} = L(A)$ be the (L)-space of all bounded valuations μ on A with $\mu[0] = 0$ ("signed measures"). Then the Riemann–Stieltjes integral

$$\int f(\mathbf{x})\, d\mu(\mathbf{x}) \quad \operatorname*{Lim}_{|\pi| \to 0} \sum f(\mathbf{x}_i)\mu(\Delta\mathbf{x}_i), \qquad \mathbf{x}_i \in \Delta\mathbf{x}_i \tag{38}$$

is a continuous linear functional on $\mathbf{M}$ for fixed μ and on $\mathbf{L}$ for fixed f. In fact $\mathbf{L}$ and $\mathbf{M}$ are *conjugate* Banach lattices, by classic theorems of F. Riesz and Steinhaus.†

Likewise, setting $\mathbf{L} = L_1(0, 1)$ and $\mathbf{M} = L_\infty(0, 1)$, we obtain another classic pair of conjugate spaces. We now give a partial generalization of this result.‡

THEOREM 29. *The dual (conjugate) $\mathbf{L}^*$ of any (L)-space $\mathbf{L}$ is an (M)-space, whose unity is the functional*

$$\epsilon(f) = \|f^+\| - \|f^-\|. \tag{39}$$

PROOF. We first note that $\epsilon(-f) = -\epsilon(f)$, since

$$\|(-f) \vee 0\| - \|(-f) \wedge 0\| = \|-(f \wedge 0)\| - \|-(f \vee 0)\| = \|f^-\| - \|f^+\|.$$

To prove that ϵ is a *linear* functional, note next that

$$f = (f + g)^+ + (f \wedge -g), \qquad g = (f + g)^- + (g \vee -f), \tag{40}$$

† F. Riesz, C. R. Acad. Sci. Paris **149** (1909), 974–7; M. H. Steinhaus, Math. Z. **5** (1918), 186–221.

‡ For the full result, see Kakutani, [**2**, p. 1021], or [**K-N**]. The proof below parallels [**LT2**, p. 256].

since $(f + g) \vee 0 = f + (g \vee -f) = f - (f \wedge -g)$, and dually. We now decompose $\mathbf{L}$ into direct summands (closed l-ideals) on which, respectively: $f \geqq 0$, $g \geqq 0$; $f < 0$, $g < 0$; $f \geqq 0 > g$, and $g \geqq 0 > f$. This is possible by Theorem 23. The linearity of ϵ in the first two cases follows from (32). In the third case, using (40) and (32), we get:

$$\|f\| = \|(f + g)^+\| + \|f \wedge (-g)\|, \qquad \|g\| = \|(f + g)^-\| + \|g \vee (-f)\|,$$

since $f \wedge (-g) = 0$ and dually. Therefore,

$$\|(f + g)^+\| - \|(f + g)^-\| = \|f\| - \|f \wedge (-g)\| - \|g\| + \|g \vee (-f)\|,$$

whose left side is $\epsilon(f) + \epsilon(g)$ since $\|f \wedge (-g)\| = \|(-f) \vee g\|$; this proves linearity.

Thirdly, if ϕ is any linear functional on $\mathbf{L}$ of norm $\|\phi\| \leqq 1$, then for any $f \geqq 0$ $\phi(f) \leqq \|f\| = \epsilon(f)$; hence ϵ is a unity for the conjugate of $\mathbf{L}$. It follows, by Lemma 3 of §15, that this conjugate is an (M)-space.

Conversely, let $\mathbf{M} = C(K)$ be the (M)-space of all continuous (bounded) real functions on a Hausdorff space K; its dual $\mathbf{M}^*$ is the (L)-space of Radon measures on K. The (natural) monomorphism $\mathbf{M} \subset (\mathbf{M}^*)^*$ has many interesting properties, for which the reader is referred to papers by S. Kaplan,† who has recently shown that $(\mathbf{M}^*)^*$ is the set of all limits of order-convergent nets of points of $\mathbf{M}$.

Abstract (L^p)-*spaces*. Intermediate to (L)-spaces and (M)-spaces are abstract (L^p)-spaces. By definition, these are Banach lattices in which

$$\text{(41)} \qquad \text{If } |f| \wedge |g| = 0, \quad \text{then} \quad \|f + g\| = \phi(\|f\|, \|g\|),$$

where $\phi(x, y)$ is a single-valued function of two variables. In an elegant paper,‡ F. H. Bohnenblust has shown that this assumption implies that $\phi(x, y) = (x^p + y^p)^{1/p}$ for some p, $1 \leqq p \leqq \infty$. Moreover, at least for finite p, any such (abstract) (L^p)-space is determined up to rigid l-isomorphism by the structure lattice A of its closed l-ideals, which is a Boolean algebra. It may be surmised that $L^p(A)$ and $L^q(A)$ are conjugate§ if $p^{-1} + q^{-1} = 1$.

Exercises for §§14–16:

1. Show that, in Theorem 24,
$$\|v\| = 2v^+[I] - v[I] = v^+[I] - v^-[I] = |f|[I] = \| |f| \|.$$

2. In any (L)-space, show that the set of positive elements of norm one (the "distributions") form a metrically closed convex set of diameter two at most.

3. Let $v[x]$ be a valuation on a relatively complemented lattice L. Show that the variation $v[a, b; \gamma]$ of v on any chain γ from a to b is equal to the variation on some such chain $v[a, b; \gamma^*]$ of length two.

4. Prove that a valuation v on a Borel algebra B is a σ-valuation if and only if $x_i \wedge x_j = 0$ for all $i \neq j$ implies
$$v\left[\bigvee_{i=1}^{\infty} x_i\right] = \sum_{i=1}^{\infty} v[x_i].$$

† Trans. AMS **86** (1957), 70–90 and **93** (1959), 320–50; also Proc. AMS **17** (1966), 401–6.

‡ Duke J. Math. **6** (1940), 627–40.

§ The case $A = B_\sigma$ of this result is classic; See F. Reisz, Math. Ann. **69** (1910), 475.

5. Show that the diameter of the set of distributions on a Boolean algebra is 0 or 2.

6. Show that every σ-additive valuation on a Borel algebra B is bounded.

7. Show that the dual of the (L)-space $L_1(0, 1)$ is the (M)-space $L_\infty(0, 1) \cong V_\sigma(B_\sigma/N)$.

*8. Show that the dual of $C[0, 1]$ is the (L)-space $V_\sigma(B_\sigma)$ of all σ-additive valuations ("signed measures") on B_σ.

9. Show that, whereas $L_\infty(0, 1)$ is separable, $V_\sigma(B_\sigma)$ is not.

PROBLEMS

129. Develop a general theory of vector lattices over ordered fields.

130. Find necessary and sufficient conditions on an (algebraic, Brouwerian) lattice for it to be the structure lattice of a suitable vector lattice.†

131. Find conditions on an integrally closed directed vector space which are necessary and sufficient for its positive linear functionals to "separate" any two distinct points.‡ (*Remark.* It is sufficient to be Banach lattice.)

132. Which integrally closed directed vector spaces V are isomorphic to their second conjugate $(V^*)^*$ ("reflexive")? For vector lattices, is completeness necessary? Which are monomorphic?

133. Prove in detail that every σ-complete vector lattice with strong unit is a subdirect product of copies of $\mathbf{R}$.

134. Extend Theorem 16 to show that any σ-complete vector lattice with strong unit can be extended to a "full" vector lattice, a "function lattice" of continuous functions on a suitable Stone space.

135. Can any σ-complete l-group be extended to a full, complete vector lattice?

136. Construct a theory of "direct integrals" of σ-complete vector lattices. Direct integrals of "full" vector lattices should be "full".

*137. *Construct* a completely meet-irreducible l-ideal in the vector lattice $L(-\infty, \infty)$, or prove that none can be "constructed". (Cf. §VIII.16.)

138. In which Banach lattices is the weak topology the new interval topology?

139. In which integrally closed directed vector spaces is the relative uniform topology (Hugh Gordon) that defined (by Arnold-Kelley Theorems) from order convergence?§

140. Which σ-complete distributive lattices (and ortholattices) admit non-trivial σ-valuations in some σ-complete vector lattice? (*Remark.* This relates to conditions for being a "measure algebra".)

141. Generalize Theorem 27 to arbitrary abstract (L)-spaces, using perhaps the Theorem of Loomis.

142. Develop a general theory of duality between (L)-spaces and (M)-spaces, relating $L(\Omega)$ to $L^\infty(\Omega)$. What is the relation of $C(K)$ with the "uniform" norm $\|x \vee y\| = \|x\| \vee \|y\|$ to $L^\infty(\Omega)$? I.e., what are $K(\Omega)$ and $\Omega(K)$?

---

† Note the necessary condition of Conrad, Harvey and C. Holland, Trans. AMS **108** (1963), 143–69, that the prime elements form a "root system".

‡ For some such conditions, see [**K–N**].

§ Correlate also with the L-topology of B. C. Rennie, Proc. Lond. Math. Soc. **52** (1951), 386–400.

CHAPTER XVI

POSITIVE LINEAR OPERATORS

1. Introduction

Many quantities are inherently *nonnegative*: mass, moles of a chemical, numbers of people, neutron density, etc. Such quantities are *conserved* by processes of convection, diffusion, (or migration), and mutation; they are often operated on *linearly* by death and unary reproduction (birth). The preceding simple observations show that one may expect positive linear operators to be very prevalent in Nature. This chapter will be devoted to deducing some of their most basic general properties. But first we will describe in some detail two typical examples, arising in neutron chain reactions and statistical mechanics, respectively.

Imagine a nuclear reactor core with time-independent physical characteristics, occupying a compact domain D in physical space. The *migration kernel* $K(\mathbf{x}, \mathbf{y})$ of the reactor is defined as the probability (per unit volume) that a neutron "born" (i.e., produced by fission) at $\mathbf{x} \in D$ will produce a fission at some other point $\mathbf{y} \in D$, after being absorbed there. Clearly $K(\mathbf{x}, \mathbf{y}) \geqq 0$, and $\int K(\mathbf{x}, \mathbf{y})\, d\mathbf{y} \leqq 1$, the "less than" sign being introduced to account for loss of neutrons by leakage and by absorption without fission.

If $f_0(\mathbf{x})$ is the initial spatial distribution of 0-th generation fission neutrons, then the spatial distribution $f_r(\mathbf{x})$ of rth generation fission neutrons will satisfy

$$f_r(\mathbf{y}) = \nu(\mathbf{y}) \int f_{r-1}(\mathbf{x}) K(\mathbf{x}, \mathbf{y})\, d\mathbf{x} = \int f_{r-1}(\mathbf{x}) P(\mathbf{x}, \mathbf{y})\, d\mathbf{x}. \tag{1}$$

where $\nu(\mathbf{y})$ is the mean neutron yield per fission at $\mathbf{y}$, and $P(\mathbf{x}, \mathbf{y}) = \nu(\mathbf{y}) K(\mathbf{x}, \mathbf{y}) \geqq 0$, on physical grounds.

In the kinetic theory of gases, one has a very different situation. One assumes that n "molecules" form a conservative (reversible) Lagrangian system, whose "state" in phase-space is described by $6n$ canonical coordinates $q_1, \cdots, q_{3n}, p_1, \cdots, p_{3n}$. The effect of time on such a system, assumed autonomous, is to produce a steady flow of the phase-space; moreover this flow is volume-conserving by a classic Theorem of Liouville.† Hence flow defines a one-parameter group of positive linear transformations T_t of the Banach lattice $L^p(X)$ for any p, $1 \leqq p \leqq \infty$. Finally, the T_t are all isometric l-automorphisms of every $L^p(X)$.

In both of the above examples, one is dealing with *nonnegative* linear operators

† See E. Hopf, *Ergodentheorie*, Springer, 1937; P. R. Halmos, *Lectures on ergodic theory*, Chelsea, 1960; K. Jacobs, *Neuere Methoden und Ergenbnisse der Ergodentheorie*, Springer, 1960; P. Caldirola, *Ergodic theories*, Academic Press, 1962. For the lattice-theoretic approach, see [**Symp**, pp. 172–8].

on some directed vector space V, with convex *positive cone* $V^+ = C$. That is, C has the properties

$$(2) \qquad C \cap -C = 0, \qquad C + (-C) = V, \qquad C + C \subset C,$$

and the operators P (or T_t) of interest are linear and *positive*: they satisfy $CP \subset C$.

The present chapter will study consequences of the preceding conditions on (V, C) and P. As shown in Chapter XV, (V, C) is bidirected by letting $f \leqq g$ mean that $(g - f) \in C$. Moreover, since P is positive and linear, it is *isotone* (i.e., $f \leqq g$ implies $fP \leqq gP$), a result which is not true for nonlinear operators.

Isotone Nonlinear Operators. A linear operator P on a partly ordered vector space is evidently *positive* if and only if it is *isotone*—i.e., if and only if

$$(1') \qquad f \leqq g \quad \text{implies} \quad fP \leqq gP.$$

Nonlinear isotone (and antitone) operators also have many interesting applications, especially to the solution of differential and integral equations by iterative and approximate methods. Surveys of such applications have been made recently by Krasnoselskii and Collatz.†

2. Hilbert's Projective Pseudo-metric

Positive linear operators on vector lattices have various remarkable properties, of which the most typical refer to dominant positive eigenvectors. The study of such positive eigenvectors is best accomplished using a special projective pseudo-metric originally invented by Hilbert.‡ This can be defined in general as follows.

Let V be any partly ordered vector space with positive cone C. Two elements $f, g \in C$ will be called *comparable* when

$$(3) \qquad \lambda f < g < \mu f \quad \text{for suitable positive scalars } \lambda, \mu.$$

Since (3) implies $\mu^{-1}g < f < \lambda^{-1}g$, the relation of comparability is symmetric; it is trivially reflexive. It is also transitive, but we shall prove a sharper result.

Namely, if f and g are comparable, then we can define $\alpha = \alpha(f, g)$ and $\beta = \beta(f, g)$ (where clearly $\alpha \leqq \beta$) by

$$(4) \qquad \alpha(f, g) = \sup \lambda \quad \text{and} \quad \beta(f, g) = \inf \mu \quad \text{such that } \lambda f < g < \mu f.$$

From this, we define the *projective distance* from f to g as

$$(5) \qquad \theta(f, g) = \ln [\beta(f, g)/\alpha(f, g)] \geqq 0.$$

Since (3) implies $\mu^{-1}g < f < \lambda^{-1}g$, and conversely, clearly $\alpha(g, f) = [\beta(f, g)]^{-1}$

† M. A. Krasnoselskii, *Positive solutions of operator equations*, Noordhoff, Groningen, Netherlands, 1964; L. Collatz, *Functional analysis and numerical analysis*, Academic Press, New York, 1966. See also papers by Albrecht and Bohl in ZaMM **46** (1966).

‡ Math. Ann. **57** (1903), 137–50. Its application to positive linear operators was first made by the author (Trans. AMS **85** (1957), 219–27; see also ibid. **104** (1962), 37–51). The present exposition uses also some unpublished ideas of A. Ostrowski (personal communication), in §§1–3.

and $\beta(g, f) = [\alpha(f, g)]^{-1}$. Hence $\theta(f, g) = \theta(g, f)$. Further, if $\lambda_n f \leqq g \leqq \mu_n f$ and $\lambda'_n g \leqq h \leqq \mu'_n g$, where $\lambda_n \uparrow \alpha$, $\mu_n \downarrow \beta$, $\lambda'_n \uparrow \alpha'$, and $\mu'_n \downarrow \beta'$, then

$$\lambda_n \lambda'_n f \leqq h \leqq \mu_n \mu'_n f, \quad \text{where } \lambda_n \lambda'_n \uparrow \alpha\alpha' \quad \text{and} \quad \mu_n \mu'_n \downarrow \beta\beta'.$$

Hence we have the triangle inequality $\theta(f, g) + \theta(g, h) \geqq \theta(f, h)$, which proves that the relation of comparability is transitive.

Finally, since $(\lambda - \epsilon)f < \lambda f + h < (\lambda + \epsilon)f$ for any $\lambda > 0$ and $f > 0$, if $h \ll f$, clearly $\theta(f, \lambda f + h) = 0$ for any such λ, f, h. Conversely, if $\theta(f, g) = 0$, then $\alpha(f, g) = \beta(f, g)$ and so we can find α such that $(\alpha - \epsilon)f < g < (\alpha + \epsilon)f$ for any $\epsilon > 0$. Therefore, $-\epsilon f < (g - \alpha f) < \epsilon f$ for all $\epsilon > 0$, so that $(g - \alpha f) \ll f$. In summary, we have proved

THEOREM 1. *The function $\theta(f, g)$ is a pseudo-metric (that is, $\theta \in [0, \infty]$, $\theta(f, f) = 0$, $\theta(f, g) = \theta(g, f)$, and the triangle inequality holds). Moreover $\theta(f, g) = 0$ if and only if $(g - \alpha f) \ll f$, for some $\alpha > 0$.*

Example 1. Let $V = \mathbf{R}^n$, and C the set of all nonnegative f. Then $f > 0$ and $g > 0$ are comparable if and only if they have the same support (set of nonzero components). Moreover $\theta(f, g) = \ln \{\sup (f_i g_j / g_i f_j)\}$, we omit the proof. In particular, when $n = 2$, $\theta(f, g) = |\ln (f_1 g_2 / g_1 f_2)|$. Hence in *projective* coordinates $x = f_2/f_1$, $\theta(x, x') = |\ln (x'/x)|$.

Example 2. In the (L)-space $L_1(-\infty, \infty)$ likewise,

$$\theta(f, g) = \ln \{\text{ess sup}_{x,y} [f(x)g(y)/g(x)f(y)]\}.$$

Two elements $f > 0$ and $g > 0$ are comparable if and only if ess sup $[f(x)/g(x)]$ and ess sup $[g(x)/f(x)]$ are both finite.

Example 3. In $V = \mathbf{R}^n$, define

$$B(x, y) = (x_1 + \cdots + x_n)(y_1 + \cdots + y_n) - (x_1 y_1 + \cdots + x_n y_n), \tag{6}$$

and let C be the set where $B(x, x) \geqq 0$ and $x_1 + \cdots + x_n \geqq 0$. Then the projective pseudo-metric in (V, C) defines Hilbert's model of non-Euclidean (i.e., hyperbolic) geometry.

We now consider *positive* linear operators on (V, C), that is, linear operators for which $f > 0$ implies $fP > 0$.

THEOREM 2. *Any positive linear operator is a contraction in the projective pseudo-metric of any directed vector space:*

$$\theta(fP, gP) \leqq \theta(f, g), \quad \text{if } \theta(f, g) < +\infty. \tag{7}$$

PROOF. If $\lambda f < g < \mu f$, then $\lambda fP < gP < \mu fP$ if P is positive. Hence $\alpha(fP, gP) \geqq \alpha(f, g)$ and $\beta(fP, gP) \leqq \beta(f, g)$, whence $\theta(fP, gP) \leqq \theta(f, g)$ as claimed.

Note that the argument applies also to *nonnegative* linear operators, provided $fP \neq 0$ and $gP \neq 0$.

Integrally closed vector spaces. The preceding ideas apply with the greatest force to "integrally closed" directed vector spaces. Thus, we recall from Chapter XV (and G. Birkhoff [7, Lemma 1]):

LEMMA 1. *In an integrally closed directed vector space, if* $\alpha_n f \geqq g$ $(f > 0, g > 0)$ *for all* $n = 1, 2, 3, \cdots$, *and* $\alpha_n \to \alpha$ *as* $n \to \infty$, *then* $\alpha f \geqq g$.

COROLLARY. *In Lemma* 1, *if* $f > 0$ *and* $g > 0$ *are linearly independent, then the plane spanned by* f *and* g *is isomorphic to* $\mathbf{R}^2$.

LEMMA 2. *In an integrally closed vector space,* $\theta(f, g) = 0$ *if and only if* $g = \alpha f$ *for some* $\alpha > 0$.

PROOF. In the paragraph preceding Theorem 1, we have $g = \alpha f$ since $(g - \alpha f) \ll f$; the converse was shown in Theorem 1.

Exercises for §§1–2:

1. Show that in Example 1, a matrix $P = \|p_{ij}\|$ defines a nonnegative linear operator if and only if all $p_{ij} \geqq 0$, and a positive linear operator if and only if all $p_{ij} \geqq 0$ and $\sum_j p_{ij} > 0$ for all i.
2. Prove that $\theta(f, g) = |\ln (f_1 g_2/g_1 f_2)|$ in Example 1, when $n = 2$.
*3. Show that Example 3 gives a model for hyperbolic n-space. (*Hint.* Consider its group of rigid motions).
4. Show that in Example 3, the projective pseudo-metric is given, for $\epsilon = (1, 0, \cdots, 0)$ and $\mathbf{x} = (1, x_1, \cdots, x_r)$, by:

$$\theta(\epsilon, \mathbf{x}) = \frac{1}{2} \ln \left(\frac{1 + r}{1 - r}\right), \quad r = \left(\sum x_i^2\right)^{1/2}, \quad \text{all} \quad x_i > 0.$$

5. If (V, C) is finite-dimensional and directed, show that C has a nonvoid interior.
*6. Show that a directed *topological* vector space V with positive cone C is "integrally closed" if and only if C is topologically closed in C.†

3. Perron's Theorem

We shall now sharpen Theorem 2 in the case of Example 1, obtaining a remarkable result of Perron on *positive* matrices,‡ generalized by Frobenius in a way to be described in §4. In this example, if $P = \|p_{ij}\|$ is positive, then for any $f, g \in C$ we have as in Example 1

$$\theta(fP, gP) = \ln \left\{ \sup_{i,j} \sum_{k,l} (f_k p_{ki} g_l p_{lj} / f_k p_{kj} g_l p_{li}) \right\}.$$

Hence the projective diameter $\Delta = \Delta(P)$ of CP is

$$\Delta = \ln \left\{ \sup_{i,j,f>0,g>0} \sum_{k,l} (f_k p_{ki} g_l p_{lj} / f_k p_{kj} g_l p_{li}) \right\}. \tag{8}$$

By choosing f and g to be suitable unit vectors $f_k = \delta_{ki}$ and $g_l = \delta_{lj}$, we can attain the limit

$$\Delta(P) = \ln \{\sup_{i,j,k,l} (p_{ki} p_{lj} / p_{kj} p_{li})\}, \tag{9}$$

which obviously cannot be exceeded since averaging (by positive weight factors $f_k g_l$) always makes ratios less extreme. In particular, for $n = 2$,

$$\Delta(P) = |\ln (p_{11} p_{22} / p_{12} p_{21})|. \tag{9'}$$

† See G. Birkhoff, J. Math. Mech. **14** (1965), 507–12, especially Theorem 1.

‡ O. Perron, Math. Ann. **64** (1907), 248–63; G. Frobenius, S.-B. Preuss. Akad. Wiss. Berlin (1912), 456–77 (and references there).

DEFINITION. A (positive) linear operator P is *uniformly positive* when the projective diameter of the transform CP of C under P is finite.

We have just shown that, in Example 1, every positive linear operator is uniformly positive. We now prove the sharpened form of Theorem 2.

THEOREM 3. *Let P be any uniformly positive linear operator on an integrally closed, directed† vector space V, and let the projective diameter of CP be Δ. Then*

$$(*)\qquad \sup_{\theta < \theta(f,g) < \infty} [\theta(fP, gP)/\theta(f, g)] = \tanh \Delta/4.$$

PROOF. The cases $\theta(f, g) = 0$ and $\theta(f, g) = \infty$ are trivial, as in Theorem 2. Hence, by the Corollary of Lemma 1, §2, it suffices to consider the case of a positive linear transformation of $V_2 = \mathbf{R}^2$ into a copy of itself. This is represented by a 2×2 matrix

$$\|p_{ij}\| = \begin{Vmatrix} a & b \\ c & d \end{Vmatrix}$$

with positive entries. Alternatively, it suffices to consider a *projective* transformation $x' = (ax + b)/(cx + d)$ (a, b, c, d positive), which maps the positive axis $[0, \infty]$ into an interval $[b/d, a/c]$ (or $[a/c, b/d]$) of projective diameter $\Delta = |\ln (ad/bc)|$, much as in (9′); see also Example 1.

The preceding remarks reduce the determination of the maximum of the *contraction ratio* $\theta(fP, gP)/\theta(f, g)$ to a straightforward calculation. In projective coordinates, the differential of "projective distance" is $|dx/x|$, as in Example 1. Hence the maximum of the contraction ratio is the maximum of

$$(10)\qquad |xdx'/x'dx| = |(ad - bc)x/(ax + b)(cx + d)|,$$

since

$$|\ln (y'/x')| = \int_{x'}^{y'} |d\xi'/\xi'| \leqq \max \left|\frac{\xi\, d\xi'}{\xi'\, d\xi}\right| \int_{x}^{y} |d\xi/\xi|.$$

This is the reciprocal of the minimum of $[acx + (ad + bc) + bdx^{-1}]/(ad - bc)$. This minimum is assumed when $x^2 = bd/ac$; substituting back into (10), we find that the contraction ratio for infinitesimal elements has the maximum

$$\tau(P) = |ad - bc|/[2(abcd)^{1/2} + (ad + bc)].$$

Dividing through the numerator and denominator by $(abcd)^{1/2}$, this gives (since $\Delta = \Delta(P) \geqq 0$):

$$(10')\qquad \begin{aligned} \tau(P) &= (e^{\Delta/2} - e^{-\Delta/2})/(e^{\Delta/2} + 2 + e^{-\Delta/2}) \\ &= (e^{\Delta/4} - e^{-\Delta/4})/(e^{\Delta/4} + e^{-\Delta/4}) = \tanh (\Delta/4). \end{aligned}$$

This proves the theorem.

† As has been observed by Professor Ostrowski, it is sufficient to assume that V is a partially ordered vector space.

Since the contraction ratio tanh $(\Delta/4) < 1$, we can then apply Picard's Fixpoint Theorem for uniform contractions of metric spaces, to prove:

THEOREM 4. *Let P be any uniformly positive operator on an integrally closed, directed vector space V. Then for any $f \in C$, the iterated transforms fP^r form a Cauchy sequence in the projective pseudo-metric $\theta(f, g)$. Moreover, for any f, g in C, $\theta(fP^r, gP^r) \to 0$.*

PROOF. By Theorem 3, the CP^r $(r = 1, 2, 3, \cdots)$ form a nested sequence of (closed) subsets of C, and the diameter of CP^r is at most $\Delta[\tanh(\Delta/4)]^{r-1}$. The stated conclusions are now obvious.

Now observe that, in the *interior* of C, by the formula of Example 1 the projective *pseudo-metric* topology in $\mathbf{R}^n$ is the natural projective topology, and indeed θ differs by a bounded factor in CP from distance on the unit sphere. Hence CP is a *complete* metric space, and the fP^r converge in Theorem 4 to a *positive eigenvector*. We thus have the

COROLLARY 1 (PERRON). *Any positive matrix P admits a unique positive eigenvector e, such that $eP = \lambda P$.*

The transpose P^T of a positive matrix P, being itself positive, likewise admits a positive eigenvector $c = (c_1, \cdots, c_n)$. Moreover the hyperplane $H: \sum c_i x_i = 0$ is invariant under P, and complementary to the one-dimensional subspace spanned by the eigenvector e of P, since $\sum c_i e_i > 0$. Finally, if μ is the spectral radius of the restriction of P to H, then Theorem 4 shows that $|\mu| \leqq \tau(P)\lambda$. This proves

COROLLARY 2. *In Corollary 1, the eigenvalue λ is simple and* dominant*: for any other eigenvalue λ_i:*

$$|\lambda_i| \leqq \tau(P)\lambda, \quad \tau(P) = \frac{\sigma - \sigma^{-1}}{\sigma + \sigma^{-1}}, \quad \sigma = \left[\sup_{i,j,k,l}\left(\frac{P_{ki}P_{lj}}{P_{kj}P_{li}}\right)\right]^{1/4}.$$

4. Primitive Nonnegative Matrices

Frobenius (op. cit.) significantly generalized the results of Perron on positive matrices, to a wide class of *nonnegative* matrices, by ingenious combinatorial considerations. His combinatorial results can be reformulated as follows.

LEMMA 1. *Under their natural order, nonnegative $n \times n$ matrices form a commutative l-monoid under matrix addition, and a noncommutative l-monoid under matrix multiplication.*

The preceding statement emphasizes the easily verified isotonicity relations

$$A \leqq B \text{ implies } A + C \leqq B + C \text{ and } C + A \leqq C + B, \tag{11}$$

$$A \leqq B \text{ implies } AC \leqq BC \text{ and } CA \leqq BA, \tag{12}$$

valid for all nonnegative matrices A, B, C. Again, just as in Theorem XIV.16, we have

LEMMA 2. *For any nonnegative $n \times n$ matrix A, let $r(A) = \|r_{ij}\|$ be the relation* *matrix obtained by replacing each nonzero (hence positive) entry of A by 1. Then the correspondence $A \to r(A)$ is an l-homomorphism:*

$$r(AB) = r(A)r(B), \qquad r(A \wedge B) = r(A) \wedge r(B), \tag{13}$$

$$r(A \vee B) = r(A + B) = r(A) \vee r(B), \qquad r(A \vee B) = r(A + B) = r(A) \vee r(B).$$

It follows that the discussion of irreducible, primitive, and semiprimitive relation matrices in §XIV.13 also applies to nonnegative matrices. In particular, some *power* Q^s of any *primitive* (i.e., irreducible acyclic) *nonnegative matrix* Q is (uniformly) *positive*. The arguments used to prove Theorem 4 and its Corollary apply with slight changes to such matrices, and so we have

THEOREM 5. *Any primitive nonnegative matrix has a positive dominant eigenvector.*

Irreducible *cyclic* nonnegative matrices behave differently. Frobenius showed that any irreducible *k-cyclic* nonnegative matrix N has a *nonnegative* eigenvector, whose eigenvalue λ_1 is simple and satisfies $\lambda_1 \geqq \lambda_j$ for all other eigenvalues. However, the eigenvector is *not* dominant: if ω is a primitive k-th root of unity, then $\omega\lambda_1, \cdots, \omega^{k-1}\lambda_1$ are also eigenvalues of N.

Theorem 5 is not "best possible." Thus, if we define a nonnegative square matrix N to be *semiprimitive* when it has the form

$$\begin{pmatrix} 0 & A \\ 0 & P \end{pmatrix}$$

where P is a primitive square submatrix, then we can prove that N has a nonnegative *dominant* eigenvector, whose positive eigenvalue λ_1 is simple and greater in magnitude than any other eigenvalue.†

During the past decades, the fundamental results of Perron and Frobenius have been extended to a wide class of "positive" and "nonnegative" linear operators on real vector spaces V with "nonnegative cones" C. However, the combinatorial approach of these authors must be essentially modified, even for finite-dimensional vector spaces. Thus, in Example 3, a rigid rotation around the axis of the "positive cone", though "positive" in the sense that $f > 0$ implies $fP > 0$, clearly does not admit a (strictly) *dominant* positive eigenvector.

Instead, it is most convenient to use concepts of *uniform* positivity and *uniform* (semi)primitivity, which are essentially dimension-free. Because of this fact, we shall now turn our attention to directed vector spaces of arbitrary dimension.

Historical note. The first generalization of Perron's Theorem to infinite-dimensional function spaces was made by Jentzsch in 1912. Far-reaching further generalizations were made by M. G. Krein and M. A. Rutman, in a series of papers written in 1938–1948.‡

† For this and related results, see G. Birkhoff and R. S. Varga, J. SIAM **6** (1958), 354–77.

‡ For the paper by Jentzsch, see Crelle's J. **141** (1912), 235–44. The results of Krein and Rutman are stated in Uspehi Mat. Nauk **3** (1948), 3–95; AMS Transl. No. 26. These authors rely on *compactness* considerations for existence proofs.

Exercises for §§3–4:

1. Show that

$$\begin{pmatrix} a & 1 \\ 0 & 1 \end{pmatrix}, \qquad a \geqq 0,$$

has a dominant nonnegative eigenvector if and only if $0 \leqq a < 1$.

2. Show that if a nonnegative matrix has a dominant eigenvector, then its eigenvalue must be positive.

*3. Show that, for a nonnegative matrix N, some power N^s of N makes CN^s have finite projective diameter if and only if N is semiprimitive.

4. Show that the *continuous* one-parameter semigroup e^{tQ}, $t \geqq 0$, consists of *positive* matrices if and only if Q is irreducible and *essentially* nonnegative, in the sense that $q_{ij} \geqq 0$ if $i \neq j$.

5. Show that, if $\Delta(P) < +\infty$ in Hilbert's projective pseudo-metric, then Theorem 4 is valid for any cone $C \subset \mathbf{R}^n$ which satisfies conditions (2), but that the condition $CP \subset C$ alone is *not* sufficient for the existence of an eigenvector in C.

*6. Let T be any linear operator on $\mathbf{R}^n = V$ such that $CT \subset C$ for some closed, proper, convex cone C with nonvoid interior. Show that C contains an eigenvector f of T whose eigenvalue is the spectral radius of T.†

5. Uniformly Semiprimitive Operators

Accordingly, we now define a linear operator P on an arbitrary directed vector space V with positive cone C to be *uniformly semiprimitive* when it is not nilpotent, and $\Delta(CP^s) < +\infty$ for some finite power of P. It is called *uniformly primitive* when it is uniformly semiprimitive *and* positive; this is equivalent to the conditions that some power of P be "uniformly positive" (§3).

To apply these definitions effectively, we must first consider more closely the geometry of directed vector spaces.

LEMMA 1. *Let P be any uniformly positive linear operator on a directed vector space (V, C). Then any nonzero $f \in CP$ is a strong unit for VP.*

PROOF. Let $(g - h)P = gP - hP$ in VP be given ($g \in C, h \in C$). By definition of uniform positivity, $gP < \beta f$ and $hP < \gamma f$ for some β, γ; hence $(g - h)P < (\beta + \gamma)f$.

We next prove a result which is geometrically obvious in the Perron–Frobenius case, of finite-dimensional V, since then the closure of any set which is bounded in the projective pseudo-metric is (pseudo-metrically) compact.

THEOREM 6. *Let (V, C) be any integrally closed directed vector space which is complete for relative uniform convergence (§XV.13). Then C is also complete in Hilbert's projective pseudo-metric.*

PROOF. Let $\{f_n\} \subset C$ be a Cauchy sequence in the pseudo-metric (5); that is, let $\theta(f_m, f_n) \to 0$ and $m, n \to \infty$. Then one can extract a subsequence $\{g_k\} = \{f_{n(k)}\}$ from $\{f_n\}$ such that $\theta(g_k, g_{k+1}) < 4^{-k}$. Hence there exist $h_k = \lambda_k g_k$, with $\lambda_1 = 1$ and all $\lambda_k > 0$, such that $h_k \leqq h_{k+1} \leqq (1 + 4^{-k})h_k$. It follows that $h_1 \leqq h_k \leqq 4h_1/3$ for all k, with

$$0 \leqq h_k - h_j \leqq 4^{1-k}h_1/3 \quad \text{if } j > k.$$

† See G. Birkhoff, Bull. AMS **16** (1965), 14–16.

Since (V, C) is complete and C closed in the relative uniform topology, the existence of a (relative uniform) limit $h \in C$ of the sequence $\{h_k\}$ follows. Moreover $0 \leqq h_k - h \leqq 4^{1-k}h_1/3$, whence $\theta(g_k, h) = \theta(h_k, h) \to 0$ as $k \to \infty$. Finally, for all k, $\theta(f_m, h) \leqq \theta(f_m, f_{n(k)}) + \theta(h_k, h)$ by the triangle inequality (since $\theta(f_{n(k)}, h_k) = 0$), and $\theta(f_m, f_{n(k)}) \to 0$; we infer that $\theta(f_m, h) \to 0$ as $m \to \infty$. This completes the proof.

COROLLARY. *In any Banach lattice (B, C), the positive cone C is a complete pseudo-metric space.*

PROOF. In (B, C), relative uniform star-convergence is equivalent to metric convergence (Theorem XV.20); hence metric completeness implies relative uniform completeness.

THEOREM 7. *Let Q be any uniformly primitive linear operator on an integrally closed, directed vector space (V, C) which is relatively uniformly complete. Then Q admits a unique positive eigenvector f. This f has a positive eigenvalue μ, and, for every $g \in C$, $\theta(gQ^r, f) \to 0$ as $r \to \infty$.*

PROOF. As in Theorem 4, the fQ^r form a Cauchy sequence for any $f \in C$ (and all $r > s$). Since C is complete (in the Hilbert pseudo-metric), the $fQ^r \to e$ for some $e \in C$. Since Q is continuous (indeed, a contraction, by (7)), it follows that $\theta(fQ^{r+1}, eQ) \to 0$, hence $\theta(e, eQ) = 0$, and $eQ = \mu e$ for some $\mu = 0$.

In the sequel, we will let $\mu = \mu(Q)$ denote the positive eigenvalue of its positive eigenvector f. We will now show that μ is the *largest* eigenvalue of Q (in magnitude), whence f is a dominant eigenvector. To this end, we define the *projective norm* of a nonnegative linear operator T as

$$\|T\| = \sup \{\theta(gT, hT)/\theta(g, h)\} \quad \text{for } 0 < \theta(g, h) < \infty. \tag{14}$$

By Theorem 2, $\|T\| \leqq 1$; and by Theorem 3, if the projective diameter of CP is Δ, then

$$\|T\| \leqq \tanh (\Delta/4). \tag{14'}$$

To get the sharpest results, it is convenient to define the (asymptotic) *dominance ratio* $\rho(T)$ of T by

$$\ln \rho(T) = \limsup_{n \to \infty} \frac{1}{n} \ln \|T^n\|. \tag{15}$$

If $T = Q$ is uniformly semiprimitive, so that some CQ^s has finite projective diameter Δ, then since CQ^r has projective diameter at most $\Delta[\tanh (\Delta/4)]^{k-1}$ for $r \geqq ks$, clearly

$$\rho(Q) \leqq (\tanh (\Delta/4))^{1/s} < 1. \tag{16}$$

From the definition of $\rho(Q)$, it follows that if $\sigma > \rho(Q)$, then for any $g \in C$ and all sufficiently large n, $|(g - \gamma f)Q^n| \leqq \sigma^n f$, where $\gamma = \gamma(g)$ is chosen to make

$$(\gamma - \epsilon)fQ^m \leqq g \leqq (\gamma + \epsilon)fQ^m,$$

for any $\epsilon > 0$, and all sufficiently large $m = m(\epsilon)$. Since any $h \in V$ can be written as a difference of positive elements, we infer as a consequence,

THEOREM 8. *In Theorem 7, there exists for each $h \in V$ a constant $\gamma = \gamma(h)$ such that, for any $\sigma > \rho(Q)$, including some $\sigma < 1$,*

$$|(h - \gamma f)Q^m| \leqq \sigma^m \mu^m f \quad \textit{for all sufficiently large } m. \tag{17}$$

Evidently, the $\gamma(h)$ in Theorem 8 is a (unique) *linear functional*, whose null-space (where $\gamma(h) = 0$) is a closed *subspace* of V complementary to the eigenspace of the eigenvalue μ. Moreover this null-space (of h such that $|hQ^m| \leqq \sigma^m\mu^m f$ for all $\sigma > \rho(Q)$) is invariant under Q. Finally, the spectral radius of the restriction of Q to this subspace is (at most) $\rho(Q)\mu$, if V is finite-dimensional. This gives the

COROLLARY. *Let Q be a uniformly semiprimitive matrix with dominant eigenvalue μ and dominance ratio $\rho(Q)$. Then the magnitudes of the other eigenvalues of Q are at most $\rho(Q)\mu$.*

6. Uniformly Semiprimitive Multiplicative Processes

The processes studied in §§2–5 were all *stationary* in time, referring to (cyclic) *semigroups* obtained by iterating a single basic nonnegative linear operator. We will now extend the ideas involved to *time-dependent multiplicative processes*, in the following sense.

DEFINITION. Let V be an integrally closed directed vector space with closed convex positive cone C satisfying (2), complete in its projective quasi-metric. A *multiplicative process* on (V, C) is a two-parameter family of nonnegative nonzero linear operators $P_{s,t}$ on (V, C), defined for all $s < t$ in an unbounded subset of $(-\infty, \infty)$, and satisfying

$$P_{s,t}P_{t,u} = P_{s,u} \quad \text{for all } s < t < u \text{ in } S. \tag{18}$$

It is *uniformly semiprimitive as $t \uparrow$* when, for some $\alpha < +\infty$, there exist for any $K > 0$ some $K < s < t$ such that $\Delta(P_{s,t}) < \alpha$. It is *uniformly semiprimitive as $t \downarrow$* when, for some $\alpha < +\infty$ and any $K < 0$, one can find $s < t < K$ such that $\Delta(P_{s,t}) < \alpha$.

THEOREM 9. *Let $P_{s,t}$ define a multiplicative process on (V, C) which is uniformly semiprimitive as $t \downarrow$. Then there is one and (up to a constant positive factor) only one function $f(s)$ with values in C which is consistent with the process in the sense that $f(s)P_{s,t} = f(t)$ for all $s, t \in S$.*

PROOF. To prove uniqueness is easy. By hypothesis, we can find an infinite sequence of negative numbers $s(k) \in S$ such that $\Delta(P_{s,t}) < \alpha$ for some $s(k) < s < t < s(k-1)$, k arbitrary. By Theorem 2 and induction, $\Delta(P_{s(k+n),s(k)}) \leqq \alpha[\tanh(\alpha/4)]^{n-1}$, for all n. But this implies that the projective diameter of the set of possible $f_{s(k)}$ is zero, as claimed.

To prove existence, the previous argument requires further elaboration. For any choice of $g_k > 0$, the function $g_kP_{s(k),t} = g_k(t)$ is consistent for all $t > s(k)$ in

S. By the preceding paragraph, however, the possible $g_k(t)$ so obtained form a Cauchy sequence in the projective quasi-metric of C. Since C is complete (by hypothesis), the limit of this sequence is the sought-for function with values in C.

THEOREM 10. *Let $P_{s,t}$ define a multiplicative process on (V, C) which is consistent as $t \uparrow$. Then there exists one and (essentially) only one positive linear time-dependent functional ϕ_s on (V, C) which is consistent with the process in the sense that $\phi_s(f(s)) = \phi_t(f(t))$ for any $f(s)$ consistent with the process.*

Thus, reversal of time *dualizes* the result, for whose proof the reader is referred to the literature.†

Exercises for §§5–6:

1. Show that, in Example 1 of the text, a primitive linear operator is necessarily uniformly primitive.

2. (a) In $V = \mathbf{R} \circ \mathbf{R} = {}^2\mathbf{R}$, show that the metric space defined by the projective pseudometric has just one point.

(b) Show that, for the positive linear operator $(x, y) \to (x, x + y)$ on ${}^2\mathbf{R}$, the largest eigenvalue is not simple.

3. Show that, for positive P in Example 1:

$$\mathrm{Min}\left(\sum_j p_{ij}\right) \leqq \lambda(P) \leqq \mathrm{Max}_i\left(\sum_j p_{ij}\right), \quad \text{and} \quad \rho(P) \leqq \ln\left[\mathrm{Max}_{j,k}\left(p_{ij}p_{lk}/p_{lj}p_{ik}\right)\right].$$

4. Show that if P is a positive matrix, and Q is nonnegative and irreducible, then PQ and QP are positive.

5. Let $K(x, y)$ be continuous and positive on $0 \leqq x, y \leqq 1$. Show that

$$f(x) = \mu \int_0^1 K(x, y) f(y)\, dy$$

has a unique positive solution, whose (positive) eigenvalue μ satisfies $\mu \leqq |\lambda_j|$ for every other eigenvalue of the equation. (Jentzsch)

6. For any uniformly primitive T, show that $\phi(n) = \ln \|T^n\|$ satisfies $\phi(m + n) \leqq \phi(m) + \phi(n)$: ϕ is subadditive (and negative). Prove that $\lim_{n\to\infty} n^{-1}\|T^n\|$ exists and equals $\ln \rho(T)$.

7. (a) Let $dx_j/dt = \sum q_{jk}(t)x_k$, where $q_{jk}(t) \geqq \epsilon > 0$ for all $j \neq k$. Show that the system has one and only one positive solution $\mathbf{x}(t)$.‡

(b) Show that, if $\psi(t) = \inf_{i\neq j}[q_{ij}(t)q_{jk}(t)]^{1/2}$, if $\mathbf{x}(t)$ and $\mathbf{y}(t)$ are any two solutions of (1) which are positive on $[0, t)$, then $\theta(\mathbf{x}(t), \mathbf{y}(t)) \leqq \theta(\mathbf{x}(0), \mathbf{y}(0)) \exp\left[-2\int_0^t \psi(s)\, ds\right]$.

7. Transition Operators

A very important special class of positive linear operators arises in ergodic theory and stationary Markov processes. A dimension-free§ definition of the relevant operators is the following.

DEFINITION. A *transition operator* on an (L)-space (V, C) is a linear operator T which carries probability distributions (i.e., $p > 0$ with $\|p\| = 1$) into probability distributions. A given probability p is *stable* under T if and only if $pT = p$.

† G. Birkhoff, J. Math. Mech. **14** (1965), 507–12.

‡ G. Birkhoff and L. Kotin, Bull. AMS **71** (1965), 771–2. For analogous results on differential-delay equations, see ibid., J. Differential Equations (to appear).

§ For the finite-dimensional case, see W. Feller, *An introduction to probability theory*···, 2d ed., Wiley, 1958, Chapter XV.

If $f > 0$ is a positive eigenvector for a transition operator T, then so is $p = f/\|f\|$, and so the eigenvalue of f (and p) must be 1 since $\|pT\| = 1$. Further, we have

THEOREM 11. *Any transition operator satisfies*

$$\|fT - gT\| \leqq \|f - g\|. \tag{19}$$

PROOF. Since $fT - gT = (f - g)T$, by linearity, we need only show that $\|hT\| \leqq \|h\|$. But $\|h^+T\| = \|h^+\|$ and $\|h^-T\| = \|h^-\|$, since T carries positive elements of norm one into positive elements of norm one, and likewise for negative elements. Therefore

$$\begin{aligned}\|hT\| &= \|h^+T + h^-T\| \leqq \|h^+T\| + \|h^-T\| \\ &= \|h^+\| + \|h^-\| = \|h\|.\end{aligned}$$

THEOREM 12. *The set F of points of any (L)-space left fixed by any transition operator T is metrically closed, a subspace, and a sublattice.*

PROOF. Since T is continuous, the set is metrically closed in L. Since T carries upper (lower) bounds into upper (lower) bounds, and is a contraction, it carries the unique least upper bound $x = f \vee g$ to f and g, which satisfies $\|f - x\| + \|x - g\| \leqq \|f - g\|$, into itself. Likewise for $f \wedge g$.

COROLLARY. *The set F is closed under the weak topology of the given (L)-space.*

For, any metrically closed subspace of any Banach space is ipso facto weakly closed (Banach [**1**, p. 133], or Dunford–Schwartz).

An especially simple case, analogous to that of "uniformly semiprimitive" linear operators (see Exercise 3 below), is that of transition operators which satisfy the following

Hypothesis of Markoff. For some s, the set of all transforms pT^s of probability distributions p has a positive lower bound $d > 0$, such that $d \leqq pT^s$ for all p.

THEOREM 13. *If Markoff's Hypothesis is satisfied, then there is a unique stable distribution p_0. Moreover, the pT^k tend to p_0 so that*

$$\|pT^k - p_0\| \leqq (1 - \|d\|)^{(k-s)/s}. \tag{20}$$

PROOF. Let there be given probability distributions $p, q \in C \cap S$ (S is the unit sphere of all p with $\|p\| = 1$). Set $h = p \wedge q$, $f = p - h$, $g = q - h$, $\mu = 1 - \|h\|$. Then $f, g \geqq 0$, $\|f\| = \|g\| = \mu$, and $\|p - q\| = \|f\| + \|g\| = 2\mu$, since any (L)-space is a metric lattice. Also, since T is additive, $pT^s - qT^s = fT^s - gT^s$. Further, since the norm is additive in (L)-spaces:

$$\|fT^s - gT^s\| = \|fT^s\| + \|gT^s\| - 2\|fT^s \wedge gT^s\| = 2\mu - 2\|fT^s \wedge gT^s\|.$$

Evidently $f = \mu p_1$ for some distribution p_1; hence, by Markoff's Hypothesis, $fT^s \geqq \mu d$; similarly, $gT^s \geqq \mu d$; therefore $fT^s \wedge gT^s \geqq \mu d$. By obvious substitutions in preceding equations, we infer

$$\|pT^s - qT^s\| = \|fT^s - gT^s\| \leqq 2\mu - 2\mu\|d\| = (1 - \|d\|)2\mu.$$

Since $2\mu = \|p - q\|$, we conclude $\|pT^s - qT^s\| \leqq (1 - \|d\|)\|p - q\|$. The inequality (20) follows by induction on the integral part of $(k - s)/s = (k/s) - 1$.

8. Ergodic Theorem

In the important *deterministic* case of classical statistical mechanics (cf. §1), $\|pT^{r+1} - pT^r\| = \|pT - p\|$ for all r. Hence Markoff's Hypothesis cannot be satisfied, and the transforms of probability distributions which are not stable initially can never converge at all, let alone to a stable distribution. However, their *time-averages* often do converge to a stable distribution; in the present section, sufficient conditions for this to happen will be given.†

Accordingly, let $\{T^r\}$ be any discrete or continuous one-parameter semigroup of linear operators (e.g., isometries) on a Banach space B. An element $f \in B$ will be called "ergodic" if and only if the *averages* $g(s) = s^{-1}(f + fT + \cdots + fT^{s-1})$, resp. $s^{-1}\int_0^s fT^t\,dt$, converge metrically to a fixpoint of T, i.e., to a point of E. There may be no nontrivial ergodic elements, as the following example shows.

Example 5. Let T be the "shift" operator on $L_1(0, \infty)$ which carries $f(x)$ into $f(x + 1)$. Then T is a transition operator, whose only fixpoint is 0.

THEOREM 14. *The set E of elements ergodic under any cyclic semigroup $\{T^r\}$ of linear isometries or contractions of any Banach space B is a weakly closed subspace of B, which contains all its images and antecedents under T^r.*

PROOF. First, E is a subspace, since if the means $g_i(s)$ of f_i converge to a_i for $i = 1, 2$, then the means of $f_1 + f_2$ and λf_1 converge to $a_1 + a_2$ resp. λa_1. Second, E is metrically closed. For, the averages $g_n(s)$ of the f_nT^r on $[0, s)$ satisfy $\|g_n(s) - g(s)\| \leqq \|f_n - f\|$. Hence, $\|f_n - f\| \to 0$ implies $\|g_n(s) - g(s)\| \to 0$ for each s, so that if $\{g_n(s)\} \to a_n$ for each n as $s \to \infty$ (i.e., if each f_n is ergodic), then the a_n form a Cauchy sequence with limit a, and $g(s) \to a$. Finally, it is obvious that E contains all its images and antecedents under T^r, since the means $g(s)$ of any fT^r have the same limit (if any) as those of f:

$$\left\| (ks)^{-1}\left[\int_0^{ks} (fT^r)T^t\,dt - \int_0^{ks} fT^t\,dt\right]\right\| \leqq 2\|f\|/k. \tag{21}$$

Since any metrically closed subspace of any Banach space is weakly closed, the proof is done.

We now obtain a general sufficient condition for an element of a Banach space B to be ergodic.

THEOREM 15. *If the means $g(s)$ of the fT^r lie in a weakly compact set, then f is ergodic.*

PROOF. By the compactness hypothesis, some subsequence $\{g(s(k))\}$ of $\{g(s)\}$ converges weakly to a limit a. On the other hand, every $f - fT^r$ is ergodic, since

† For references to original sources, see [**LT2**, p. 267, footnote 19].

is $s > r$, then as in (21):

$$(21') \qquad \left\| \frac{1}{s} \sum_{k=0}^{s-1} (f - fT^r)T^k \right\| = \frac{1}{s} \left\| \sum_{k=0}^{r-1} fT^k - \sum_{k=s}^{s+r-1} fT^k \right\| \leqq 2r\|f\|/s,$$

so that the means converge metrically to 0.

Hence, writing $g_k = g(s(k))$ for short, every difference $f - g_k$ belongs to the set E of ergodic elements, being a mean of $f - fT^r$. Since E is weakly closed in B, $(f - a) = \mathrm{Lim}_{k\to\infty} (f - g_k) \in E$. It remains to show that $a \in E$; $f = a + (f - a) \in E$ will follow by Theorem 14. But for any linear functional ϕ on B and every k:

$$|\phi(a - aT)| \leqq |\phi(a - g_k)| + |\phi(g_k - g_kT)| + |\phi(g_kT - aT)|.$$

As $k \to \infty$, the first and last summands on the right tend to zero since $g_k \to a$; the middle summand does by (21'). Hence $\phi(a) = \phi(aT)$ for all $\phi \in B'$ (the dual of B); this implies $a = aT$, whence $a \in E$. Q.E.D.

Corollary 1. *If $a \leqq fT^r \leqq b$ for all r, and B is the Banach lattice L_1 or l_1, then f is ergodic.*

In words, f is ergodic if its transforms under the T^r are uniformly order-bounded. This follows from Theorem 14 and the fact [**D-S**, p. 293] that closed intervals $[a, b]$ in the vector lattices specified are weakly compact.

Corollary 2. *If the function $f(x) \equiv 1$ in L_1 is invariant under T, then every element is ergodic.*

For, the condition $f \in [-M, M]$ is invariant under T, since transition operators are linear and order preserving; hence by Corollary 1 every bounded function $f \in L_1$ is ergodic. But bounded functions are dense in L_1; hence the proof is completed by appeal to Theorem 14.

In particular, Corollary 2 applies to any dynamical systems whose phase-space has *finite total measure*, e.g., to compact phase-spaces.

We now use the Kakutani–Yosida representation theorem for abstract (L)-spaces (Theorem XV.27). Let T^n be any discrete semigroup of transition operators on an abstract (L)-space L. For any $f \in L$, the fT^n generate a *separable* closed subspace $F \subset L$ under finite linear combination and metric closure. By the theorem of Kakutani–Yosida, this can be embedded isomorphically and isometrically in $L_1[0, 1]$. Hence the fT^n, *if* uniformly order-bounded in F, are similarly bounded and lie in a weakly compact subset of L. Hence Corollary 1 above applies also to discrete semigroups of transition operators on any abstract (L)-space. To cover the case of continuous semigroups, we need merely transfer our attention to the means of $f_1 = \int_0^1 fT^s\,ds$ under the action of the discrete semigroup $\{T, T^2, T^3, \cdots\}$. This proves

Theorem 16. *Any element f of any abstract (L)-space L, whose transforms under a cyclic semigroup of transition operators are order-bounded, is ergodic.*

The preceding result may be called the Mean Ergodic Theorem for transition operators: in the case of $L_1(M)$, M any measure space, it asserts the convergence *in the mean* of the averages $g(s)$ of the fT^r, *provided* that the latter constitute an order-bounded set.

Exercises for §§7–8:

1. Show that for a square matrix A with spectral radius one:

(a) If diagonalizable, $(f + fA + \cdots + fA^{n-1})/n \to Ef$, where E projects f onto the subspace of fixpoints of A.

(b) This need not be true if A is not diagonalizable.

2. Show that a nonnegative square matrix P represents a transition operator on $\mathbf{R}^n$ if and only if $\sum_j p_{ij} = 1$ for all i. Such nonnegative matrices are called *stochastic*.

3. Show that for stochastic matrices T and probability vectors p, the following conditions on T are equivalent: (a) Uniform semiprimitivity, (b) Markoff's Hypothesis, (c) $\|pT^n - p_0\| \leqq M\rho^n$ for some fixed $\rho < 1$ and p_0, and all p, (d) for some $d > 0$ and finite r, $\|pT^r - qT^r\| \leqq (1 - d)\|p - q\|$ for all p, q.

4. Show that the results of Exercise 3 are not true for transition operators on infinite-dimensional spaces.

5. Show that, if an element f of a Banach space is ergodic, then its means lie on a weakly compact set.

6. Show that, in any abstract (L)-space Σ, the set of all elements which are ergodic for a given transition operator is a closed l-ideal of Σ.

7. Show that, if a subsequence $\{g(s(k))\}$ of the means $g(s)$ of the fT^r converges weakly to a limit a in a Banach space, then a is a fixpoint of T.

*8. A nonnegative square matrix P such that P and its transpose P^T are both stochastic is called *doubly* stochastic. Prove that every doubly stochastic matrix is a weighted mean of permutation matrices.†

9. Metric Transitivity; Pointwise Ergodic Theorem

The ergodic theorem of §8 was concerned with the *existence* of fixpoints of a given transition operator T, and with the convergence of first Cesaro means $(f + fT + \cdots + fT^{n-1})/n$ of the transforms of a given f to such fixpoints. For applications to statistical mechanics, the *uniqueness* of such an invariant probability distribution is equally important. Indeed, in statistical mechanics it is usually *assumed* (on physical grounds) that there is just one invariant ("stable") probability distribution; this is called the Hypothesis of Metric Transitivity.

Metric transitivity is not very interesting when applied to deterministic processes with finite phase-spaces, for the following reasons.

A stochastic matrix represents a *deterministic* process if and only if, by definition, each i is followed by a unique $j(i)$ so that

$$P_{i,j(i)} = 1, \quad \text{and} \quad P_{ik} = 0 \quad \text{if } k \neq j(i). \tag{22}$$

To be "reversible", a deterministic process must be defined by a permutation matrix (so as to have an inverse). In any case, since the set of matrices which satisfy (22) is finite for any n, deterministic processes on finite phase-spaces must

† For continuous analogs of this result, see J. W. Ryff, Trans. AMS **117** (1965), 92–100. For infinite matrix analogs, see D. G. Kendall, J. London Math. Soc. **35** (1960), 81–4, and J. R. Isbell, Canadian Math. Bull. **5** (1962), 1–4.

be *discrete* in time. Moreover, since every permutation matrix is a direct sum of cyclic permutation matrices, a reversible deterministic transition operator can only be *metrically transitive* if it is a *cyclic* permutation matrix of order n. Finally, its stable distribution cannot ever be "dominant"!

On the other hand Markoff's Hypothesis (and likewise the hypothesis of uniform semiprimitivity) both imply metric transitivity, as we have seen earlier.

Pointwise Ergodic Theorem. Theorem 15 is often called the Mean Ergodic Theorem, because it refers to convergence in the mean of "time-averages" under iteration. A stronger result is the following, first proved for measure-preserving flows such as arise in dynamics by G. D. Birkhoff.†

THEOREM 17. *Let T be any transition operator on $L_1(X)$, X a probability measure space. Then, with probability one (for almost all $x \in X$):*

$$\lim \frac{1}{n} (f + fT + \cdots + fT^{n-1}) \tag{23}$$

exists.

PROOF. Since lim inf and lim sup exist for any sequence of measurable function, the set where lim inf < lim sup is measurable. If it had positive measure, we would have

$$\liminf \frac{1}{n} (f + \cdots + fT^{n-1}) < a < b < \limsup \frac{1}{n} (f + \cdots + fT^{n-1}) \tag{23'}$$

on a set of positive measure m on X. This is however incompatible with the following Maximal Ergodic Theorem, due to E. Hopf‡ and recently given a very short proof by A. M. Garsia.‡

MAXIMAL ERGODIC THEOREM. *Let P be any nonnegative linear operator on $L_1(X)$ such that $\|gP\| \leqq \|g\|$ for all g. Let $fS_n = f + fP + \cdots + fP^{n-1}$, and let R_n be the sublinear operator:*

$$fR_n \equiv 0 \vee f \vee (f + fP) \vee \cdots \vee fS_n = 0 \vee \bigvee_{k=1}^{n} fS_k. \tag{24}$$

Then, if M_n is the set where $fR_n > 0$ (some $fS_k > 0$).

$$\int_{M_n} f(x)\, dm(x) \geqq 0. \tag{24'}$$

PROOF. Since P is nonnegative, from (24) we get

$$fR_nP \geqq \bigvee_{k=1}^{n} fS_kP, \tag{25a}$$

whence

$$f + fR_nP \geqq f + \bigvee_{k=1}^{n} fS_kP. \tag{25b}$$

---

† Proc Nat. Acad. Sci. **17** (1931), 650–65. See [**LT2**, pp. 266–7], for a discussion of the literature.

‡ E. Hopf, J. Rat. Mech. Anal. **3** (1954), 13–45; A. M. Garsia, J. Math. Mech. **14** (1965), 381–2. See also G.-C. Rota, Proc. AMS **14** (1963), 722–3.

But by definition of S_k, $f + \bigvee_{k=1}^{n} fR_kP \geqq \bigvee_{k=1} fS_k$. Hence

(26) $$f + fR_nP \geqq \bigvee_{k=1}^{n} S_k.$$

Again, on the set M_n where† $fR_n > 0$, by (24) we have $fR_n = \bigvee_{k=1}^{n+1} fS_k$, since $fS_0 \geqq 0$; hence (26) implies

(26′) $$f \geqq fR_n - fR_nP \quad \text{on } M_n.$$

Integrating (26′), we get the first inequality of

(27) $$\int_{M_n} f\,dm \geqq \int_{M_n} (fR_n - fR_nP)\,dm \geqq \int_X (fR_n - fR_nP)\,dm.$$

The second inequality follows since, by construction, $fR_n \leqq 0$ on the complement of M_n and (P being positive) $fR_nP \geqq 0$ on X. Finally, in the last term of (26), $\int_X fR_n\,dm = \|fR_n\|$, while $\int_X (-fR_nP)\,dm = \|fR_nP\| \leqq \|fR_n\|$, P being a contraction. Hence the last term in (27) is nonnegative, which completes the proof of (24). We omit the proof that (24′) implies (23′); for this, see G.-C. Rota, op. cit. supra.

PROBLEMS

153. Establish a bijection between the notion of a (consistent) one-parameter family of transition operators on $L^1(\mathbf{R})$ with that of a "stochastic process" in the sense of Doob.‡

154. A linear operator Q is called *essentially positive* when $e \wedge f = 0$ implies $e \wedge fQ = 0$. Prove that, if Q is *bounded* (i.e., $fQ \leqq \lambda f$ for some fixed $\lambda = \lambda(Q)$), and acts on a complete vector lattice, then $\{e^{tQ}\}$ is a semigroup of positive linear operators.

155. Establish existence and uniqueness theorems for positive distributions which are consistent with bidirected groups (and directed semigroups) of positive linear operators.§

156. Let $K(\mathbf{x}, \mathbf{u}; \mathbf{y}, \mathbf{v})$ be the kernel for "elastic" scattering of neutrons. Show that the resulting "migration kernel" is uniformly primitive in a bounded reactor.¶

† See [**LT2**, p. 250, Theorem 13] for an extension of this concept to abstract (L)-spaces.

‡ J. L. Doob, *Stochastic processes*, Wiley, New York, 1954. Cf. [**LT2**, Problem 110].

§ That is, do for positive linear operators what was done for transition operators in G. Birkhoff and L. Alaoglu, Ann. of Math. **41** (1940), 293–309.

¶ Cf. G. Birkhoff, Rend. Mat. **22** (1963), 102–26.

CHAPTER XVII

LATTICE-ORDERED RINGS

1. *po*-rings and *l*-rings

Many rings are partly ordered, or even lattice-ordered, in a way which satisfies the following conditions.

DEFINITION. A *po-ring* is a ring R which is also a poset under a relation $\geqq$, in which

$$\text{(1)} \qquad x \geqq y \quad \text{implies} \quad a + x \geqq a + y \quad \text{for all } a \in R,$$

$$\text{(2)} \qquad x \geqq 0 \quad \text{and} \quad y \geqq 0 \quad \text{imply} \quad xy \geqq 0 \quad \text{in } R.$$

An *l-ring* is a *po*-ring R which is a lattice under $\geqq$.

Remark. In a *po*-ring without zero-divisors, (2) is also equivalent to the condition

$$\text{(2')} \qquad x > 0 \quad \text{and} \quad y > z \quad \text{imply} \quad xy > xz.$$

Since any ring is a commutative group under addition, we see that under $+$, $\geqq$, any *po*-ring is an Abelian *po*-group, and any *l*-ring is an Abelian *l*-group. Conversely, any Abelian *po*-group (or *l*-group) which is also a ring is a *po*-ring (resp. *l*-ring) if (2) holds. In particular, any *l*-group G becomes an *l*-ring if one sets $ab \equiv 0$, thus making G into a zero ring.

Example 1. Let R be any commutative associative† ring in which no sum of nonzero squares is zero. Then, by defining $a \geqq 0$ to mean that a is a sum of squares, we make R into a *po*-ring.

Example 2. Let $M_n(K)$ be the *full matrix ring* of all $n \times n$ matrices $A = \|a_{ij}\|$ having entries in an ordered field K. Then, by defining $A \geqq B$ to mean that $a_{ij} \geqq b_{ij}$ for all $i, j = 1, \cdots, n$, we get an *l*-ring.

Example 3. Let $C[0, 1]$ be the vector lattice of all continuous functions on $[0, 1]$. Then, under the usual definitions of $f + g$ and fg, $C[0, 1]$ is an *l*-ring.

Note that the vector lattices $L_p[0, 1]$ with $1 \leqq p < \infty$ are *not* *l*-rings, since the classes of functions which they include are not closed under multiplication.

po-rings and *l*-rings are of many diverse types, which have few common properties. However, we have

LEMMA 1. *Condition* (2) *is equivalent in any po-group to*

$$\text{(3)} \qquad x \geqq 0 \quad \textit{and} \quad y \geqq z \quad \textit{imply} \quad xy \geqq xz \quad \textit{and} \quad yx \geqq zx.$$

† In Chapter XVII, rings are not assumed to be associative unless this is specifically stated.

PROOF. By (1) and the elementary theory of *po*-groups, $y \geqq z$ if and only if $y - z \geqq 0$, while $xy \geqq xz$ if and only if $x(y - z) = xy - xz \geqq 0$—and similarly for $yx \geqq zx$. The equivalence now follows.

THEOREM 1. *In any l-ring R*:

$$a \geqq 0 \quad \textit{implies} \quad a(b \vee c) \geqq ab \vee ac, \quad (a \vee b)c \geqq ac \vee bc, \tag{4}$$

$$a(b \wedge c) \leqq ab \wedge ac, \quad \textit{and} \quad (a \wedge b)c \leqq ac \wedge bc,$$

$$|ab| \leqq |a| \cdot |b|. \tag{5}$$

PROOF. The inequalities (4) follow from Lemma 1, since $b \vee c \geqq b, c$ etc. The inequality (5) also follows from Lemma 1, since

$$\begin{aligned} -|a| \cdot |b| &= -a^+b^+ + a^+b^- + a^-b^+ - a^-b^- \\ &\leqq a^+b^+ + a^+b^- + a^-b^+ + a^-b^- = ab \\ &\leqq a^+b^+ - a^+b^- - a^-b^+ + a^-b^- = |a| \cdot |b|. \end{aligned}$$

THEOREM 2. *The class of l-rings is equationally definable; so is the class of commutative l-rings.*

PROOF. The class of Abelian *l*-groups is equationally definable (Theorem XIII.2, Corollary 1); so is the class of rings. But as remarked above, an *l*-ring is just an Abelian *l*-group which is a ring, in which (2) holds. But in such a system, the implication (2) is equivalent to the identity

$$[(a \vee 0)(b \vee 0)] \wedge 0 = 0;$$

hence the class of *l*-rings is equationally definable. The second assertion of Theorem 2 is a trivial corollary.

2. Ordered Rings and Fields

There is an extensive theory of (simply) ordered rings and fields. An excellent exposition of this theory.may be found in [**Fu**, Chapters VII–VIII]. We will here present only a few simple results. First, we have

LEMMA 2. *In any ordered ring, the inequalities of Theorem 1 can be replaced by equalities.*

PROOF. Ad (4), $b \vee c$ is just the larger of b and c, etc. Ad (5), $|a| = \pm a$. We omit the obvious details.

Next, let us define a *po*-ring to be *Archimedean* when it is Archimedean considered as an additive *po*-group. We then have the following result of Pickert and Hion.†

THEOREM 3. *An Archimedean ordered ring R is either a zero-ring, or l-isomorphic with a unique subring of the real field* **R**.

† Ya. V. Hion, Uspehi Mat. Nauk **9** (1954), 237–42.

PROOF. By Hölder's Theorem XIII.12, R is isomorphic with an additive subgroup of $\mathbf{R}$ under $\geqq, t$; moreover every $e > 0$ in R is a strong unit. Furthermore, either $e^2 = 0$ (in which case R is a zero-ring), or $e^2 > 0$ satisfies for some positive scalar α:

$$(6) \qquad ne^2 \geqq me \quad \text{if and only if } m/n \geqq \alpha.$$

It follows, referring back to Hölder's construction, that the mapping $\lambda e \to \lambda\alpha$ is an *isomorphism* of R with an ordered subring of $\mathbf{R}$; we omit the details.

Non-Archimedean ordered fields are easy to construct as fields of *formal power series* with exponents in $\mathbf{Z}$ or some other ordered group Γ.

Example 4. Let Γ be any ordered group, and K any ordered field. Define $K[[\Gamma]]$ as the set of all "formal power series" of the form

$$(7) \qquad f = \sum_W a_\gamma x^\gamma, \qquad a_\gamma \in K, \qquad \gamma \in W,$$

for some *well-ordered* subset W of Γ. Sums and products are defined by the rules

$$(7a) \qquad \sum_W a_\gamma x^\gamma + \sum_W b_\gamma x^\gamma = \sum_W (a_\gamma + b_\gamma) x^\gamma,$$

$$(7b) \qquad \left(\sum_W a_\gamma x^\gamma\right)\left(\sum_W b_\gamma x^\gamma\right) = \sum_{W+W} (a_\gamma b_{\gamma'}) x^{\gamma+\gamma'},$$

inserting "dummy" terms with zero coefficients as needed. Since $W \cup W_1$ is well ordered if W and W_1 are, we have closure under $+$. The ordering is according to the first nonzero coefficient a_γ, as in $[^\Gamma K]$ defined in §VIII.14; hence $K[[\Gamma]]$ is an ordered (Abelian) group. Again, the set of γ such that $\gamma + \gamma' = \delta$ (fixed) is well ordered and (setting $\gamma = -\gamma' + \delta$) dually; hence it is finite. Therefore (7b) is well defined.

LEMMA 3. *$K[[\Gamma]]$ is an ordered ring.*

The proof will be left to the reader.

THEOREM 4 (HAHN [**1**]). *The ring $K[[\Gamma]]$ is a field.*†

PROOF. It suffices to show that, given $f \neq 0$ in (7), the equation $fg = 1 = 1x^0$ has a solution. Since $f \neq 0$, we can find $\sum_B a_\alpha^{-1} x^{-\alpha} = h$ such that

$$fh = 1 + \sum_B b_\beta x^\beta, \quad \text{all } \beta > 0,$$

where B is a *well-ordered* subgroup of positive elements of Γ. We can then solve $(1 + \sum_B b_\beta x^\beta)(1 + \sum_B c_\beta x^\beta) = 1 = 1 + \sum 0 \cdot x^\beta$ by induction on β, for the well-ordered set of c_β; we omit the details.

In an ordered field or division ring (skew field) F, one can define the relation

† The preceding construction has been extended to division rings (noncommutative rings) by B. H. Neumann, Trans. AMS **66** (1949), 202–52. See [**Fu**, pp. 134–8]. For a special case, see [**LT2**, p. 227, Example 11].

$a \ll b$ just as for ordered groups. Moreover, since the mapping $a \to |a|$ is a morphism for multiplication, we can restrict attention to positive elements in considering "orders of magnitude" in F (i.e., equivalence classes of elements $a \sim b$). We have

THEOREM 5. *In any ordered field or division ring, the orders of magnitude of nonzero elements form an ordered group.*

PROOF. Since $nx \leqq y$ if and only if $n(ax) = a(nx) \leqq ay$ and symmetrically, the equivalence relation $\sim$ and the ordering $\ll$ are preserved under multiplications $x \to axb$ for all nonzero a, b.

In Example 4, the group in question is Γ. In an Archimedean field, it is the trivial one-element group.

Exercises for §§1–2:

1. Show that, in any (simply) ordered ring, any square is positive.
2. Show that the real field **R** and the rational field **Q** can be made into ordered fields in just one way, but that this is not true of $\mathbf{Q}(\sqrt{2})$.

Define an *l-algebra* as an l-ring which is a real vector lattice under a scalar multiplication on its additive l-group.

3. Show that, in any two-dimensional l-algebra A, the set A^+ of all $a > 0$ is a sector of one of the two following types: (i) $\alpha \leqq \arg a \leqq \beta$, $0 < \beta - \alpha < \pi$ (Archimedean case), (ii) $\alpha < \arg a \leqq \alpha + \pi$ or $\alpha \leqq \arg a < \alpha + \pi$ (non-Archimedean case).
4. Show that the complex field **C** cannot be made into an l-algebra.
5. Let A be the algebra of real *dual numbers* $a = x + ye$ $(x, y \in \mathbf{R})$, where $e^2 = 0$; let A^+ be the sector $|(\pi/4) - \arg a| \leqq \pi/8$. Show that A is an l-algebra.
6. Construct an l-algebra with unity 1, where $1 \not> 0$. (*Hint.* See Exercise 5)

*7. Determine all two-dimensional l-algebras. (*Hint.* See Birkhoff–Pierce [**1**, §4, Examples 9a–9g].)

*8. Show that a field F is "simply orderable" if and only if it is "formally real", in the sense that equations of the form $\sum x_i^2 = -1$ have no solution in F.†

3. *L*-ideals; the Radical

Let R be any l-ring, and let θ be any congruence relation on R. Since R is an l-group, θ is determined by the l-ideal J of $x \equiv 0$ (θ) in R. Moreover, by the theory of rings, the l-ideals which correspond to congruence relations for multiplication as well as those for which $x \in J$ and $a \in R$ imply $ax \in J$ and $xa \in J$. The preceding observations can be summarized as follows.

DEFINITION. An L-ideal of an l-ring R is a nonvoid subset $J \subset R$ such that: (i) $x \in J$ and $y \in J$ imply $x \pm y \in J$, (ii) $x \in J$ and $|t| \leqq |x|$ imply $t \in J$, (iii) $x \in J$ and $a \in R$ imply $ax \in J$ and $xa \in J$.

THEOREM 6. *The congruence relations on any l-ring R are the partitions of R into the cosets of its different L-ideals.*

An l-ring R without proper L-ideals (i.e., whose only L-ideals are 0 and R) may therefore be called *simple*. Using Zorn's Lemma (Chapter VIII), it is easy to show that any l-ring $R \neq 0$ with multiplicative unity 1 has at least one *maximal*

† E. Artin and O. Schreier, Hamb. Abh. **5** (1926), 83–100. For later work, see D. W. Dubois, Proc. AMS **7** (1956), 918–30.

proper L-ideal M, such that R/M is simple. We define the *L-radical* of an l-ring R as the intersection $\bigcap M_\alpha$ of its maximal proper L-ideals. Clearly, $R/\bigcap M_\alpha$ is a subdirect union of simple l-rings, and conversely; such an l-ring is called *semisimple.*

It is easily verified that the l-ring of Example 2, §1, is simple, while that of Example 3 is semisimple. For any $x \in [0, 1]$, the set of all $f \in C[0, 1]$ such that $f(x) = 0$ is a maximal proper ideal M_x, and $\bigcap M_x = 0$. We will now show that $C[0, 1]$ has no other maximal proper l-ideals, let alone maximal proper L-ideals.

LEMMA. *Let J be any proper l-ideal of $C[0, 1]$. Define the "support" of J as the set $S(J)$ of all x such that $f(x) \neq 0$ for some $f \in J$. Then $S(J) < [0, 1]$.*

PROOF. Suppose $S(J) = [0, 1]$. Then for each $x \in [0, 1]$, there would exist an $f \in J$ such that $f(x) \neq 0$; hence an $h = |f| \in J^+$ such that $h(x) > 0$. By the Heine–Borel covering theorem, we could therefore find a finite set $h_1, \ldots, h_n \in J^+$ such that the union of the open sets S_i where $h_i(y) > 0$ covered $[0, 1]$. The sum $h^* = h_1 + \cdots + h_n \in J^+$ would therefore be identically positive on $[0, 1]$, and therefore a strong unit of $C[0, 1]$. This would imply $J = C[0, 1]$, contrary to hypothesis. Q.E.D.

Now let M be a *maximal* proper l-ideal of $C[0, 1]$. By the preceding lemma, there exists some $x \in [0, 1]$ such that $f(x) = 0$ for all $f \in M$. Hence $M \subset M_x$—but since M is maximal, this implies $M = M_x$. We have proved the

COROLLARY. *In $C[0, 1]$, the maximal proper l-ideals are precisely the sets M_x of functions such that $f(x) = 0$ for some fixed $x \in [0, 1]$; they are all L-ideals.*

Another kind of radical is the l-radical, defined as follows in *associative* rings.

DEFINITION. The l-radical of an l-ring R is the set N of all $a \in R$ such that, for some positive integer $n = n(a)$ and all $x_0, \cdots, x_n \in R$:

$$x_0 |a| x_1 |a| x_2 \cdots x_{n-1} |a| x_n = 0. \tag{8}$$

THEOREM 7. *The l-radical of R is an L-ideal, which is the union of the nilpotent L-ideals of R. Every element of N is nilpotent.*

PROOF. If J is a nilpotent L-ideal with $J^n = 0$, then evidently (8) holds for all $a \in J$. Conversely, if $a \in N$ satisfies (8), then

$$(|a|\, R)^{n+1} = (R\, |a|)^{n+1} = (R\, |a|\, R)^n = (|a| + |a|\, R + R\, |a| + R\, |a|\, R)^{2n+1} = 0;$$

hence a is contained in a nilpotent L-ideal. Finally, (8) trivially implies $|a^{2n+1}| = 0$: every $a \in R$ is nilpotent.

For various corollaries of Theorem 7, see G. Birkhoff and R. S. Pierce [**1**, §3].

4. Representations; Regular l-rings

In many l-rings, the "positive cone" has important properties not implied by (2). One of these properties is *regularity*, to be defined below. This property is closely related to the concept of l-representation, which will be introduced first. To define it we need

Example 5. Let G be any additive Abelian *po*-group. In the ring $E(G)$ of all group-endomorphisms of G, define $\theta \geqq 0$ to mean that $g \geqq 0$ implies $g\theta \geqq 0$ in G. This makes $E(G)$ into a *po*-ring. (For instance, if $G = \mathbf{R}^n$, then $E(G)$ is the l-ring of Example 2, §1.)

DEFINITION. By an *l-representation* of an l-ring R, is meant a morphism of l-rings from R into the *po*-ring $E(G)$ of all endomorphisms of some l-group G.

For any l-ring $R = (R, +, \cdot, \vee)$, consider the ring-morphism† which maps each $a \in R$ onto the endomorphism $\theta_a: x \to xa$ of $(R, +)$, the so-called *regular* representation. The mapping $a \to \theta_a$ is a *p-representation*, in the sense that $a \geqq 0$ in R implies $\theta_a \geqq 0$ in $E(R, +, \vee)$. It is faithful (i.e., one-one) if R has a unity, or more generally if $0{:}R = 0$.

DEFINITION. The l-ring R is *right*-regular when its regular p-representation is a faithful l-representation; it is *left*-regular when its opposite is right-regular; R is *regular* when it is both right- and left-regular.

Thus, in a right-regular ring, we have

$$\theta_a + \theta_b = \theta_{a+b}, \qquad \theta_a\theta_b = \theta_{ab}, \qquad \theta_a \wedge \theta_b = \theta_{a\wedge b}. \tag{9}$$

Setting $b = 0$, the last identity implies

$$\text{If } xa \geqq 0 \text{ for all } x \geqq 0, \text{ then } a \geqq 0 \text{ in } R. \tag{10}$$

It seems unlikely that conversely, (10) implies right-regularity. Similar remarks hold for left-regular rings. Since $\theta_1 > \theta_0$ in $E(R)$, clearly any right-unity 1 is positive in any right-regular l-ring.

Although the majority of l-rings are not "regular" in the preceding sense,‡ apparently most significant l-rings *are* regular. For example, we have

LEMMA 1. *The full matrix ring $M_n(K)$ of Example 2 is regular, for any ordered field K.*

PROOF. Let E_{hk} be the matrix with (h, k)-entry 1 and all other entries 0. Let $A = \sum a_{hk}E_{hk}$ and $B = \sum b_{hk}E_{hk}$ be any elements of $M_n(K)$. Then:

$$E_{hk}A \wedge E_{hk}B = \left(\sum a_{kj}E_{hj}\right) \wedge \left(\sum b_{kj}E_{hj}\right) > 0 \quad \text{if and only if, for some } j,\ a_{kj} \wedge b_{kj} > 0. \tag{11}$$

It follows that, if $A \wedge B = 0$, then $E_{hk}A \wedge E_{hk}B = 0$. Hence, if $A \wedge B = 0$ and $\theta \leqq \theta_A \wedge \theta_B$ is in $E(M_n(K))$, then $\theta \leqq 0$ in $E(M_n(K))$. That is

$$A \wedge B = 0 \ \text{ in } M_n(K) \ \text{ implies } \ \theta_A \wedge \theta_B = 0 \ \text{ in } E(M_n(K)). \tag{12}$$

The proof that $M_n(K)$ is right-regular is completed by

LEMMA 2. *An l-ring R is right-regular if and only if $a \wedge b = 0$ in R implies $\theta_a \wedge \theta_b = 0$ in $E(R)$.*

---

† See for example N. Jacobson, *The theory of rings*, AMS Survey #2, 1943, p. 16. By the "opposite" of R is meant the ring R obtained by reversing the order of multiplication.

‡ See Birkhoff–Pierce [**1**, §4]. The present discussion reworks the material of §§6–7, op. cit.

PROOF. The necessity is trivial, by (9). To prove sufficiency, observe that $a \wedge b = c$ in R implies $\theta_a \wedge \theta_b = \theta_c$ in $E(R)$ if:

$$(12') \quad (a - c) \wedge (b - c) = 0 \quad \text{in } R \quad \text{implies} \quad \theta_{a-c} \wedge \theta_{b-c} = 0 \quad \text{in } E(R),$$

by homogeneity (the invariance of order in R and in $E(R)$ under group-translations).

Finally, since $M_n(K)$ is its own opposite (under the correspondence $A \to A^T$) its right-regularity implies also left-regularity. Q.E.D.

We note also that, since the correspondences $a \to -a$ and $\theta \to -\theta$ are dual automorphisms in R and in $E(R)$ alike, we have

LEMMA 3. *An l-ring is right-regular if and only if the mapping $a \to \theta_a$ is an l-homomorphism of R into $E(R)$.*

It follows that the positive cone of any regular l-ring is an l-semigroup under multiplication and joins. The converse is presumably not true.

Exercises for §§3–4:

1. (a) Show that the l-ideals of any l-ring R form a complete Brouwerian lattice.
   (b) Prove the same result for the L-ideals of R.
2. Show that, if the lattice of all l-ideals of an l-ring R is Noetherian, then R can be uniquely represented as a direct "product" (sum) of indecomposable factors.
3. Let all $c_{ij}^k \geqq 0$, and let $\sum \gamma_i e_i \geqq 0$ mean that every $\gamma_i \geqq 0$. Show that the set of all $\sum \gamma_i e_i$, with the multiplication $(\sum \lambda_i e_i)(\sum \mu_j e_j) = \sum \lambda_i \mu_j c_k^{ij} e_k$, defines an l-algebra.
4. Let R be an l-ring whose L-ideals satisfy the ascending or descending chain condition. Show that the l-radical of R is nilpotent.
5. (a) Show that, if N is the l-radical of a *commutative* l-ring R, then the l-radical of R/N is $N/N \cong 0$.
   (b) Show that this is not true in all l-rings.†

The *i-radical* of a (real) vector lattice with strong unit u is the set $i(V)$ of all $x \in V$ such that $n|x| \leqq u$ for every positive integer n. (N.B. The choice of u is immaterial.)

6. (a) Let A be a finite-dimensional real l-algebra. Show that $i(A) \subset N$, where N is the l-radical of A.
   (b) Infer that, if $N = 0$, then A is Archimedean and isomorphic to one of the l-algebras of Exercise 3.

5. Function Rings

Lattice-ordered rings of functions with values in a (simply) ordered field have a characteristic property which is much stronger than "regularity": namely, their closed l-ideals are also L-ideals. Though this property implies regularity, as will be shown below, the "regular" l-ring $M_n(K)$ does not have it, hence it is not implied by regularity.

DEFINITION. A function ring or *f-ring* is an l-ring in which

$$(13) \qquad a \wedge b = 0 \quad \text{and} \quad c \geqq 0 \quad \text{imply} \quad ca \wedge b = ac \wedge b = 0.$$

Clearly any (simply) ordered ring is an f-ring, since $a \wedge b = 0$ implies $a = 0$ or $b = 0$ in any ordered ring; in either case, for any $c \geqq 0$, one easily verifies that $ca \wedge b = 0$ and $ac \wedge b = 0$.

† Proofs of this and most of the other results stated in the Exercises for §§1–5 may be found in Birkhoff–Pierce [1].

LEMMA 1. *In any f-ring, we have*

(14) $$a \wedge b = 0 \quad \textit{implies} \quad ab = 0.$$

PROOF. By (13), $a \wedge b = 0$ implies $ab \wedge b = 0$ and hence $ab \wedge ab = 0$; (14) is now obvious.

An important class of f-rings is obtained from the following easily proved lemma.

LEMMA 2. *Let R be an l-ring with unity 1. If 1 is a strong unit, then R is an f-ring. Conversely, if R is an f-ring, then 1 is a weak unit.*

PROOF. Let $a \wedge b = 0$ and $c \geqq 0$. If 1 is a strong unit, then $0 \leqq c \leqq n1$ for some positive integer n. Hence, by the theory of l-groups $0 \leqq ca \wedge b \leqq na \wedge b = 0$. Similarly, $ac \wedge b = 0$. Conversely, in any f-ring, $a \wedge 1 = 0$ implies $0 = a \wedge (1a) = a \wedge a = a$.

LEMMA 3. *An l-ring is an f-ring if and only if its elements satisfy the identities:*

(15) $$(a \vee 0) \wedge [(-a \vee 0)(c \vee 0)] = 0 = (a \vee 0) \wedge [(c \vee 0)(-a \vee 0)].$$

PROOF. The set of pairs, a, b with $a \wedge b = 0$ is the same as the set of pairs $d \vee 0, -d \wedge 0$, by the formulas of §XIII.4 (set $d = a - b$).

As a corollary of Lemma 3 and Theorem 2, we have

THEOREM 8. *The class of f-rings is equationally definable.*

COROLLARY. *Any f-ring is a subdirect product of subdirectly irreducible f-rings.*

LEMMA 4. *Any subdirectly irreducible f-ring F is simply ordered.*

PROOF. First consider F as an additive Abelian l-group. The argument used to prove Theorem XIII.21 can be repeated. Either F is simply ordered, or it contains two elements $a > 0$ and $b > 0$ with $a \wedge b = 0$. Let A be the closed l-ideal $b^\perp$ and let $B = A^\perp = (b^\perp)^\perp$. Then $A \cap B = 0, a \in A$, and $b \in B$. Moreover A and B are both L-ideals, being intersections of L-ideals by (13). Hence, unless simply ordered F is subdirectly reducible. Q.E.D.

From the preceding corollary, it follows that every f-ring is a subdirect product of ordered rings. Conversely, (13) holds trivially in any ordered ring; hence by Theorem 8 it holds in every subdirect product of ordered rings. In summary, we have proved

THEOREM 9. *An l-ring is an f-ring if and only if it is a subdirect product of (simply) ordered rings.*

COROLLARY 1. *In any f-ring, the following identities are satisfied:*

(16) $$\textit{If} \quad a \geqq 0, \quad \textit{then} \quad a(b \vee c) = ab \vee ac, \qquad a(b \wedge c) = ab \wedge ac,$$
$$(b \vee c)a = ba \vee ca, \qquad (b \wedge c)a = ba \wedge ca.$$

(17) $$|ab| = |a| \cdot |b|.$$

COROLLARY 2. *Any f-ring is regular.*

Again, it is of interest to note that any l-ring R with l-radical 0 which satisfies the identities (16) is an f-ring. More generally (Birkhoff–Pierce, [**1**, Theorem 14]), if $0 .\!\cdot R = 0 \cdot\!. R = 0$ and (16) holds, then R is an f-ring.

THEOREM 10. *Any Archimedean f-ring is commutative and associative.*

PROOF. We first prove that if $a \geqq 0$, $b \geqq 0$ in any f-ring R, then

$$n|ab - ba| \leqq a^2 + b^2, \quad \text{for any positive integer } n. \tag{18}$$

By Theorem 9, it suffices to prove (18) for ordered rings, in which we can suppose $a \geqq b$. Then, for some nonnegative integer k, $nb = ka + r$, where $0 \leqq r < a$. Hence $n|ab - ba| = |ka^2 + ar - ka^2 - ra| = |ar - ra| \leqq a^2 \leqq a^2 + b^2$. If R is Archimedean, then (18) implies $ab - ba = 0$ by definition, whence any Archimedean f-ring is commutative.

We next recall that associativity of multiplication was not assumed in proving Theorem 9. Further, by an argument similar to that used in proving (18), one can prove—still without assuming associativity—that if a, b, c are nonnegative elements of an f-ring R, and n is any positive integer, then:

$$\begin{aligned} n|(ab)c - a(bc)| \leqq\ & ab(a + b + ab) + a(a + a^2 + ba) \\ & + cb(c + b + cb) + c(c + c^2 + bc). \end{aligned} \tag{19}$$

Hence, if R is Archimedean, $(ab)c - a(bc) = 0$. This completes the proof.

For additional properties of l-radicals of f-rings, see Birkhoff–Pierce, [**1**, §10]; also R. S. Pierce, Duke Math. J. **23** (1956), 253–61, D. G. Johnson, Acta Math. **104** (1960), 163–215, and [**Fu**, §IX.3]. In particular, we note the following

THEOREM OF FUCHS. *An l-ring is an f-ring if and only if all its closed l-ideals are L-ideals.*

6. Almost f-rings

An associative l-ring which satisfies (14) may be called an *almost f-ring*. Such an "almost" f-ring need not be an f-ring (see Birkhoff–Pierce, [**1**, p. 62, Example 16]). In this section, we shall characterize the class of "almost f-rings" with positive unity 1. We first prove

THEOREM 11. *Let R be an l-ring with a positive unity* 1 *which is a weak order unit. If $a > 0$ and a is nilpotent, then $a \ll 1$.*

PROOF. Let $B(R)$ be the set of all "bounded" elements $x \in R$ such that $|x| \leqq n$ for some integral multiple of 1. Clearly, $B(R)$ is an l-ideal and a subring of R. Moreover, by Lemma 2 of §5, $B(R)$ is an *f-ring* with strong unit 1. Now let $a > 0$ be nilpotent; clearly $a \wedge 2 \in B(R)$.

Suppose it were *not* true that $a \wedge 2 \leqq 1$. Then, by Theorem 9, we could find an l-epimorphism $b \to b'$ of $B(R)$ onto an ordered ring in which $a' \wedge 2' > 1'$, whence $(a \wedge 1)' = 1'$ would not be nilpotent. This is impossible, since $a \wedge 1$ is nilpotent; hence $a \wedge 2 \leqq 1$.

From this it follows that $(a - 1) \wedge 1 \leqq 0$ (subtracting 1 from all terms), or $0 = 0 \vee [(a-1) \wedge 1] = [0 \vee (a-1)] \wedge 1$. Since 1 is a weak unit by hypothesis, this implies $0 \vee (a - 1) = 0$; or $a - 1 \leqq 0$, whence $a \leqq 1$.

Finally, if $a > 0$ is nilpotent, the same is true of $na = n|a|$ for every positive integer n; hence $na \leqq 1$ for every such n by the preceding argument. That is, $a \ll 1$. Q.E.D.

Using Theorem 11, a rather involved argument which will not be reproduced here leads to the following result (Birkhoff–Pierce [**1**, Theorem 15]).

Theorem 12. *Let R be an (associative) l-ring with positive unity 1. Then R satisfies (14) if and only if 1 is a weak order unit.*

We now justify the name "almost f-ring", by showing that almost f-rings which are not f-rings must have some very special properties.

Lemma 1. *If R is an almost f-ring which contains no nonzero, positive, nilpotent elements, then R is an f-ring.*

Proof. If $a \wedge b = 0$ and $c \geqq 0$, then

$$0 \leqq (ac \wedge b)^2 \leqq b(ac) = (ba)c = 0,$$

by (14). Thus $ac \wedge b = 0$. Similarly, $ca \wedge b = 0$.

Corollary. *Let R be an Archimedean associative l-ring with (positive) ring-unity 1 which is a weak order unit. Then R is an f-ring.*

Proof. Since R is Archimedean, it can contain no positive nilpotent element, by Theorem 11. Again, R satisfies (14) (is an almost f-ring), by Theorem 12. Hence, by Lemma 1, R is an f-ring.

Lemma 2. *In an almost f-ring, every square is positive.*

Proof. For any a,

$$a^2 = (a^+ - (-a)^+)^2 = (a^+)^2 + ((-a)^+)^2 = (a^+)^2 + (-a^-)^2,$$

since $a^+ \wedge (-a)^+ = 0$, whence $a^+(-a)^+ = (-a)^+a^+ = 0$.

Exercises for §§5–6:

1. Show that one can make any vector lattice into an l-ring, by defining $ab \equiv 0$ (zero-ring).
2. Prove that an l-ring is an f-ring if and only if all its closed l-ideals are L-ideals. [**Fu**, §IX. 3]
3. Show, that, if $a \wedge b \geqq 0$, then $|ab - ba| \ll a^2 \vee b^2$ in any f-ring.†
4. For an f-ring R, show that the following three conditions are equivalent: (i) R is a subdirect product of a finite set of (simply) ordered rings, (ii) the lattice $L_c(R)$ of all closed L-ideals of R satisfies the ascending chain condition, (iii) $L_c(R)$ satisfies the descending chain condition.

† Exercise 3 states a result of S. J. Bernau, Proc. Cambridge Philos. Soc. **61** (1965), 613–16. Exercise 4 states results of F. W. Anderson, Proc. AMS **13** (1962), 715–21.

7*. Complete l-rings

We have already proved in §XV.10 that one can represent general σ-complete l-groups with suitable units as additive groups of continuous functions on appropriate Stone spaces. These representation theorems have analogs for (σ-complete) l-rings, which will be stated below without proof. We first note the following result.

THEOREM 13. *Let R be any σ-complete (associative) l-ring with positive unity* 1 *which is a weak order unit. Then R is a commutative f-ring with zero l-radical.*

PROOF. Being σ-complete, it is Archimedean. By the preceding corollary, it is therefore an f-ring. By Theorem 10, it is consequently commutative. (For the proof that its l-radical is zero, see Birkhoff–Pierce [**1**, §10].)

The preceding results can be used to prove representation theorems for σ-complete l-rings having a ring-unity 1 which is a weak unit. As stated above, the situation is analogous to that for σ-complete l-groups and vector lattices. As in these cases, simpler results are available if 1 is a strong unit.

THEOREM 14 (STONE–VON NEUMANN). *Let R be a σ-complete l-ring with ring-unity* 1 *which is a strong order unit. Let X be the Stone space of the Boolean algebra of components of* 1. *Then, for some unique closed subspace S of X, R is isomorphic to the f-ring $C(X, S)$ of all continuous, real-valued functions on X which are integer-valued on S. If R is complete, then S is open and closed.*

For the case that the ring unity 1 is a weak unit, see Birkhoff–Pierce, [**1**, Theorem 21].†

8. **Averaging Operators**

The concept of a (weighted) *average* of a bounded Borel function $f(x)$ on a space X is classic. It is a function $A: B(X) \to \mathbf{R}$ defined by

$$A(f) = \int_X f(x)\, dm(x), \tag{18}$$

for some probability measure on X. Obviously, A is *linear* and satisfies:

A1. $$A1 = 1.$$

A2. $$f \geqq 0 \quad \text{implies} \quad Af \geqq 0.$$

Since A is linear and its range is $\mathbf{R}$, there follows

A3. $$A(fAg) = (Af)(Ag),$$

since $Af = \lambda$, $Ag = \mu$ imply $A(fAg) = A(f\mu) = \lambda\mu = (Af)(Ag)$. We shall now study a far-reaching generalization of the notion of an averaging operator, due to Kampé de Fériet.‡

† For other recent results on complete l-rings and l-algebras, see M. Henriksen and D. G. Johnson, Fund. Math. **50** (1961), 73–94, and Henriksen-Isbell-Johnson, ibid., 107–17.

‡ Ann. Soc. Sci. Bruxelles **59** (1939), 145–54. For a discussion of my own ideas, see [**Symp**, pp. 163–72].

DEFINITION. An *averaging operator* on an Archimedean f-algebra F with unity 1 is a linear operator $A: F \to F$ which satisfies A1–A3.

For many basic results about averaging operators, the positivity hypothesis A2 is superfluous (as we shall now see). It is sufficient that F be a commutative and associative ring with unity, and that A satisfy A1 and A3. (We recall that, by Theorem 10, any Archimedean f-ring is commutative and associative.)

LEMMA 1. *Any averaging operator is idempotent:*

$$A^2f = Af \quad \text{for all } f \in F. \tag{19}$$

PROOF. By A1, A3: $A(Af) = A(1Af) = (A1)(Af) = Af$.

COROLLARY. *The domain F of any averaging operator is the direct sum of the range $R = AF$ of A and its null-space $N = 0:A$.*

This follows from elementary linear algebra (over any field); moreover the range of A is the set of its fixpoints.

LEMMA 2. *The range R of any averaging operator A is a subring of F which contains* 1.

PROOF. By A1, $1 \in R$. Also, if $Af = f$ and $Ag = g$, then $A(fg) = A(fAg) = (Af)(Ag) = fg$, by hypothesis and A3.

THEOREM 15. *In any f-algebra F with* 1, *every averaging operator satisfies the Reynolds identity*†

$$A(fg) = AfAg + A((f - Af)(g - Ag)). \tag{20}$$

PROOF. Expanding the last term of (20), we get by commutativity:

$$A(fg) - A(fAg) - A(gAf) + A(AfAg) = A(fg) - AfAg - AgAf + A(A(fAg)).$$

But $A(A(fAg)) = A(fAg)$ (by Lemma 1) $= AfAg$. Hence the last term of (20) is equal to $A(fg) - AgAf$. Since $AgAf = AfAg$, the Reynolds identity is obvious.

THEOREM 16. *In any f-algebra F with* 1, *let e be a proper idempotent such that $Ae = e$. Then F is the direct sum (as f-ring with operator A) of the subring E of f such that $ef = f$ and the subring E' of $f \in F$ such that $ef = 0$.*

PROOF. Since F is commutative, evidently

$$E = eF = 0: (1 - e) \quad \text{and} \quad E' = (1 - e)F = 0: e, \tag{21}$$

so that $E \cap E' = 0$, $E + E' = F$. Also:

$$[ea + (1 - e)a][eb + (1 - e)b] = (ea)(eb) + (1 - e)a(1 - e)b,$$

† For the general theory of "Reynolds' operators" satisfying (20), see G.-C. Rota, Proc. AMS Sympos. Appl. Math. Vol. XVI, 1963, pp. 70–83. Also F. V. Atkinson, J. Math. Anal. Appl. **7** (1963), 1–30; and J. B. Miller, J. f. Math. **218** (1965), 1–16, and J. Math. Anal. Appl. **14** (1966), 527–48; N. Starr, Trans. AMS **121** (1966), 90–116.

since $e(1 - e) = 0$. Hence, as a ring, F is the direct sum of E and E'. Also, for any $f \in F$

$$A(fe) = A(fAe) = (Af)(Ae) = (Af)e \in E$$

and likewise $A(f - fe) = (Af)(1 - e) \in E'$. Hence E and E' are mapped into themselves by F.

To complete the proof, it suffices to recall that, by Lemma 2 of §6, $e \geqq 0$ and $(1 - e) \geqq 0$; hence $f \geqq 0$ in F if and only if $ef \geqq 0$ and $(1 - e)f \geqq 0$.

LEMMA 3. *In* $\mathbf{R}^n$, *the lattice of l-ideals is a Boolean algebra, crypto-isomorphic with the Boolean ring of idempotents under the crypto-isomorphism of* §II.12.

We omit the proof, which is more relevant to Chapter VI. We prove instead a closely related result, needed for the purpose in hand.

LEMMA 4. *Let* A *be an averaging operator on* $\mathbf{R}^n$, *and suppose* $Af = g$ *is not a "scalar"* $(\lambda, \cdots, \lambda)$. *Then* A *admits an invariant proper idempotent.*

PROOF. Let $\lambda_1, \cdots, \lambda_m$ be the *distinct* values of $f_1, \cdots, f_n$. Then (cf. Lagrangian interpolation)

$$e^{(j)} = \prod_{i \neq j} (f_i - \lambda_j) / \prod_{i \neq j} (\lambda_i - \lambda_j)$$

is an idempotent, with $e_i^{(j)} = 1$ if $f_i = \lambda_j$ and $e_i^{(j)} = 0$ elsewhere.

THEOREM 17. *The most general averaging operator* A *on the f-ring* $\mathbf{R}^n$ *is given as follows. For some partition* $\pi = \pi(A)$ *of the subscripts into (disjoint) subsets* $S_1, \cdots, S_m$, *and set of weighting coefficients* w_k^j *for each* S_j *such that*

$$w_k^j \geqq 0, \qquad \sum_k w_k^j = 1, \qquad w_k^j = 0 \quad \textit{unless } k \in S_j, \tag{22}$$

we have

$$(Af)_i = \sum_{S_{j(i)}} w_k^j f_k, \quad \textit{where } i \in S_{j(i)}. \tag{23}$$

In other words, there exists a direct decomposition of $\mathbf{R}^n$ into A-invariant subrings (L-ideals), on each of which A is a *scalar* averaging operator of the form (18).

Generalizations. Using the Stone–Weierstrass Theorem, one can extend the preceding result to averaging operators on the f-ring $C(K)$ of continuous functions on any compact Hausdorff space K. We omit the proof;† the essential idea is to decompose the space K into closed "T-reducing subsets" (there are not usually enough idempotents).

Important further extensions of the ideas of §8 have been made by Mme. Dubreil and her collaborators, and by Rota (op. cit.). We quote only two results.

† See §§9–10 of G. Birkhoff, *Moyennes des fonctions bornées*, pp. 143–54 of "Algebre et théorie des nombres", Paris, 1950. J. Sopka, Ph.D. Thesis, Harvard, 1950.

Rota's Theorem. *Let R be any continuous operator on (the M-space and f-ring) $L_\infty(S, \Sigma, \mu)$. Then R is an averaging operator if and only if it has closed range.*

Corollary. *Let $L_\infty(S, \Sigma, \mu)$ be finite-dimensional. Then every Reynolds operator is an averaging operator.*

Problems

157. Is associativity redundant in the hypotheses of Theorem 12?

158. Develop a theory of Jordan l-algebras and directed algebras, which yields the spectral theory of bounded symmetric operators on real Hilbert space as a corollary.

159. Can the real field be made into an l-ring by some ordering other than the usual one? Can the complex field be made into an l-ring?

160. In how many ways can the quaternions be made into an l-ring? an l-algebra? a directed algebra?

161. Solve the word problem for the free, commutative, l-algebra with n generators.

162. Same problem for the free l-ring with n generators.

163. Same problem for the free f-ring with n generators.

164. Is every quotient-ring of a regular l-ring regular?

165. Is every l-ring with zero l-radical, having a ring unity which is a weak order unit, an f-ring?

166. Which abstract rings are ring-isomorphic to l-rings? (Fuchs)

BIBLIOGRAPHY

I. BOOKS. An attempt is made to identify each book by a mnemonic letter sequence. For example, **LT1** and **LT2** refer to the first two editions of the present book. Other books especially relevant include:

[A-H] P. Alexandroff and H. Hopf, *Topologie*, Springer, 1935.

[Ba] S. Banach, *Theorie des opérations linéares*, Warsaw, 1933.

[B-M] G. Birkhoff and S. Mac Lane, *A survey of modern algebra*, 3d ed., Macmillan, 1965.

[Bo] G. Boole, *An investigation into the laws of thought*, London, 1854 (reprinted by Open Court Publishing Co., Chicago, 1940).

[Ca] C. Carathéodory, *Mass und Integral und ihre Algebraisierung*, Birkhäuser, 1956. Also, *Measure and Integration*, Chelsea, 1963.

[CG] J. von Neumann, *Continuous geometry*, Princeton, 1960.

[D-S] N. Dunford and J. Schwartz, *Theory of linear operators*, Part I, New York, 1958.

[Fu] L. Fuchs, *Partially ordered algebraic systems*, Pergamon Press, 1963.

[GA] E. H. Moore, *Introduction to a form of general analysis*, AMS Colloq. Publ. no. 2, New Haven, 1910.

[Hal] P. R. Halmos, *Lectures on Boolean algebras*, Van Nostrand, 1963.

[Hau] F. Hausdorff, *Grundzüge der Mengenlehre*, Leipzig, 1914.

[KaVP] L. V. Kantorovich, B. C. Vulich and A. G. Pinsker, *Functional analysis in partially ordered spaces* (in Russian), Moscow-Leningrad, 1950.

[Kel] J. L. Kelley, *General topology*, Van Nostrand, 1955.

[KG] F. Maeda, *Kontinuierliche Geometrien*, Springer, 1958.

[K-N] J. L. Kelley, I. Namioka and others, *Linear topological spaces*, Van Nostrand, Princeton, 1963.

[Li] L. R. Lieber, *Lattice theory*, Galois Institute of Mathematics and Art, Brooklyn, N.Y., 1959.

[MA] B. L. van der Waerden, *Moderne Algebra*, 2 vols., Springer, 1930.

[Sch] E. Schröder, *Algebra der Logik*, 3 vols., Leipzig, 1890–5.

[Sik] R. Sikorski, *Boolean Algebras*, 2d ed., Springer, 1964.

[Sk] L. A. Skornjakov, *Complemented modular lattices and regular rings*, Gos. Izd.-Mat. Lit., Moscow, 1961.

[Su] M. Suzuki, *Structure of a group and the structure of its lattice of subgroups*, Springer, 1956.

[Symp] *Lattice theory*, Proc. Sympos. Pure Math. vol. 2, AMS, Providence, R.I., 1961.

[Sz] G. Szasz, *Introduction to lattice theory*, 3d ed., Academic Press and Akademiai Kiado, 1963.

[Ta] A. Tarski, *Cardinal algebras*, Oxford Univ. Press, 1949.

[Tr] M. L. Dubreil-Jacotin, L. Lesieur, and R. Croisot, *Leçons sur la théorie des treillis*, Gauthier-Villars, Paris, 1953.

[UA] P. M. Cohn, *Universal algebra*, Harper and Row, 1965.

[W-R] A. N. Whitehead and B. Russell, *Principia Mathematica*, 3 vols., Cambridge Univ. Press, 1925, 1927.

[Da] M. M. Day, *Normed linear spaces*, 2d ed., Springer, 1962. (Contains material on Banach lattices.)

[Pe] A. L. Peressini, *Ordered topological vector spaces*, Harper and Row, 1967. (See attached summary of contents.)

[Sch] H. H. Schaefer, *Topological vector spaces*, Macmillan, 1966. (Chapter V is on vector lattices; Appendix on the spectra of positive linear operators.)

[Vu] B. Z. Vulikh, *Introduction to the theory of partially ordered spaces*, Noordhoff.

II. PAPERS

G. Birkhoff, [1] *On the combination of subalgebras*, Proc. Cambridge Philos. Soc. 29 (1933), 441–64; [2] *Combinatorial relations in projective geometries*, Ann. of Math. 36 (1935), 743–8; [3] *On the structure of abstract algebras*, Proc. Cambridge Philos. Soc. 31 (1935), 433–54; [4] *Moore-Smith convergence in general topology*, Ann. of Math. 38 (1937), 39–56; [5] *Generalized arithmetic*, Duke Math. J. 9 (1942), 283–302; [6] *Lattice-ordered groups*, Ann. of Math. 43 (1942), 298–331; [7] *Uniformly semi-primitive multiplicative processes*, Trans. AMS 104 (1962), 37–51.

G. Birkhoff and O. Frink, [1] *Representations of lattices by sets*, Trans. AMS 64 (1948), 299–316.

G. Birkhoff and J. von Neumann, [1] *On the logic of quantum mechanics*, Ann. of Math. 37 (1936), 823–43.

G. Birkhoff and R. S. Pierce, [1] *Lattice-ordered rings*, An. Acad. Brasil. Ci. 28 (1956), 41–69.

J. Certaine, [1] *Lattice-ordered groupoids and some related problems*, Harvard Doctoral Thesis (unpublished), 1943.

M. M. Day, [1] *Arithmetic of ordered systems*, Trans. AMS 58 (1945), 1–43.

R. Dedekind, [1] *Über Zerlegungen von Zahlen durch ihre grössten gemeinsamen Teiler*, Festschrift Techn. Hoch. Branuschweig (1897), and Ges. Werke, Vol. 2, 103–48; [2] *Über die von drei Moduln erzeugte Dualgruppe*, Math. Ann. 53 (1900), 371–403, and Ges. Werke, Vol. 2, 236–71.

R. P. Dilworth, [1] *Non-commutative residuated lattices*, Trans. AMS 46 (1939), 426–44; [2] *Lattices with unique complements* Trans. AMS 57 (1945), 123–54.

R. P. Dilworth and Peter Crawley, [1] *Decomposition theory for lattices without chain condition*, Trans. AMS 96 (1960), 1–22.

C. J. Everett and S. Ulam, [1] *On ordered groups*, Trans. AMS 57 (1945), 208–16.

H. Freudenthal, [1] *Teilweise geordnete Moduln*, Proc. Akad. Wet. Amsterdam 39 (1936), 641–51.

O. Frink, [1] *Topology in lattices*, Trans. AMS 51 (1942), 569–82; [2] *Complemented modular lattices and projective spaces of infinite dimension*, ibid., 60 (1946), 452–67.

H. Gordon, [1] *Relative uniform convergence*, Math. Ann. 153 (1964), 418–27.

H. Hahn, [1] Über die nichtarchimedische Grössensysteme, S.-B. Akad. Wiss. Wien IIa 116 (1907), 601–55.

A. Hales, [1] *On the non-existence of free complete Boolean algebras*, Fundamental Math. 54 (1964), 45–66.

J. Hashimoto [1] *Ideal theory for lattices*, Math. Japonica 2 (1952), 149–86.

M. Henrikson and J. R. Isbell, [1] *Lattice-ordered rings and function rings*, Pacific J. Math. 12 (1962), 533–65.

O. Hölder, [1] *Die Axiome der Quantität und die Lehre vom Mass*, Leipzig Ber., Math.-Phys. Cl. 53 (1901), 1–64.

S. S. Holland, Jr., [1] *A Radon-Nikodym theorem for dimension lattices*. Trans. AMS 108 (1963), 66–87.

E. V. Huntington, [1] *Sets of independent postulates for the algebra of logic*, Trans. AMS 5 (1904), 288–309.

B. Jonsson, [1] *Representations of complemented modular lattices*, Trans. AMS 97 (1960), 64–94.

B. Jonsson and A. Tarski, [1] *Direct decompositions of finite algebraic systems*, Notre Dame Mathematical Lectures, No. 5 (1947), 64 pp.

R. V. Kadison, [1] *A representation theorem for commutative topological algebra*, Mem. AMS #7, (1951), 39 pp.

S. Kakutani, [1] *Concrete representation of abstract* (L)*-spaces and the mean ergodic theorem*, Ann. of Math. 42 (1941), 523–37; [2] *Concrete representation of* (M)*-spaces*, ibid., 994–1024.

L. Kantorovich, [1] *Lineare halbgeordnete Räume*, Mat. Sb. 2 (44) (1937), 121–68.

I. Kaplansky, [1] *Any orthocomplemented complete modular lattice is a continuous geometry*, Ann. of Math. 61 (1955), 524–41.

J. L. Kelley, [1] *Convergence in topology*, Duke Math. J. 17 (1950), 277–83.

M. G. Krein and M. A. Rutman, [1] *Linear operators leaving invariant a cone in a Banach space*, Uspehi Mat. Nauk 3 (1948), 3–95. (AMS Translations No. 26.)

L. Lesieur, [1] *Sur les demigroupes reticulés satisfaisant à une condition de chaine*, Bull. Soc. Math. France 83 (1955), 161–93.

L. Loomis, [1] *On the representation of σ-complete Boolean algebras*, Bull. AMS 53 (1947), 757–60; [2] *The lattice theoretic background of the dimension theory*, Mem. AMS #18, A 1955.

S. Mac Lane, [1] *A lattice formulation for transcendence degrees and p-bases*, Duke Math. J. 4 (1938), 455–68.

H. MacNeille, [1] *Partially ordered sets*, Trans. AMS 42 (1937), 416–60.

F. Maeda, [1] *Matroid lattices of infinite length*, J. Sci. Hiroshima Univ. Ser. A-15 (1952), 177–82.

K. Menger, F. Alt, and O. Schreiber, [1] *New foundations of projective and affine geometry*, Ann. of Math. 37 (1936), 456–82.

E. H. Moore, [1] The New Haven Mathematical Colloquium, *Introduction to a form of general analysis*, AMS Colloq. Publ. vol. 2, New Haven, 1910.

I. Namioka, [1] *Partially ordered linear topological spaces*, Mem. AMS #24 (1957).

J. von Neumann, [1] *Continuous geometries* and *Examples of continuous geometries*, Proc. Nat. Acad. Sci. U.S.A. 22 (1936), 707–13.

M. H. A. Newman, [1] *A characterization of Boolean lattices and rings*, J. London Math. Soc. 16 (1941), 256–72.

T. Ogasawara, [1] *Theory of vector lattices*, I, J. Sci. Hiroshima Univ. 12 (1942), 37–100; [2] *Theory of vector lattices*, II, ibid. 13 (1944), 41–161. (See Math. Revs. 10 (1949), 545–6.)

O. Ore, [1] *On the foundations of abstract algebra*, I, Ann. of Math. 36 (1935), 406–37; [2] *On the foundations of abstract algebra*, II, ibid., 37 (1936), 265–92.

C. S. Peirce, [1] *On the algebra of logic*, Amer. J. Math. 3 (1880), 15–57.

L. Rieger, [1] *On the free $\aleph_\xi$-complete Boolean algebras. . .*, Fund. Math. 38 (1951), 35–52.

F. Riesz, [1] *Sur la théorie générale des opérations linéaires*, Ann. of Math. 41 (1940), 174–206.

G.-C. Rota, [1] *On the foundations of combinatorial theory. I. Theory of Möbius functions*, Zeits, f. Wahrscheinlichkeitzrechnung 2 (1964), 340–68.

Th. Skolem, [1] *Om konstitutionen av den identiske kalkuls grupper*, Third Scand, Math. Congr. (1913), 149–63.

M. H. Stone, [1] *The theory of representations for Boolean algebras*, Trans. AMS 40 (1936), 37–111; [2] *Applications of the theory of Boolean rings to general topology*, ibid., 41 (1937), 375–481; [3] *Topological representations of distributive lattices and Brouwerian logics*, Cas. Mat. Fys. 67 (1937), 1–25; [4] *A general theory of spectra*, Proc. Nat. Acad. Sci. U.S.A. 27 (1941), 83–7.

A. Tarski, [1] *Sur les classes closes par rapport à certaines opérations élémentaires*, Fund. Math. 16 (1929), 181–305; [2] *Zur Grundlagung der Booleschen Algebra*. I, ibid., 24 (1935), 177–98; [3] *Grundzüge des Systememkalkuls*, ibid., 25 (1936), 503–26 and 26 (1936), 283–301.

M. Ward and R. P. Dilworth, [1] *Residuated lattices*, Trans. AMS 45 (1939), 335–54.

H. Whitney, [1] *The abstract properties of linear dependence*, Amer. J. 57 (1935), 507–33.

P. Whitman, [1] *Free lattices*, Ann. of Math. 42 (1941), 325–9; [2] *Free lattices*. II, ibid. 43 (1942), 104–15.

K. Yosida and S. Kakutani, [1] *Operator-theoretical treatment of Markoff's process and mean ergodic theorem*, Ann. of Math. 42 (1941), 188–228.

INDEX

BCDEFGHIJ–AMS–8987654